The Universal Book
of Mathematics

The Universal Book of Mathematics

From **A**bracadabra to **Z**eno's Paradoxes

David Darling

WILEY

John Wiley & Sons, Inc.

Published by John Wiley & Sons, Inc., Hoboken, New Jersey
Published simultaneously in Canada

For general information about our other products and services, please contact
our Customer Care Department within the United States at (800) 762-2974,
outside the United States at (317) 572-3993 or fax (317) 572-4002.

Wiley also publishes its books in a variety of electronic formats. Some content
that appears in print may not be available in electronic books. For more
information about Wiley products, visit our web site at www.wiley.com.

Library of Congress Cataloging-in-Publication Data:

Darling, David J.
   The universal book of mathematics : from abracadabra to Zeno's paradoxes /
David Darling.
      p. cm.
Includes bibliographical references and index.
  ISBN 0-471-27047-4 (cloth: alk. paper)
  1. Mathematics—Encyclopedias.   I. Title.
QA5 .D27 2004
510'.3—dc22   2003024670
Printed in the United States of America

10 9 8 7 6 5 4 3 2 1

Mathematics is not a careful march down a well-cleared highway, but a journey into a strange wilderness, where the explorers often get lost. Rigor should be a signal to the historian that the maps have been made, and the real explorers have gone elsewhere.
—William S. Anglin

But leaving those of the Body, I shall proceed to such Recreation as adorn the Mind; of which those of the Mathematicks are inferior to none.
—William Leybourn (1626–1700)

The last thing one knows when writing a book is what to put first.
—Blaise Pascal (1623–1662)

Contents

Acknowledgments

Many people have helped me enormously in assembling this collection of mathematical oddities, delights, whimsies, and profundities. Thanks especially go to Jan Wassenaar (www.2dcurves.com) for drawing many of the plane curves that are featured in the book; Robert Webb (www.software3d.com) for numerous photos of his wonderful, homemade polyhedra; Jos Leys (www.josleys.com) for his mesmerizing fractal artwork; Xah Lee (www.xahlee.org) for a variety of ingenious digital imagery; Sue and Brian Young at Mr. Puzzle Australia (www.mrpuzzle.com.au) and Kate and Dick Jones at Kadon Enterprises (www.gamepuzzles.com) for their advice and photos of puzzles from their product lines and personal collections; Gideon Weisz (www.gideonweisz .com) and Istvan Orosz for stunning recursive and anamorphic art images; my good friend Andrew "Dogs" Barker for stimulating discussions and the solution to one of the problems; William Waite for pictures from his antique math collection; and Peter Cromwell, Lord & Lady Dunsany, Peter Knoppers, John Lienhard, John Mainstone, David Nicholls, Paul and Colin Roberts, Anders Sandberg, John Sullivan, and others for their valuable contributions.

I'm greatly indebted to Stephen Power, senior editor, and to Lisa Burstiner, senior production editor, at John Wiley & Sons, for their encouragement and unfailing attention to detail, and even proffering of alternative, clever solutions to some of the problems in the book. Any errors that remain are entirely my own responsibility. Thanks also to my marvelous agent, Patricia Van der Leun. And last but most of all, thanks to my family for letting me pursue a career that is really a fantasy.

Introduction

You are lost in a maze: How do you find your way out? You want to build a time machine, but is time travel logically possible? How can one infinity be bigger than another? Why can't you drink from a Klein bottle? What is the biggest number in the world to have a proper name, and how can you write it? Who claimed he could see in the fourth dimension? And what does "iteration" mean? And what does "iteration" mean?

Mathematics was never my strong point in school, but because I wanted to become an astronomer, I was told to stick with it. Fortunately, in my last two years before heading off to university, I had a wonderful old-fashioned, eccentric teacher (he actually wore a black gown when teaching), called Mr. Kay (known to one and all as "Danny"), who would suddenly divert from the chalk and blackboard to ask, "But how did the universe come to be *asymmetric*–that's what I want to know," or "These imaginary numbers are *very* interesting; in part, because they are so remarkably real." During lunch-break, Danny and the senior chemistry teacher, Mr. Erp (whose nickname I need hardly spell out), would always meet in the chemistry prep room for a game of chess. They looked and acted very much like characters from a Wellsian science fiction tale, and I sometimes imagined them musing on formulas for invisibility or doorways to higher dimensions. At any rate, though I was never a shining student, I realize what a profound effect those two deeply imaginative, thoughtful men had on my future career. I did become an astronomer. I did persevere with math to a certain level of competence. But, much more than that, my curiosity was fired by the wonderful and weird possibilities of these subjects: curved space, Möbius bands, parallel universes, patterns in the heart of chaos, alternative realities. These strange possibilities, and a thousand others, make up the stuffing of this book. If you want a comprehensive, academic dictionary of mathematics, look elsewhere. If you want rigor and proof, try the next shelf. Herein you will find only the unusual and the outrageous, the fanciful and the fantastic: a compendium of the mathematics they *didn't* teach you in school.

Entries range from short definitions to lengthy articles on topics of major importance or unusual interest. These are arranged alphabetically according to the first word of the entry name and are extensively cross-referenced. Terms that appear in **bold** type have their own entries. A number of puzzles are included for the reader to try; the answers to these can be found at the back of the book. Also at the back are a comprehensive list of references and a category index. Readers are invited to visit the author's Web site at www.daviddarling.info for the latest news in mathematics and related subjects.

abacus

A counting frame that started out, several thousand years ago, as rows of pebbles in the desert sands of the Middle East. The word appears to come from the Hebrew *âbâq* (dust) or the Phoenician *abak* (sand) via the Greek *abax*, which refers to a small tray covered with sand to hold the pebbles steady. The familiar frame-supporting rods or wires, threaded with smoothly running beads, gradually emerged in a variety of places and mathematical forms.

In Europe, there was a strange state of affairs for more than 1,500 years. The Greeks and the Romans, and then the medieval Europeans, calculated on devices with a **place-value system** in which **zero** was represented by an empty line or wire. Yet the written notations didn't have a symbol for zero until it was introduced in Europe in 1202 by **Fibonacci**, via the Arabs and the Hindus.

abacus   A special form of the Chinese abacus (c. 1958) consisting of two abaci stacked one on top of the other. *Luis Fernandes*

The Chinese *suan pan* differs from the European abacus in that the board is split into two decks, with two beads on each rod in the upper deck and five beads, representing the digits 0 through 4, on each rod in the bottom. When all five beads on a rod in the lower deck are moved up, they're reset to the original position, and one bead in the top deck is moved down as a carry. When both beads in the upper deck are moved down, they're reset and a bead on the adjacent rod on the left is moved up as a carry. The result of the computation is read off from the beads clustered near the separator beam between the upper and lower decks. In a sense, the abacus works as a 5-2-5-2-5-2 . . . –based number system in which carries and shifts are similar to those in the decimal system. Since each rod represents a digit in a decimal number, the capacity of the abacus is limited only by the number of rods on the abacus. When a user runs out of rods, she simply adds another abacus to the left of the row.

The Japanese *soroban* does away with the dual representations of fives and tens by having only four counters in the lower portion, known as "earth," and only one counter in the upper portion, known as "heaven." The world's largest abacus is in the Science Museum in London and measures 4.7 meters by 2.2 meters.

Abbott, Edwin Abbott (1838–1926)

An English clergyman and author who wrote several theological works and a biography (1885) of Francis Bacon, but is best known for his standard *Shakespearian Grammar* (1870) and the pseudonymously written *Flatland: A Romance of Many Dimensions* (by A Square, 1884).[1]

ABC conjecture

A remarkable **conjecture**, first put forward in 1980 by Joseph Oesterle of the University of Paris and David Masser of the Mathematics Institute of the University of Basel in Switzerland, that is now considered one of the most important unsolved problems in **number theory**. If it were proved correct, the proofs of many other famous conjectures and theorems would follow immediately—in some cases in just a few lines. The vastly complex current proof of **Fermat's last theorem**, for example, would reduce to less than a page of mathematical reasoning. The ABC conjecture is disarmingly simple compared to most of the deep questions in number theory and, moreover, turns out to be equivalent to all the main problems that involve **Diophantine equation**s (equations with integer coefficients and integer solutions).

Only a couple of concepts need to be understood to grasp the ABC conjecture. A *square-free number* is an integer that isn't divisible by the square of any number. For example, 15 and 17 are square-free, but 16 (divisible by 4^2) and 18 (divisible by 3^2) are not. The *square-free part* of an integer n, denoted sqp(n), is the largest square-free number that can be formed by multiplying the prime factors of n. For $n = 15$, the prime factors are 5 and 3, and $3 \times 5 = 15$, a square-free number, so that sqp(15) = 15. On the other hand, for $n = 16$, the prime factors are all 2, which means that sqp(16) = 2. In general, if n is square-free, the square-free part of n is just n; otherwise, sqp(n) represents what is left over after all the factors that create a square have been eliminated. In other words, sqp(n) is the product of the distinct **prime number**s that divide n. For example, sqp(9) = sqp(3×3) = 3 and sqp(1,400) = sqp($2 \times 2 \times 2 \times 5 \times 5 \times 7$) = $2 \times 5 \times 7$ = 70.

The ABC conjecture deals with pairs of numbers that have no common factors. Suppose A and B are two such numbers that add to give C. For example, if $A = 3$ and $B = 7$, then $C = 3 + 7 = 10$. Now, consider the square-free part of the product $A \times B \times C$: sqp(ABC) = sqp($3 \times 7 \times 10$) = 210. For most values of A and B, sqp(ABC) > C, as in the prior example. In other words, sqp(ABC)/C > 1. Occasionally, however, this isn't true. For instance, if $A = 1$ and $B = 8$, then $C = 1 + 8 = 9$, sqp(ABC) = sqp($1 \times 8 \times 9$) = sqp($1 \times 2 \times 2 \times 2 \times 3 \times 3$) = $1 \times 2 \times 3 = 6$, and sqp(ABC)/C = $^6\!/_9$ = $^2\!/_3$. Similarly, if $A = 3$ and $B = 125$, the ratio is $^{15}\!/_{64}$.

David Masser proved that the ratio sqp(ABC)/C can get arbitrarily small. In other words, given any number greater than zero, no matter how small, it's possible to find integers A and B for which sqp(ABC)/C is smaller than this number. In contrast, the ABC conjecture says that [sqp(ABC)]^n/C reaches a minimum value if n is any number greater than 1—even a number such as 1.0000000001, which is only barely larger than 1. The tiny change in the expression results in a huge difference in its mathematical behavior. The ABC conjecture in effect translates an infinite number of Diophantine equations (including the equation of Fermat's last theorem) into a single mathematical statement.[144]

Abel, Niels Henrik (1802–1829)

The divergent series are the invention of the devil, and it is a shame to base on them any demonstration whatsoever. By using them, one may draw any conclusion he pleases and that is why these series have produced so many fallacies and so many paradoxes.

A Norwegian mathematician who, independently of his contemporary Évariste **Galois**, pioneered **group** theory and proved that there are no algebraic solutions of the general **quintic** equation. Both Abel and Galois died tragically young—Abel of tuberculosis, Galois in a sword fight.

While a student in Christiania (now Oslo), Abel thought he had discovered how to solve the general quintic algebraically, but soon corrected himself in a famous pamphlet published in 1824. In this early paper, Abel showed the impossibility of solving the general quintic by means of radicals, thus laying to rest a problem that had perplexed mathematicians since the mid-sixteenth century. Abel, chronically poor throughout his life, was granted a small stipend by the Norwegian government that allowed him to go on a mathematical tour of Germany and France. In Berlin he met Leopold Crelle (1780–1856) and in 1826 helped him found the first journal in the world devoted to mathematical research. Its first three volumes contained 22 of Abel's papers, ensuring lasting fame for both Abel and Crelle. Abel revolutionized the important area of elliptic integrals with his theory of **elliptic functions**, contributed to the theory of **infinite series**, and founded the theory of commutative groups, known today as **Abelian group**s. Yet his work was never properly appreciated during his life, and, impoverished and ill, he returned to Norway unable to obtain a teaching position. Two days after his death, a delayed letter was delivered in which Abel was offered a post at the University of Berlin.

Abelian group

A **group** that is **commutative**, that is, in which the result of multiplying one member of the group by another is independent of the order of multiplication. Abelian groups, named after Niels **Abel**, are of central importance in modern mathematics, most notably in **algebraic topology**. Examples of Abelian groups include the **real numbers** (with addition), the nonzero real numbers (with multiplication), and all cyclic groups, such as the **integers** (with addition).

abracadabra

A word famously used by magicians but which started out as a cabalistic or mystical charm for curing various ailments, including toothache and fever. It was first mentioned in a poem called "Praecepta de Medicina" by the Gnostic physician Quintus Severus Sammonicus in the second century A.D. Sammonicus instructed that the letters be written on parchment in the form of a triangle:

```
A B R A C A D A B R A
 A B R A C A D A B R
  A B R A C A D A B
   A B R A C A D A
    A B R A C A D
     A B R A C A
      A B R A C
       A B R A
        A B R
         A B
          A
```

This was to be folded into the shape of a cross, worn for nine days suspended from the neck, and, before sunrise, cast behind the patient into a stream running eastward. It was also a popular remedy in the Middle Ages. During the Great Plague, around 1665, large numbers of these amulets were worn as safeguards against infection. The origin of the word itself is uncertain. One theory is that it is based on Abrasax, the name of an Egyptian deity.

PUZZLE

A well-known puzzle, proposed by George Polya (1887–1985), asks how many different ways there are to spell abracadabra in this diamond-shaped arrangement of letters:

```
          A
         B B
        R R R
       A A A A
      C C C C C
     A A A A A A
      D D D D D
       A A A A
        B B B
         R R
          A
```

Solutions begin on page 369.

---

abscissa

The *x*-coordinate, or horizontal distance from the *y*-axis, in a system of **Cartesian coordinates**. Compare with **ordinate**.

absolute

Not limited by exceptions or conditions. The term is used in many different ways in mathematics, physics, philosophy, and everyday speech. Absolute space and absolute time, which, in Newton's universe, form a unique, immutable frame of reference, blend and become deformable in the **space-time** of Einstein. See also

absolute zero. In some philosophies, the absolute stands behind the reality we see–independent, transcendent, unconditional, and all-encompassing. The American philosopher Josiah Royce (1855–1916) took the absolute to be a spiritual entity whose self-consciousness is imperfectly reflected in the totality of human thought. Mathematics, too, reaches beyond imagination with its absolute **infinity**. See also **absolute value**.

absolute value

The value of a number without regard to its sign. The absolute value, or *modulus*, of a **real number**, *r*, is the distance of the number from zero measured along the real **number line**, and is denoted $|r|$. Being a distance, it can't be negative; so, for example, $|3| = |-3| = 3$. The same idea applies to the absolute value of a **complex number** $a + ib$, except that, in this case, the complex number is represented by a point on an **Argand diagram**. The absolute value, $|a + ib|$, is the length of the line from the origin to the given point, and is equal to $\sqrt{(a^2 + b^2)}$.

absolute zero

The lowest possible temperature of a substance, equal to 0 Kelvin (K), −273.15°C, or −459.67°F. In classical physics, it is the temperature at which all molecular motion ceases. However, in the "real" world of **quantum mechanics** it isn't possible to stop all motion of the particles making up a substance as this would violate the Heisenberg uncertainty principle. So, at 0 K, particles would still vibrate with a certain small but nonzero energy known as the *zero-point energy*. Temperatures within a few billionths of a degree of absolute zero have been achieved in the laboratory. At such low temperatures, substances have been seen to enter a peculiar state, known as the *Bose-Einstein condensate*, in which their quantum wave functions merge and particles lose their individual identities. Although it is possible to approach ever closer to absolute zero, the third law of thermodynamics asserts that it's impossible to ever attain it. In a deep sense, absolute zero lies at the asymptotic limit of low energy just as the speed of light lies, for particles with mass, at the asymptotic limit of high energy. In both cases, energy of motion (kinetic energy) is the key quantity involved. At the high energy end, as the average speed of the particles of a substance approaches the speed of light, the temperature rises without limit, heading for an unreachable ∞ K.

abstract algebra

To a mathematician, real life is a special case.
                                        —Anonymous

Algebra that is not confined to familiar number systems, such as the **real numbers**, but seeks to solve equations that may involve many other kinds of systems. One of its aims, in fact, is to ask: What other number systems are there? The term *abstract* refers to the perspective taken on the subject, which is very different from that of high school algebra. Rather than looking for the solutions to a particular problem, abstract algebra is interested in such questions as: When does a solution exist? If a solution does exist, is it unique? What general properties does a solution possess? Among the structures it deals with are **group**s, **ring**s, and **field**s. Historically, examples of such structures often arose first in some other field of mathematics, were specified rigorously (axiomatically), and were then studied in their own right in abstract algebra.

Abu'l Wafa (A.D. 940–998)

A Persian mathematician and astronomer who was the first to describe geometrical constructions (see **constructible**) possible only with a straightedge and a fixed compass, later dubbed a "rusty compass," that never alters its radius. He pioneered the use of the tangent function (see **trigonometric function**), apparently discovered the secant and cosecant functions, and compiled tables of sines and tangents at 15′ intervals—work done as part of an investigation into the orbit of the Moon.

abundant number

A number that is smaller than the sum of its **aliquot parts** (proper divisors). Twelve is the smallest abundant number; the sum of its aliquot parts is $1 + 2 + 3 + 4 + 6 = 16$, followed by 18, 20, 24, and 30. A *weird number* is an abundant number that is not *semiperfect;* in other words, n is weird if the sum of its divisors is greater than n, but n is not equal to the sum of any subset of its divisors. The first few weird numbers are 70, 836, 4,030, 5,830, and 7,192. It isn't known if there are any odd weird numbers. A *deficient number* is one that is greater than the sum of its aliquot parts. The first few deficient numbers are 1, 2, 3, 4, 5, 8, and 9. Any divisor of a deficient (or perfect) number is deficient. A number that is not abundant or deficient is known as a **perfect number**.

Achilles and the Tortoise paradox

See **Zeno's paradoxes**.

Ackermann function

One of the most important **functions** in computer science. Its most outstanding property is that it grows astonishingly fast. In fact, it gives rise to **large numbers** so quickly that these numbers, called *Ackermann numbers,* are written in a special way known as **Knuth's up-arrow notation**. The Ackermann function was discovered and studied by Wilhelm Ackermann (1896–1962) in 1928.

Ackermann worked as a high school teacher from 1927 to 1961 but was also a student of the great mathematician David **Hilbert** in Göttingen and, from 1953, served as an honorary professor in the university there. Together with Hilbert he published the first modern textbook on mathematical logic. The function he discovered, and that now bears his name, is the simplest example of a well-defined and total function that is also computable but not primitive recursive (PR). "Well-defined and total" means that the function is internally consistent and doesn't break any of the rules laid down to define it. "Computable" means that it can, in principle, be evaluated for all possible input values of its variables. "Primitive recursive" means that it can be computed using only *for loops*—repeated application of a single operation a predetermined number of times. The recursion, or feedback loop, in the Ackermann function overruns the capacity of any *for loop* because the number of loop repetitions isn't known in advance. Instead, this number is itself part of the computation, and grows as the calculation proceeds. The Ackermann function can only be calculated using a *while loop,* which keeps repeating an action until an associated test returns false. Such loops are essential when the programmer doesn't know at the outset how many times the loop will be traversed. (It's now known that everything computable can be programmed using while loops.)

The Ackermann function can be defined as follows:

$$A(0, n) = n + 1 \text{ for } n = 0$$
$$A(m, 0) = A(m - 1, 1) \text{ for } m = 1$$
$$A(m, n) = A(m - 1, A(m, n - 1)) \text{ for } m, n = 1.$$

Two positive integers, m and n, are the input and $A(m, n)$ is the output in the form of another positive integer. The function can be programmed easily in just a few lines of code. The problem isn't the complexity of the function but the awesome rate at which it grows. For example, the innocuous-looking $A(4,2)$ already has 19,729 digits! The use of a powerful large-number shorthand system, such as the up-arrow notation, is indispensable as the following examples show:

$$A(1, n) = 2 + (n + 3) - 3$$
$$A(2, n) = 2 \times (n + 3) - 3$$
$$A(3, n) = 2{\uparrow}(n + 3) - 3$$
$$A(4, n) = 2{\uparrow}(2{\uparrow}(2{\uparrow}(\ldots {\uparrow}2))) - 3 \ (n + 3 \text{ twos})$$
$$= 2{\uparrow}{\uparrow}(n + 3) - 3$$
$$A(5, n) = 2{\uparrow}{\uparrow}{\uparrow}(n + 3) - 3, \text{ etc.}$$

Intuitively, the Ackermann function defines generalizations of multiplication by 2 (iterated additions) and exponentiation with base 2 (iterated multiplications) to iterated exponentiation, iteration of this operation, and so on.[84]

acre
An old unit of **area**, equal to 160 square rods, 4,840 square yards, 43,560 square feet, or 4,046.856 square meters.

acute
From the Latin *acus* for "needle" (which also forms the root for *acid, acupuncture,* and *acumen*). An acute **angle** is less than 90°. An acute **triangle** is one in which all three angles are acute. Compare with **obtuse**.

adjacent
Next to. *Adjacent angles* are next to each other, and thus share one side. *Adjacent sides* of a **polygon** share a vertex.

affine geometry
The study of properties of geometric objects that remain unchanged after parallel projection from one plane to another. During such a projection, first studied by Leonhard **Euler**, each point (x, y) is mapped to a new point $(ax + cy + e, bx + dy + f)$. Circles, angles, and distances are altered by affine transformations and so are of no interest in affine geometry. Affine transformations do, however, preserve collinearity of points: if three points belong to the same straight line, their images (the points that correspond to them) under affine transformations also belong to the same line and, in addition, the middle point remains between the other two points. Similarly, under affine transformations, parallel lines remain parallel; concurrent lines remain concurrent (images of intersecting lines intersect); the ratio of lengths of line segments of a given line remains constant; the ratio of areas of two triangles remains constant; and ellipses, parabolas, and hyperbolas continue to be ellipses, parabolas, and hyperbolas.

age puzzles and tricks
Problems that ask for a person's age or, alternatively, when a person was a certain age, given several roundabout facts. They go back at least 1,500 years to the time of Metrodorus and **Diophantus's riddle**. A number of distinct types of age puzzles sprang up between the sixteenth and early twentieth centuries, in most cases best solved by a little algebra. One form asks: if X is now a years old and Y is now b years old, when will X be c times as old as Y? The single unknown, call it x, can be found from the equation $a + x = c(b + x)$. Another type of problem takes the form: if X is now a times as old as Y and after b years X will be c times as old as Y, how old are X and Y now? In this case the trick is to set up and solve two simultaneous equations: $X = aY$ and $X + b = c(Y + b)$.

Various mathematical sleights of hand can seem to conjure up a person's age as if by magic. For example, ask a person to multiply the first number of his or her age by 5, add 3, double this figure, add the second number of his or her age to the figure, and tell you the answer. Deduct 6 from this and you will have their age.

Alternatively, ask the person to pick a number, multiply this by 2, add 5, and multiply by 50. If the person has already had a birthday this year and it's the year 2004, she should add 1,754, otherwise she should add 1,753. Each year after 2004 these numbers need to be increased by 1. Finally, the person should subtract the year they were born. The first digits of the answer are the original number, while the last two digits are the person's age.

Here is one more trick. Take your age, multiply it by 7, then multiply again by 1,443. The result is your age repeated three times. (What you have actually done is multiplied by 10,101; if you multiply by 1,010,101, the repetition is fourfold, and so on.)

Agnesi, Maria Gaetana (1718–1799)
An Italian mathematician and scholar whose name is associated with the curve known as the *Witch of Agnesi.* Born in Milan, Maria was one of 24 children of a professor of mathematics at the University of Bologna. A child prodigy, she could speak seven languages, including Latin, Greek, and Hebrew, by the age of 11 and was solving difficult problems in geometry and ballistics by her early teens. Her father encouraged her studies and her appearance at public debates. However, Maria developed a chronic illness, marked by convulsions and headaches, and, from the age of about 20, withdrew socially and devoted herself to mathematics. Her *Instituzioni analitiche ad uso della gioventu italiana,* published in 1748, became a standard teaching manual, and in 1750, she was appointed to the chair of mathematics and natural philosophy at Bologna. Yet she never fulfilled her early promise in terms of making new

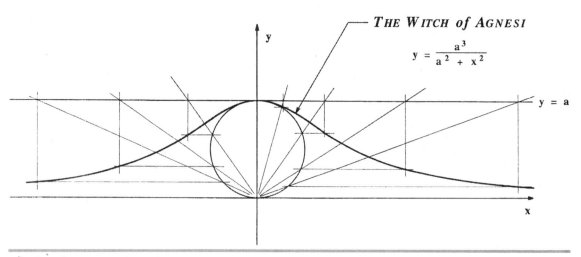

$$y = \frac{a^3}{a^2 + x^2}$$

THE WITCH of AGNESI

$y = a$

Agnesi, Maria Gaetana The Witch of Agnesi curve. *John H. Lienhard*

breakthroughs. After the death of her father in 1752, she moved into theology and, after serving for some years as the directress of the Hospice Trivulzio for Blue Nuns at Milan, joined the sisterhood herself and ended her days in this austere order.

The famous curve that bears her name had been studied earlier, in 1703, by Pierre de **Fermat** and the Italian mathematician Guido Grandi (1671–1742). Maria wrote about it in her teaching manual and referred to it as the *aversiera*, which simply means "to turn." But in translating this, the British mathematician John Colson (1680–1760), the fifth Lucasian professor of mathematics at Cambridge University, confused *aversiera* with *avversiere* which means "witch," or "wife of the devil." And so the name of the curve came down to us as the Witch of Agnesi. To draw it, start with a circle of diameter a, centered at the point $(0, a/2)$ on the y-axis. Choose a point A on the line $y = a$ and connect it to the origin with a line segment. Call the point where the segment crosses the circle B. Let P be the point where the vertical line through A crosses the horizontal line through B. The Witch is the curve traced by P as A moves along the line $y = a$. By a happy coincidence, it does look a bit like a witch's hat! In Cartesian coordinates, its equation is

$$y = a^3/(x^2 + a^2).$$

Ahmes papyrus
See **Rhind** papyrus.

Ahrens, Wilhelm Ernst Martin Georg (1872–1927)
A great German exponent of recreational mathematics

whose *Mathematische Unterhaltungen und Spiele*[6] is one of the most scholarly of all books on the subject.

Alcuin (735–804)
A leading intellectual of his time and the probable compiler of *Propositiones ad Acuendos Juvenes* (Problems to sharpen the young), one of the earliest collections of recreational math problems. According to David **Singmaster** and John Hadley: "The text contains 56 problems, including 9 to 11 major types of problem which appear for the first time, 2 major types which appear in the West for the first time and 3 novel variations of known problems. . . . It has recently been realized that the **river-crossing problems** and the crossing-a-desert problem, which appear here for the first time, are probably the earliest known combinatorial problems."

Alcuin was born into a prominent family near the east coast of England. He was sent to York, where he became a pupil and, eventually, in 778, the headmaster, of Archbishop Ecgberht's School. (Ecgberht was the last person to have known the Venerable Bede.) Alcuin built up a superb library and made the school one of the chief centers of learning in Europe. Its reputation became such that, in 781, Alcuin was invited to become master of Charlemagne's Palace School at Aachen and, effectively, minister of education for Charlemagne's empire. He accepted and traveled to Aachen to a meeting of the leading scholars. Subsequently, he was made head of Charlemagne's Palace School and there developed the Carolingian minuscule, a clear, legible script that became the basis of how letters of the present Roman alphabet are written.

Before leaving Aachen, Alcuin was responsible for the most prized of the Carolingian codices, now called the

Golden Gospels: a series of illuminated masterpieces written largely in gold, on white or purple vellum. The development of Carolingian minuscule had, indirectly, a major impact on the history of mathematics. Because it was a far more easily readable script than the older unspaced capital, it led to many mathematical works being newly copied into this new style in the ninth century. Most of the works of the ancient Greek mathematicians that have survived did so because of this transcription. Alcuin lived in Aachen from 782 to 790 and again from 793 to 796. In 796, he retired from Charlemagne's Palace School and became abbot of the Abbey of St. Martin at Tours, where he and his monks continued to work with the Carolingian minuscule script.

aleph

The first letter of the Hebrew alphabet, ℵ. It was first used in mathematics by Georg **Cantor** to denote the various orders, or sizes, of **infinity**: $\aleph_0$ (aleph-null), $\aleph_1$ (aleph-one), etc. An earlier (and still used) symbol for infinity, ∞, was introduced in 1655 by John **Wallis** in his *Arithmetica infinitorum* but didn't appear in print until the *Ars conjectandi* by Jakob Bernoulli, published posthumously in 1713 by his nephew Nikolaus Bernoulli (see **Bernoulli Family**).

Alexander's horned sphere A sculpture of a five-level Alexander's horned sphere. *Gideon Weisz, www.gideonweisz.com*

Alexander's horned sphere

In **topology,** an example of what is called a "wild" structure; it is named after the Princeton mathematician James Waddell Alexander (1888–1971) who first described it in the early 1920s. The horned sphere is topologically equivalent to the **simply connected** surface of an ordinary hollow **sphere** but bounds a region that is not simply connected. The horns-within-horns consist of a recursive set—a **fractal**—of interlocking pairs of orthogonal rings (rings set at right angles) of decreasing radius. A rubber band around the base of any horn couldn't be removed from the structure even after infinitely many steps. The horned sphere can be embedded in the plane by reducing the interlock angle between ring pairs from 90° to 0°, then weaving the rings together in an over-under pattern. The sculptor Gideon Weisz has modeled a number of approximations to the structure, one of which is shown in the photograph.

algebra

A major branch of mathematics that, at an elementary level, involves applying the rules of arithmetic to numbers, and to letters that stand for unknown numbers, with the main aim of solving equations. Beyond the algebra learned in high school is the much vaster and more profound subject of **abstract algebra**. The word itself comes from the Arabic *al-jebr*, meaning "the reunion of broken parts." It first appeared in the title of a book, *Al-jebr w'al-muqabalah* (The science of reduction and comparison), by the ninth-century Persian scholar **al-Khowarizmi**—probably the greatest mathematician of his age, and as famous among Arabs as Euclid and Aristotle are to the Western world.

algebraic curve

A curve whose equation involves only *algebraic functions*. These are functions that, in their most general form, can be written as a sum of **polynomial**s in x multiplied by powers of y, equal to zero. Among the simplest examples are straight lines and **conic section**s.

algebraic fallacies

Misuse of **algebra** can have some surprising and absurd results. Here, for example, is a famous "proof" that $1 = 2$:

$$\text{Let } a = b.$$
$$\text{Then } a^2 = ab$$
$$a^2 + a^2 = a^2 + ab$$
$$2a^2 = a^2 + ab$$
$$2a^2 - 2ab = a^2 + ab - 2ab$$
$$2a^2 - 2ab = a^2 - ab$$
$$2(a^2 - ab) = 1(a^2 - ab).$$
$$\text{Dividing both sides by } a^2 - ab$$
$$2 = 1.$$

Where's the mistake? The problem lies with the seemingly innocuous final division. Since $a = b$, dividing by $a^2 - ab$ is the same as dividing by **zero**–the great taboo of mathematics.

Another false argument runs as follows:

$$(n + 1)^2 = n^2 + 2n + 1$$
$$(n + 1)^2 - (2n + 1) = n^2.$$

Subtracting $n(2n + 1)$ from both sides and factorizing gives

$$(n + 1)^2 - (n + 1)(2n + 1) = n^2 - n(2n + 1).$$

Adding $\tfrac{1}{4}(2n + 1)^2$ to both sides yields

$$(n + 1)^2 - (n + 1)(2n + 1) + \tfrac{1}{4}(2n + 1)^2$$
$$= n^2 - n(2n + 1) + \tfrac{1}{4}(2n + 1)^2.$$

This may be written:

$$[(n + 1) - \tfrac{1}{2}(2n + 1)]^2 = [(n - \tfrac{1}{2}(2n + 1)]^2.$$

Taking square roots of both sides,

$$n + 1 - \tfrac{1}{2}(2n + 1) = n - \tfrac{1}{2}(2n + 1).$$

Therefore,

$$n = n + 1.$$

The problem here is that there are *two* square roots for any positive number, one positive and one negative: the square roots of 4 are 2 and −2, which can be written as ±2. So the penultimate step should properly read:

$$\pm(n + 1 - \tfrac{1}{2}(2n + 1)) = \pm(n - \tfrac{1}{2}(2n + 1))$$

algebraic geometry
Originally, the geometry of **complex number** solutions to **polynomial** equations. Modern algebraic geometry is also concerned with algebraic varieties, which are a generalization of the solution sets found in the traditional subject, as well as solutions in fields other than complex numbers, for example *finite fields*.

algebraic number
A **real number** that is a **root** of a **polynomial** equation with integer coefficients. For example, any **rational number** a/b, where a and b are nonzero integers, is an algebraic number of degree one, because it is a root of the linear equation $bx - a = 0$. The **square root of two** is an algebraic number of degree two because it is a root of the quadratic equation $x^2 - 2 = 0$. If a real number is not algebraic, then it is a **transcendental number**. Almost all real numbers are transcendental because, whereas the set of algebraic numbers is countably infinite (see **countable**

set), the set of transcendental numbers is uncountably infinite.

algebraic number theory
The branch of **number theory** that is studied without using methods such as **infinite series** and **convergence** taken from **analysis**. It contrasts with **analytical number theory**.

algebraic topology
A branch of **topology** that deals with invariants of a topological space that are algebraic structures, often **group**s.

algorithm
A systematic method for solving a problem. The word comes from the name of the Persian mathematician, **al-Khowarizmi**, and may have been first used by Gottfried **Liebniz** in the late 1600s. It remained little known in Western mathematics, however, until the Russian mathematician Andrei Markov (1903–1987) reintroduced it. The term became especially popular in the areas of math focused on computing and computation.

algorithmic complexity
A measure of **complexity** developed by Gregory **Chaitin** and others, based on Claude **Shannon**'s **information theory** and earlier work by the Russian mathematicians Andrei **Kolmogorov** and Ray Solomonoff. Algorithmic complexity quantifies how complex a system is in terms of the shortest computer program, or set of **algorithm**s, needed to completely describe the system. In other words, it is the smallest model of a given system that is necessary and sufficient to capture the essential patterns of that system. Algorithmic complexity has to do with the mixture of repetition and innovation in a complex system. At one extreme, a highly regular system can be described by a very short program or algorithm. For example, the bit string 010101010101010101 . . . follows from just three commands: print a zero, print a one, and repeat the last two commands indefinitely. The complexity of such a system is very low. At the other extreme, a totally **random** system has a very high algorithmic complexity since the random patterns can't be condensed into a smaller set of algorithms: the program is effectively as large as the system itself. See also **compressible**.

Alhambra
The former palace and citadel of the Moorish kings of Granada, and perhaps the greatest monument to Islamic mathematical art on Earth. Because the Qur'an consid-

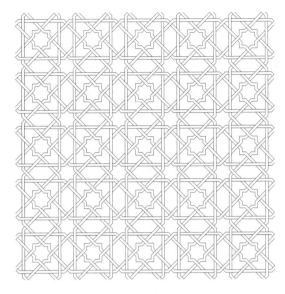

Alhambra Computer-generated tilings based on Islamic tile designs such as those found in the Alhambra. *Xah Lee, www.xahlee.org*

ers the depiction of living beings in religious settings blasphemous, Islamic artists created intricate patterns to symbolize the wonders of creation: the repetitive nature of these complex geometric designs suggests the limitless power of God. The sprawling citadel, looming high above the Andalusian plain, boasts a remarkable array of mosaics with tiles arranged in intricate patterns. The Alhambra **tiling**s are *periodic;* in other words, they con-

sist of some basic unit that is repeated in all directions to fill up the available space. All 17 different groups of isometries—the possible ways of repeatedly tiling the plane—are used at the palace. The designs left a deep impression on Maurits **Escher**, who came here in 1936. Subsequently, Escher's art took on a much more mathematical nature, and over the next six years he produced 43 colored drawings of periodic tilings with a wide variety of symmetry types.

aliquot part

Also known as a *proper divisor,* any divisor of a number that isn't equal to the number itself. For instance, the aliquot parts of 12 are 1, 2, 3, 4, and 6. The word comes from the Latin *ali* ("other") and *quot* ("how many"). An *aliquot sequence* is formed by taking the sum of the aliquot parts of a number, adding them to form a new number, then repeating this process on the next number and so on. For example, starting with 20, we get $1 + 2 + 4 + 5 + 10 = 22$, then $1 + 2 + 11 = 14$, then $1 + 2 + 7 = 10$, then $1 + 2 + 5 = 8$, then $1 + 2 + 4 = 7$, then 1, after which the sequence doesn't change. For some numbers, the result loops back immediately to the original number; in such cases the two numbers are called **amicable numbers.** In other cases, where a sequence repeats a pattern after more than one step, the result is known as an *aliquot cycle* or a *sociable chain.* An example of this is the sequence 12496, 14288, 15472, 14536, 14264, . . . The aliquot parts of 14264 add to give 12496, so that the whole cycle begins again. Do all aliquot sequences end either in 1 or in an aliquot cycle (of which amicable numbers are a special case)? In 1888, the Belgian mathematician Eugène Catalan (1814–1894) conjectured that they do, but this remains an open question.

al-Khowarizmi (c. 780–850)

An Arabic mathematician, born in Baghdad, who is widely considered to be the founder of modern day **algebra**. He believed that any math problem, no matter how difficult, could be solved if broken down into a series of smaller steps. The word *algorithm* may have derived from his name.

Allais paradox

A **paradox** that stems from questions asked in 1951 by the French economist Maurice Allais (1911–).[8] Which of these would you choose: (A) an 89% chance of receiving an unknown amount and 11% chance of $1 million; or (B) an 89% chance of an unknown amount (the same amount as in A), a 10% chance of $2.5 million, and a 1% chance of nothing? Would your choice be the same if the unknown amount was $1 million? What if the unknown amount was zero?

Most people don't like risk and so prefer the better chance of winning $1 million in option A. This choice is firm when the unknown amount is $1 million, but seems to waver as the amount falls to nothing. In the latter case, the risk-averse person favors B because there isn't much difference between 10% and 11%, but there's a big difference between $1 million and $2.5 million. Thus the choice between A and B depends on the unknown amount, even though it is the same unknown amount independent of the choice. This flies in the face of the so-called *independence axiom*, that rational choice between two alternatives should depend only on how those two alternatives differ. Yet, if the amounts involved in the problem are reduced to tens of dollars instead of millions of dollars, people's behavior tends to fall back in line with the axioms of rational choice. In this case, people tend to choose option B regardless of the unknown amount. Perhaps when presented with such huge numbers, people begin to calculate qualitatively. For example, if the unknown amount is $1 million the options are essentially (A) a fortune guaranteed or (B) a fortune almost guaranteed with a small chance of a bigger fortune and a tiny chance of nothing. Choice A is then rational. However, if the unknown amount is nothing, the options are (A) a small chance of a fortune ($1 million) and a large chance of nothing, and (B) a small chance of a larger fortune ($2.5 million) and a large chance of nothing. In this case, the choice of B is rational. Thus, the Allais paradox stems from our limited ability to calculate rationally with such unusual quantities.

almost perfect number

A description sometimes applied to the powers of 2 because the **aliquot part**s (proper divisors) of 2^n sum to $2^n - 1$. So a power of 2 is a deficient number (one that is less than the sum of its proper divisors), but only just. It isn't known whether there is an odd number n whose divisors (excluding itself) sum to $n - 1$.

alphamagic square

A form of **magic square**, introduced by Lee **Sallows**,[278–280] in which the number of letters in the word for each number, in whatever language is being used, gives rise to another magic square. In English, for example, the alphamagic square:

5 (five)	22 (twenty-two)	18 (eighteen)
28 (twenty-eight)	15 (fifteen)	2 (two)
12 (twelve)	8 (eight)	25 (twenty-five)

generates the square:

4	9	8
11	7	3
6	5	10

A surprisingly large number of 3×3 alphamagic squares exist—in English and in other languages. French allows just one 3×3 alphamagic square involving numbers up to 200, but a further 255 squares if the size of the entries is increased to 300. For entries less than 100, none occurs in Danish or in Latin, but there are 6 in Dutch, 13 in Finnish, and an incredible 221 in German. Yet to be determined is whether a 3×3 square exists from which a magic square can be derived that, in turn, yields a third magic square—a magic triplet. Also unknown is the number of 4×4 and 5×5 language-dependent alphamagic squares. Here, for example, is a four-by-four English alphamagic square:

26	37	48	59
49	58	27	36
57	46	39	28
38	29	56	47

alphametic

A type of **cryptarithm** in which a set of words is written down in the form of a long addition sum or some other mathematical problem. The object is to replace the letters of the alphabet with decimal digits to make a valid arithmetic sum. The word *alphametic* was coined in 1955 by James **Hunter**. However, the first modern alphametic, published by Henry **Dudeney** in the July 1924 issue of *Strand Magazine*, was "Send more money," or, setting it out in the form of a long addition:

SEND
MORE
MONEY

and has the (unique) solution:

9567
1085
10652

PUZZLES
The reader is invited to try to solve the following elegant examples:
1. Earth, air, fire, water: nature. (Herman Nijon)
2. Saturn, Uranus, Neptune, Pluto: planets. (Peter J. Martin)
3. Martin Gardner retires. (H. Everett Moore)

Solutions begin on page 369.

Two rules are obeyed by every alphametic. First, the mapping of letters to numbers is one-to-one; that is, the same letter always stands for the same digit, and the same digit is always represented by the same letter. Second, the digit zero isn't allowed as the left-most digit in any of the numbers being added or in their sum. The best alphametics are reckoned to be those with only one correct answer.

Altekruse puzzle

A symmetrical 12-piece **burr puzzle** for which a patent was granted to William Altekruse in 1890. The Altekruse family is of Austrian-German origin and, curiously, the name means "old cross" in German, which has led some authors to incorrectly assume that it was a pseudonym. William Altekruse came to the United States as a young man in 1844 with his three brothers to escape being drafted into the German army. The Altekruse puzzle has an unusual mechanical action in the first step of disassembly by which two halves move in opposition to each other, unlike the more familiar burr types that have a key piece or pieces. Depending on how it is assembled, this action can take place along one, two, or all three axes independently but not simultaneously.

alternate

A mathematical term with several different meanings: (1) *Alternate angles* are angles on opposite sides and opposite ends of a line that cuts two parallel lines. (2) A well-known theorem called the *alternate segment theorem* involves the segment on the opposite side of a given chord of a circle. (3) An *alternate hypothesis* in statistics is the alternative offered to the **null hypothesis**. (4) To alternate is to cycle backward and forward between two different values, for example, 0, 1, 0, 1, 0, 1, ...

altitude

A perpendicular line segment from one **vertex** of a figure or solid to an edge or face opposite to that vertex. Also the length of such a line segment.

ambiguous figure

An **optical illusion** in which the subject or the perspective of a picture or shape may suddenly switch in the mind of the observer to another, equally valid possibility. Often the ambiguity stems from the fact that the figure and ground can be reversed. An example of this is the vase/profile illusion, made famous by the Danish psychologist Edgar John Rubin (1886–1951) in 1915, though earlier versions of the same illusion can be

ambiguous figure   The Rubin vase illusion: one moment a vase, the next two people face to face.

found in many eighteenth-century French prints depicting a variety of vases, usually in a naturalistic setting, and profiles of particular people. The same effect can be created in three dimensions with a suitably shaped solid vase. In some ambiguous figures, the features of a person or of an animal can suddenly be seen as different features of another individual. Classic examples include the old woman–young woman illusion and the duck-rabbit illusion. **Upside-down picture**s involve a special case of dual-purpose features in which the reversal is accomplished not mentally, by suddenly "seeing" the alternative, but physically, by turning the picture 180°. Ambiguity can also occur, particularly in some geometric drawings, when there is confusion as to which are the front and the back faces of a figure, as in the **Necker cube**, the **Thiery figure**, and **Schröder's reversible staircase**.

ambiguous connectivity

See **impossible figure**.

Ames room

The famous distorted room illusion, named after the American ophthalmologist Adelbert Ames Jr. (1880–1955), who first constructed such a room in 1946 based on a concept by the German physicist Hermann Helmholtz in the late nineteenth century. The Ames room looks cubic when seen with one eye through a specially positioned peephole; however, the room's true shape is trapezoidal. The floor, ceiling, some walls, and the far windows are trapezoidal surfaces; the floor appears level but is actually at an incline (one of the far corners being

Ames room Misleading geometry makes these identical twins appear totally different in size. *Technische Universitat, Dresden*

much lower than the other); and the walls are slanted outward, though they seem perpendicular to the floor. This shape makes it look as if people or objects grow or shrink as they move from one corner of the room to another. See also **distortion illusion**.[142, 178]

amicable numbers

A pair of numbers, also known as *friendly numbers,* each of whose **aliquot parts** add to give the other number. (An aliquot part is any divisor that doesn't include the number itself.) The smallest amicable numbers are 220 (aliquot parts 1, 2, 4, 5, 10, 11, 20, 22, 44, 55, and 110, with a sum of 284) and 284 (aliquot parts 1, 2, 4, 71, and 142, with a sum of 220). This pair was known to the ancient Greeks, and the Arabs found several more. In 1636, Pierre de **Fermat** rediscovered the amicable pair 17,296 and 18,416; two years later René **Descartes** rediscovered a third pair, 9,363,584 and 9,437,056. In the eighteenth century, Leonhard **Euler** drew up a list of more than 60. Then, in 1866, B. Nicolò Paganini (not the violinist), a 16-year-old Italian, startled the mathematical world by announcing that the numbers 1,184 and 1,210 were amicable. This second-lowest pair of all had been completely overlooked! Today, the tally of known amicable numbers has grown to about 2.5 million. No amicable pair is known in which one of the two numbers is a square. An unusually high proportion of the numbers in amicable pairs ends in either 0 or 5. A *happy amicable pair* is an amicable pair in which both numbers are **happy numbers**; an example is 10,572,550 and 10,854,650. See also **Harshad number**.

amplitude

Size or magnitude. The origin of the word is the same Indo-European *ple* root that gives us *plus* and *complement.* The more immediate Latin source is *amplus* for "wide." Today, **amplitude** is used to describe, among other things, the distance a **periodic** function varies from its central value, and the magnitude of a **complex number**.

anagram

The rearrangement of the letters of a word or phrase into another word or phrase, using all the letters only once. The best anagrams are meaningful and relate in some way to the original subject; for example, "stone age" and "stage one." There are also many remarkable examples of long anagrams. " 'That's one small step for a man; one giant leap for mankind.' Neil Armstrong" becomes "An 'Eagle' lands on Earth's Moon, making a first small permanent footprint."

PUZZLES

The reader is invited to untangle the following anagrams that give clues to famous people:
1. A famous German waltz god.
2. Aha! Ions made volts!
3. I'll make a wise phrase.

Solutions begin on page 369.

An *antonymous anagram,* or *antigram,* has a meaning opposite to that of the subject text; for example, "within earshot" and "I won't hear this." *Transposed couplets,* or *pairagrams,* are single word anagrams that, when placed together, create a short meaningful phrase, such as "best bets" and "lovely volley." A rare *transposed triplet,* or *trianagram,* is "discounter introduces reductions." See also **pangram**.

anallagmatic curve

A curve that is invariant under inversion (see **inverse**). Examples include the **cardioid**, **Cassinian ovals**, **limaçon of Pascal**, **strophoid**, and **Maclaurin trisectrix**.

analysis

A major branch of mathematics that has to do with *approximating* certain mathematical objects, such as numbers or **functions**, in terms of other objects that are easier to understand or to handle. A simple example of analysis is the calculation of the first few decimal places of **pi** by writing it as the **limit** of an **infinite series**. The origins of analysis go back to the seventeenth century, when people such as Isaac **Newton** began investigating how to approximate locally—in the neighborhood of a point—the behavior of quantities that vary continuously. This led to an intense study of limits, which form the basis of understanding infinite series, differentiation, and integration.

Modern analysis is subdivided into several areas: real analysis (the study of derivatives and integrals of real-valued functions); functional analysis (the study of spaces of functions); **harmonic analysis** (the study of **Fourier series** and their abstractions); **complex analysis** (the study of functions from the complex plane to the complex plane that are complex differentiable); and **nonstandard analysis** (the study of **hyperreal number**s and their functions, which leads to a rigorous treatment of infinitesimals and of infinitely large numbers).

analytical geometry

Also known as *coordinate geometry* or *Cartesian geometry,* the type of geometry that describes points, lines, and shapes in terms of **coordinate**s, and that uses **algebra** to prove things about these objects by considering their coordinates. René **Descartes** laid down the foundations for analytical geometry in 1637 in his *Discourse on the Method of Rightly Conducting the Reason in the Search for Truth in the Sciences,* commonly referred to as *Discourse on Method.* This work provided the basis for **calculus**, which was introduced later by Isaac **Newton** and Gottfried **Leibniz**.

analytical number theory

The branch of **number theory** that uses methods taken from **analysis**, especially **complex analysis**. It contrasts with **algebraic number theory**.

anamorphosis

The process of distorting the perspective of an image to such an extent that its normal appearance can only be restored by the observer completely changing the way he looks at the image. In *catoptric anamorphosis,* a curved mirror, usually of cylindrical or conical shape, is used to restore an anamorphic picture to its undistorted form. In other kinds of anamorphism, the observer has to change her viewing position—for example, by looking at the picture almost along its surface. Some anamorphic art adds deception by concealing the distorted image in an

anamorphosis   "Self-portrait with Albert" is a clever example of anamorphic art by the Hungarian artist Istvan Orosz. The artist's hands over his desk and a small round mirror in which the artist's face is reflected can be seen in the etching. *Istvan Orosz*

(continued)

anamorphosis *(continued)* A cylindrical mirror is placed over the circle. *Istvan Orosz*

The mirror reveals a previously unsuspected aspect of the picture. The distorting effect of the curved mirror is to undistort a face hidden amid the shapes on the desk: the face of Albert Einstein. Orosz created this etching for an exhibition in Princeton, where the great scientist lived. *Istvan Orosz*

otherwise normal looking picture. At one time, artists who had the mathematical knowledge to create anamorphic pictures kept their calculations and grids well-guarded secrets. Now it is relatively easy to create such images by computer.

angle

My geometry teacher was sometimes acute, and sometimes obtuse, but he was always right.

—Anonymous

The opening between two lines or two planes that meet; the word comes from the Latin *angulus* for "sharp bend." Angles are measured in **degrees**. A right angle has 90°, an **acute** angle less than 90°, and an **obtuse** angle has between 90° and 180°. If an angle exceeds the straight angle of 180°, it is said to be convex. *Complementary angles* add to 90°, and *supplementary angles* make a total of 180°.

angle bisection

See **bisecting an angle.**

angle trisection

See **trisecting an angle.**

animals' mathematical ability

Many different species, including rats, parrots, pigeons, raccoons, and chimpanzees, are capable of doing simple calculations. Tests on dogs have shown that they have a basic grasp of cardinality—the number of things on offer. If they're shown a pile of treats and then shown the pile again after it has been concealed and the number of treats changed slightly, they will react differently than if there's been no change. However, not all purported animal math talents stand the test of time. At the turn of the century, a horse named Clever Hans wowed audiences with his counting skills. His trainer would pose a problem, and the horse would tap out the answer. In the end, though, it was found that Hans couldn't really add or subtract but was instead responding to subtle, unintended clues from its trainer, who would visibly relax when the horse reached the correct number.

annulus

The region between the smaller and the larger of two circles that share a common center.

antigravity houses and hills

The House of Mystery in the Oregon Vortex, Gold Hill, Oregon, built during the Great Depression in the 1930s, can claim to be the first "antigravity house." It spawned

antigravity houses and hills   Visitors to a "house of mystery," believing that the floor of the house is horizontal, may be astonished by the apparent gravity-defying effects (right). All these effects are easily understood, however, when it is realized that the entire house tilts at the same angle as a hill on which it is built.

many imitators around the United States and in other parts of the world. Such buildings give rise to some spectacular visual effects, which seem bewildering until the underlying cause is revealed. Of course, the visitor guides are not forthcoming about what is really going on and make fantastic claims about magnetic or gravitational anomalies, UFOs, or other weird and wonderful phenomena. The fact is that all the stunning effects stem from clever construction and concealment that make an incline seem like a horizontal in the mind of the visitor. All antigravity houses are built on hills, with a typical incline of about 25°. But unlike a normal house on the side of a hill, an antigravity house is built so that its walls are perpendicular to the (inclined) ground. In addition, the area around the house is surrounded by a tall fence that prevents the visitor from establishing a true horizontal. Thus compelled to fall back on experience, the visitor assumes that the floor of the house is horizontal and that the walls are vertical with respect to Earth's gravity. All the stunning visual shenanigans follow from this.

In addition to man-made antigravity illusions, there are also a number of remarkable natural locations around the world where gravity seems to be out of kilter. One example is the "Electric Brae," known locally as Croy Brae, in Ayrshire, Scotland. This runs the quarter-mile from the bend overlooking the Croy railway viaduct in the west (86 meters above sea level) to the wooded Craigencroy Glen (92 meters above sea level) to the east. While there is actually a slope of 1 in 86 (a rise of 1 meter for every 86 meters horizontally) upward from the bend at the Glen, the configuration of the land on either side of the road creates the illusion that the slope runs the other way. The author is among countless folk who have parked their cars with the brakes off on this stretch of road and been amazed to see it roll apparently uphill. See also **distortion illusion**.

antimagic square

An $n \times n$ square arrangement of the numbers 1 to n^2 such that the totals of the n rows and n columns and two long diagonals form a sequence of $(2n + 2)$ consecutive integers. There are no antimagic squares of size 2×2 and 3×3 but plenty of them for larger sizes. Here is a 4×4 example:

1	13	3	12
15	9	4	10
7	2	16	8
14	6	11	5

See also **magic square**.

antiprism

A **semi-regular polyhedron** constructed from two *n-sided* **polygon**s and $2n$ triangles. An antiprism is like a **prism** in that it contains two copies of any chosen regular polygon, but is unlike a prism in that one of the copies is given a slight twist relative to the other. The polygons are connected by a band of triangles pointing alternately up and down. At each vertex, three triangles and one of the chosen polygons meet. By spacing the two polygons at the proper distance, all the triangles become equilateral. Antiprisms are named square antiprisms, pentagonal antiprisms, and so on. The simplest, the triangular antiprism, is better known as the **octahedron**.

aperiodic tiling

A **tiling** made from the same basic elements or tiles that can cover an arbitrarily large surface without ever exactly repeating itself. For a long time it was thought that whenever tiles could be used to make an aperiodic tiling, those same tiles could also be fitted together in a different way to make a **periodic tiling**. Then, in the 1960s, mathematicians began finding sets of tiles that were uniquely aperiodic. In 1966, Robert Berger produced the first set of 20,426 aperiodic tiles, and soon lowered this number to 104. Over the next few years, other mathematicians reduced the number still further.

antiprism　A pentagonal antiprism. *Robert Webb, www.software3d.com; created using Webb's Stella program*

In 1971, Raphael Robinson found a set of six aperiodic tiles based on notched squares; then, in 1974, Roger **Penrose** found a set of two colored aperiodic tiles (see **Penrose tilings**). The coloring can be dispensed with if the pieces are notched. There is a set of three convex (meaning no notches) aperiodic tiles, but it isn't known if there is a set of two such tiles or even a single tile (see **Einstein problem**). In three dimensions, Robert Ammann found two aperiodic polyhedra, and Ludwig Danzer found four aperiodic tetrahedra.

Apéry's constant
The number defined by the formula $\zeta/(3) = \sum_{n=1}^{\infty} 1/n^3$, where ζ is the **Riemann zeta function**: It has the value $1.202056\ldots$ and gives the odds (1 in $1.202056\ldots$) of any three positive integers, picked at random, having no common divisor. In 1979, the French mathematician Roger Apéry (1916–1994) stunned the mathematical world with a proof that this number is irrational.[11] Whether it is a **transcendental number** remains an open question.

apex
The **vertex** of a cone or a pyramid.

apocalypse number
See **beast number**.

Apollonius of Perga (c. 255–170 B.C.)
A highly influential Greek mathematician (born in a region of what is now Turkey), known as the "Great Geometer," whose eight-part work *On Conics* introduced such terms as *ellipse, parabola,* and *hyperbola.* **Euclid** and others had written earlier about the basic properties of **conic section**s but Apollonius added many new results, particularly related to **normal**s and **tangent**s to the various conic curves. One of the most famous questions he raised is known as the **Apollonius problem**. He also wrote widely on other subjects including science, medicine, and philosophy. In *On the Burning Mirror* he showed that parallel rays of light are not brought to a focus by a spherical mirror (as had been previously thought), and he discussed the focal properties of a parabolic mirror. A few decades after his death, Emperor Hadrian collected Apollonius's works and ensured their publication throughout his realm.

Apollonius problem
A problem first recorded in *Tangencies*, written around 200 B.C. by **Apollonius of Perga**. Given three objects in the plane, each of which may be a circle C, a point P (a degenerate circle), or a line L (part of a circle with infinite

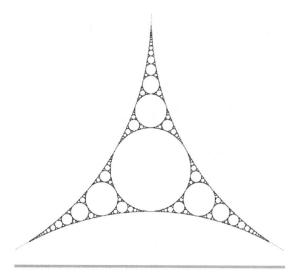

Apollonius problem   The Apollonian gasket.

radius), find another circle that is tangent to (just touches) each of the three. There are ten cases: *PPP, PPL, PLL, LLL, PPC, PLC, LLC, LCC, PCC, CCC.* The two easiest involve three points or three straight lines and were first solved by **Euclid**. Solutions to the eight other cases, with the exception of the three-circle problem, appeared in *Tangencies;* however, this work was lost. The most difficult case, to find a tangent circle to any three other circles, was first solved by the French mathematician François Viète (1540–1603) and involves the simultaneous solution of three **quadratic** equations, although, in principle, a solution could be found using just a compass and a straightedge. Any of the eight circles that is a solution to the general three-circle problem is called an *Apollonius circle.* If the three circles are mutually tangent then the eight solutions collapse to just two, which are known as **Soddy circles**. A **fractal** is produced by starting with three mutually tangent circles and creating a fourth–the inner Soddy circle–that is nested between the original three. The process is repeated to yield three more circles nested between sets of three of these, and then repeated again indefinitely. The points that are never inside a circle form a fractal set called the *Apollonian gasket,* which has a fractional dimension of about 1.30568.

apothem
Also known as a *short radius,* the perpendicular distance from the center of a **regular polygon** to one of its sides. It is the same as the radius of a circle inscribed in the polygon.

apotome

One of Euclid's categories of **irrational number**s. An apotome has the form $\sqrt{(\sqrt{A} - \sqrt{B})}$. The corresponding number with a "+" sign is called a *binomial* in Euclid's scheme.

applied mathematics

Mathematics for the sake of its use to science or society.

Arabic numeral

A **numeral** written with an Arabic digit alone: 0, 1, 2, 3, 4, 5, 6, 7, 8, 9, or in combination: 10, 11, 12, . . . 594, ...

arbelos

A figure bounded by three semicircles *AB*, *BC*, and *AC*, where *ABC* is a straight line. Archimedes (about 250 B.C.) called it an *arbelos*–the Greek word for a knife of the same shape used by shoemakers to cut and trim leather–and wrote about it in his *Liber assumptorum* (Book of lemmas). Among its properties are that the sum of the two smaller arc lengths is equal to the larger; the area of the arbelos is $\pi/4$ times the product of the two smaller diameters (*AB* and *BC*); and the area of the arbelos is equal to the area of a circle whose diameter is the length of a perpendicular segment drawn from the tangent point *B* of the two smaller semicircles to the point *D*, where it meets the larger semicircle. The circles inscribed on each half of *BD* of the arbelos (called *Archimedes's circles*) each have a diameter of (*AB*)(*BC*)/(*AC*). Furthermore, the smallest circumcircle of these two circles has an area equal to that of the arbelos. **Pappus of Alexandria** wrote on the relations of the chain of circles, C_1, C_2, C_3, . . . (called a *Pappus chain* or an *arbelos train*) that are mutually tangent to the two largest semicircles and to each other. The centers of these circles lie on an ellipse and the diameter of the *n*th circle is $(1/n)$ times the base of the perpendicular distance to the base of the semicircle.

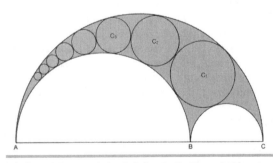

arbelos A Pappus chain of circles, C_1, C_2, C_3, . . ., inside an arbelos (shaded region).

arc

Any part of a curved line or part of the circumference of a circle; the word comes from the Latin *arcus* for a bow, which also gives rise to *arch*. *Arc length* is the distance along part of a curve.

arch

A strong, curved structure, traditionally made from wedge-shaped elements, that may take many different forms and that provides both an opening and a support for overlying material. Two common forms are the semicircular arch, first used by the Romans, and the pointed Gothic arch. The semicircular arch is the weaker of the two because it supports all the weight on the top and tends to flatten at its midpoint. It also requires massive supporting walls since all the stress on the arch acts purely downward. The pointed arch, by contrast, directs stresses both vertically and horizontally, so that the walls can be thinner, though buttressing may be required to prevent the walls from collapsing sideways. See also **Vesica Piscis**.

Archimedean dual

See **Catalan solid**.

Archimedean solid

A convex **semi-regular polyhedron**; a solid made from regular polygonal sides of two or more types that meet in a uniform pattern around each corner. (A regular polyhedron, or **Platonic solid**, has only one type of polygonal side.) There are 13 Archimedean solids (see table "Archimedian Solids"). Although they are named after their discoverer, the first surviving record of them is in the fifth book of the *Mathematical Collection* of **Pappus of Alexandria**. The **dual**s of the Archimedean solids (made by replacing each face with a vertex, and each vertex with a face) are commonly known as **Catalan solid**s. Apart from the Platonic and Archimedean solids, the only other convex uniform polyhedra with regular faces are **prism**s and **antiprism**s. This was shown by Johannes **Kepler**, who also gave the names generally used for the Archimedean solids. See also **Johnson solid**.

Archimedean spiral

A **spiral**, like that of the groove in a phonograph record, in which the distance between adjacent coils, measured radially out from the center, is constant. **Archimedes** was the first to study it and it was the main subject of his treatise *On Spirals*. The Archimedean spiral has a very simple equation in **polar coordinates** (r, θ):

$$r = a + b\theta$$

where *a* and *b* can be any real numbers. Changing the parameter *a* turns the spiral, while *b* controls the distance

Archimedean Solids

Name	Vertices	Number of	
		Faces	Edges
Truncated tetrahedron	8 = 4 + 4	12	18
Truncated cube	14 = 8 + 6	24	36
Truncated octahedron	14 = 6 + 8	24	36
Truncated dodecahedron	32 = 20 + 12	60	90
Truncated icosahedron	32 = 12 + 20	60	90
Cuboctahedron	14 = 8 + 6	12	24
Icosidodecahedron	32 = 20 + 12	30	60
Snub dodecahedron	92 = 80 + 12	60	150
Rhombicuboctahedron	26 = 8 + 18	24	48
Great rhombicosidodecahedron	62 = 30 + 20 + 12	120	180
Rhombicosidodecahedron	62 = 20 + 30 + 12	60	120
Great rhombicuboctahedron	26 = 8 + 12 + 6	48	72
Snub cube	38 = 32 + 6	24	60

Archimedean solid The complete set of Archimedean solids, starting far left and going clockwise: truncated cube, small rhombicuboctahedron, great rhombicuboctahedron, snub cube, snub dodecahedron, great rhombicosidodecahedron, small rhombicosidodecahedron, truncated dodecahedron, truncated icosahedron (soccer ball), icosidodecahedron, truncated tetrahedron, cuboctahedron, and truncated octahedron. *Robert Webb, www.software3d.com; created using Webb's Stella program*

between the arms. The Archimedean spiral is distinguished from the **logarithmic spiral** by the fact that successive arms have a fixed distance (equal to $2\pi b$ if θ is measured in radians), whereas in a logarithmic spiral these distances form a **geometric sequence**. Note that the Archimedean spiral has two possible arms that coil in opposite directions, one for $\theta > 0$ and the other for $\theta < 0$. Many examples of spirals in the man-made world, such as a watch spring or the end of a rolled carpet, are either Archimedean spirals or another curve that is very much like it, the **circle involute**.

Archimedean tessellation

Also known as a *semiregular tessellation,* a **tiling** that uses only regular **polygon**s arranged so that two or more different polygons are around each vertex and each vertex involves the same pattern of polygons. There are eight such tessellations, two involving triangles and squares, two involving triangles and hexagons, and one each involving squares and octagons; triangles and dodecagons; squares, hexagons, and dodecagons; and triangles, squares, and hexagons.

Archimedes of Syracuse (c. 287–212 B.C.)

One of the greatest mathematicians and scientists of all time. He became a popular figure because of his involvement in the defense of Syracuse against the Roman siege in the first and second Punic Wars when his war machines helped keep the Romans at bay. He also devised a scheme to move a full-size ship, complete with crew and cargo, by pulling a single rope, and invented the irrigation device known as the Archimedean screw. According to one of many legends about him, he is said to have discovered the principle of buoyancy while taking a bath and then ran into the street naked shouting "eureka" ("I found it!").

In his book *The Sand-Reckoner,* he described a positional **number system** and used it to write the equivalent of numbers up to 8×10^{64}–the number of grains of sand he thought it would take to fill the universe. He devised a rule-of-thumb method to do private calculations that closely resembles integral calculus (2,000 years before its "discovery"), but then switched to geometric proof for his results. He demonstrated that the ratio of a circle's perimeter to its diameter is the same as the ratio of the circle's area to the square of the radius. Although he didn't call this ratio **"pi,"** he showed how to work it out to arbitrary accuracy and gave an approximation of it as "exceeding 3 in less than $^1/_7$ but more than $^{10}/_{71}$."

Archimedes was the first, and possibly the only, Greek mathematician to introduce mechanical curves (those traced by a moving point) as legitimate objects of study, and he used the **Archimedean spiral** to square the circle. He proved that the area and volume of the **sphere** are in the same ratio to the area and volume of a circumscribed straight cylinder, a result that pleased him so much that he made it his epitaph. Archimedes is probably also the first mathematical physicist on record, and the best before Galileo and Isaac **Newton**. He invented the field of statics, enunciated the law of the lever, the law of equilibrium of fluids, and the law of buoyancy, and was the first to identify the concept of center of gravity. He is also, perhaps erroneously, credited with the invention of a square **dissection** puzzle known as the **loculus of Archimedes**. Many of his original works were lost when the library at Alexandria burned down and they survive only in Latin or Arabic translations. Plutarch wrote of him: "Being perpetually charmed by his familiar siren, that is, by his geometry, he neglected to eat and drink and took no care of his person; that he was often carried by force to the baths, and when there he would trace geometrical figures in the ashes of the fire, and with his finger draws lines upon his body when it was anointed with oil, being in a state of great ecstasy and divinely possessed by his science."

Archimedes's cattle problem

A fiendishly hard problem involving very **large numbers** that **Archimedes** presented in a 44-line letter to **Eratosthenes,** the chief librarian at Alexandria. It ran as follows:

If thou art diligent and wise, O stranger, compute the number of cattle of the Sun, who once upon a time grazed on the fields of the Thrinacian isle of Sicily, divided into four herds of different colors, one milk white, another a glossy black, a third yellow and the last dappled. In each herd were bulls, mighty in number according to these proportions: Understand, stranger, that the white bulls were equal to a half and a third of the black together with the whole of the yellow, while the black were equal to the fourth part of the dappled and a fifth, together with, once more, the whole of the yellow. Observe further that the remaining bulls, the dappled, were equal to a sixth part of the white and a seventh, together with all of the yellow. These were the proportions of the cows: The white were precisely equal to the third part and a fourth of the whole herd of the black; while the black were equal to the fourth part once more of the dappled and with it a fifth part, when all, including the bulls, went to pasture together. Now the dappled in four parts were equal in number to a fifth part and a sixth of the yellow herd. Finally the yellow were in number equal to a sixth part and a seventh of the white herd. If thou canst accurately tell, O stranger, the number of cattle of the Sun, giving separately the number of well-fed bulls and again the number of females according to each color, thou wouldst not be called unskilled or ignorant of numbers, but not yet shalt thou be numbered among the wise.

But come, understand also all these conditions regarding the cattle of the Sun. When the white bulls mingled their number with the black, they stood firm, equal in depth and breadth, and the plains of Thrinacia, stretching far in all ways, were filled with their multitude. Again, when the yellow and the dappled bulls were gathered into one herd they stood in such a manner that their number, beginning from one, grew slowly greater till it completed a triangular figure, there being no bulls of other colors in their midst nor none of them lacking. If thou art able, O stranger, to find out all these things and gather them together in your mind, giving all the relations, thou shalt depart crowned with glory and knowing that thou hast been adjudged perfect in this species of wisdom.

The answer to the first part of the problem—the smallest solution for the total number of cattle—turns out to be 50,389,082. But when the extra two constraints in the second part are factored in, the solution is vastly larger. The approximate answer of 7.76×10^{202544} was found in 1880 by A. Amthor, having reduced the problem to a form called a **Pell equation**.[9] His calculations were continued by an ad hoc group called the Hillsboro Mathematical Club, of Hillsboro, Illinois, between 1889 and 1893. The club's three members (Edmund Fish, George Richards, and A. H. Bell) calculated the first 31 digits and the last 12 digits of the smallest total number of cattle to be

7760271406486818269530232833**209** . . . 719455081800

though the two digits in bold should be 13.[31] In 1931, a correspondent to the *New York Times* wrote: "Since it has been calculated that it would take the work of a thousand men for a thousand years to determine the complete [exact] number [of cattle], it is obvious that the world will never have a complete solution." But *obvious* and *never* are words designed to make fools of prognosticators. Enter the computer. In 1965, with the help of an IBM 7040, H. C. Williams, R. A. German, and

C. R. Zarnke reported a complete solution to the cattle problem, though it was 1981 before all 202,545 digits were published, by Harry Nelson, who used a Cray-1 supercomputer to generate the answer, which begins: $7.760271406486818269530232833213 . . . \times 10^{202544}$.[341]

Archimedes's square
See **loculus of Archimedes**.

area
A measure of surface extension in two-dimensional space. *Area* is the Latin word for a vacant piece of level ground and still carries this common meaning. The French shortened form *are* denotes a square of land with a side length of 10 meters, that is, an area of 100 square meters. A *hectare* is a hundred are.

area codes
North American telephone area codes seem to have been chosen at random. But there was a method to their selection. In the mid-1950s when direct dialing of long-distance calls first became possible, it made sense to assign area codes that took the shortest time to dial to the larger cities. Almost all calls were from rotary dials. Area codes such as 212, 213, 312, and 313 took very little time for the dial to return to its starting position compared, for example, to numbers such as 809, 908, 709. The quickest-to-dial area codes were assigned to the places expected to receive the most direct-dialed calls. New York City got 212, Chicago 312, Los Angeles 213, and Washington, D.C., 202, which is a little longer to dial than 212, but much shorter than others. In order of decreasing size and estimated amount of telephone traffic, the numbers grew larger: San Francisco got 415, Miami 305, and so on. At the other end of the spectrum came places like Hawaii (the last state annexed in 1959) with 808, Puerto Rico with 809, and Newfoundland with 709. The original plan (still in use until about 1993) was that area codes had a certain construction to the numbers: the first digit is 2 through 9, the second digit is 0 or 1, and the third digit is 1 through 9. Three-digit numbers with two zeros are special codes, that is, 700, 800, or 900. Three-digit numbers with two ones are for special local codes such as 411 for local directory assistance, 611 for repairs, and so forth.

Argand diagram
A way of representing **complex number**s as points on a coordinate plane, also known as the *Argand plane* or the *complex plane,* using the x-axis as the real axis and the y-axis as the imaginary axis. It is named for the French amateur mathematician Jean Robert Argand

(1768–1822) who described it in a paper in 1806.[14] John **Wallis** suggested a similar method 120 years earlier and Casper **Wessel** extensively developed it. But Wessel's paper was published in Danish and wasn't circulated in the languages more common to mathematics at that time. In fact, it wasn't until 1895 that his paper came to the attention of the mathematical community—long after the name "Argand diagram" had stuck.

argument

(1) The input for a **function**. (2) The angle between ZO, where Z is the point representing a complex number on an **Argand diagram** and O is the origin, and the real axis. (3) A mathematical proof, possibly an informal one.

Aristotle's wheel

A **paradox** mentioned in the ancient Greek text *Mechanica,* whose author is unknown but is suspected by some to have been Aristotle. The paradox concerns two concentric circles on a wheel, as shown in the diagram. A **one-to-one** correspondence exists between points on the larger circle and those on the smaller circle. Therefore, the wheel should travel the same distance regardless of whether it is rolled from left to right on the top straight line or on the bottom one. This seems to imply that the two circumferences of the different-sized circles are equal, which is impossible. How can this apparent contradiction be resolved? The key lies in the (false) assumption that a one-to-one correspondence of points means that two curves must have the same length. In fact, the cardinalities of points in a line segment of any length (or even an infinitely long line or an infinitely large *n*-dimensional Euclidean space) are all the same. See also **infinity**.

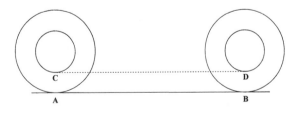

Aristotle's wheel   The outer circle turns once when going from A to B, as does the inner circle when going from C to D. Yet AB is the same length as CD. How can this be, since the circles are a different size?

arithmetic

A branch of mathematics concerned with doing calculations with numbers using addition, subtraction, multiplication, and division.

arithmetic mean

The sum of *n* given numbers divided by *n*. See also **geometric mean** and **harmonic mean**.

arithmetic sequence

Also known as an *arithmetic progression,* a finite **sequence** of at least three numbers, or an infinite sequence, whose terms differ by a constant, known as the *common difference.* For example, starting with 1 and using a common difference of 4 we can get the finite arithmetic sequence: 1, 5, 9, 13, 17, 21, and also the infinite sequence 1, 5, 9, 13, 17, 21, 25, 29, . . . , $4n + 1$, ... In general, the terms of an arithmetic sequence with the first term a_0 and common difference d, have the form $a_n = dn + a_0$ ($n = 1, 2, 3, . . .$). Does every increasing sequence of integers have to contain an arithmetic progression? Surprisingly, the answer is no. To construct a counterexample, start with 0. Then for the next term in the sequence, take the *smallest possible integer* that doesn't cause an arithmetic progression to form in the sequence constructed thus far. (There must be such an integer because there are infinitely many integers beyond the last term, and only finitely many possible progressions that the new term could complete.) This gives the nonarithmetic sequence 0, 1, 3, 4, 9, 10, 12, 13, 27, 28, ...

If the terms of an arithmetic sequence are added together the result is an *arithmetic series,* $a_0 + (a_0 + d) + . . . + (a_0 + (n - 1)d)$, the sum of which is given by:

$$S_n = n/2 \ (2a_0 + (n - 1)d) = n/2 \ (a_0 + a_n).$$

See also **geometric sequence**.

around the world game

See **Icosian game**.

array

A set of numbers presented in a particular pattern, usually a grid. Matrices (see **matrix**) and **vector**s are examples of arrays.

Arrow paradox

The oldest and best-known **paradox** related to voting. The American economist Kenneth Arrow (1921–) showed that it is impossible to devise a perfect democratic voting system. In his book *Social Choice and Individual Values,*[16] Arrow identified five conditions that are universally regarded as essential for any system in which social decisions are based on individual voting preferences. The Arrow paradox is that these five conditions are logically inconsistent: under certain conditions, at least one of the essential conditions will be violated.

arrowhead

See **dart**.

artificial intelligence (AI)

The subject of "making a machine behave in ways that would be called intelligent if a human were so behaving," according John McCarthy, who coined the term in 1955. How can we tell if a computer has acquired AI at a human level? One way would be to apply the **Turing test**, though not everyone agrees that this test is foolproof (see **Chinese room**). Certainly, AI has not developed at nearly the rate many of its pioneers expected back in the 1950s and 1960s. Meanwhile, progress in fields such as **neural network**s and **fuzzy logic** continues to be made, and most computer scientists have no doubt that it is only a matter of time before computers are outperforming their biological masters in a wide variety of tasks beyond those that call for mere number-crunching ability.

artificial life

A lifelike pattern that may emerge from a **cellular automaton** and appear organic in the way it moves, grows, changes shape, reproduces, aggregates, and dies. Artificial life was pioneered by the computer scientist Chris Langton, and has been researched extensively at the Santa Fe Institute. It is being used to model various **complex system**s such as ecosystems, the economy, societies and cultures, and the immune system. The study of artificial life, though controversial, promises insights into natural processes that lead to the buildup of structure in self-organizing (see **self-organization**) systems.

associative

Three numbers, x, y, and z, are said to be *associative under addition* if

$$x + (y + z) = (x + y) + z,$$

and to be *associative under multiplication* if

$$x \times (y \times z) = (x \times y) \times z.$$

In general, three elements a, b, and c of a set S are associative under the binary operation (an operation that works on two elements at a time) * if

$$a * (b * c) = (a * b) * c.$$

The word incorporates the Greek root *soci*, from which we also get *social*, and may have been first used in the modern mathematical sense by William **Hamilton** around 1850. Compare with **distributive** and **commutative**.

astroid

A **hypocycloid**—the path of a point on a circle rolling inside another circle—for which the radius of the inner circle is four times smaller than that of the larger circle; this ratio results in the astroid having four cusps. The astroid was first studied by the Danish astronomer Ole Römer in 1674, in his search for better shapes for gear teeth, and later by Johann Bernoulli (1691) (see **Bernoulli family**), Gottfried **Leibniz** (1715), and Jean **d'Alembert** (1748). Its modern name comes from the Greek *aster* for "star" and was introduced in a book by Karl Ludwig von Littrow published in Vienna in 1836; before this, the curve had a variety of names, including tetracuspid (still used), cubocycloid, and paracycle. The astroid has the Cartesian equation

$$x^{2/3} + y^{2/3} = r^{2/3}$$

where r is the radius of the fixed outer circle, and $r/4$ is the radius of the rolling circle. Its area is $3\pi r^2/8$, or $3/2$ times that of the rolling circle, and its length is $6r$. The astroid is a sextic curve and also a special form of a **Lamé curve**. It has a remarkable relationship with the quadrifolium (see **rose curve**): the radial, pedal, and orthoptic of the astroid are the quadrifolium, while the cata**caustic** of the quadrifolium is the astroid. The

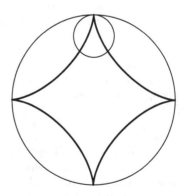

astroid   As a small circle rolls around the inside of a larger one with exactly four times its circumference, a point on the rim of the small circle traces out an astroid. © *Jan Wassenaar, www.2dcurves.com*

astroid is also the catacaustic of the **deltoid** and the **evolute** of the ellipse.

asymptote

A curve that gets closer and closer to a fixed straight line without ever actually touching it. Imagine facing along the direction of a great wall that is just a meter to your left. Every second, you walk forward a meter and at the same time move sideways slightly so that you halve the distance between you and the wall. The path you follow is an asymptote. The word comes from the Greek roots *a* (not), *sum* (together), and *piptein* (to fall), so that it literally means "not falling together" and was originally used in a wider sense to describe any two curves that don't intersect. **Proclus** writes about both asymptotic lines and symptotic lines (those that do cross). Nowadays, "symptotic" is almost never heard, and "asymptote" is used mainly to denote lines that serve as a limiting barrier for some curve as one of its parameters approaches plus or minus infinity. The "~" symbol is often used to show that one **function** is asymptotic to another. For example, $f(x) \sim g(x)$ indicates that the ratio of the functions $f(x)$ to $g(x)$ approaches 1 as x tends to infinity. Asymptotes are not always parallel to the x- and y-axes, as shown by the graph of $x + 1/x$, which is asymptotic to both the y-axis and the diagonal line $y = x$.

Atiyah, Michael Francis (1929–)

An English mathematician who has contributed to many topics in mathematics, notably dealing with the relationships between geometry and **analysis**. In **topology**, he developed *K-theory*. He proved the *index theorem* on the number of solutions of elliptic differential equations, linking **differential geometry**, topology, and analysis—a theorem that has been usefully applied to quantum theory. Atiyah was influential in initiating work on gauge theories and applications to nonlinear differential equations, and in making links between topology and quantum field theory. Theories of superspace and supergravity, and string theory, were all developed using ideas introduced by him.

Atomium, the

A giant steel monument in Heysel Park, Brussels, Belgium, consisting of 9 spheres that represent the body-centered cubic structure of an iron crystal magnified 150 billion times. Designed by the architect André Waterkeyn and built for the 1958 World's Fair, the 103-meter-high Atomium was originally meant to stand for only 6 months. It may be the world's largest cube. Each of its spheres have a diameter of 18 meters and are connected by escalators. Three of the upper spheres have no vertical support, and so for safety reasons are not open to the public. However, the top sphere offers a panoramic view of Brussels through its windows, and the lower spheres contain various exhibitions.

attractor

A trajectory, or set of points in **phase space**, toward which nearby orbits converge, and which is stable. Specific types of attractor include **fixed-point attractor**, **periodic attractor**, and **chaotic attractor**.

Aubel's theorem

Given a quadrilateral and a square drawn on each side of it, the two lines connecting the centers of the squares on opposite sides are perpendicular and of equal length.

autogram

See **self-enumerating sentence**.

automorphic number

Also known as an *automorph*, a number n whose square ends in n. For instance 5 is automorphic, because $5^2 = 25$, which ends in 5. A number n is called *trimorphic* if n^3 ends in n. For example $49^3 = 117,649$, is trimorphic. Not all trimorphic numbers are automorphic. A number n is called *tri-automorphic* if $3n^2$ ends in n; for example 6,667 is tri-automorphic because $3 \times 667^2 = 133,346,667$ ends in 7.

automorphism

An **isomorphism** from a **set** onto itself. An *automorphism group* of a **group** G is the group formed by the automorphisms of G (bijections from G to itself that preserve the multiplication). Similarly, one can consider the automorphism groups of other structures such as **rings** and **graph**s, by looking at bijections that preserve their mathematical structure.

Avagadro constant

One of the best known examples of a **large number** in science. It is named after the Italian physicist Amedio Avagadro (1776–1856) and is defined as the number of carbon atoms in 12 grams of pure carbon, or, more generally as the number of atoms of n grams in an element with atomic weight n. It has the value $6.02214199 \times 10^{23}$.

average

A vague term that usually refers to the **arithmetic mean** but can also signify the **median**, **mode**, **geometric mean**, or weighted mean. The word stems from a commercial practice of the shipping age. The root *aver* means to declare, and the shippers of goods would declare the value of their goods. When the goods were sold, a deduction was made from each person's share, based on their declared value, for a portion of the loss or "average."

axiom

A statement that is considered to be true without need of **proof**. The term *axiom* comes from the Greek *axios* meaning "worthy" and was used by many Greek philosophers and mathematicians, including **Aristotle**. Curiously, **Euclid**, whose axioms are best known of all, seems to have favored a more general phrase meaning "common notion."

axiom of choice

An **axiom** in **set theory** that is one of the most controversial axioms in mathematics; it was formulated in 1904 by the German mathematician Ernst Zermelo (1871–1953) and, at first, seems obvious and trivial. Imagine there are many—possibly an unlimited number of—boxes in front of you, each of which has at least one thing in it. The axiom of choice (AC) says simply that you can always choose one item out of each box. More formally, if S is a collection of nonempty sets, then there exists a set that has exactly one element in common with every set S of S. Put another way, there exists a function f with the property that, for each set S in the collection, $f(S)$ is a member of S. Bertrand **Russell** summed it up neatly: "To choose one sock from each of infinitely many pairs of socks requires the Axiom of Choice, but for shoes the Axiom is not needed." His point is that the two socks in a pair are identical in appearance, so, to pick one of them, we have to make an arbitrary choice. For shoes, we can use an explicit rule, such as "always choose the left shoe." Russell specifically mentions *infinitely many* pairs, because if the number is finite then AC is superfluous: we can pick one member of each pair using the definition of "nonempty" and then repeat the operation finitely many times using the rules of formal logic.

AC lies at the heart of a number of important mathematical arguments and results. For example, it is equiv-

alent to the *well-ordering principle,* to the statement that for any two **cardinal number**s m and n, then $m < n$ or $m = n$ or $m > n$, and to *Tychonoff's theorem* (the product of any collection of compact spaces in topology is compact). Other results hinge upon it, such as the assertion that every infinite set has a denumerable subset. Yet AC was strongly attacked when it was first suggested, and still makes some mathematicians uneasy. The central issue is what it means to *choose* something from the sets in question and what it means for the choosing function to *exist*. This problem is brought into sharp focus when S happens to be the collection of all nonempty subsets of the **real number**s. No one has ever found a suitable choosing function for this collection, and there are good reasons to suspect that no one ever will. AC just mandates that there *is* such function. Because AC conjures up sets without offering workable procedures, it is said to be *nonconstructive,* as are any theorems whose proofs involve AC. Another reason that some mathematicians aren't greatly enamored with AC is that it implies the existence of some bizarre counterintuitive objects, the most famous and notorious example of which is the **Banach-Tarski paradox**. The main reason for accepting AC, as the majority of mathematicians do (albeit often reluctantly), is that it is useful. However, as a result of work by Kurt **Gödel** and, later, by Paul Cohen, it has been proven to be independent of the remaining axioms of set theory. Thus there are no contradictions in choosing to reject it; among the alternatives are to adopt a contradictory axiom or to use a completely different framework for mathematics, such as **category theory**.

axis

A line with respect to which a curve or figure is drawn, measured, rotated, and so forth. The word comes from the Greek root *aks* for a point of turning or rotation and seems to have first been used in English by Thomas Digges around 1570 in reference to the rotational axis of a right circular cone.

Babbage, Charles (1791–1871)

On two occasions I have been asked [by members of Parliament], "Pray, Mr. Babbage, if you put into the machine wrong figures, will the right answers come out?" I am not able rightly to apprehend the kind of confusion of ideas that could provoke such a question.

An English mathematician who served as Lucasian Professor of Mathematics at Cambridge (1828–1839) and became the most important figure in the prehistory of computers. Babbage noted that astronomical and other mathematical tables of the period were riddled with errors because all the calculations had to be done by hand. This gave him the idea of building a machine that would do the tedious work of computation more accurately, faster, and without ever getting tired.

In 1822, Babbage wrote a letter to one of the top British scientists of the day, Humphrey Davy, in which he talked about the design of an automatic calculator. Shortly after, he was given a grant by the British government to build this device—an elaborate symphony of rods and interlocking gear teeth—which Babbage called the Difference Engine. Construction started but never finished. Despite heroic efforts to construct a working model, the critical tolerances were beyond what engineers could provide in the first half of the eighteenth century (though the resultant gear-making skills gave Britain an edge in precision machinery for several decades and even contributed to the qualitative superiority of the British navy in World War I). The government had spent £17,000, and Babbage contributed a similar amount of his own money, on the project, when Babbage set his sights on something even more ambitious. He grasped

Babbage, Charles   The realization of Babbage's dream: the Manchester Mark 1 computer at Manchester University, England, in 1948. This was the first computer that could store both data and programs electronically. *Ferranti Electronics Ltd.*

that the basic mechanisms of the Difference Engine could be generalized to an all-purpose calculating machine, programmable by a punched-card mechanism like that of a Jacquard loom. This vastly more powerful machine was called the Analytical Engine and would have been the world's first true computer. But it never got off the ground. "He was ill-judged enough," wrote the secretary of the Royal Astronomical Society, "to press the consideration of this new machine upon the members of Government, who were already sick of the old one." Prime Minister Robert Peel was less than enthusiastic: "I would like a little previous consideration before I move in a thin house of country gentlemen a large vote for the creation of a wooden man to calculate tables from the formula $x^2 + x + 41$." The government's eventual withdrawal of support for his schemes left Babbage a disappointed and embittered man. However, his ideas survived and proved to be the forerunner of modern computers. Parts of his uncompleted mechanisms are on display in the London Science Museum. In 1991, working from Babbage's original plans, a Difference Engine was completed—and functioned perfectly. Among Babbage's many lesser known accomplishments was his cracking of the Vigenère cipher, a discovery that helped English military campaigns but wasn't published for several years, by which time the credit had gone instead to Friedrich Kasiski, who broke the code some years after Babbage.[329] See also **Byron, Ada**.

Bachet de Méziriac, Claude-Gaspar (1581–1638)

A poet and early mathematician of the French Academy, best known for his 1621 translation of **Diophantus's** *Arithmetica,* the book that Pierre de **Fermat** was reading when he inscribed the margin with his famous last theorem. Bachet is also remembered as a collector of mathematical puzzles, many of which he published in *Problèmes plaisans et délectables qui font par les nombres* (1612) (Pleasant and delightful problems that involve numbers), including **river-crossing problems, measuring and weighing puzzles,** number tricks, and **magic squares**. One of the puzzles is to find the least number of weights that can be used on a scale pan to weigh any integral number of pounds from 1 to 40 inclusive, if the weights can be placed in either of the scale pans. The answer is four: 1, 3, 9, and 27 pounds. On a slightly more serious note, Bachet observed that apparently every positive number can be expressed as a sum of at most four squares; for example, $5 = 2^2 + 1^2$, $6 = 2^2 + 1^2 + 1^2$, $7 = 2^2 + 1^2 + 1^2 + 1^2$, $8 = 2^2 + 2^2$, and $9 = 3^2$. The case of 7 shows that sometimes three squares wouldn't be enough. Bachet said he had checked this for more than 300 numbers but didn't know how to prove it. It wasn't until the late eighteenth century that Joseph **Lagrange** supplied a complete proof.[339]

backgammon

A gambling game for two in which each player seeks to get a set of pieces from one side of the board to the other, while trying to prevent the other player from doing the same. The distance that a piece can be moved at each turn is determined by the throw of dice.

Backgammon has roots stretching back 5,000 years. From Mesopotamia, versions of it spread to Greece and Rome as well as to India and China. The rules of the modern form of the game were largely established in England in 1743 by Edmond Hoyle but benefited from a crucial modification that emerged in American gambling clubs in the 1920s. This final innovation, which added a new level of subtlety, is known as the *doubling cube.*

Backgammon is played with two sets of 15 checkers: one player has black, the other white. The players' checkers move in opposite directions on a board with 24 spaces or *points*. Each player's goal is to be the first to bring all their own checkers "home" (into their own quarter of the board) and then "bear them off" (remove them from the board altogether). The movement of the checkers follows the outcome of a roll of two dice, the numbers on the two dice constituting separate moves.

The actual amount that changes hands at the end of the game can be more than the initial stake. For instance, in certain winning positions called *gammon* and *backgammon,* the stake is doubled or trebled, respectively. The other way the stake can change is by means of the doubling cube. If one of the players thinks that she is in a winning position, she can turn the doubling cube and *announce a double,* which means that the total stake will be doubled. If her opponent *refuses the double,* he immediately loses his (undoubled) stake and the game is finished. If he *accepts the double,* the stakes are doubled and, as a compensation, the doubling cube is handed over to him and he gets the exclusive right to announce the next double. (He is now said to *own the cube.*) If the luck of the game changes so that he later judges that he is now winning, he'll be in a position to announce a *redouble,* which means that the stake is doubled again. If the first player refuses the double, she now loses the doubled stake; if she accepts, the game goes on with a redoubled stake, four times the original value. There's no limit to how many times the stake can be doubled, but the right to announce a double switches from one player to the other every time it's exercised. (Initially either player can double—no one owns the cube.) This aspect of the game adds greatly to the variety of tactical possibilities and problems.

baker's dozen
See **thirteen**.

Bakhshali manuscript

An early mathematical manuscript, written on birch bark and found in the summer of 1881 near the village of Bakhshali in the Yusufzai subdivision of the Peshawar district (now in Pakistan). A large part of the manuscript had been destroyed and only about 70 leaves of birch bark, of which a few were mere scraps, survived to the time of its discovery. Although its date is uncertain, it is most commonly put at about the third or fourth century A.D. and appears to be a commentary on an earlier mathematical work. Among the rules and techniques it sets out for solving problems, mostly in arithmetic and algebra, but also to a lesser extent in geometry and mensuration, is this formula (stated here in modern terms) for calculating the square root of a nonsquare number Q:

$$\sqrt{Q} = \sqrt{(A^2 + b)} = A + b/2A - (b/2A)^2/[2(A + b/2A)]$$

If $Q = 41$ (so that $A = 6$ and $b = 5$) this gives $\sqrt{Q} = 6.403138528$, which compares very favorably with the correct result of 6.403124237.

ball

Mathematicians, unlike the rest of the human race, draw a sharp distinction between a **sphere** and a ball. A sphere (in mathematics) is only a surface, whereas a ball is everything inside, and possibly including, that surface—the filling of the sphere. An *open ball* consists of all the points that are less than a given distance (the radius) away from a given point (the center); a *closed ball* consists of all the points that are less than or equal to the radius.

Mathematical balls can also exist in any number of dimensions. A one-dimensional ball of radius r is just a line segment. It consists of all the points on a line between $-r$ and r, or, in the case of a one-dimensional *unit ball* (a ball with a radius of 1), between -1 and 1. A 1 – d unit ball thus has a length, or "1 – d volume," of 2. A 2 – d unit ball, which is the filling of a unit circle, has an area, or 2 – d volume, of π. The volume of a unit ball in 3 – d is $4/3\pi$. In 4 – d it is $\pi^2/2$. Apparently, as the number of dimensions increases, so does the volume of the unit ball. What does this volume tend to do as the dimension tends to infinity? Intuitively, it might seem that in higher and higher dimensions there's more and more "room" in the unit ball, allowing its volume to become larger and larger. Does the volume become infinite, or does it approach a sufficiently large constant as the dimension increases? The answer is surprising and shows how our intuition is often misleading. Using a technique called *multivariable calculus* the volume of the unit ball in n dimensions, $V(n)$, can be shown to be $\pi^{n/2}/\Gamma(n/2 + 1)$, where Γ is the **gamma function** that generalizes the factorial function (i.e., $\Gamma(z + 1) = z!$). For n even, say $n = 2k$, the volume of the unit ball is thus given by $V(n) = \pi^k / k!$.

Ball, Walter William Rouse (1850–1925)

A British mathematician who lectured at Trinity College, Cambridge University, from 1878 to 1905, but is best known as a historian and as the author of the timeless classic *Mathematical Recreations and Essays*.[24] It was first published in 1892 and ran to fourteen editions, the last four with revisions by the great geometer Harold **Coxeter**.

Banach, Stefan (1892–1945)

A great Polish mathematician who founded functional **analysis** and also made important contributions to the understanding of **vector spaces**, **measure theory**, and **set theory**. His name is associated with *Banach space*, *Banach algebra*, the *Hahn-Banach theorem*, and the remarkable **Banach-Tarski paradox**. Largely self-taught in mathematics, Banach was "discovered" by Hugo **Steinhaus** and when World War II began was president of the Polish Mathematical Society and a full professor at Lvov University. Being on good terms with Soviet mathematicians, he was allowed to hold his chair during the Soviet occupation of Lvov. The German occupation of the city in 1941 resulted in the mass murder of Polish academics. Banach survived, but the only way he could earn a living was by feeding lice with his blood in a German institute where typhoid fever research was conducted. His health declined during the occupation, and Banach died before he could be repatriated from Lvov, which was incorporated into the Soviet Union and returned to Poland after the war. *Théorie des opérations linéaires* (The theory of linear operations) is regarded as his most influential work.

Banach-Tarski paradox

> *There are more things in heaven and earth, Horatio, than are dreamt of in your philosophy.*
> —William Shakespeare

A seemingly bizarre and outrageous claim that it is possible to take a ball, break it into a number of pieces, and then reassemble those pieces to make two identical copies of the ball. The claim can be made even stronger: it is possible to decompose a ball the size of a marble and then reassemble the pieces to make another ball the size of Earth, or, indeed, the size of the known universe!

Before writing off Banach and Tarski as being either very bad mathematicians or very good practical jokers, it's important to understand that this is not a claim about what can actually be done with a real ball, a sharp knife, and some dabs of glue. Nor is there any chance of some entrepreneur being able to slice up a gold ingot and assemble in its place two new ones like the original. The Banach-Tarski paradox tells us nothing new about the physics of the world around us but a great deal about how volume, space, and other familiar-sounding things can assume unfamiliar guises in the strange abstract world of mathematics.

Stefan **Banach** and Alfred **Tarski** announced their startling conclusion in 1924, having built on earlier work by Felix **Hausdorff**, who proved that it's possible to chop up the unit interval (the line segment from 0 to 1) into countably many pieces, slide these bits around, and fit them together to make an interval of length 2. The Banach-Tarski paradox, which mathematicians often refer to as the *Banach-Tarski decomposition* because it's really a proof not a paradox, highlights the fact that among the infinite **set** of points that make up a mathematical **ball**, the concepts of volume and measure can't be defined for all possible subsets. What this boils down to is that quantities that can be measured in any familiar sense are not necessarily preserved when a ball is broken down into subsets and then those subsets reassembled in a different way using just translations (slides) and rotations (turns). These unmeasurable subsets are extremely complex, lacking reasonable boundaries and volume in the ordinary sense, and thus are not attainable in the real world of matter and energy. In any case, the Banach-Tarski paradox doesn't give a prescription for *how* to produce the subsets: it only proves their *existence* and that there must be at least five of them to produce a second copy of the original ball. The fact that the Banach-Tarski paradox depends on the **axiom of choice** (AC), yet is so strongly counterintuitive, has been used by some mathematics to suggest that the AC must be wrong; however, the benefits of adopting the AC are so great that such black sheep of the mathematical family as the paradox are generally tolerated.[324, 340]

Bang's theorem
If all the faces of a **tetrahedron** have the same perimeter, then the faces are all **congruent** triangles.

banker's rounding
For banking or scientific purposes it's often considered correct to round something 0.5 to the nearest *even* number (not always upward). For instance, 5.5 rounds to 6, but 12.5 rounds to 12. This method avoids introducing a bias to a large set of numbers, by rounding up more or less as often as rounding down. Unfortunately, at a lower level, it is often taught to round something 0.5 upward all the time. See also **round-off error**.

Barbaro, Daniele (1513–1570)
A Venetian geometer whose book, *La Practica della Perspectiva* (1568–9), presents the techniques of perspective, illustrated in part with a range of **polyhedra**. Partly based on the methods and writings of the great artist Piero della Francesca (1416–1492), but written in a more readable and humanistic style, it includes the earliest drawing of the *truncated icosidodecahedron* and one of the earliest representations, along with that of the German goldsmith Wenzel Jamnitzer (1508–1585), of a *rhombicosidodecahedron*. *La Practica* was one of the most respected texts on perspective in the sixteenth century, comparable to Albrecht **Dürer**'s *Painter's Manual*.

barber paradox
See **Russell's paradox**.

Barbier's theorem
See **curve of constant width**.

Barlow, Peter (1776–1862)
A self-educated English mathematician who wrote several important books on the subject but is best known for *New Mathematical Tables* (generally known as *Barlow's Tables*), a compendium of factors, squares, cubes, square roots, reciprocals, and hyperbolic logarithms of all numbers from 1 to 10,000, and his invention of a special telescope lens. "Barlows" are popularly used by amateur astronomers to this day to multiply the power of other lenses. Barlow also worked on the design of bridges and was appointed as royal commissioner for railways, conducting experiments to see if the limitation on gradients and radius of curvature proposed by George Stephenson was correct.

Barnsley's fern
A **fractal** shape, first explored by Michael F. Barnsley at the Georgia Institute of Technology in the 1980s, that has many geometric features in common with a natural fern,

Barnsley's fern   *David Nicholls*

most notably the appearance of frondlike forms at different scales. As in the case of real ferns, Barnsley's fern reveals smaller prominences along the edge of each frond that are miniature versions of the overall figure. Along these small prominences are still smaller protuberances, and so on. Barnsley's fern is created by the repetitive application of four relatively simple mathematical rules and is a type of fractal, introduced by Barnsley, known as an *iterated function system* (IFS).[26]

base
(1) The flat plane or straight line upon which a shape or a solid rests. (2) The number upon which a **number system** is based; this is also the number of different characters or figures needed by the number system. The base, or *radix*, of our familiar decimal system is 10. Thus, there are ten symbols, 0, 1, 2, 3, 4, 5, 6, 7, 8, and 9, and a decimal number is written right to left in terms of units, tens, hundreds, and so on. Each move to the left represents a jump by a **power** of 10. The decimal number 375, for example, equals $(3 \times 10^2) + (7 \times 10) + (5 \times 1)$. This can easily be written to another base. Decimal 375_{10} becomes in octal (base 8) $567_8 = (5 \times 8^2) + (6 \times 8) + (7 \times 1)$, or in binary (base 2) 101111001_2.

basin of attraction
The set of all points in **phase space** that are under the influence of an **attractor**, or, more generally, the initial conditions of a system that evolve into the range of behavior allowed by the attractor. If one imagines a **complex system** as a sink, then the attractor can be considered the drain at the bottom, and the basin of attraction is the sink's basin.

basis
In mathematics, usually associated with linear algebra; a minimal set of **vectors** that spans a **vector space**.

Bayes, Thomas (1702–1761)
An English mathematician and theologian, remembered chiefly for the theorem named after him (see **Bayes's theorem**), and the technique of **Bayesian inference** that arises from it. Bayes wrote on probability theory, the logical basis of calculus, and asymptotic series.

Bayesian inference
Statistical inference in which probabilities are interpreted not as frequencies or proportions, but rather as degrees of belief. A prior distribution for a certain random variable is assumed; then this is modified, in the light of experimentation, using **Bayes's theorem**. Pierre **Laplace** applied Bayesian inference to estimate the mass of Saturn and in a variety of other problems.

Bayes's theorem
Also known as *Bayes's rule,* a result in **probability theory**, named after Thomas **Bayes**, who proved a special case of it. It is used in statistical inference to update estimates of the probability that different hypotheses are true, based on observations and a knowledge of how likely those observations are, given each hypothesis. In fact, it is habitually used by scientists in preference to the principle of **induction**. Bayes's theorem says that if an instance X is actually observed, then the probability of a hypothesis H must be multiplied by the following ratio:

$$\frac{\text{probability of observing } X \text{ if } H \text{ is true}}{\text{probability of observing } X}$$

In other words, the probability of a hypothesis H conditional on a given body of data X is equal to the ratio of the unconditional probability of the conjunction of the hypothesis with the data to the unconditional probability of the data alone.

Beale cipher
One of the greatest unsolved puzzles in **cryptography**— or a mere hoax. About a century ago, a fellow by the name of Thomas Beale supposedly buried two wagonloads of pots filled with silver coins in Bedford County, near Roanoke, Virginia. Local rumors claim the treasure was buried near Bedford Lake. Beale wrote three encoded letters telling what was buried, where it was buried, and to whom it belonged. He entrusted these three letters to a friend, went west, and was never heard from again. Several years later, someone examined the letters and was able to crack the code in the second one, which turned out to be based on the text from the Declaration of Independence. A number in the letter indicated which word in the document was to be used. The first letter of that word replaced the number. For example, if the first four words of the document were "We hold these truths," the number 3 in the letter would represent the letter *t*. The second letter translated as follows:

I have deposited in the county of Bedford about four miles from Bufords in an excavation or vault six feet below the surface of the ground the following articles belonging jointly to the parties whose names are given in number three herewith. The first deposit consisted of ten hundred and fourteen pounds of gold and thirty eight hundred and twelve pounds of silver deposited Nov eighteen nineteen. The second was made Dec eighteen twenty one and consisted of nineteen hundred and seven pounds of gold and twelve hundred and eighty eight of silver, also jewels obtained in St. Louis in exchange to save transportation and valued at thirteen [t]housand dollars. The above is securely packed i[n] [i]ron pots with iron cov[e]rs. Th[e] vault is roughly lined with stone and the vessels rest on solid stone and are covered [w]ith others. Paper number one describes th[e] exact locality of the va[u]lt so that no difficulty will be had in finding it.

One of the remaining letters supposedly contains directions on how to find the treasure but, to date, no one has solved the code. One theory is that both the remaining letters are encoded using either the same document in a different way, or another very public document. Or, of course, this could all be an elaborate, but entertaining, wheeze. Those interested may wish to contact the Beale Cypher Association, P.O. Box 975, Beaver Falls, PA 15010.

Beal's conjecture

In 1997, the Texan financier Andrew Beal offered $75,000, later increased to $100,000, to the first person who could prove or provide a counterexample to the following conjecture:

If $x^m + y^n = z^r$, where x, y, z $m, n,$ and r are all positive integers, and m, n and r are greater than two, then $x, y,$ and z have a common factor (greater than one).

Fermat's last theorem, which was proved in 1994, is a special case of Beal's conjecture. However, no one has yet been able to use this fact to prove or disprove the conjecture, nor has anyone been able to come up with a counterexample as a disproof. It is known that for any set of three exponents $m, n,$ and $r,$ each greater than two, there can be at most finitely many solutions. But is this finite number zero? The prize remains unclaimed.

beast number (666)

Also known as the *Apocalypse number,* the "number of the beast" mentioned in the Bible's book of Revelation, the relevant verse of which (Rev. 13:18) is often quoted as:

Here is wisdom. Let him that hath understanding count the number of the beast: for it is the number of a man; and his number is six hundred threescore and six.

Leaving aside the thorny issue of what this actually means, the number 666 does have some interesting mathematical properties. Most notably, it is the sum of the first 36 natural numbers (all the numbers on a roulette wheel): $1 + 2 + 3 + \ldots + 36$, which makes it the thirty-sixth **triangular number**. It is also the sum of the squares of the first seven **prime numbers**, $2^2 + 3^2 + 5^2 + 7^2 + 11^2 + 13^2 + 17^2$. Other curious representations of "the beast" include:

$$1^3 + 2^3 + 3^3 + 4^3 + 5^3 + 6^3 + 5^3 + 4^3 + 3^3 + 2^3 + 1^3$$
$$3^6 - 2^6 + 1^6$$
$$6 + 6 + 6 + 6^3 + 6^3 + 6^3.$$

Furthermore, 666 is one member of a **Pythagorean triplet** (216, 630, 666), which can be written in the remarkable form:

$$(6 \times 6 \times 6)^2 + (666 - 6 \times 6)^2 = 666^2.$$

In Roman numerals, 666 represents all the numbers from 500 in descending order, namely D (500) + C (100) + L (50) + X (10) + V (5) + I (1), or DCLXVI. In fact, it's been suggested that the Roman representation of 666 may have something to do with the biblical reference. DCLXVI was often used as a generic way of referring to any unspecified or unknown large number–the Roman equivalent of our "zillion." Thus, the writer of Revelation might simply have been using "666" to mean "big but unspecified."

Beatty sequences

Suppose R is an **irrational number** greater than 1, and let S be the number satisfying the equation $1/R + 1/S = 1$. Let $[x]$ denote the *floor function* of $x,$ that is, the greatest integer less than or equal to x. Then the sequences $[nR]$ and $[nS]$, where n ranges through the set N of positive integers, are the Beatty sequences determined by R. The interesting thing about them is that they partition N; in other words, every positive integer occurs exactly once in one sequence or the other. For example, when R is the **golden ratio** (about 1.618), the two sequences begin with

1, 3, 4, 6, 8, 9, 11, 12, 14, 16, 17, 19, 21, . . . , and
2, 5, 7, 10, 13, 15, 18, 20, 23, 26, 28, 31, 34 . . . .

Beatty sequences are named after the American mathematician Samuel Beatty (1881–1970), who introduced them in 1926 in a problem in the *American Mathematical Monthly*. Beatty was the first person to receive a Ph.D. in mathematics from a Canadian university, and later became the chairman of the mathematics department and chancellor of the University of Toronto.

beauty and mathematics

Many mathematicians and scientists have commented on the beauty they find in the structure and symmetry of the equations that underpin their work, and that beauty is often the first sign of truth. In *A Mathematician's Apology,*[151] G. H. **Hardy** wrote:

The mathematician's patterns, like the painter's or the poet's must be beautiful; the ideas, like the colors or the words must fit together in a harmonious way. Beauty is the first test: there is no permanent place in this world for ugly mathematics.

The physicist Paul **Dirac** went even further:

I think that there is a moral to this story, namely that it is more important to have beauty in one's equations than to have them fit experiment. If [Erwin] Schrödinger had been more confident of his work, he could have published it some months earlier, and he could have published a more accurate equation. It seems that if one is working from the point of view of getting beauty in one's equations, and if one has really a sound insight, one is on a sure line of progress. If there is not complete agreement between the results of one's work and experiment, one should not allow oneself to be too discouraged, because the discrepancy may well be due to minor features that are not properly taken into account and that will get cleared up with further development of the theory.[176]

The architect Richard Buckminster Fuller also saw beauty as an acid test of truth: "When I'm working on a problem, I never think about beauty. I think only how to solve the problem. But when I have finished, if the solution is not beautiful, I know it is wrong."

Bell, Eric Temple (1883–1960)
A Scottish-born mathematician and writer who, from 1903, spent most of his life in the United States, teaching at the University of Washington from 1912 until 1926, then serving as professor of mathematics at the California Institute of Technology. He did work in **number theory** but is best remembered for his books, including *Algebraic Arithmetic* (1927) and *The Development of Mathematics* (1940), which became classics, and, at a more popular level, *Men of Mathematics* (1937)[32] and *Mathematics, Queen and Servant of Science* (1951). He was also a prolific writer of science fiction under the penname John Taine.

bell curve
The characteristic shape of the graph of a normal (Gaussian) distribution.

Bell number
Named after Eric **Bell,** one of the first to analyze them in depth, the number of ways that n distinguishable objects (such as differently colored balls) can be grouped into sets (such as buckets) if no set can be empty. For example, if there are three balls, colored red (R), green (G), and blue (B), they can be grouped in five different ways: (RGB), (RG)(B), (RB)(G), (BG)(R), and (R)(G)(B), so that the third Bell number is 5. The sequence of Bell numbers, 1, 2, 5, 15, 52, 203, 877, 4,140, 21,147, . . . , can be built up in the form of a triangle, as follows. The first row has just the number one. Each successive row begins with the last number of the previous row and continues by adding the number just written down to the number immediately above and to the right of it.

```
            1
          1   2
        2   3   5
      5   7  10  15
   15  20  27  37  52
52 . . .
```

The Bell numbers appear down the left-hand side of the triangle. These normal Bell numbers contrast with *ordered Bell numbers,* which count the number of ways of placing n distinguishable objects (balls) into one or more *distinguishable* sets (buckets). The ordered Bell numbers are 1, 3, 13, 75, 541, 4,683, 47,293, 545,835, ... Bell numbers are related to the **Catalan numbers**.

Benford's law
If a number is chosen at random from a large table of data or statistics, such as stock quotations, populations of towns in Germany, or half-lives of radioactive atoms, the chance that the first digit is 1 is about 30.1%, that the first digit is 2 is about 17.6%, that it is 3 is 12.4%, . . . , and that it is 9 is 4.5%. These figures fit the rule that the probability that the first digit is d is $\log_{10}(1 + 1/d)$. This rule is called Benford's law after the American physicist Frank Benford, who publicized his findings in 1938.[34] The same discovery had been made 57 years earlier by the astronomer and mathematician Simon Newcomb, who noticed that the front pages of logarithm tables tended to be more dog-eared than pages later on.[232]

Benford tested thousands of different collections of data, including the surface areas of 335 rivers, specific heats and molecular weights of thousands of chemicals, baseball statistics, and the street addresses of the first 342 people listed in the book *American Men of Science.* All these seemingly unrelated sets of numbers followed the same first-digit probability pattern as the worn pages of logarithm tables suggested. In all cases, the number 1 showed up as the first digit about 30% of the time, more often than any other, and seven times more often than the number 9.

It seems extraordinary. Why shouldn't the numbers 1 to 9 take equal turns to be the first digit? Benford's findings have been verified by other researchers. The larger and more varied the sampling of numbers from different data sets, it has been found, the more closely the distribution of numbers approaches what Benford's law predicts. Moreover these probabilities are *scale invariant* and *base invariant.* For example, it doesn't matter whether the numbers are based on the dollar prices of stocks or their prices in yen or euros, nor does it matter if the numbers are in terms of stocks per dollar; provided there are enough numbers in the sample, Benford's law will hold.[164, 263]

Benham's disk

Benham's disk

A disk, marked with a black and white pattern, which, when spun around, causes people to see colors. Benham's disk, also known as *Benham's wheel* and *Benham's top,* was invented in 1894 by the toy maker C. E. Benham and originally sold through Messrs. Newton and Co. as the Artificial Spectrum Top. It is one of a number of spinning disk color illusions first described by Gustav **Fechner** in 1838. For this reason, the illusory colors are known as *Fechner colors.* From the beginning, it was realized that the root of the illusion probably lay in the variation of retinal response time with wavelength. An online animated version of the disk can be seen at http://www.michaelbach.de/ot/col_benham/index.html.

Bernoulli family

An extraordinary Swiss family from Basle that produced eight outstanding mathematicians within three generations. Together with Isaac **Newton**, Gottfried **Leibniz**, Leonhard **Euler**, and Joseph **Lagrange**, the Bernoulli family dominated mathematics and physics in the seventeenth and eighteenth centuries, making important contributions to differential calculus, geometry, mechanics, ballistics, thermodynamics, hydrodynamics, optics, elasticity, magnetism, astronomy, and probability theory. Unfortunately, the Bernoullis were as conceited and arrogant as they were brilliant, and engaged in bitter rivalries and rows with one another.

The patriarchs of this mathematical dynasty were Jakob I (1654–1705) and his brother Johann I (1667–1748). (The Roman numerals are used to tell fathers, brothers, sons, and cousins apart, as the same Christian names were used repeatedly.) Next came Jakob's son, Nikolaus I, and Johann's three sons, Nikolaus II, Daniel (1700–1772), and Johann II. Finally, came Johann II's mathematical offspring, Johann III and Jakob II.

Jakob I developed a passion for science and mathematics after meeting Robert Boyle during a trip to England in 1676. He largely taught himself in these subjects and went on to lecture in experimental physics at the University of Basle. He also secretly introduced his younger brother to mathematics, against the wishes of his parents who wanted the younger brother to go into commerce. The cooperation between the two brothers soon degenerated, however, into vitriolic argument. Irked by Johann's bragging, Jakob publicly claimed that his younger brother had copied his own results. Later, having been appointed to the chair of mathematics at Basle, Jakob succeeded in blocking his brother's appointment to the same department, forcing Johann to take a teaching job at the University of Groningen instead. Johann proposed the so-called **brachistrochrone problem** and, along with Newton, Leibniz, l'Hospital, and Jakob, managed to solve it—but only after he first came up with a faulty proof and then tried to substitute one of Jakob's in its place! Eventually, Johann was offered a post at Basle as, of all things, the department head of Ancient Greek. But, en route to Basle, Johann learned that Jakob had died of tuberculosis. Upon his arrival he set about lobbying for the vacant position and, in less than two months, got his way. Jakob's most important work, his *Ars Conjectandi* (The art of conjecture), was published posthumously and formed the basis of **probability theory**.

Sadly, Johann I repeated his father's mistake and tried to force the most mathematically talented of his three sons, Daniel, into an unwanted career as a merchant. When the attempt failed, Johann let Daniel study medicine, in order to prevent his son from becoming a competitor. But all three sons followed their father's path, and Daniel, while studying medicine, was tutored in math by his older brother Nikolaus II. In 1720, Daniel went to Venice to work as a physician but won such a big reputation for his work in physics and mathematics that Peter the Great of Russia offered him a chair at the Academy of Science in St. Petersburg. Daniel went, along with Nikolaus II, who was also offered a position at the Academy. However, after just eight months, Nikolaus came down with a fever and died. Upset, Daniel wanted to return to Basle, but Johann I didn't want his son—a potential rival—back home. Instead he sent one of his pupils, none other than the great Leonhard **Euler**, to St. Petersburg to keep Daniel company. The two Swiss mathematicians became good friends, and the six years they spent together in St. Petersburg were the most productive of Daniel's life.

When Daniel finally returned to Basle, quarrels within the family flared up again after he won the prize of the Parisian Academy of Science with a paper, co-authored with his father, on astronomy. Jealous of Daniel's success, Johann threw him out of the family house. And worse was to come. In 1738 Daniel published his magnum opus, *Hydrodynamica*. Johann I read the book, hastily wrote one of his own called *Hydraulica*, back-dated it to 1732, and claimed to be the inventor of fluid dynamics! The plagiarism was soon uncovered, and Johann was ridiculed by his colleagues, but his son never recovered from the blow. See also **St. Petersburg paradox**.

Bernoulli number

A number of the type defined by Jakob **Bernoulli** in connection with evaluating sums of the form $\sum i^k$. The sequence $B_0, B_1, B_2, \ldots$ can be generated using the formula

$$x/(e^x - 1) = \sum (B_n x^n)/n!$$

though various different notations are used for them. The first few Bernoulli numbers are: $B_0 = 1$, $B_1 = -\frac{1}{2}$, $B_2 = \frac{1}{6}$, $B_4 = -\frac{1}{30}$, $B_6 = \frac{1}{42}, \ldots$. They crop up in many diverse areas of mathematics including the series expansions of $\tan(x)$ and **Fermat's last theorem**.

Berry's paradox

A **paradox**, devised by G. G. Berry of the Bodleian Library at Oxford University in 1906, that involves statements of the form: "The smallest number not nameable in under ten words." At first sight, there doesn't seem anything particularly mysterious about this sentence. After all, there are only so many sentences that have less than ten words, and only a set **S** of these specify unique numbers; so there is clearly some number N that is the smallest integer not in **S**. The trouble is, the Berry sentence itself is a specification for that number *in only nine words!* Berry's paradox shows that the concept of name-ability is inherently ambiguous and a dangerous concept to be used without qualification. A similar air of the paradoxical swirls around the notion of **interesting numbers**.[60]

Bertrand's box paradox

A problem, similar to the **Monty Hall problem**, that was published by the French mathematician Joseph Bertrand (1822–1900) in his 1889 text *Calcul des Probabilités*. Suppose there are three desks, each with two drawers. One desk contains a gold medal in each drawer, one contains a silver medal in each drawer, and one contains one of each, but you don't know which desk is which. If you open a drawer and find a gold medal, what are the chances that the other drawer in that desk also contains gold? This comes down, then, to figuring out the probability that you've picked the gold-gold desk instead of the gold-silver desk. Many people quickly jump to the conclusion that there are two possibilities, and since the selection was random, it must be 50-50. But this is wrong. Think of the initial selection as picking from among six drawers:

Before			After		
S	S	G		G	G
S	G	G			G
1	2	3	1	2	3

So, we have it narrowed down to three drawers, with an equal probability of each one being the one that was picked. One of the drawers is in desk 2, so there's a one-third chance that desk 2 was picked. Two of the drawers are in desk 3, so there are two one-third chances (i.e., a two-third chance) that desk 3 was picked.

Bertrand's postulate

Also known as *Betrand's conjecture*, if n is an integer greater than 3, then there is at least one **prime number** between n and $2n - 2$. This postulate (which should now be called a theorem) is named after the French mathematician Joseph Bertrand (1822–1900) who, in 1845, showed it was true for values of n up to 3 million. The Russian Pafnuty Chebyshev (1821–1894) gave the first complete proof in 1850, so that it is sometimes called *Chebyshev's theorem* (although another theorem also goes by this name). In 1932, Paul **Erdös** gave a more elegant proof, using the **binomial coefficients**, which is the one that appears in most modern textbooks. Bertrand's postulate implies that the nth prime p_n is at most 2^n.

Bessel, Friedrich Wilhelm (1784–1846)

A German astronomer and mathematician who became director of the observatory at Königsberg (see **bridges of Königsberg**). Much of Bessel's work dealt with **perturbation**s (wobbles) in the motion of planets and stars caused by the gravitational influence of other bodies. To help analyze these perturbations he developed certain mathematical **functions** that are now known as *Bessel functions* and are used widely in physics.

beta function

The function

$$B(m, n) = \int_0^1 x^{m-1} (1 - x)^{n-1} \, dx.$$

It can be defined in terms of the **gamma function** by

$$B(m, n) = \frac{\Gamma(m)\Gamma(n)}{\Gamma(m + n)}.$$

Many integrals can be reduced to the evaluation of beta functions.

Betti number

An important topological property of a surface, named after the Italian mathematician Enrico Betti (1823–1892). The Betti number is the maximum number of cuts that can be made without dividing the surface into two separate pieces. If the surface has edges, each cut must be a "crosscut," one that goes from a point on an edge to another point on an edge. If the surface is closed, like a sphere, so that it has no edges, each cut must be a "loop cut," a cut in the form of a simple closed curve. The Betti number of a square is 0 because it is impossible to crosscut without leaving two pieces. However, if the square is folded into a tube, its topology changes–it now has two disconnected edges–and its Betti number changes to 1. A **torus**, or donut shape, has a Betti number of 2. See also **chromatic number**.

bicorn

Also known as the *cocked-hat,* any of a collection of **quartic** curves studied by James **Sylvester** in 1864 and by Arthur **Cayley** in 1867. The bicorn has the Cartesian equation

$$y^2(a^2 - x^2) = (x^2 + 2ay - a^2)^2.$$

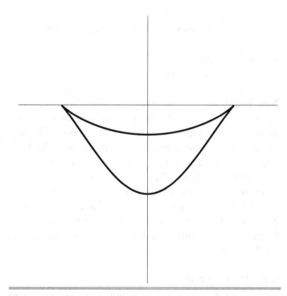

bicorn The bicorn curve © *Jan Wassenaar, www.2dcurves.com*

bicuspid curve

The **quartic** curve given by the equation

$$(x^2 - a^2)(x - a)^2 + (y^2 - a^2)^2 = 0.$$

Bieberbach conjecture

A celebrated conjecture made by the German mathematician Ludwig Bieberbach (1886–1982) in 1916, that was finally proved, after many partial results by others, by Louis de Branges of Purdue University in 1984.[54] Bierberbach is infamous in the history of mathematics because of his outspoken anti-Semitism during the Nazi era. Following the dismissal of Edmund Landau (1877–1938) from the University of Göttingen, Bierberbach wrote: "This should be seen as a prime example of the fact that representatives of overly different races do not mix as students and teachers. . . . The instincts of the Göttingen students felt that Landau was a type who handled things in an un-German manner."

Bieberbach's conjecture (BC) stemmed from the Riemann mapping theorem (RMT), which makes a claim about any region of a plane that is simply connected (in other words, any region, however complicated, that doesn't have any holes). The RMT says there must be some function, or mapping, such that every point in the arbitrary region is associated with one and only one point inside a circle with unit radius. Complex functions are best suited to plane-to-plane mappings and are often easier to work with if they can be represented as a power series. For example, given the complex number z, the function e^z can be expressed as the infinite series $1 + z + z^2/2! + z^3/3! + \dots$. Bieberbach guessed that there is a link between the conditions imposed on a function by RMT and the numerical coefficients of the terms in a power series that represents the function. The BC says that if a function gives a one-to-one association between points in the unit circle and points in a simply connected region of the plane, the coefficients of the power series that represents the function are never larger than the corresponding power. In other words, given that $f(z) = a_0 + a_1 z + a_2 z^2 + a_3 z^3 + \dots$, then $|a_n| \le n|a_1|$ for each n.

bifurcation

The value of a smoothly varying control parameter, or the point in *parameter space,* at which the behavior of a **dynamical system** undergoes a qualitative change. For example, a simple equilibrium, or a **fixed-point attractor,** might give way to a periodic oscillation as the stress on a system increases. Similarly, a **periodic attractor** might become unstable and be replaced by a **chaotic attractor**. To give a real-world example, drops fall individually at equal intervals, from a dripping faucet at low pressure. As the pressure is increased, however, the pattern of dripping abruptly changes so that two drops fall close together, with a longer interval before the next pair fall. In this case, a simple periodic process has given way to a periodic process with twice the period, a process described as "period doubling." If the flow rate of water through the faucet is increased still further, beyond the bifurcation point, often an irregular dripping is found and the behavior can become chaotic. See also **chaos**.

bilateral
Having two sides, or relating to the right and left sides of an object. *Bilateral symmetry* is a symmetrical arrangement, as of an organism or a body part, along a central axis, so that the body is divided into equivalent right and left halves by only one plane. See also **mirror reversal problem**.

bilateral diagram
A device invented by Lewis **Carroll** to represent the different logical states that two objects with two properties can take. Each cell in a four-square array represents one of the four possible object/property states and is covered by a red counter if the state is present and by a gray counter if it is absent.

billion
See **large numbers**.

bimagic square
A **magic square** that becomes a new magic square when each integer is squared. If, in addition to being bimagic, the integers in the square can be cubed and the resulting square is still magic, the square is said to be *trimagic*. To date the smallest bimagic square seems to be of order 8, while the smallest trimagic square is of order 32.

binary

> *There are 10 kinds of people in the world, those who understand binary math, and those who don't.*
>
> —Anonymous

The simplest positional **number system** and the natural one to use when dealing with computers; it employs just two symbols, 0 and 1, which correspond to the possible states of an off-on switch. Each place to the left in a binary number represents the next highest power of two. The binary number 10110_2, for example, means $1 \times 2^4 + 0 \times 2^3 + 1 \times 2^2 + 1 \times 2^1 + 0 \times 2^0$, or 22_{10} in the familiar decimal notation. Nonintegers can be represented by using negative powers, which are set off from the other digits by means of a *radix point* (called a *decimal point* in base 10). The binary number 11.01_2 thus means $1 \times 2^1 + 1 \times 2^0 + 0 \times 2^{-1} + 1 \times 2^{-2}$ which equals 3.25_{10}. A number that terminates in a decimal doesn't necessarily do so in binary (e.g., $0.3_{10} =$

$0.0100110011001 \ldots _2$), and vice versa. An irrational number, however, is nonperiodic in both systems (e.g., **pi**, $= 3.1415926 \ldots _{10} = 11.001001000011111 \ldots _2$). Binary arithmetic was first investigated by Gottfried **Leibniz** in 1672, though he didn't publish anything about it until 1701.

binary operation
An operation that involves two operands. For example, addition and subtraction are binary operations.

binomial
An expression containing two terms, joined by + or −. The **binomial theorem** gives the result of raising a binomial expression to a power; the expansion and the series it leads to are called the *binomial expansion* and the *binomial series*. A *binomial distribution* is described by a formula related to the binomial expansion. A binomial equation is a particular kind of equation that contains two terms.

binomial coefficient
A **coefficient** of x in the expansion of $(x + y)^n$. The binomial coefficient $_nC_m$ or $\binom{m}{n}$ gives the number of ways of picking m unordered outcomes from n possibilities, and is also known as a **combination**. It has the value $n!/(n - m)!m!$ The binomial coefficients form the rows of **Pascal's triangle**.

binomial theorem
The result that allows the expansion of a **binomial** expression:

$$(x + y)^n = x^n + a_{n-1}x^{n-1}y + a_{n-2}x^{n-2}y^2 + \ldots + y^n$$

where the coefficients a_i are called **binomial coefficients**.

Birkhoff, George David (1884–1944)
The foremost American mathematician of the early twentieth century and the first prominent dynamicist in the New World. He is known for his work on linear **differential equations** and **difference equations**, and was also deeply interested in and made contributions to the analysis of **dynamical systems**, celestial mechanics, the **four-color map problem**, and function spaces. Although a geometer at heart, he discovered new symbolic solution methods. He saw beyond the theory of oscillations, created a rigorous theory of **ergodic** behavior, and foresaw dynamical models for **chaos**. In addition he wrote on the foundations of relativity and quantum mechanics and, in *Aesthetic Measure* (1933), on art and music.

birthday paradox
The fact—not really a **paradox**—that you need a group of only 23 people for there to be a better than 50/50

chance that two of these people will have the same birthday. This seems surprising because we are used to comparing our particular birthday with others and only rarely finding a perfect match. The probability of any two individuals having the same birthday is just $1/365$. Even if you were to ask 20 people, the probability of finding someone with your birthday is still less than $1/20$. But the odds improve dramatically when a group of people *ask each other* about their birthdays because then there are many more opportunities for a matchup. One way to calculate the probability of a birthday match is to count the pairs of people involved. In a room of 23 people, there are $(23 \times 22)/2$, or 253, possible pairs. Each pair has a probability of success of $1/365 = 0.00274$ (0.274%), and thus a probability of failure of $(1 - 0.00274) = 0.99726$ (99.726%). The probability of no match among any of the pairs of people is 0.99726 to the 253rd power, which is 0.499 (49.9%). So the probability of a successful match is $(1 - 0.499)$, or slightly better than even (50/50). With 42 people, the probability of a birthday match climbs to 90%.

birthday surprise

Here is a way to learn someone's birthday by doing a little simple math. Ask a person to take the month number (January = 1, February = 2, and so forth) of their birthday, multiply by 5, add 6, multiply the total by 4, add 9, and multiply the new total by 5 again. Finally, have her add the number of the day on which she was born and give you the total. In your head, subtract 165 and you will have the month and day of her birth. How does this work? If M is the month number and D the day number, then after the seven steps the expression for their calculation is: $5(4(5M + 6) + 9) + D = 100M + D + 165$. Thus, if you subtract the 165, what will remain will be the month in hundreds plus the day.

bisect

To cut in half.

bisecting an angle

Splitting an angle exactly in two. The ancient Greeks knew how to easily do it using only a pair of compasses and a straightedge. Here's how: Put the point of the compass at a point O and draw a circle so that it cuts the two lines coming out from the angle. Call these intersection points A and B. Now put the point of the compass at A and draw an arc that follows within the opening of the angle. Without changing the radius at which the compass is set, move its point to B and draw another arc. Join the point where the two arcs cross, P, to O using the straight edge: angle POB is half of angle AOB. See also **trisecting an angle**.

bishops problem

To find the maximum number of bishops (chess pieces capable of moving any number of spaces along diagonals of their own color) that can be placed on an $n \times n$ chessboard in such a way that no two are attacking each other. The answer is $2n - 2$, which gives the solution 14 for a standard (8×8) chessboard. The numbers of distinct maximal arrangements for $n = 1, 2, \ldots$ bishops are 1, 4, 26, 260, 3,368, ...

bistromathics

The revolutionary new (and totally fictitious) field of mathematics in restaurants, as described by Douglas Adams in his book *Life, the Universe and Everything*:[4]

> Numbers written on restaurant bills within the confines of restaurants do not follow the same mathematical laws as numbers written on any other pieces of paper in any other parts of the Universe. This single statement took the scientific world by storm. . . . So many mathematical conferences got held in such good restaurants that many of the finest minds of a generation died of obesity and heart failure and the science of math was put back by years.

Adams explains that just as Einstein found that space and time are not an absolute but depend on the observer's movement, so numbers are not absolute, but depend on the observer's movement in restaurants:

> The first non-absolute number is the number of people for whom the table is reserved. This will vary during the course of the first three telephone calls to the restaurant, and then bear no apparent relation to the number of people who actually turn up, or to the number of people who subsequently join them after the show/match/party/gig, or to the number of people who leave when they see who else has turned up. The second non-absolute number is the given time of arrival, which is now known to be one of the most bizarre of mathematical concepts, a recipriversexcluson, a number whose existence can only be defined as being anything other than itself. In other words, the given time of arrival is the one moment of time at which it is impossible that any member of the party will arrive. . . . The third and most mysterious piece of nonabsoluteness of all lies in the relationship between the number of items on the bill, the cost of each item, the number of people at the table and what they are each prepared to pay for.

See also **large numbers**.

bit

A **binary** digit, either 0 or 1. See also **byte**.

blackjack

Also known as *twenty-one,* the most popular casino game in the world and the only such game with a *fluctuating* probability: the odds of winning change with the makeup of the deck. The cards two to nine have a numerical value equal to the number printed on the card. Tens and all face cards (jack [J], queen [Q], and king [K]) have the value of 10. Aces may be counted as either 11 or one. A dealer plays against one to seven players. Every player and the dealer initially receive two cards each, dealt by the dealer. Each player's hand is played against the dealer's hand only. If a player's hand has a value closer to 21 (without going over) than the dealer's hand, the player wins. The best possible hand is known as a *blackjack* (21 in the first two cards) and consists of an ace and a ten-valued card (10, J, Q, K). The payout for a blackjack is 3-to-2: the player is paid three chips for every two chips bet. When both the player and the dealer have blackjacks, it is a normal *tie (push)* situation and the player retains the initial bet. The player has several choices after receiving the first two cards: (1) *Hit* or *draw:* take one or more cards to add-up to a better hand. (2) *Stand:* stop taking more cards. (3) *Double down:* double the initial amount (in cases considered more favorable). (4) *Split pairs:* if the two cards are equal in value they may be played in two separate hands. The dealer must draw until his hand adds up to 17 or more. Both the player and the dealer can go over 21, a situation known as *bust;* the player loses the bet immediately. The dealer plays his hand last, after all the players at the table. This rule creates the so-called *house edge.* John Scarne[282] was the first to calculate the house advantage at blackjack as 5.9%. However, the house edge can be cut to around 1% if the player follows certain rules. The set of rules known as *basic strategy* make blackjack one of the fairest games of any kind, almost as fair as coin tossing.

In 1962 Edward O. Thorp, an IBM computer scientist, published *Beat the Dealer,*[333] which introduced a winning method called *card counting.* This method considered the 10-valued cards and the aces as positive, and the cards 2 to 6 as negative. If the net value of the remaining deck was positive, the player must increase the bet accordingly. The method had visible results when only one deck was used and very few cards remained in the deck. Casinos responded by changing the rules dramatically. The *penetration* was introduced: not all the cards in the deck are played. Shuffling is done unexpectedly. Also, most casinos introduced *multiple-deck blackjack.*

Blanche's dissection

The simplest **dissection** of a square into rectangles of the same areas but different shapes. It is composed of seven pieces; the square is 210 units on a side, and each rectangle has area of $210^2/7 = 6,300$.

Bólyai, János (1802–1860)

A Hungarian mathematician who was one of the founders of **non-Euclidean geometry**, independently coming to almost the same conclusions as Nikolai **Lobachevsky**. He was initially taught by his father, Farkas, also a mathematician, then he studied at the Royal Engineering College in Vienna from 1818 to 1822. Between 1820 and 1823 he prepared his treatise on a complete system of non-Euclidean geometry, commenting, "Out of nothing I have created a strange new universe." It was published in 1832 as an appendix to an essay by his father. Carl **Gauss**, on reading the appendix, wrote to a friend saying, "I regard this young geometer Bólyai as a genius of the first order." It was not until 1848 that Bólyai learned that Lobachevsky had produced a similar piece of work in 1829. Although he never published more than the 24 pages of the appendix, Bólyai left more than 20,000 pages of manuscript of mathematical work when he died. He was an accomplished linguist, speaking nine foreign languages including Chinese and Tibetan.

book-stacking problem

How much of an overhang can be achieved by stacking books on a table before the books overbalance and fall off? Assume each book is one unit long. To balance one book on a table, the center of gravity of the book must be somewhere over the table; to achieve the maximum overhang, the center of gravity should be just over the table's edge. The maximum overhang with one book is

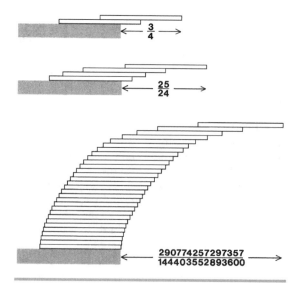

book-stacking problem   The solution to the book-stacking problem.

obviously ½ unit. For two books, the center of gravity of the first should be directly over the edge of the second, and the center of gravity of the stack of two books should be directly above the edge of the table. The center of gravity of the stack of two books is at the midpoint of the books' overlap, or $(1 + ½)/2$, which is ¾ unit from the far end of the top book. It turns out that the overhangs are related to the harmonic numbers H_n, (see **harmonic sequence**), which are defined as $1 + ½ + ⅓ + \ldots + 1/n$: the maximum overhang possible for n books is $H_n/2$. With four books, the overhang $(1 + ½ + ⅓ + ¼)/2$ exceeds 1, so that no part of the top book is directly over the table. With 31 books, the overhang is 2.0136 book lengths.

Boole (Stott), Alicia (1860–1940)

The third daughter of George **Boole** and an important mathematician in her own right. At the age of 18, she was introduced to a set of wooden cubes devised by her brother-in-law Charles **Hinton** as an aid to visualization of the **fourth dimension**. Despite having had no formal education, she surprised everyone by becoming adept with the cubes and developing an amazing feel for four-dimensional geometry. She introduced the word *polytope* to describe a four-dimensional convex solid, and went on to explore the properties of the six regular polytopes and to make 12 beautiful card models of their three-dimensional central cross sections. She sent photographs of these models to the Dutch mathematician Pieter Schoute (1846–1923), who had done similar work and with whom she subsequently published two papers. The models themselves are now housed in the Department of Pure Mathematics and Mathematical Statistics at Cambridge University.

Boole, George (1815–1864)

An English mathematician and philosopher who is regarded as one of the founders of computer science. His great contribution was to approach logic in a new way, reducing it to a simple algebra and thus incorporating logic into mathematics. He pointed out the analogy between algebraic symbols and those that represent logical forms; his algebra of logic became known as **Boolean algebra** and is now used in designing computers and analyzing logic circuits. Although he never studied for a degree, Boole was appointed to the chair of mathematics at Queens College, Cork, Ireland, in 1849. One day in 1864 he walked the two miles in pouring rain from his home to the college and then lectured in wet clothes. A fever followed but whether this alone would have caused his demise is unknown. Certainly his condition wasn't helped by his wife, Mary (a niece of Sir George Everest, after whom the mountain is

named), who, following the principle that remedy should resemble cause, put Boole to bed and threw buckets of water over him. He expired shortly after. See also **Boole (Stott), Alicia**.

Boolean

Taking only 0/1, true/false, yes/no values.

Boolean algebra

An **algebra** in which the **binary** operations are chosen to model the **union** and **intersection** operations in **set theory**. For any set A, the subsets of A form a Boolean algebra under the operations of union, intersection, and complement.

Borel, Emile (1871–1956)

A French mathematician who worked on divergent series, the theory of functions, probability, and **game theory**, and was the first to define games of strategy. He also founded **measure theory**, which applies the theory of sets to the theory of functions, and thus became an originator, with Henri **Lebesgue** and René Louis Baire (1874–1932), of the modern theory of functions of a real variable.

Borges, Jorge Luis (1899–1986)

An Argentinian author, essayist, and poet, many of whose short stories explore paradoxes and other strange avenues of mathematics, logic, philosophy, and **time**. For example, the possibility of branches in time is dealt with in "The Garden of Forking Paths," while the strange notion of the **Universal Library** is the subject of "The Library of Babel." Borges was profoundly influenced by European culture, English literature, and such thinkers as George Berkeley.

Borromean rings

Three rings linked in such a way that although they can't be separated, no two rings are linked; remove any one ring, however, and the other two fall apart. Named after the Italian family of Borromeo whose family crest has borne the rings since the fifteenth century, the design has been used in many places and times as a symbol of strength in unity. A form of the Borromean link known as Odin's triangle or the walknot ("knot of the slain") was used by the Norse folk of Scandinavia in two variants: a set of Borromean triangles and a unicursal curve that makes a trefoil knot. A motif of three interlaced crescent moons, similar to the Borromean rings, can be seen at the Palace of Fontainebleau. Designed by the architect Philibert de l'Orme, it is based on the moon emblem used by Diane de Poitiers (1499–1566), mistress of King Henry II of France. A similar pattern, but with three

Borromean rings A flowerpot on Isola Bella, an island on Lake Maggiore, near Arona in northern Italy, that bears the Borromeo family crest. *Peter Cromwell*

interlaced snakes in place of the crescent moons, occurs at various sites in Wales, including Bangor Cathedral. The Borromean rings are commonly used to symbolize the Christianity Trinity. An early source for this was a thirteenth-century French manuscript, now lost, in which the word *unitas* appears in the center, inside all the circles, and the three syllables of "tri-ni-tas" are distributed in the outer sectors. Borromean rings can also be found on Japanese family emblems, at a Japanese Shinto shrine north of Sakurai in the province of Nara, and in the sculptures of the Australian artist John Robinson. In North America, the design is known as the Ballantine rings after the New Jersey brewing company P. Ballantine and Sons who use it as a trademark with the rings labeled Purity, Body, and Flavor.

The rings first appeared in mathematics in the earliest work on knots by Peter **Tait** in 1876. The pattern of circles can be interlaced by replacing each of the six vertices by a crossing that shows how the circles pass over and under one another. Since there are two choices for each crossing, there are $2^6 = 64$ possible interlaced patterns. However, after taking symmetry into account, these 64 reduce to only 10 geometrically distinct patterns. Two patterns are considered to be the same if one can be obtained from the other by applying one or more of the following operations: rotation by 120°, reflection, and reflection in the plane of the pattern. The last symmetry operation means that the sense of all the crossings is switched. The rings can also be analyzed from the viewpoint of **topology**, which means that the designs are thought of as links made from a flexible and elastic material. If two links can be manipulated and deformed to look like one another (without breaking and joining the rings) then they are topologically equivalent. The 10 geometrically distinct patterns boil down to only five distinct topological types.

Borromean rings   Three alternating rings carved in a panel in a walnut door of the church of San Sigismondo in Cremona, Italy. The emblem is one of several belonging to the Sforza family. *Peter Cromwell*

Borsuk-Ulam theorem

One of the most important and profound statements in **topology**: if there are *n* regions in *n*-dimensional space, then there is some hyperplane that cuts each region exactly in half, measured by volume. All kinds of interesting results follow from this. For example, at any given moment on Earth's surface, there must exist two antipodal points—points on exactly opposite sides of Earth—with the same temperature and barometric pressure! One way to see that this must be true is to think of two opposite points *A* and *B* on the equator. Suppose *A* starts out warmer than *B*. Now move *A* and *B* together around the equator until *A* moves into *B*'s original position, and simultaneously *B* into *A*'s original position. *A* is now cooler than *B*, so somewhere in between they must have been the same temperature. The Borsuk-Ulam theorem implies the **Brouwer fixed-point theorem** and also the **ham sandwich theorem**.

bottle sizes

Wine and champagne come in various standard bottle sizes, as shown in the table "Wine Bottle Sizes." These follow a **geometric sequence**, doubling in size with each step, up to the double-magnum, but thereafter increase in a more complicated way. There are also regional variations (for example, a Nebuchadnezzar may hold from 12 to 15 liters) and differences depending on the type of drink held.

Wine Bottle Sizes

Name of Size	Region	Capacity (liters)	Standard Bottle Equivalents
Baby/split	All	0.1875	0.25
Half-bottle	All	0.375	0.5
Bottle	All	0.75	1
Magnum	All	1.5	2
Double-magnum	All	3	4
Jeroboam	Burgundy, Champagne	5	6.67
Jeroboam	Bordeaux, Cabernet S.	4.5	6
Rehoboam	Burgundy, Champagne	4.5	6
Imperial	Bordeaux, Cabernet S.	6	8
Methuselah	Burgundy, Champagne	6	8
Salmanazar	Burgundy, Champagne	9	12
Balthazar	Burgundy, Champagne	12	16
Nebuchadnezzar	Burgundy, Champagne	15	20

boundary condition

The value of a **function** at the edge of the range of some of its variables. Recognizing the boundary conditions of an unknown function helps in its identification since other unknowns, such as variables in integrations, can then be eliminated.

boundary value problem

An **ordinary differential equation** or a **partial differential equation** given together with **boundary conditions** to ensure a unique solution.

Bourbaki, Nicholas

Not an individual but a collective mathematician. In the 1930s, the Bourbaki group, made up of some of the brightest mathematicians in France, began as a club, holding secret meetings in Strasbourg to update university lectures and texts in the wake of World War I, which had essentially wiped out a generation of young talent. In time, Bourbaki authored encyclopedic accounts of all areas of mathematics, and its influence became widespread.

Its origins can be traced to 1934 and to André Weil and Henri Cartan who were maîtres de conférences (the equivalent of assistant professors) at the University of Strasbourg. One of their duties was to teach differential and integral calculus but they found the standard text on this subject, *Traité d'Analyse* by E. Goursat, wanting. Following a suggestion by Weil to write a new "Treatise on Analysis," a group of about 10 mathematicians began to meet regularly to plan the new work. Quickly, it was decided that the work should be collective, without any acknowledgment of individual contributions, and this became a feature of Bourbaki's output. In the summer of 1935, the pen name Nicholas Bourbaki was chosen, and the initial membership consisted of Weil, Cartan, Claude Chevalley, Jean Delsarte, and Jean Dieudonné—all former students at the École Normale Supérieure in Paris. Over the years, the membership varied; some people in the first group dropped out quickly, others were added, and later there was a regular process of addition and retirement (mandatory by the age of 50).

Bourbaki adopted rules and procedures that to outsiders often seemed eccentric and even bizarre. For example, during meetings to review and revise drafts for the various books the group developed, anyone could express his opinion as loudly as he wanted at any time, so it was not uncommon for several distinguished mathematicians to be on their feet at the same time shouting monologues at the top of their voices. Somehow out of this mayhem emerged work of extreme precision, to the point of pedantry and dryness. Bourbaki would have nothing to do with geometry or any attempt at visualization, and believed that mathematics should distance itself from the sciences. However, despite its tendency to be boring and long-winded, Bourbaki did achieve its goal: to set down in writing what was no longer in doubt in modern mathematics.

brachistochrone problem

A problem with which Johann Bernoulli (see **Bernoulli Family**) challenged his contemporaries in *Acta Eruditorum* in June 1696:

> Following the example set by Pascal, Fermat, etc., I hope to gain the gratitude of the whole scientific community by placing before the finest mathematicians of our time a problem which will test their methods and the strength of their intellect. If someone communicates to me the solution of the proposed problem, I shall publicly declare him worthy of praise. . . . Given two points A and B in a vertical plane, what is the curve traced out by a point acted on only by gravity, which starts at A and reaches B in the shortest time?

Isaac **Newton** reportedly solved the problem between four in the evening and four in the morning after a hard day at the Royal Mint, later commenting: "I do not love to be dunned [pestered] and teased by foreigners about mathematical things. . . ." Other correct solutions came from Gottfried **Leibniz**, the Frenchman Guillaume **de L'Hôpital**, and Johann's brother Jakob. They, like Johann, realized that the solution to the brachistochrone problem, as it was also to the **tautochrone problem**, was a curve known as the **cycloid**.

Brahmagupta (A.D. 598–after 665)

A Hindu astronomer and mathematician who became the head of the observatory at Ujjain—the foremost mathematical center of ancient India. His main work, *Brahmasphutasiddhanta* (The opening of the universe), written in 628, contains some remarkably advanced ideas, including a good understanding of the mathematical role of **zero**, rules for manipulating both positive and **negative numbers**, a method for computing square roots, methods of solving linear and some **quadratic** equations, and rules for summing series. His contributions to astronomy were equally ahead of their time. *Brahmagupta's theorem* states that in a cyclic quadrilateral (a four-sided shape whose corners lie on a circle) having perpendicular diagonals, the perpendicular to a side from the point of intersection of the diagonals always bisects the opposite side. **Brahmagupta's formula** for the area of a cyclic quadrilateral with sides of length a, b, c, and d is $\sqrt{(S-a)(S-b)(S-c)(S-d)}$, where $S = (a+b+c+d)/2$. As d goes to zero, this reduces to **Heron's formula**.

braid

A collection of lines or strings that are plaited together and whose ends are attached to two parallel straight lines.

Braid theory was pioneered by the Austrian mathematician Emil Artin (1898–1962) and is related to **knot** theory. It also has other applications: for instance, if we consider the way the roots of a **polynomial** move as one of the polynomial's coefficients changes, this motion can be thought of as a braid.

Brianchon's theorem

Given a **conic section**, if we circumscribe a hexagon about it, then the major diagonals of the hexagon are concurrent.

bridges of Königsberg

A famous routing problem that was analyzed and solved by Leonhard **Euler** in 1736, and that helped spur the development of **graph** theory. The old city of Königsberg, once the capital of East Prussia, is now called Kaliningrad. It falls within a tiny part of Russia known as the Western Russian Enclave, between Poland and Lithuania, which (to the surprise even of many modern Russians) is not connected with the rest of the country! Königsberg lay some four miles from the Baltic Sea on rising ground on both sides of the river Pregel (now the Pregolya), which flowed through the town in two branches before uniting below the Grune Brocke (Green Bridge). Seven bridges (numbered in the diagram) crossed the Pregel and connected various parts of the city (letters A to D), including Kneiphof Island (B), the site of Königsberg University and the grave of its most famous son, the great philosopher Emmanuel Kant (1724–1804).

A question arose among the town's curious citizens: Was it possible to make a journey across all seven bridges *without having to cross any bridge more than once?* No one had been able to do it, but was there a solution? Euler, who was in St. Petersburg, Russia, at the time, heard

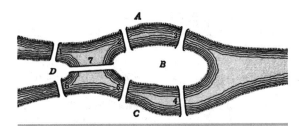

bridges of Königsberg The essential layout of the bridges. *B* represents Kneiphof Island.

about this puzzle and looked into it. In 1736, he published a paper called *Solutio problematis ad geometriam situs pertinentis* (The solution of a problem relating to the geometry of position) in which he gave his answer. Euler reasoned that, for such a journey to be possible, each land mass would need to have an even number of bridges connected to it, or, if the journey began at one land mass and ended at another then those two land masses alone could have an odd number of connecting bridges while all the other land masses would have to have an even number of connecting bridges. Since the Königsberg bridges violated this layout, a grand tour that involved only one crossing per bridge was impossible. Euler's paper was important because it solved not just the Königsberg conundrum but the much more general case of any network of points, or *vertices*, that are connected by lines, or *arcs*. What is more, the words *geometry of position* in the title show that Euler realized that he was dealing with a different type of geometry where distance is irrelevant. So this work can be seen as a prelude to the subject of **topology**. See also **Euler path**.

Briggs, Henry (1561–1630)

An English mathematician who introduced common **logarithms** (to base 10) and was largely responsible for getting scientists to use them. Although a well-regarded mathematician in his own right, holding the Savilian chair of geometry at Oxford, Briggs was most important as a contact and as a public relations man for his field.

Brocard problem

The problem of finding the integer solutions of the equation

$$n! + 1 = m^2.$$

These solutions are called *Brown numbers*, and only three of them are known: (5, 4), (11, 5), and (71, 7). Paul **Erdös** conjectured that there are no others.

broken chessboard

See **polyomino**.

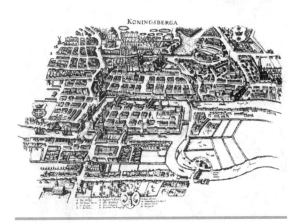

bridges of Königsberg An old map of the city showing the seven bridges.

Bronowski, Jacob (1908–1974)

A Polish mathematician who worked first on operation theory and its application to military strategy, but later on the ethics of science. He is remembered for writing and narrating the television series *The Ascent of Man* in 1973.

Brouwer, Luitzen Egbertus Jan (1881–1966)

A Dutch mathematician who opposed the logicist school of Bertrand **Russell** and established the intuitionist school of mathematical thought. He was also one of the founders of **topology**, doing most of his work in this field between 1909 and 1913.

Brouwer fixed-point theorem

An amazing result in **topology** and one of the most useful theorems in mathematics. Suppose there are two sheets of paper, one lying directly on top of the other. Take the top sheet, crumple it up, and put it back on top of the other sheet. Brouwer's theorem says that there must be at least one point on the top sheet that is in exactly the same position relative to the bottom sheet as it was originally. The same idea works in three dimensions. Take a cup of coffee and stir it as much as you like. Brouwer's theorem insists that there must be some point in the coffee that is in exactly the same spot as it was before you started stirring (though it might have moved around in between). Moreover, if you stir again to move that point out of its original position, you can't help but move another point back into its original position! Not surprisingly, the formal definition of Brouwer's theorem makes no mention of sheets of paper or cups of coffee. It states that a continuous function from an *n*-ball into an *n*-ball (that is, any way of mapping points in one object that is topologically the same as the filling of an *n*-dimensional sphere to another such object) must have a fixed point. Continuity of the function is essential: for example, if you rip the paper in the previous example then there may not be a fixed point.

Brownian motion

The most common type of continuous **random** motion of a particle, one in which the particle's vibrations have more energy at short length and time scales. It models the motion of a particle in a fluid, fluctuation of stock prices, and many other processes. Brownian motion is named after the Scottish botanist Robert Brown (1773–1858) who first studied it.

Brun's constant

See **twin primes**.

bubbles

Whether alone or in groups joined together, bubbles get their shape by following one simple rule: soap film always tries to form a **minimal surface**. The mathematical study of bubbles and films began in earnest in the 1830s with the experiments of Joseph **Plateau**. A single bubble will always try to form a sphere because this shape, as proposed by **Archimedes** and proved by Hermann Amandus Schwarz (1843–1921) in 1884, is the minimal surface enclosing a single volume. For a long time, mathematicians believed that the minimal surface for enclosing two separate volumes of air is a double bubble separated by a third surface, which meets the other two along a circle at 120° and is flat if the bubbles enclose the same volume and, otherwise, is a spherical surface that bulges a little into the larger of the two. The *double bubble conjecture* was finally confirmed by a team of four mathematicians in 2000.

Another great bubble mystery that has recently been solved is why the bubbles in a glass of Guinness appear to sink rather than rise. Bubbles that rise, like those in a saucepan or those breathed out by a diver, are a familiar sight and easy to explain: gas bubbles are lighter than liquid and experience a buoyancy force that drives them up toward the liquid surface. But many of the bubbles in a glass of Guinness can be seen heading downward. Researchers have found that large bubbles in the center of the glass move upward relatively quickly and drag liquid with them. Since the amount of liquid in the glass stays the same (unless someone drinks it!) the liquid moving upward near the center must eventually move back down near the walls of the glass. This downward-

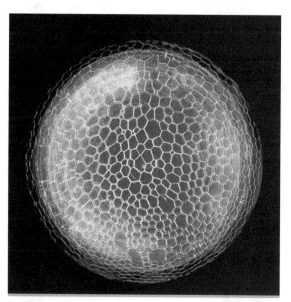

bubbles A honeycomb-like arrangement of tightly packed bubbles.
Australian National University

moving liquid has a dragging effect on the bubbles. Larger bubbles are more buoyant than smaller bubbles, and continue to move upward. Smaller bubbles (less than 0.05 mm in diameter) aren't buoyant enough to resist this drag force, and move downward with the liquid near the sides of the glass. Since Guinness is quite opaque, and these downward-moving small bubbles are close to the side of the glass, it often looks as if almost all the bubbles are moving down. See also **Plateau problem**.

buckyball

Also known as a *fullerene*, a large molecule made of carbon atoms arranged in the form of a convex polyhedral cage. Buckyballs are named after the architect Richard Buckminster Fuller because they look like the geodesic domes that he invented. The first buckyball to be discovered (by accident) was C_{60}, in which 60 carbon atoms are arranged at each of the vertices of a truncated **icosahedron**. This shape, which looks like a soccer ball, has 32 faces, of which 20 are regular hexagons and 12 are regular pentagons. Many

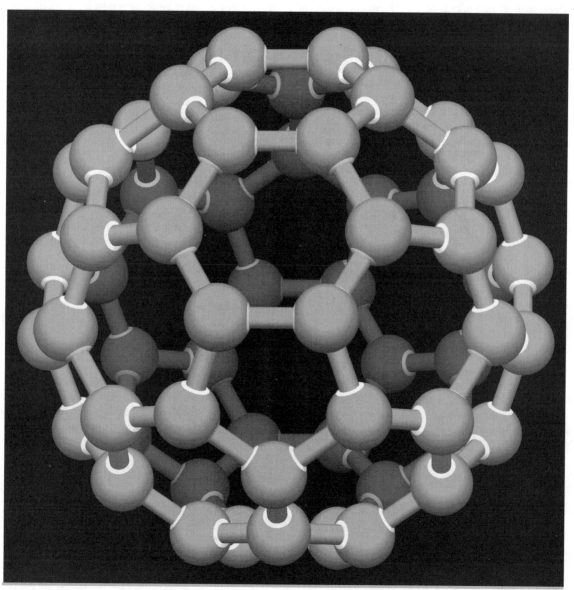

buckyball   A molecule of Buckminster fullerene—a buckyball. *Nick Wilson*

different types of buckyballs are known. Common ones have 70, 76, and 84 carbon atoms, but all are built up exclusively from hexagonal and pentagonal faces in arrangements that follow **Euler's formula**. This formula ensures that while the number of hexagons can vary from one type of fullerene to another, every fullerene has exactly 12 pentagons. (In fact, buckyballs with heptagonal faces have been seen, but these faces are concave and are regarded as defects.)

Buffon's needle

An early problem in geometrical probability (see **probability theory**) that was investigated experimentally in 1777 by the French naturalist and mathematician Comte Georges Louis de Buffon (1707–1788). It involves dropping a needle repeatedly onto a lined sheet of paper and finding the probability of the needle crossing one of the lines on the page. The result, surprisingly, is directly related to the value of **pi**.

Consider a simple case in which the lines are 1 cm apart and the needle is 1 cm in length. After many drops the probability of the needle lying across a line is found to be very close to $2/\pi$. Why? There are two variables: the distance from the center of the needle to the closest line, d,

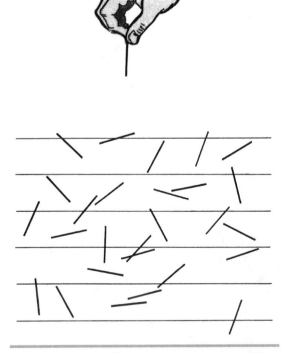

Buffon's needle Needles are dropped randomly onto a lined surface.

which can vary between 0 and 0.5 cm, and the angle, θ, at which the needle falls with respect to the lines, which can vary between 0 and 180°. The needle will hit a line if $d \leq \frac{1}{2} \sin\theta$. In a plot of d against $\frac{1}{2} \sin\theta$, the values on or below the curve represent a hit; thus, the probability of a success is the ratio of the area below the curve to the area of entire rectangle. The area below the curve is given by the integral of $\frac{1}{2} \sin\theta$ from 0 to π, which is 1. The area of the rectangle is $\pi/2$. So, the probability of a hit is $1/(\pi/2)$ or $2/\pi$ (about 0.637). Dropping a needle many times onto lined paper gives an interesting (but slow) way to find π. This kind of probabilistic means of performing calculations is the basis of a technique known as the **Monte Carlo method**.

bundle

A map between two **topological space**s A and B, where the sets $f^{-1}(b)$ for elements b of B (known as fibers), are all **homeomorphic** to a single space. The simplest example is the **Möbius band**, for which A is the Möbius band, B is a circle, and the fibers are homeomorphic to an interval on the real number line.

Burali-Forti paradox

An argument that shows that the collection of **ordinal numbers** (numbers that give the position of objects) do not, unlike the **natural numbers**, form a **set**. Each ordinal number can be defined as the set of all its predecessors. Thus:

0 is defined as {}, the **empty set**

1 is defined as {0} which can be written as {{}}

2 is defined as {0, 1} which can be written as {{}, {{}}}

3 is defined as {0, 1, 2} which can be written as {{}, {{}}, {{}, {{}}}} · · ·

in general, n is defined as {0, 1, 2, . . . $n - 1$}

If the ordinal numbers formed a set, this set would then be an ordinal number greater than any number in the set. This contradicts the assertion that the set contains all ordinal numbers. Although the ordinal numbers don't form a set, they can be regarded as a collection called a class.

Buridan's ass

A paradox of medieval logic concerning the behavior of an ass that is placed equidistantly from two piles of food of equal size and quality. Assuming that the behavior of the ass is entirely rational, it has no reason to prefer one pile to the other. Thus lacking a basis to decide which pile to eat first, it remains in its original position and starves. With one pile it would have lived; having two identical piles it dies. How can this make sense? The paradox is named after the French philosopher Jean Buridan (c. 1295–1356).

burr puzzle

An interlocking wooden puzzle that, when put together, typically looks like three rectangular blocks crossing one another at right-angles. Little is known about its early history, but it was certainly produced both in Asia and Europe in the eighteenth century. It acquired the name *Chinese puzzle*, probably because so many were produced in the Orient from the early 1900s. In 1928, Edwin Wyatt published *Puzzles in Wood*,[355] the first book devoted to the subject, and introduced the term *burr puzzle* because of the likeness of the assembled toy to a seed burr.

Burr puzzles consist of three (the smallest number), six (the most common number), twelve, or other numbers of pieces that are notched in various ways so as to pose a challenge to the would-be assembler. The earliest known reference to the popular six-piece burr appears in a Berlin catalog of 1790, but not until 1917 was a patent taken out on a particular design. In 1977, William **Cutler** proved that 25 possible notchable pieces can be used to make solid six-piece burrs and that they can be put together in 314 ways. (Pieces are considered notchable if they can be made by a sequence of notches that are produced by chiseling out the space between two saw cuts.) Cutler also proved there are 369 general pieces from which solid burrs can be made and that these can be assembled in 119,979 ways. One particular form of burr has six identical pieces, all of which move outward or inward together. Another form, with flat notched pieces, has one piece with an extra notch or an extended notch that allows it to fit in last, either by sliding or twisting, although this isn't initially obvious. This form is sometimes made with equal pieces so that it can only be assembled by force, perhaps after steaming.

butterfly effect

One of the more sensational and loudly touted claims of **chaos** theory: a butterfly beating its wings could, by an intricate chain of causes and effects, give rise to a hurricane. The gist of the argument is that minuscule disturbances can be amplified unpredictably into major phenomena. However, the overwhelming likelihood is that any effect as small as the beating of a butterfly's wing would be quickly dampened out and play no significant part in future events. See also **causality**.

butterfly theorem

Let *M* be the midpoint of a **chord** *PQ* of a circle, through which two other chords *AB* and *CD* are drawn. If *AD* intersects *PQ* at *X* and *CB* intersects *PQ* at *Y*, then *M* is also the midpoint of *XY*. The theorem gets its name from the shape of the resulting figure.

Byron, (Augusta) Ada (1815–1852)

Also known as Lady Lovelace, the daughter of Lord Byron and one of the most picturesque characters in computer prehistory. Her parents separated five weeks after her birth and she was raised by her mother (neé Annabella Milbanke), whom Lord Byron had called his "Princess of Parallelograms" because of her interest in mathematics. She was determined that Ada would become a mathematician and scientist, not a poet like her father. Ada, however, managed to combine both worlds, blending her science with poetical vision and her mathematics with metaphor.

At the age of 17 she was introduced to Mary Somerville, a remarkable woman who translated **Laplace**'s works into English, and whose texts were used at Cambridge (where a

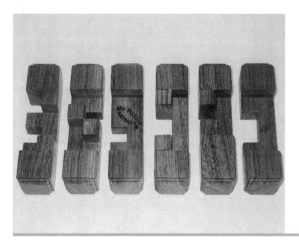

burr puzzle   A difficult six-piece burr with 19 false assemblies and a final piece that requires 10 directional moves. *Mr. Puzzle Australia/William Cutler*

women's college is now named after her). It was at one of Mary Somerville's dinner parties, in November 1834, that Ada first heard of Charles **Babbage**'s ideas for a new calculating machine, the Analytical Engine, and was immediately intrigued. In 1843, Ada, married and the mother of three children, translated a French article about the Engine and showed it to Babbage. He suggested that she add her own notes, which turned out to be three times the length of the original piece and included prescient comments about how such a machine might be used to compose complex music, produce graphics, and solve scientific problems. A regular correspondence ensued between Ada and Babbage, during which Ada suggested to Babbage a plan for how the engine might calculate **Bernoulli numbers**—a plan now regarded as the first computer program. In recognition of this, a software language developed by the U.S. Department of Defense was named "Ada" in 1979. Like her father, she died at the age of 36, following a series of illnesses.[335]

byte
A string of 8 **bits**, used to represent a character.

C

caduceus

In mathematics, a pair of curves in space, each of which is a **helix** and which twist in opposite directions around one another. In mythology, the caduceus is the wing-topped staff, wound about by two snakes, carried by Hermes, the Greek messenger of the gods. The snakes became entwined after Hermes threw his staff at them to stop their fighting. A caduceus was carried by Greek officials and became a Roman symbol for truce and neutrality. Since the sixteenth century it has also served as a symbol of medicine. Before modern medicine, people infected by parasitic worms were treated by physicians using a stick and a knife. A slit would be cut in the patient's skin in front of the worm, and as the parasite crawled out of the incision, the worm would be wound around a stick until it was totally removed. The medical treatment of parasitic worm infection by knife and stick is believed to be the inspiration for the original caduceus. It was used as a promotional sign for physicians of that period.

Cage, John (1912–1992)

The American avant-garde composer perhaps best known for the quietest piece of music ever written. His piano composition *4'33"* calls for the player to sit in silence for 273 seconds–this being the number of degrees below

caduceus *Auckland Medical Research Foundation*

zero on the centigrade scale of **absolute zero** at which molecular motion stops. *4'33"* was inspired by Cage's visit to Harvard University's anechoic chamber about which he wrote:

> There is no such thing as empty space or empty time. There is always something to hear or something to see. In fact, try as we might to make a silence, we cannot. For certain engineering purposes, it is desirable to have as silent a situation as possible. Such a room is called an anechoic chamber, its walls made of special materials, a room without echoes. I entered one at Harvard University . . . and heard two sounds, one a high and one a low. When I described them to the engineer in charge, he informed me that the high one was my nervous system and the low one was my blood circulation.

Cage's *4'33"* breaks traditional boundaries by shifting attention from the stage to the audience and even beyond the concert hall. The listener becomes aware of all sorts of sound, from the mundane to the profound, from the expected to the surprising, from the intimate to the cosmic: shifting in seats, riffling programs, breathing, a creaking door, passing traffic, a recaptured memory. Is sitting quietly alone for 273 seconds equivalent to a private performance (and audience) of the piece? Or, in the final analysis, is it all pretentious nonsense? In his essay on "Nothing" Martin Gardner wrote: "I have not heard *4'33"* performed, but friends who have tell me it is Cage's finest composition."

Caesar cipher

The simplest and oldest known type of **substitution cipher**, attributed to Julius Caesar, who used it to send government messages. In it, each letter in the alphabet is replaced by another letter using a predefined rule that shifts the alphabet a uniform amount to the right or left. For example, a shift of three units to the right, would turn the "This is secret" into "Wklv lv vhfuhw."

cake-cutting

How can a group of people cut up a cake so that each gets what they consider to be a fair share? In its modern mathematical form, this classic problem of *fair division* dates from World War II, when Hugo **Steinhaus** tackled it using a **game theory** approach.[315] Any number of "players" are allowed. They agree on rules for dividing the cake, and then everyone follows those rules. In the end, each player must get what he or she perceives to be a fair share. In the simplest case, involving just two people, there is an easy, well-known strategy: one cuts, the other chooses. Can this method be extended to three people? Can it be extended so that each person, according to his own judgment, receives the *biggest* piece? Steinhaus was able to prove that a so-called *envy-free division,* where everyone believes they got the best deal exists in every case, for any number of players. However, it was left for others to find actual algorithms that worked for three or more players.

A three-person envy-free method was first devised by John Selfridge and John **Conway**. Suppose the players are called Alice, Bob, and Carol. The method goes like this: (1) Alice cuts the cake into what she thinks are thirds. (2) Bob trims one piece to create a two-way tie for largest, and sets the trimmings aside. (3) Carol picks a piece, then Bob, then Alice. Bob has to take a trimmed piece if Carol does not. Call the person who took the trimmed piece T, and the other (of Bob and Carol) NT. (4) To deal with the trimmings, NT cuts them into what she thinks are thirds. (5) The players pick pieces in this order: T, Alice, NT. The key to the success of the Selfridge-Conway strategy is that for the trimmings, Alice has an "irrevocable advantage" with respect to T, since Alice will never envy T even if T gets all the trimmings. Thus Alice can pick after T and allow the method to end in a finite number of steps.

For four or more cake-cutters, envy-free solutions are very complex and can take arbitrarily long to resolve. However, a general solution of the problem of fair, envy-free division was eventually found in 1992 by the Americans Steven Brams, a political scientist at New York University, and Alan Taylor, a mathematician at Union College in Schenectady, New York.[51, 52] With two players, the first player cuts the cake in half. With three players, the first player cuts the cake into thirds. With four players, Brams and Taylor showed, the first player, say Bob, cuts the cake into five equal-looking pieces. He passes them to Carol, who trims two at most to create a three-way tie for largest in her eyes. She sets the trimmings aside and gives the five pieces to Don, who trims one at most to create a two-way tie for largest in his eyes. Alice, the fourth player, now selects the piece she likes best. Choosing proceeds in the reverse order from cutting, with the proviso that anyone who trimmed one or more pieces must take one of them if any are still available when it's his or her turn to choose. The extra piece to begin with assures that no player gets second-best. If someone takes a piece she likes before it's her turn to choose, an equivalent piece or better always remains on the table. According to a formula Brams and Taylor developed, Bob must cut the cake into at least $2^{(n-2)+1}$ pieces at the start. This amounts to nine pieces for five players, 17 pieces for six, and so on. Bob has to cut all these extra pieces to make sure that, when he finally gets to choose at the end, there will be a piece left that hasn't been either trimmed or chosen by one of the many other players. With 22 players, Bob has to divide the cake into over a million pieces—small crumbs of comfort in the quest for a fairer world.[268]

Calabi-Yau space

A type of mathematical space that enters into **string theory**, where the geometry of the universe is held to consist of at least 10 dimensions—the four familiar dimensions of **space-time** and six compact dimensions of Calabi-Yau space. These extra dimensions are so tightly curled up that they aren't noticed. Although the main application of Calabi-Yau spaces is in theoretical physics, they are also interesting from a purely mathematical standpoint.

calculus

The calculus is the greatest aid we have to the application of physical truth in the broadest sense of the word.
—William Fogg Osgood (1864–1943)

The branch of mathematics that deals with (1) the rate of change of quantities (which can be interpreted as the slopes of curves), known as *differential calculus,* and (2) the length, area, and volume of objects, known as *integral calculus.* Calculus was one of the most important developments in mathematics and also in physics, much of which involves studying how quickly one quantity changes with respect to another. It is no coincidence that one of the founders of calculus was the brilliant English physicist Isaac **Newton**; another was Gottfried **Leibniz**. Although students nowadays learn differential calculus first, integral calculus has older roots.

calculus of variations

Calculus problems, especially **differentiation** and *maximization,* that involve **functions** on a set of functions of a real variable. An example is to find the shape of a cable suspended from both ends.

calendar curiosities

The earliest event in human history for which a definite date is known is a battle between the Lydians (allies of the Greek Spartans) and the Medes (ruled by the Persian king Cyrus) who had been locked in a war for five years. As the two sides faced each other for a crucial daytime confrontation, a solar eclipse occurred. This was taken as a sign of the gods' disapproval and the Lydians and Medes agreed to end the fighting then and there. The dates of solar eclipses can be figured out with great accuracy, and this one is known to have taken place on May 28, 586 B.C.

Much less certain is the birth date of Christ. It was not until A.D. 440 that Christmas was celebrated on December 25. This date was chosen because it coincided with the birth date of Mithras, the Persian sun-god, and was close to the pagan festival of Yule. In A.D. 534, Dionysius Exiguus (also known as Dennis the Little) created the system, still used today, of counting the years from the birth of Christ. Unfortunately, he slipped up in his calculations. No one knows exactly when Jesus was born, but it was probably around 6 B.C. and certainly before the death of Herod the Great in 4 B.C.

As for the future, there's no shortage of predictions about the end of the world. According to the Mayan "long count" linear calendar, it will happen on June 5, 2012. Other calendric curiosities: February 1865 is the only month in recorded history not to have a full moon, and months that begin on a Sunday will always have a Friday the 13th.

Caliban

A pseudonym of Hubert **Phillips**.

Caliban puzzle

A logic puzzle in which one is asked to infer one or more facts from a set of given facts.

cannonball problem

The mathematical analysis of stacks of cannonballs (or of spheres in general) that has its roots in a question posed by Sir Walter Raleigh, explorer, introducer of the potato and tobacco to Britain, and part-time pirate on the high seas. Raleigh asked his mathematical assistant, Thomas Harriot, how he could quickly figure out the number of cannonballs in a square pyramidal stack without having to count them individually. Harriot solved this problem without difficulty. If k is the number of cannonballs along the side of the bottom layer, the number of cannonballs in the pyramid n is equal to $\frac{1}{6}k(1+k)(1+2k)$. For example, if $k = 7$, $n = 420$. A more specific form of the cannonball problem asks what is the smallest number of balls that can first be laid out on the ground as an $n \times n$

cannonball problem   Cannonballs stacked in Narbonne, France. *Australia National University*

square, then piled into a square pyramid k balls high? In other words, what is the smallest square number that is also a square pyramidal number? This answer is the smallest solution to the **Diophantine equation**

$$\tfrac{1}{6} k(1 + k)(1 + 2k) = n^2$$

and turns out to be $k = 24$, $n = 70$, corresponding to 4,900 cannonballs. The ultimate form of the cannonball problem is to ask if there are any other, larger solutions. In 1875 Edouard **Lucas** conjectured that there weren't, and in 1918 G. N. Watson proved that Lucas was right.[345]

Returning to Elizabethan times, Thomas Harriot's interest in spheres extended far beyond piles of cannonballs. Harriot was an atomist, in the classical Greek sense, and believed that understanding how spheres pack together was crucial to understanding how the basic constituents of nature are arranged. Harriot also carried out numerous experiments in optics and was far ahead of his time in this field. So when, in 1609, Johannes **Kepler** wanted some advice on how to give his own theories on optics a stronger scientific underpinning who better to turn to than the Englishman? Harriot supplied Kepler with important data on the behavior of light rays passing through glass, but he also stimulated the German's interest in the sphere-packing problem. In response, Kepler published a little booklet titled *The Six-Cornered Snowflake* (1611) that would influence the science of crystallography for the next two centuries and that contained what has come to be known as **Kepler's conjecture** about the most efficient way to pack spheres.

canonical form
A form of any given **polyhedron** distorted so that every edge is **tangent** to the unit sphere and the center of gravity of the tangent points is the origin.

Cantor, Georg Ferdinand Ludwig Philipp (1845–1918)
A Russian-born German mathematician who founded **set theory** and introduced the concept of **transfinite numbers**. His shocking and counterintuitive ideas about **infinity** drew widespread criticism before being accepted as a cornerstone of modern mathematical theory.

Cantor was 11 when his family moved from St. Petersburg to Germany. Despite attempts to push him into the more lucrative field of engineering, he eventually won his father's approval to study math at the Polytechnic of Zurich. The following year, 1863, his father died and Cantor switched to the University of Berlin where he studied under some of the greats of the day, including Karl **Weierstrass** and Leopold **Kronecker**. After receiving his doctorate in 1867, he had trouble finding a good job and was forced to accept a position as an unpaid lecturer and later

as an assistant professor at the backwater University of Halle. In 1872, he achieved his first breakthrough—and a promotion—by proving that if a **function** is continuous (in other words, its graph is smooth) throughout an interval, it can be represented by a unique trigonometric series. This work, suggested to him by his colleague Heinrich Heine, was crucial because it led Cantor to think about the relations between points, represented by **real numbers**, that make up an unbroken line—the so-called *continuum*. Cantor realized that **irrational numbers** can be represented as infinite sequences of **rational numbers**, so that they can be understood as geometric points on the real-number line, just as rational numbers can. He was now in uncharted territory and at odds with mathematical orthodoxy, which frowned on the idea of actual **infinity**; however, he found like-minded friends in Richard **Dedekind** and, later, Gösta **Mittag-Leffler**.

In 1873 to 1874 Cantor proved that the rational numbers could be paired off, one by one, with the natural numbers and were therefore countable, but that there was no such one-to-one correspondence with the real numbers. He then went on to show, incredibly, that there are exactly the same number of points on a short line as there are on an indefinitely long line, or on a plane, or in any mathematical space of higher dimensions. On this, he wrote to Dedekind: "I see it, but I don't believe it!"

By 1883, Cantor had abandoned his earlier reticence about dealing with irrationals only as sequences of rationals and started to think in terms of a new type of number—the **transfinite numbers**. The sets of natural numbers and of real numbers were, he reasoned, just two elements of a series of different kinds of infinity. This dramatic extension of the number system to allow for legitimate mathematics of the infinite was violently opposed. Henri **Poincaré** said that Cantor's theory of infinite sets would be regarded by future generations as "a disease from which one has recovered." Kronecker went further and did all he could to ridicule Cantor's ideas, suppress publication of his results, and block Cantor's ambition of gaining a position at the prestigious University of Berlin. In the spring of 1884 Cantor suffered the first of several attacks of depression, exacerbated if not induced by the negative reaction of his contemporaries. In between these attacks, he published further results but was increasingly troubled by his failure to prove the **continuum hypothesis**—his belief that the order of infinity of the real numbers came next after that of the natural numbers. Although his later years were spent in and out of a sanatoria, he lived long enough to see his ideas on set theory vindicated and be described by David **Hilbert** as "the finest product of mathematical genius and one of the supreme achievements of purely intellectual human activity."

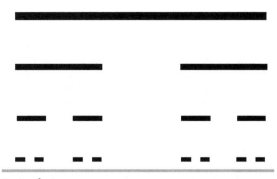

Cantor dust.

Cantor dust

Also known as the *Cantor set,* possibly the first pure **fractal** ever found. It was detected by Georg **Cantor** around 1872. To produce Cantor dust, start with a line segment, divide it in three equal smaller segments, take out the middle one, and repeat this process indefinitely. Although Cantor dust is riddled with infinitely many gaps, it still contains uncountably many points. It has a **fractal dimension** of log 2/log 3, or approximately 0.631. See also **Sierpinski carpet**.

cap

The symbol ∩ used to denote the **union** between two **sets**.

Cardano, Girolamo (1501–1570)

A celebrated Renaissance mathematician, physician, astrologer, and gambler, whose writings on the use of **negative numbers** are the earliest known in Europe. As a physician, he gave one of the first clinical descriptions of typhoid fever. The illegitimate child of a mathematically gifted lawyer who was a friend of Leonardo da Vinci, he entered the University of Pavia in 1520 and later studied medicine at Padua. His eccentric and confrontational style earned him few friends, and he had trouble finding work. Eventually, he won a reputation as a physician and his services were highly valued at the courts.

Today, Cardano is remembered mostly for his achievements in algebra. He published the solutions to the **quartic** and **cubic equations** in his book *Ars magna* (1545). The solution to the cubic was communicated to him by Niccoló **Tartaglia** (who later claimed that Cardano had sworn not to reveal it, and became embroiled with Cardano in a decade-long fight), and the quartic was solved by Cardano's student Lodovico Ferrari. Both were acknowledged in the foreword of the book. Cardano was notoriously short of money and kept himself afloat by being an accomplished gambler and chess player. A book by him about games of chance, *Liber de Ludo Aleae* (Book on games of chance), written in the 1560s but published posthumously in 1663,

contains the first systematic treatment of **probability theory**, as well as a section on effective cheating methods. Cardano invented several mechanical devices including the combination lock, the *Cardano suspension* (consisting of three concentric circles that allow a supported compass to rotate freely), and the *Cardan shaft,* which allows the transmission of rotary motion at various angles and is used in vehicles to this day. He made several contributions to hydrodynamics and claimed that perpetual motion is impossible, except in celestial bodies. He published two encyclopedias of natural science that contain a wide variety of inventions, facts, and occult superstitions.

Cardano led a beleaguered life. His elder and favorite son was executed in 1560 after he confessed to having poisoned his mercenary, cuckolding wife. Cardano's daughter was allegedly a prostitute who died from syphilis, prompting him to write a treatise about the disease. His younger son was a gambler who stole money from him. And Cardano himself was accused of heresy in 1570 because he computed the horoscope of Jesus Christ. Apparently, his own son contributed to the prosecution. Cardano was arrested and had to spend several months in prison, then was forced to abjure and had to give up his professorship. He moved to Rome, received a lifetime annuity from Pope Gregory XIII, and finished his not-uneventful autobiography. He died on the day he had (supposedly) astrologically predicted. See also **Chinese rings**.

Cardan's rings

See **Chinese rings**.

cardinal number

A number, often called simply a *cardinal,* that is used to count the objects or ideas in a **set** or collection: 0, 1, 2, . . . , 83, and so on. The *cardinality* of a set is just the number of elements the set contains. For finite sets this is always a **natural number**. To compare the sizes of two sets, X and Y, all that's necessary is to pair off the elements of X with those of Y and see if there are any left over. This concept is obvious in the case of finite sets but leads to some strange conclusions when dealing with infinite sets (see **infinity**). For example, it is possible to pair off all the natural numbers with all the even numbers, with none left over; thus the set of natural numbers and the set of even numbers have the same cardinality. In fact, an infinite set can be *defined* as any set that has a proper subset of the same cardinality. Every **countable set** that is infinite has a cardinality of **aleph**-null; the set of **real numbers** has cardinality aleph-one. See also **ordinal number**.

cardioid

A heart-shaped curve first studied in 1674 by the Danish astronomer Ole Römer who was trying to find the

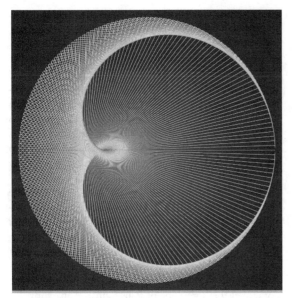

cardioid A cardioid curve spun by thread on a computer loom. *Jos Leys www.josleys.com*

best shape for gear teeth; the curve appears to have been named by Giovanni Salvemini de Castillon (1708–1791). When a circle rolls around another circle of the same size, any point on the moving circle traces out a cardioid. The Greeks used this fact when attempting to describe the motions of the planets. The cardioid is also the envelope of all circles with centers on a fixed circle, passing through one point on the fixed circle. In polar coordinates, it has the equation $r = 2a(1 − \cos\theta)$. It can also be described as an **epicycloid** with one cusp.

cards

The standard deck of 52 cards can be ordered in 52! (see **factorial**), or $8.065817517094 \times 10^{67}$ ways. There are various ways to **shuffle** cards in order to randomize them or to perform tricks with them. Each of the four kings in a deck represents a great leader from history: Charlemagne (hearts), Alexander the Great (clubs), Julius Caesar (diamonds), King David (spades). The king of hearts is the only one without a moustache. See also **blackjack**.

Carmichael number

Also known as an *absolute pseudoprime*, a number n that is a Fermat **pseudoprime** to any **base**, that is, it divides $(a^n − a)$ for any a. Another way of saying this is that a Carmichael number is actually a **composite number** even though **Fermat's little theorem** suggests it is probably a **prime number**. (Fermat's little theorem says

that if P is a prime number, then for any number a, $(a^P − a)$ must be divisible by P. Carmichael numbers satisfy this condition to any base despite being composite.) There are only seven Carmichael numbers under 10,000 (they are 561, 1,105, 1,729, 2,465, 2,821, 6,601, and 8,911), and less than a quarter of a million of them under 10^{16}. Nevertheless, in 1994 it was proved that there are infinitely many of them. All Carmichael numbers are the product of at least three distinct primes, for example, $561 = 3 \times 11 \times 17$.

Carroll, Lewis (1832–1898)

The pen name of Charles Lutwidge Dodgson (obtained by anglicizing the Latin translation, "Carolus Lodovicus," of his first two names), an English mathematician, logician, and writer. Carroll was born at the Old Parsonage, Newton-by-Daresbury, Cheshire, his father being the vicar of All Saints Church, Daresbury. There is a commemorative window in the church, and a "Wonderland" weathervane, showing the Mad Hatter, the White Rabbit, and Alice, on the local primary school. Carroll was educated at Rugby School (the student mathematics society there is still called the Dodgson Society in his honor) and then at Christ Church, Oxford, at which college he was to spend the rest of his life, employed mainly as a lecturer.

Carroll's most famous book, *Alice's Adventures in Wonderland* (1865), grew out of a story he told on the

Carroll, Lewis Lewis Carroll's Chess Wordgame, a game based on notation in Carroll's diaries with rules devised by Martin Gardner. *Kadon Enterprises, Inc., www.gamepuzzles.com*

hot summer afternoon of July 4, 1862, when out rowing with the three young daughters of the Greek scholar H. G. Liddell, dean of Christ Church. Alice, named after Alice Liddell (later Hargreaves, 1852–1934), continued her adventures in *Through the Looking-Glass* (1871). Carroll wrote other books for children, including a long poem, "The Hunting of the Snark" (1876) and published several mathematical works, but was not distinguished academically. He stammered badly, never married, and seemed to find greatest pleasure in the company of little girls, with whom he lost his shyness. He was also an amateur pioneer in photography and an inventor of puzzles, games, ciphers, and mnemonics.[57–59] Carroll was a master of fantasy and his stories have their own logic. Carroll used puns and coined neologisms, including what he called "portmanteau words" like chortle (combining chuckle and snort). He played games with idioms, using such expressions as "beating time" (to music) in a literal sense. He reshaped such animals of fable or rhetoric as the Gryphon, the March Hare (said to have been inspired by a carved hare carrying a satchel located at St. Mary's Church, Beverly, Humberside, where Carroll visited), and the Cheshire Cat, and invented new ones, including the Bandersnatch and the Boojum.

PUZZLES
Here are a few examples of puzzles invented by Carroll:
1. You are given two glasses. One contains 50 tablespoons of milk, the other 50 tablespoons of water. Take one tablespoon of milk and mix it with the water. Now take one tablespoon of the water/milk mixture and mix it with the pure milk to obtain a milk/water mixture. Is there more water in the milk/water mixture or more milk in the water/milk mixture?
2. If you paint the faces of a cube with six different colors, how many ways are there to do this if each face is painted a different color and two colorings of the cube are considered equivalent if you can rotate one to get the other? What if we drop the restriction that the faces be painted different colors?
3. Make a word-ladder from FOUR to FIVE. (Every step in a word ladder differs from the previous step in exactly one letter and each step in the ladder is an English word.)
4. Why is a raven like a writing desk?

Solutions begin on page 369.

CARROLLIAN QUOTES
From *Alice's Adventures in Wonderland:*
- "The different branches of Arithmetic—Ambition, Distraction, Uglification, and Derision."
- "Then you should say what you mean," the March Hare went on.
  "I do," Alice hastily replied; "at least I mean what I say, that's the same thing, you know."
  "Not the same thing a bit!" said the Hatter. "Why, you might just as well say that 'I see what I eat' is the same thing as 'I eat what I see!'"
- "Take some more tea," the March Hare said to Alice, very earnestly.
  "I've had nothing yet," Alice replied in an offended tone, "so I can't take more."
  "You mean you can't take *less*," said the Hatter. "It's very easy to take *more* than nothing."
From *The Hunting of the Snark:*
- "What I tell you three times is true."
From *Alice through the Looking Glass*
- "Can you do addition?" the White Queen asked. "What's one and one and one and one and one and one and one and one and one and one?"
  "I don't know," said Alice. "I lost count."
- "It's very good jam," said the Queen.
  "Well, I don't want any to-day, at any rate."
  "You couldn't have it if you did want it," the Queen said. "The rule is jam tomorrow and jam yesterday but never jam to-day."
  "It must come sometimes to 'jam to-day,'" Alice objected.
  "No it can't," said the Queen. "It's jam every other day; to-day isn't any other day, you know."
  "I don't understand you," said Alice. "It's dreadfully confusing."
- "When I use a word," Humpty Dumpty said, in a rather scornful tone, "it means just what I choose it to mean—neither more nor less."
  "The question is," said Alice, "whether you can make words mean so many different things."
  "The question is," said Humpty Dumpty, "which is to be master—that's all."

Cartesian geometry
See **analytical geometry**.

Cartesian coordinates
An ordered set of **real number**s that defines the position of a point in terms of its projection onto mutually perpendicular number lines. In the plane, each point is defined by two such projections, one onto the *x*-axis and

one onto the *y*-axis, and is written as an **ordered pair** of real numbers (*x, y*). The same system works equally in spaces of three or more dimensions.

Cartesian oval

A curve that actually consists of two ovals, one inside the other. It is the locus of a point whose distances *s* and *t* from two fixed points *S* and *T* satisfy the equation *s* + *mt* = *a*. When *c* is the distance between *S* and *T* then the curve can be expressed in the form:

$$((1 - m^2)(x^2 + y^2) + 2m^2cx + a^2 - m^2c^2)^2 = 4a^2(x^2 + y^2)$$

The curves were first studied by René **Descartes** in 1637 and are sometimes called the *ovals of Descartes;* they were also investigated by Isaac **Newton** in his classification of **cubic curve**s. If *m* = ±1, then the Cartesian oval is a central conic. If *m* = *a/c*, then it becomes a **limaçon of Pascal**, in which case the inside oval touches the outside one. Cartesian ovals are **anallagmatic curve**s.

Cassinian ovals

Also known as *Cassini's ovals,* a family of curves, each member of which is defined as follows: given two points *A* and *B* and a constant *c²*, the locus of points *P* with *PA* × *PB* = *c²*. The locus has the equation $(x^2 + y^2)^2 - 2a^2(x^2 - y^2)^2 - a^4 + c^4 = 0$, where *a* = *AB*. Equivalently, Cassinian ovals can be thought of as the set of curves produced when a circular **torus** is sliced at every possible point parallel to its axis. If *c* = *a*, then the curve is a special case known as the **lemniscate of Bernoulli** (a figure-eight type curve). The ovals are named after the Italian-born astronomer Giovanni Cassini (1625–1712) who first investigated them in 1680 while studying the relative motions of Earth and the Sun. Cassini

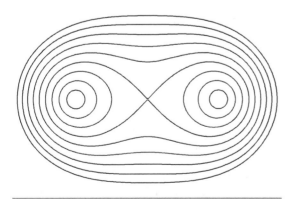

Cassinian ovals The many different ways to slice a doughnut. *Xah Lee, www.xahlee.org*

thought that the Sun traveled round Earth on one of these curves (rather than the **ellipse**, as correctly proposed in Kepler's heliocentric scheme), with Earth at one focus.

casting out nines

A method for checking arithmetic that uses the idea of the **digital root** of a number. Let the digital root of a number *n* be *r(n)*; for example, *r(7,586)* = 8. For any two numbers *a* and *b*: *r(a + b)* = *r(r(a) + r(b))* and *r(a × b)* = *r(r(a) × r(b))*. These rules allow checks on addition and multiplication as the following examples show. Does 7,586 + 9,492 = 16,978? *r(r(7,586) + r(9,492))* = *r(8 + 6)* = 5; *r(16,978)* = 4; so the sum given is incorrect. Does 7,586 × 9,492 = 72,006,312. *r(r(7,586) × r(9,492))* = *r(8 × 6)* = *r(48)* = 3; *r(72,006,312)* = *r(21)* = 3; which suggests that the product given is likely to be correct. The name "casting out nines" comes from the fact that nines need not be included in the calculation of the digital roots, since they have no effect on the final result. This a direct outcome of the fact that we use a **decimal** number system. If we calculated instead in octal (base eight), say, then the process would be one of "casting out sevens." This kind of checking will pick up most errors, but not all. For example, an interchange of two digits will not be detected, nor will replacing a nine by a zero or vice versa. The method appears in the work of ninth-century Arab mathematicians but may have originated earlier with the Greeks and, possibly, the Hindus.

Catalan number

Any number, u_n, from the *Catalan sequence* defined by

$$u_n = (2n)! / (n + 1)!n!.$$

It begins: 1, 2, 5, 14, 42, 132, 429, 1,430, 4,862, 16,796, 58,786, 208,012, 742,900, ... The values of u_n represent the number of ways a **polygon** with *n* + 2 sides can be cut into *n* triangles using straight lines joining vertices (see **vertex**). Catalan numbers are named after the Belgian mathematician Eugène Catalan (1814–1894). They also arise in other counting problems, for example in determining how many ways 2*n* beans can be divided into two containers if one container can never have less than the second.

Catalan solid

A **polyhedron** that is a dual of an **Archimedean solid**. (A dual of a polyhedron is obtained by replacing each face with a vertex, and each vertex with a face.) Catalan solids are named after the Belgian mathematician Eugène Catalan (1814–1894) who first described them in 1865. See also **Platonic solid** and **Johnson solid**. (See table, "Catalan solids.")

Catalan solid   Two of the Catalan solids: the rhombic tricontahedron (right) and the disdyakistriacontahedron (left). *Robert Webb, www.software3d.com; created using Webb's Stella program*

Catalan Solids

Name	Corresponding Archimedean Solid
Triakis tetrahedron	Truncated tetrahedron
Rhombic dodecahedron	Cuboctahedron
Triakis octahedron	Truncated cube
Tetrakis hexahedron	Truncated octahedron
Deltoidal icositetrahedron	Small rhombicuboctahedron
Disdyakis dodecahedron	Great rhombicuboctahedron
Pentagonal icositetrahedron	Snub cube
Rhombic triacontahedron	Icosidodecahedron
Triakis icosahedron	Truncated dodecahedron
Pentakis dodecahedron	Truncated icosahedron
Deltoidal hexecontahedron	Rhombicosidodecahedron
Disdyakis triacontahedron	Great rhombicosidodecahedron
Pentagonal hexecontahedron	Snub dodecahedron

Catalan's conjecture

The hypothesis, put forward by the Belgian mathematician Eugène Catalan (1814–1894) in 1844, that 8 (= 2^3) and 9 (= 3^2) are the only pair of consecutive powers. In other words, the *Catalan equation* for **prime numbers** p and q and positive integers x and y

$$x^p - y^q = 1$$

has only the one solution

$$3^2 - 2^3 = 1.$$

In 1976 R. Tijdeman took the first major step toward showing this is true by proving that for any solution, y^q is less than **e** to the power e to the power e to the power e to the power 730 (a huge number!). Since then, this bound has been reduced many times, and it is now know that the larger of p and q is at most 7.78×10^{16} and the smaller is at least 10^7. On April 18, 2002, the Romanian number theorist Preda Mihailescu sent a manuscript to several mathematicians with a proof of the entire conjecture together with an analysis by Yuri Bilu. It is expected that as soon as this work is completely reviewed by other mathematicians that Catalan's conjecture will have been proved.[266]

Solutions to Catalan's conjecture and **Fermat's last theorem** are special cases of the *Fermat-Catalan equation:*

$$x^p + y^q = z^r,$$

where x, y, and z are positive, **coprime** integers and the exponents are all primes with

$$1/p + 1/q + 1/r \leq 1.$$

The *Fermat-Catalan conjecture* is that there are only finitely many solutions to this system. These solutions include:

$$1^p + 2^3 = 3^2 \ (p \geq 2); \ 2^5 + 7^2 = 3^4$$
$$13^2 + 7^3 = 2^9;$$
$$33^8 + 1{,}549{,}034^2 = 15{,}613^3$$
$$\text{and } 43^8 + 96{,}222^3 = 30{,}042{,}907^2.$$

Catalan's constant

A constant that crops up regularly in combinatorial problems, especially in the evaluation of certain infinite series and integrals. For example, it is equal to

$\int_0^1 \arctan(x) / x \, dx$, and

$1 - 1/3^2 + 1/5^2 - 1/7^2 + 1/9^2 - \ldots$

It is also the solution to the following problem as n becomes arbitrarily large: If you have a $2n \times 2n$ checkerboard and a supply of $2n^2$ dominoes that are just large enough to cover two squares of the checkerboard, how many ways are there to cover the whole board with the dominoes? Catalan's constant has the value $0.915965\ldots$; it is not known if it's an **irrational number**.

catastrophe theory

A theory, developed by the French mathematician René Thom (1923–2003), that attempts to explain the behavior of complex **dynamical systems** by relating it to **topology**. The evolution of such systems consists of steady continuous change interspersed with sudden major jumps, or "catastrophes," when the topology of the set changes. Catastrophe theory has been applied, with varying degrees of success, to phenomena as diverse as earthquakes, stock market crashes, prison riots, and human conflicts, at the personal, group, and societal level. The theory was first developed by Thom in a paper published in 1968 but became well known through his book *Structural Stability and Morphogenesis* (1972).[331] Many mathematicians took up the study of catastrophe theory and it was in tremendous vogue for a while, yet it never achieved the success that its younger cousin **chaos** theory has because it failed to live up to its promise of useful predictions. Late in his career, the surrealist Salvador Dali painted *Topological Abduction of Europe: Homage to René Thom* (1983), an aerial view of a seismically fractured landscape juxtaposed with the equation that strives to explain it.

catch-22

A situation in which a person is frustrated by a paradoxical rule or set of circumstances that preclude any attempt to escape from them. The name comes from the title of a novel by Joseph Heller (1923–1999), based on his personal experiences, about an American airman's attempts to survive the madness of World War II. Heller wrote:

> There was only one catch and that was Catch-22, which specified that concern for one's own safety in the face of dangers that were real and immediate was the process of a rational mind. Orr was crazy and could be grounded. All he had to do was ask; and as soon as he did, he would no longer be crazy and would have to fly more missions. Orr would be crazy to fly more missions and sane if he didn't, but if he was sane he had to fly them. If he flew them he was crazy and didn't have to; but if he didn't want to he was sane and had to.

category theory

The study of abstracted collections of mathematical objects, such as the category of **sets** or the category of **vector spaces**, together with abstracted operations sending one object to another, such as the collection of **functions** from one set to another or linear **transformations** from one vector space to another.

catenary

The shape that a rope or telephone cable makes, under the influence of gravity, when suspended between two points. The word comes from the Latin *catena*, meaning "chain," and was first used by Christiaan **Huygens** while studying the form of suspended chains. Galileo thought the shape would be a **parabola**. In fact, near the **vertex**, a parabola and a catenary do look very similar. When x is slightly greater than three, however, the catenary begins to rapidly outgrow the value of the parabola. The two shapes are related in another way. If a parabola is rolled along a straight line, the **focus** of the parabola moves along a catenary curve. Surprisingly, too, if a bicycle with square (or any polygon-shaped) wheels is ridden along a road made of upturned catenaries the wheels will roll smoothly and the rider will stay at the same height! The St. Louis Arch, which is 192 meters wide at the base and 192 meters tall, follows the form of a catenary, the exact formula for which is displayed inside the arch: $y = 68.8 \cosh(0.01x - 1)$, where cosh is the hyperbolic cosine function.

The general equation of a catenary can be written

$$y = k \cosh(x/k),$$

where k is a constant, or, in terms of the **exponential** function,

catenary The catenary curve. © Jan Wassenaar, www.2dcurves.com

$$y = k(e^{x/k} + e^{-x/k})/2.$$

In the special case where $k = 1$, these reduce to

$$y = \cosh(x) = (e^x + e^{-x})/2.$$

In terms of a polynomial series

$$y = 0.5 \, (1 + x + x^2/2! + x^3/3! + x^4/4! + \ldots + 1 - x + x^2/2! \\ - x^3/3! + x^4/4! - \ldots) \\ = 1 + x^2/2! + x^4/4! + x^6/6! + \ldots .$$

For small values of x the terms beyond $x^2/2!$ are very small, so that the equation closely approximates that of a parabola, as we have already seen.

catenoid
The **surface of revolution** produced when a **catenary** rotates about its central axis. The catenoid was first described by Leonhard **Euler** in 1740 and is the oldest known **minimal surface** (a shape of least area when bounded by a given closed space). It is the minimal surface connecting two parallel circles of unequal diameter on the same axis; soap film between two circular rings takes this form (see also **bubbles**). The catenoid is the only known minimal surface that is also a surface of revolution, and is one of only four minimal surfaces that have the topological properties of being unbounded, embedded, and non-periodic; the others are the simple plane, the **helicoid**, and *Costa's surface*.

cathetus
A line that is perpendicular to another line. Usually, it refers to one of the lines in a right triangle that is not the **hypotenuse**.

Cauchy, Augustin Louis, Baron (1789–1857)
A French mathematician who founded **complex analysis** by discovering the *Cauchy-Riemann equations* and wrote 789 papers—an output surpassed only by Leonhard **Euler**, George **Cayley**, and Paul **Erdös**. He coined the name for the **determinant** and systematized its study and gave nearly modern definitions of **limit**, **continuity**, and **convergence**.

causality
The relationship between causes and effects. An event or state of affairs A is the cause of an event B if A is the reason that brings about the effect B. For instance, one might say, "my pushing the gas pedal *caused* the car to go faster." An important question in philosophy and other fields is how (and if) causes can bring about effects. In a strict reading, if A causes B, then A must *always* be followed by B. In this sense, for example, smoking doesn't cause cancer. In everyday usage, we therefore often take "A causes B" to mean "A causes an increase in the probability of B." The establishment of cause and effect, even with this relaxed reading, is notoriously difficult. The Scottish philosopher David Hume held that causes and effects are not real, but instead are imagined by our minds to make sense of the observation that A often occurs together with or slightly before B. All we can actually observe are *correlations,* not causations. This is also expressed in the logical fallacy, "correlation implies causation." For instance, the observation that smokers have a dramatically increased lung cancer rate doesn't establish that smoking must be the *cause* of that increased cancer rate: maybe there exists a certain genetic defect which both causes cancer and a yearning for nicotine.[194]

caustic
The envelope of rays of light reflected (or refracted) by a given curve from a given point source of light; a *catacaustic* results from reflection, a *diacaustic* from refraction. Among the caustic curves of a circle are a lima, if the light source is nearby, a **nephroid**, if the source is at infinity, and a **cardioid**, if the source is on the circle.

Cavalieri's principle

If two solids have the same height and the same cross-sectional area at every level, then they have the same volume. This principle is named after the Italian mathematician Bonaventura Cavalieri (1598–1647).

Cayley, Arthur (1821–1895)

A British mathematician who made important contributions to **non-Euclidean geometry** and the algebra of matrices (see **matrix**). The former eventually found its way into the study of the **space-time** continuum, the latter into a formulation of **quantum mechanics** by the German physicist Werner Heisenberg. Cayley was also far ahead of his time in pioneering the idea of abstract **group**s.

Cayley number

See **octonion**.

Cayley's mousetrap

A **permutation** problem invented by Arthur **Cayley**. Write the numbers $1, 2, \ldots, n$ on a set of cards and shuffle the deck. Start counting using the top card. If the card chosen does not equal the count, move it to the bottom of the deck and continue counting forward. If the card chosen *does* equal the count, discard the chosen card and begin counting again at 1. The game is won if all cards are discarded, and lost if the count reaches $n + 1$. The number of ways the cards can be arranged such that at least one card is in the proper place for $n = 1, 2, \ldots$, are 1, 1, 4, 15, 76, 455, $\ldots$.

Cayley's sextic

A sinusoidal spiral curve described by the Cartesian equation

$$4(x^2 + y^2 - ax)^3 = 27a^2(x^2 + y^2)^2.$$

It was discovered by Colin **Maclaurin** but was first studied in detail by Arthur **Cayley** and named after him by R. C. Archibald in 1900.

ceiling

The largest value that something can take. The *ceiling function* of a number x is the smallest integer that is not smaller than x.

cell

(1) A three-dimensional object that is part of a higher-dimensional object, such as a **polychoron**. A cell is related to higher-dimensional objects in the way that a face, or (two-dimensional) **polygon**, is related to higher-dimensional objects. For example, a cell is to a Four-dimensional polytope, or polychoron, what a face is to a three-dimensional polytope, or **polyhedron**. Often polytopes are classified simply by how many cells they have. For example, the **tesseract** has eight cells, each one of which is a cube. (2) The fundamental spatial unit operated on by the rules of a **cellular automaton** during one generation.

cellular automaton

An array of cells that evolves according to a set of rules based on the states of surrounding cells; for example, a cell might be "on" if its four neighbor cells (east, west, north, and south) are also on. The entire array can self-organize into global patterns that may move around the screen. These patterns can be quite complex even though they emerge from just a few very simple rules governing the connections among the cells. Cellular automata are the simplest models of spatially distributed processes. They were first investigated by John **von Neumann** in about 1952. Von Neumann incorporated a cellular model into his "universal constructor" and also proved that an automaton consisting of cells with four orthogonal neighbors and 29 possible states would be capable of simulating a **Turing machine** for some configuration of about 200,000 cells. The best-known cellular automaton is Life (see **Life, Conway's game of**). Another example is **Langton's ant**. The study of cellular automata and their patterns has led to insights into the way structure is built up in biological and other complex systems, and for this reason forms part of the subject of **artificial life**.

celt

Also known as a *rattleback,* a simple ancient toy that behaves in a very counterintuitive way. When spun one way about its vertical axis, the celt spins for a long time. When spun the other way, however, a wobble quickly sets in that halts the rotation and then, incredibly, reverses it. In his 1986 paper on the subject, the British physicist Hermann Bondi wrote: "Many people, even trained scientists, find it hard to understand that the behaviour of the toy doesn't violate the principle of conservation of angular momentum."[47] The celt's remarkable antics stem from three factors: a curved base that has two different radii—one long radius for the lengthwise curve and one shorter radius for the tighter curve across the width; axes of symmetry that are skewed slightly from the principal axes of inertia; and a different distribution of mass about each of the two horizontal axes of inertia. To understand how the celt can switch direction halfway through its performance, think of the frictional force that acts at the point of contact between the celt and the surface. One component of the friction creates a torque (twisting force) that tends to rotate the celt about its vertical axis. The point of contact is moving all the time and the torque changes. If the inertial and symmetrical axes coincided, the average torque over a single oscillation would be zero. But for the celt, there's a net torque in one direction, which reverses the angular momentum. See also **Tippee Top**.

center of perspective

The point where the lines joining corresponding points of two figures that are in perspective meet.

centillion

See **large number**.

central angle

The angle subtended at the center of a **circle** by an **arc** or a **chord**; in other words, the angle between two radii.

centroid

For a **triangle**, the point of intersection of the **medians**. For any other shape, the point where coordinates are the average of the coordinates of the shape's vertices (see **vertex**). The centroid is the *center of mass* of a figure.

century

A period of 100 years. The original Latin *centuria* means simply "one hundred" and was used to describe any collection of 100 items. In the Roman army, a century was a group of 100 men, each known as a *centurion*. One of the few modern examples of "century" being used other than to denote a period of time is in the game of cricket where a batsman who scores 100 runs in an inning is said to have "made a century."

Ceva, Giovanni (1647–1734)

A Jesuit-trained Italian mathematician who specialized in **geometry**. His greatest discovery, now known as *Ceva's theorem*, can be stated as follows. Given a triangle with vertices (corners), *A, B,* and *C* and points *D, E,* and *F* on the opposite sides, the lines *AD, BE,* and *CF* will intersect at a single point if $BD \times CE \times AF = DC \times EA \times FB$. The term *cevian line* was coined by French geometers around the end of the eighteenth century to honor Ceva. It is defined as any line joining a vertex of a triangle to a point on the opposite side. The **median**, **altitude**, and angle **bisect**or are all examples of cevians. The perpendicular bisector, however, in most cases, is not a cevian because it doesn't usually pass through a vertex.

chained arrow notation

See **Conway's chained arrow notation**.

Chaitin, Gregory (1947–)

An American mathematician and computer scientist at IBM's T. J. Watson Research Center who is the chief architect of a new subject known as *algorithmic information theory,* which has profound consequences for our ideas about **random**ness. In particular, because of the limitations of computers and the programs they run, Chaitin has shown that there is an inherent uncertainty or unknowability in mathematics that is similar to the uncertainty principle in physics. Although there are an infinite number of mathematical facts, they are, for the most part, unrelated to each other and impossible to tie together with unifying theorems. His powerful message is that most of mathematics is true for no particular reason; math is true by accident. See **Chaitin's constant**.

Chaitin's constant

A **real number**, represented by capital omega (Ω) and also known as the *Halting probability,* whose digits are distributed so **random**ly that no rule can be found to predict them. Discovered by Gregory **Chaitin**, Ω is definable but not computable. It has no pattern or structure to it whatsoever, but consists instead of an infinitely long string of zeros and ones in which each digit is as unrelated to its predecessor as one coin toss is to the next. Although called a constant, it is not a constant in the sense that, for example, **pi** is, since its definition depends on the arbitrary choice of computation model and programming language. For each such model or language, Ω is the probability that a randomly produced string will represent a program that, when run, will eventually halt. To derive it, Chaitin considered all the possible programs that a hypothetical computer known as a **Turing machine** could run, and then looked for the probability that a program, chosen at random from among all the possible programs, will halt. He eventually showed that this halting probability turns Turing's question of whether a program ever stops into a real number, somewhere between zero and one. He further showed that, just as there are no computable instructions for deciding in advance whether a computer will halt, there are also no instructions for determining the digits of Ω. Omega is uncomputable and unknowable: we don't know its value for any programming language and we never will. This is extraordinary enough in itself, but Chaitin has found that Ω permeates the whole of mathematics, placing fundamental limits on what we can know.

And Ω is just the beginning. There are more disturbing numbers called Super-Omegas, whose degree of randomness is vastly greater even than that of Ω. If there were an omnipotent computer that could solve the halting problem and evaluate Ω, this mega-brain would have its own unknowable halting probability called Ω'. And if there were a still more godlike machine that could find Ω', its halting probability would be Ω''. These higher Omegas, it has been recently discovered, are not meaningless abstractions. Ω', for instance, gives the probability that an infinite computation produces only a finite amount of output. Ω'' is equivalent to the probability that, during an infinite computation, a computer will fail to produce an output—for example, get no result from a computation and move on to the next one—and that it will do this only a finite number of times. Omega and the Omega hierarchy are revealing to mathematicians an unsettling truth: the problems that we can hope ever to solve form a tiny archipelago in a vast ocean of undecidability.[130]

Champernowne's number

The first known **normal number**. It was discovered in 1933 by the English mathematician David G. Champernowne and consists of a decimal fraction in which the decimal integers are written down in increasing order: 0.12345678910111213 . . . . Champernowne's number has been proven to be a normal number in base 10 and also to be an **irrational number**. However, although its digits appear with equal frequency, the sequence of its digits are not unpredictable. An example of a number whose sequence of digits *is* unpredictable is **Chaitin's constant**.

chance

See **probability theory**.

change ringing

The art of change ringing is peculiar to the English, and, like most English peculiarities, unintelligible to the rest of the world. To the musical Belgian, for example, it appears that the proper thing to do with a carefully tuned ring of bells is to play a tune upon it. By the English campanologist . . . the proper use of the bells is to work out mathematical permutations and combinations.

—Dorothy L. Sayers, *The Nine Tailors*

The ringing of a set of bells in a precise relationship to one another to produce a pleasing sound. Bells are numbered 1, 2, 3, 4, 5 . . . from lightest (highest-pitched) to heaviest. After each sequence, or *round*, the order of the bells is changed slightly in a predetermined way. With 5 bells, there are $5 \times 4 \times 3 \times 2 \times 1$, or 120, possible changes, which take about 4 minutes to ring. With 6, 7, or 8 bells, the number of unique changes is 720, 5,040, and 40,320, respectively. To produce pleasing variations in the sound, bells are made to change places with adjacent bells in the row, for example:

```
1  2  3  4  5  6  7  8
2  1  4  3  6  5  8  7
```

These rows are the musical notation of change ringing. No bell moves more than one place in the row at a time, although more than one pair may change in the same

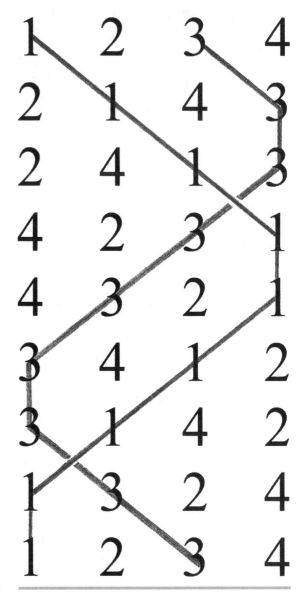

change ringing The "Plain Hunt Minimus" for four bells. The sequences for bells 1 and 3 are shown by lines.

row. In order to ring a different row with each pull of the rope, ringers have devised methods for changing pairs in orderly ways. In ringing a method, the bells begin in rounds, ring changes according to the method, and return to rounds without repeating any row along the way. These place changes produce musical patterns, with the sounds of the bells weaving in and out. For example, a "Plain Hunt Minimus" with four bells is rung as shown in the diagram.

Experienced ringers test and extend their abilities by ringing peals: 5,000 or more changes without breaks or repeating a row. Peals customarily last about three hours. The first peal was rung in England in 1715. Chiming bells (swinging them through a short arc using a rope and a lever) goes well back into the Middle Ages, but it wasn't until the seventeenth century that ringers developed the full wheel, which allowed enough control for orderly ringing. In 1668 Fabian Stedman published *Tintinnalogia* (The art of change ringing), containing all the available information on systematic ringing. The theory of change ringing set forth by Stedman has been refined in later years but remains essentially unchanged today. Bells for change ringing are hung in stout frames that allow the bells to swing through 360°. Each bell is attached to a wooden wheel with a handmade rope running around it and takes about 2 seconds to rotate. The bells are arranged in the frame so their ropes hang in a circle in the ringing chamber below. Into each rope is woven a tuft of brightly colored wool (a *sally*), which marks where the ringer must catch the rope while ringing. Bells are rung from the "mouth up" position. With a pull of the rope, the bell swings through a full circle to the up position again. With the next pull it swings back in the other direction. The plot of Dorothy Sayers's *The Nine Tailors* (1934), considered one of her best works, revolves around the art of change ringing.

chaos

We adore chaos because we love to produce order.
—M. C. Escher

A phenomenon shown by some **dynamical system**s, which consists of a curious, infinitely complex pattern of behavior that lies just beyond the edge of total order. A system is chaotic if it is predictable in principle and yet is unpredictable in practice over long periods because its behavior depends very sensitively on initial conditions. Despite this unpredictability, however, there are certain constants, such as **Feigenbaum's constant**, and certain structures, such as **chaotic attractor**s, that are fixed and susceptible to analysis. The weather, the movements of a metal pendulum moving over fixed magnets, and the orbits of closely spaced moons are all examples of chaotic systems. Although the ideas behind modern chaos theory were actively studied at some level throughout most of the twentieth century, the word as a mathematical term dates only from an article in *American Mathematical Monthly* in 1975 called "Period Three Implies Chaos."

In everyday language, chaos has come to mean the exact opposite of order. But the Greek root *khaox* means "empty space" and this meaning still persists in archaic usage where it refers to a canyon or abyss. The evolution

of the word to mean disorder seems to come from reference to the time before God created the universe. Empty space was formless and the creation filled the emptiness and established order. Mathematical chaos represents an unexpected third state: a deterministic system subject to simple rules that nevertheless displays infinitely complex behavior.[135]

chaos tiles

See **Penrose tiling**.

chaotic attractor

Also known as a *strange attractor,* a type of **attractor** (i.e., an attracting set of states) in a complex **dynamical system**'s **phase space** that shows **sensitivity** to initial conditions. Because of this property, once the system is on the attractor, nearby states diverge from each other exponentially fast. Consequently, small amounts of noise are amplified. Once sufficiently amplified the noise determines the system's large-scale behavior and the system is then unpredictable. Chaotic attractors themselves are markedly patterned, often having elegant, fixed geometric structures, despite the fact that the trajectories moving within them appear unpredictable. The chaotic attractor's geometric shape is the order underlying the apparent chaos. It functions in much the same way as someone kneading dough. The local separation of trajectories corresponds to stretching the dough and the global attraction property corresponds to folding the stretched dough back onto itself. One result of the stretch-and-fold aspect of chaotic attractors is that they are **fractals**; that is, some cross section of them reveals similar structure on all scales.

character theory

The study of the traces (sums of the diagonal elements) of the **matrix** representations of a **group**. The information gained is listed in *character tables,* the properties of which give insight into the group's properties.

chess

A game of strategy for two players that probably originated in India, though the earliest documentary references are in Chinese and Persian texts in about A.D. 600. Each player has 16 pieces, either black or white, consisting of eight pawns, two rooks (also known as castles), two knights, two bishops, a queen, and a king. The object is to lay siege to the opposing king in such a way that it cannot escape attack—a position known as *checkmate* (from the Persian phrase *Shah Mat,* meaning "the king is dead"). There are 400 first-move combinations—20 for white × 20 for black (though only 64 of these are regarded as strong), 318,979,564,000 ways of playing the first four moves, and 169,518,829,100,544,000 trillion ways of playing the first

chess An illumination from the *Cantigas de Santa Maria* (a thirteenth-century collection of songs) showing a chess game in progress.

10 moves. The total number of possible board configurations is estimated at 10^{120}; for comparison, that of **Go** is generally put at about 10^{174}.

A standard chessboard is a square plane divided into 64 smaller squares by straight lines at right angles. Originally, it wasn't checkered (that is, made with its rows and columns alternately of dark and light colors), and this feature was introduced merely to help the eye in actual play. In many puzzles based on chess the utility of checkering is questionable, and the board may be generalized to any $n \times n$ size.

One of the first puzzles to use a chessboard was the **wheat and chessboard problem**, posed in 1256 by the Arabic mathematician Ibn Kallikan. Among the earliest problems to involve chess *pieces*, proposed by Guarini di Forli in 1512, asks how two white and two black knights can be interchanged, using normal knight's moves, if they are placed at the corners of a 3×3 board. The unusual L-shaped movement of the knight is what makes one of the best known chess puzzles, the **knight's tour**, such a challenge. Other standard puzzles, often called simply the **kings problem**, the **queens puzzle**, the **rooks problem**,

the **bishops problem**, and the **knights problem**, ask for the greatest number of each of these pieces that can be placed on an 8×8 board or on a generalized $n \times n$ board without attacking each other, and/or the smallest number of each of these pieces that are needed to occupy or attack every square. *Fairy chess* is any variant on the standard game, which may involve a change in the form of the board, the rules of play, or the pieces used. For example, the normal rules of chess can be used but with a cylindrical or **Möbius band** connection of the edges.

Chinese cross
See **burr puzzle**.

Chinese remainder theorem
If there are n numbers, a_1 to a_n, that have no **factor**s in common (i.e., are pairwise relatively prime), then any integer greater than or equal to 0 and less than the product of all the numbers n can be uniquely represented by a series consisting of the remainders of division by the numbers n. For example, if $a_1 = 3$ and $a_2 = 5$, the Chinese remainder theorem (CRT) says that every integer from 0

to 14 will have a unique set of remainders when divided separately by (**modulo**) 3 and 5. Listing out all the possibilities shows that this is true:

0 has a remainder of 0 modulo 3 and a remainder of 0 modulo 5.

1 has a remainder of 1 modulo 3 and a remainder of 1 modulo 5.

2 has a remainder of 2 modulo 3 and a remainder of 2 modulo 5.

3 has a remainder of 0 modulo 3 and a remainder of 3 modulo 5.

4 has a remainder of 1 modulo 3 and a remainder of 4 modulo 5.

5 has a remainder of 2 modulo 3 and a remainder of 0 modulo 5.

6 has a remainder of 0 modulo 3 and a remainder of 1 modulo 5.

7 has a remainder of 1 modulo 3 and a remainder of 2 modulo 5.

8 has a remainder of 2 modulo 3 and a remainder of 3 modulo 5.

9 has a remainder of 0 modulo 3 and a remainder of 4 modulo 5.

10 has a remainder of 1 modulo 3 and a remainder of 0 modulo 5.

11 has a remainder of 2 modulo 3 and a remainder of 1 modulo 5.

12 has a remainder of 0 modulo 3 and a remainder of 2 modulo 5.

13 has a remainder of 1 modulo 3 and a remainder of 3 modulo 5.

14 has a remainder of 2 modulo 3 and a remainder of 4 modulo 5.

CRT enables problems such as the following to be solved: Find the two smallest counting numbers that will each have the remainders 2, 3, and 2 when divided by 3, 5, and 7, respectively. It is said that the ancient Chinese used a variant of this theorem to count their soldiers by having them line up in rectangles of 7 by 7, 11 by 11, and so forth. After counting only the remainders, they solved the associated system of equations for the smallest positive solution.

Chinese rings

One of the oldest known **mechanical puzzle**s, the object of which is to remove all n rings from a horizontal loop of stiff wire, and/or put them back on the loop. On the first move it is possible to take up to two rings off the left end of the wire. One or both of those can then be slipped through the wire loop (from top to bottom). If both are removed then the fourth ring can be slipped over the end. If just one of the first two is removed, then the next step is to slip the third ring over the end. Subsequently, rings must be put back on to the wire loop in order to remove other rings, and this procedure is repeated over and over again. In general, the minimum number of moves needed is $(2^{n+1} - 2)/3$ if n is even and $(2^{n+1} - 1)/3$ if n is odd. For example, with seven rings the solution takes 85 moves. Most of the solution is easy, as each move normally involves going forward or back to the previous state. The key to a correct solution is the first step: if n is even, you must remove two rings; if n is odd, you must remove only one. The solution is similar to that of the **Tower of Hanoi**. In fact, Edouard **Lucas**, who invented the Tower of Hanoi, gave an elegant solution to the Chinese rings that uses **binary** arithmetic.

Stewart Cullin, the noted nineteenth-century ethnologist, relates that the puzzle was invented by the famous Chinese general Chu-ko Liang (A.D. 181–234), in the second century, as a present to his wife so that, in trying to solve it, she would have something to do while he was away at the wars. However, this is anecdotal and its origins remain obscure. The earliest reference to it in Europe may be in about 1500 in the form of Problem 107 of the manuscript *De Viribus Quantitatis* by Luca **Pacioli** in which the description appears: "*Do Cavare et Mettere una Strenghetta Salda in al Quanti Anelli Saldi, Difficil Caso*" (Remove and put a little bar joined in some joined rings,

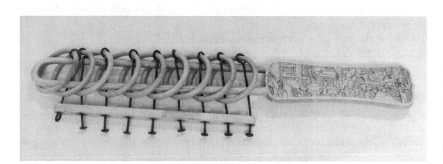

Chinese rings An unusual example of Chinese rings in ivory, dating from the mid-nineteenth century. *Sue & Brian Young/Mr. Puzzle Australia, www.mrpuzzle.com.au*

difficult case). It was also mentioned by Girolamo **Cardan** in the 1550 edition of his book *De Subtililate* from which comes the name *Cardan's rings,* and was treated at length in mathematical terms by John **Wallis** in about 1685. By the end of the seventeenth century, it had become popular in many European countries. French peasants used it to lock chests and called it *baguenaudier,* or "time-waster."

Chinese room

An argument first put forward by the American philosopher John Searle (1932–) in 1980 in an attempt to show that the human mind is not a computer and that the **Turing test** is not adequate to prove that a machine can have strong **artificial intelligence** (strong AI)—in other words, can think in a humanlike way.[292] In the Chinese room scenario, a person who understands no Chinese sits in a room into which written Chinese characters are passed. The person uses a complex set of rules, established ahead of time, to manipulate these characters, and pass other characters out of the room. The idea is that a Chinese-speaking interviewer would pass questions written in Chinese into the room, and the corresponding answers would come out of the room in Chinese. Searle maintains that if such a system could indeed pass a Turing test, the person who manipulated the symbols would obviously not understand Chinese any better than he did before entering the room.

Searle proceeds systematically to refute the claims of strong AI by positioning himself as the one who manipulates the Chinese symbols. The first claim is that a system able to pass the Turing test understands the input and output. Searle replies that as the "computer" in the Chinese room, he gains no understanding of Chinese by simply manipulating the symbols according to the formal program (the complex translation rules). The operator in the room need not have any understanding of what the interviewer is asking, or of the replies that he is producing. He may not even know that there is a question-and-answer session going on outside the room.

The second claim of strong AI to which Searle objects is the claim that the system explains human understanding. Searle asserts that since the system is functioning—in this case passing the Turing Test—and yet there is no understanding on the part of the operator, then the system does not understand and therefore could not explain human understanding.

chiral

Having different left-hand and right-hand forms; not mirror symmetric. For example, the *snub cube* (one of the **Archimedean solid**s) is chiral, where as the ordinary **cube** is not.

Chladni, Ernst Florens Friedrich (1756–1827)

A German lawyer, musician (he was born in Leipzig in the same year as Mozart and died in the same year as Beethoven), and amateur scientist who founded the science of acoustics. While investigating musical tones, he had the inspired idea of making the sounds visible in a solid material. He spread fine sand over a glass or metal plate and set it into vibration with the bow of a violin by scraping the bow along one edge of the plate. The bow alternately stuck and slipped in rapid succession on the edge of the plate creating waves that moved across the plate and were reflected from the edges. These reflected waves became superimposed on the new waves coming from the bow edge, resulting in symmetrical patterns of nodal lines where the plate wasn't moving. The type of pattern produced on a Chladni plate depends on a variety of factors, including the point or points of support and their location; the point where the bow touches the plate; the frequency of the vibration, which is influenced by the speed the bow; and the shape and other properties of the plate itself.

chord

A straight line that joins two points on a curve. Most commonly, *chord* is used to mean a straight line segment joining, and included between, two points on a **circle**. In this more restricted sense it first appears in English in 1551 in Robert **Recorde**'s *The Pathwaie to Knowledge:* "*Defin.,* If the line goe crosse the circle, and passe beside the centre, then is it called a corde, or a stryngline."

Some surprising results emerge from moving chords. For example, take a chord in a circle *C,* and slide the chord around the circle so that the midpoint of the chord traces out a smaller concentric circle. Call the area between the two circles $A(C)$. Now do the same thing with a larger circle C' but with the same length chord. Is $A(C')$ larger or smaller than $A(C)$? Surprisingly, they are the same. In other words $A(C)$ doesn't depend on what circle you start with, only the length of the chord. An even more amazing fact is that if you slide a chord of fixed length around *any convex shape C* so that the chord's midpoint traces out another figure D, the area between C and D doesn't depend on what shape you started with.

chromatic number

(1) In **graph theory**, the minimum number of colors needed to color (the vertices of) a **connected graph** so that no two adjacent vertices are colored the same. In the case of simple graphs, this so-called *coloring problem* can be solved by inspection. In general, however, finding the chromatic number of a large graph (and, similarly, an optimal coloring) is an **NP-hard problem**. (2) In **topology**, the maximum number of regions that can be drawn

on a surface in such a way that each region has a border in common with every other region. If each region is given a different color, each color will border on every other color. The chromatic number of a square, tube, or sphere, for example, is 4; in other words, it is impossible to place more than four differently colored regions on one of these figures so that any pair has a common boundary. "Chromatic number" also indicates the least number of colors needed to color any finite map on a given surface. Again, this is 4 in the case of the plane, tube, and sphere, as was proved quite recently in the solution to the **four-color map problem**. The chromatic number, in both senses just described, is 7 for the **torus**, 6 for the **Möbius band**, and 2 for the **Klein bottle**. See also **Betti number**.

chronogram

A phrase or sentence in which certain letters represent, cryptically, a date, epoch, or, in rare cases, a non-date number. For example, the chronogram "My Day Is Closed In Immortality" commemorates the death of Queen Elizabeth the First of England: the capital letters can be rearranged to give MDCIII, or 1603, the year in which she died.

Church, Alonzo (1903–1995)

An American logician and professor at Princeton University who was an early pioneer of theoretical computer science. He is best known for his development, in 1934, of the so-called **lambda calculus**, a model of computation, and his discovery, in 1936, of an "undecidable problem" within it. This result preceded Alan **Turing**'s famous work on the **halting problem**, which also pointed out the existence of a problem unsolvable by mechanical means. Church and Turing then showed that the lambda calculus and the **Turing machine**, which is used in the halting problem, are equivalent in capability. They also demonstrated a variety of alternative "mechanical processes for computation" with equivalent computational abilities. See also **Church-Turing thesis**.

Church-Turing thesis

A logical/mathematical postulate, independently arrived at by Alan **Turing** and Alonzo **Church**, which asserts that as long as a procedure is sufficiently clear-cut and mechanical, there is some **algorithm**ic way of solving it (such as via computation on a **Turing machine**). Thus, there are some processes or problems that are computable according to some set of algorithms, and other processes or problems that are not. A strong form of the Church-Turing thesis claims that all neural and psychological processes can be simulated as computational processes on a computer.

cipher

(1) A cryptographic system (see **cryptography**) in which units of plain text of regular length, usually letters, are arbitrarily transposed (see **transposition cipher**) or substituted (see **substitution cipher**) according to a predetermined code, or a message written or transmitted in such a system. See also **Caesar cipher** and **Beal cipher**. (2) The mathematical symbol (0) for **zero**.

circle

The set of all points in a plane at a given distance, called the *radius,* from a fixed point, called the *center.* A circle is a simple **closed** curve that divides the plane into an interior and exterior. It has a perimeter, called a *circumference,* of length $2\pi r$ and encloses an area of πr^2. In coordinate geometry a circle with center (x_0, y_0) and radius r is the set of all points (x, y) such that:

$$(x - x_0)^2 + (y - y_0)^2 = r^2$$

"Circle" comes from the Latin *circus,* which refers to a large round or rounded oblong enclosure in which the famous Roman chariot races were held.

A line cutting a circle in two places is called a *secant.* The *segment* of a secant bound by the circle is called a *chord,* and the longest chord is that which passes through the center and is known as a *diameter.* The ratio of the circumference to the diameter of a circle is **pi**. The length of a circle between two radii is called an *arc;* the ratio between the length of an arc and the radius defines the *angle* between two radii in *radians.* The area bounded by two radii and an arc is known as a *sector.* A line touching a circle in one place is called a *tangent.* Tangent lines are perpendicular to radii. In **affine geometry** all circles and ellipses become **congruent**, and in **projective geometry** the other **conic section**s join them. A circle is a conic section with eccentricity zero. In **topology** all simple closed curves are **homeomorphic** to circles, and the word *circle* is often applied to them as a result. The three-dimensional analog of the circle is the **sphere,** and the four-dimensional analog is the **hypersphere**.

circle involute

The simplest kind of **spiral** to draw and understand. It is the path that a goat, tethered to a post, would follow if it walked around and around in the same direction, keeping its tether taught until it wound its way to the center. The radial distance between adjacent loops of the spiral is equal to the circumference of the central circle. Except for the innermost loop, the circle involute is hard to distinguish from the **Archimedean spiral**, though the two curves are never identical.

circular cone

A **cone** whose base is a **circle**.

circular helix
See **helix**.

circular prime
A **prime number** that remains prime on any cyclic rotation of its digits. An example (in the decimal system) is 1,193 because 1,931, 9,311, and 3,119 are also prime. Any one-digit prime is circular by default. In base ten, any circular prime with two or more digits can only contain the digits 1, 3, 7, and 9; otherwise when 0, 2, 4, 5, 6, or 8 is rotated into the units place, the result can be divided by 2 or 5. The only circular primes known, listing just the smallest representative from each cycle, are: 2, 3, 5, 7, 11, 13, 17, 37, 79, 113, 197, 199, 337, 1,193, 3,779, 11,939, 19,937, 193,939, 199,933, R_{19}, R_{23}, R_{317}, R_{1031} and possibly R_{49081}. These last five are the known **rep-unit** primes and probable primes. It's generally believed that there are infinitely many rep-unit primes, so there should be infinitely many circular primes. But it's very likely that all circular primes not on the list above are rep-units.

circumcenter
The center of a **circle** that passes through the vertices (see **vertex**) of a given **polygon**, usually a triangle. For a triangle, it is the same as the point of intersection of the perpendicular **bisector**s of the three sides.

circumcircle
The **circle** that passes through all three vertices (see **vertex**) of a given triangle. It is said to *circumscribe* the triangle.

circumference
The distance around the outside of a **circle**. The word comes the Latin *circus* ("circle") and *ferre* ("to carry"), thus means "to carry around."

cissoid
Given a fixed point A and two curves C and D, the cissoid of the two curves with respect to A is constructed as follows: pick a point P on C, and draw a line l through P and A. This cuts D at Q. Let R be the point on l such that $AP = QR$. The **locus** of R as P moves on C is the cissoid. The name *cissoid*, meaning "ivy-shaped," first appears in the work of Geminus in the first century B.C.

A special case of this curve, now known as the *cissoid of Diocles*, was first explored by **Diocles** in his attempt to solve the classical problem of **duplicating the cube**. Later investigators of the same curve include Pierre de **Fermat**, Christiaan **Huygens**, John **Wallis**, and Isaac **Newton**. The cissoid of Diocles is traced out by the **vertex** of a **parabola** as it rolls, without slipping, on a second parabola of the same size. It has the Cartesian equation

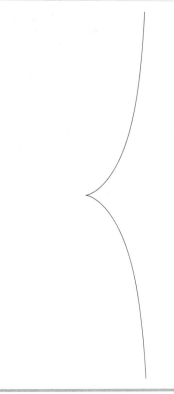

cissoid   The cissoid of Diocles. © *Jan Wassenaar, www.2dcurves.com*

$$y^2 = x^3/(2a - x).$$

Interestingly, Diocles investigated the properties of the focal point of a parabola in *On Burning Mirrors* (a similar title appears in the works of Archimedes). The problem, then as now, is to find a mirror surface such that when it is placed facing the Sun, it focuses the maximum amount of heat.

classification
The goal in a branch of mathematics of providing an exhaustive list of some type of mathematical object with no repetitions. For example, the classification of 3-**manifold**s is one of the outstanding problems in **topology**. With the advent of computers, one weak but precise way to state a classification problem is to ask whether there is an **algorithm** to determine whether two given objects are equivalent.

clelia
Also known as a *clelie curve*, the **locus** of a point P that moves on the surface of a **sphere** in such a way that ϕ/θ is constant, where ϕ and θ are the longitude and colatitude (the angular distance from a pole).

Clifford, William Kingdon (1845–1879)

An English mathematician who studied **non-Euclidean geometry** and **topology**. In 1870, he wrote *On the Space Theory of Matter* in which he argued that energy and matter are simply different types of curvature of space–a remarkably advanced idea that would come to fruition in Einstein's general **relativity theory**. Although small of build Clifford was remarkably strong and able to do one-armed chin-ups. His death at an early age was the result of overwork and exhaustion.

clock puzzles

The earliest known clock problem was posed in 1694 by Jacques **Ozanam** in his *Récréations mathématiques et physiques*.

PUZZLES

Here are two clock puzzles invented by Lewis **Carroll**:

1. A clock has hour and minute hands of the same length and no numerals on its face. At what time between 6 and 7 o'clock will the time on the clock appear to be the same as the time read on the reflection of the clock in a mirror?

2. Which has a better chance of giving the right time: a clock that has stopped or one that loses a minute every day?

And here is another from Henry **Dudeney**'s *Amusements in Mathematics* called "The Club Clock":

3. One of the big clocks in the Cogitators' Club was found the other night to have stopped just when the second hand was exactly midway between the other two hands. One of the members proposed to some of his friends that they should tell him the exact time when (if the clock had not stopped) the second hand would next again have been midway between the minute hand and the hour hand. Can you find the correct time that it would happen?

Solutions begin on page 369.

---

closed

A *closed curve* is one that has no endpoints so that it completely encloses a certain area. A *closed interval*, which corresponds to a *closed set*, is an interval that includes its endpoints.

cochleoid

A **spiral** curve that was first studied by J. Peck in 1700 and Bernoulli in 1726. Its name, meaning "snail-form" (*kochlias* is Greek for "snail"), was coined by Benthan and Falkenburg in 1884. It can be constructed starting from a point O on the *y*-axis. For all circles through O (tangent

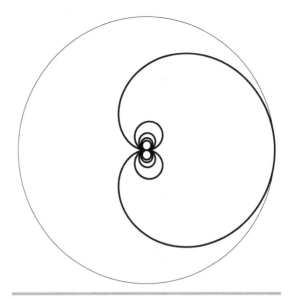

cochleoid   A cochleoid inside the circle used to construct it.
© *Jan Wassenaar, www.2dcurves.com*

---

to the *y*-axis), pace a constant distance on the circle. The collection of those points is the cochleoid. In Cartesian coordinates, it is given by the formula

$$(x^2 + y^2)\tan^{-1}(y/x) = ay$$

and in polar coordinates by

$$r = a\sin\theta \,/\, \theta.$$

The points of contact of parallel tangents to the cochleoid lie on a **strophoid**.

code

See **cipher**.

codimension

In general, if a mathematical object sits inside or is associated with another object of **dimension** *n*, then it is said to have codimension *k* if it has dimension $n - k$.

coding theory

The branch of mathematics concerned with sending data across noisy channels and recovering the message. Whereas **cryptography** is about making messages *hard* to read, coding theory focuses on making messages *easy* to read. The basic problem is that messages, in the form of binary digits or *bits* (strings of 0 or 1) have to be sent along a channel (such as a phone line) in which errors occur randomly, but at a predictable overall rate. To compensate for the errors, more bits have to be sent than are contained in the original message. The easiest way to detect errors in binary data is the **parity** code, which inserts an extra parity bit after every 7 bits from the source message. To correct as well as detect errors, the data has to be retransmitted. A simple way to do this is to repeat each bit a set number of times. The recipient sees which value, 0 or 1, occurs more often and assumes that to be the intended bit. This method can cope with error rates up to 1 error in every 2 bits transmitted but it means that an awful lot of extra bits have to be sent.

In 1948, Claude **Shannon** at Bell Labs began the subject of coding theory by proving the minimum number of extra bits that had to be transmitted to encode messages but without showing ways to find these optimal codes. Two years later, Richard Hamming, also at Bell Labs, gave details of error-correcting codes with information transmission rates more efficient than simple repetition. His first code, in which four data bits were followed by three check bits, allowed not only the detection but the correction of a single error.

While Shannon and Hamming were involved with information transmission in the United States, John Leech devised similar codes while working on **group** theory at Cambridge University. This research also took in the sphere **packing** problem and culminated in the amazing, 24-dimensional *Leech lattice,* the study of which proved crucial to understanding and classifying finite symmetry groups. The value of error-correcting codes for information transmission, both on Earth and from space, was immediately grasped, and a variety of codes were constructed that boosted both economy of transmission and error-correction capacity. Between 1969 and 1973 the NASA Mariner probes used a powerful *Reed-Muller code* capable of correcting 7 errors out of 32 bits transmitted. A less obvious application of error-correcting codes came with the development of the compact disk on which the signal is encoded digitally. To guard against scratches and other damage, two interleaved codes that can correct up to 4,000 consecutive errors are used. By the late 1990s the goal of finding explicit codes that reach the limits predicted by Shannon's original work had been achieved.

codomain

For a given **function** or mapping, a **set** within which the values of the function lie. This is different from the set of values, known as the **range**, that the function actually takes.

coefficient

A number or other factor that multiplies a variable. For example, in the equation $3x - 4ky = 8$, the 3 and $4k$ are coefficients of the variables x and y. The word combines three elements, the Latin *facere* ("to do"), and the prefixes *ex* ("out") and *co* ("with"), to give the overall meaning of joining two things together to bring about a result. The sixteenth-century mathematician Francois Vieta may

have coined the word, but it was not commonly used until around the beginning of the eighteenth century.

Coffin, Stewart T.

A leading designer of mechanical puzzles. He is also the author of *The Puzzling World of Polyhedral Dissections,*[64] one of the most significant works produced on this subject.

cohomology

A subject that involves calculating algebraic **invariant**s of **topological space**s that are formally dual to **homology**. The invariants obtained are in general more powerful than those given by homology and usually have more algebraic structure. *Generalized cohomology theories,* both for topological spaces and for purely algebraic structures, have been developed that have some of the formal properties of cohomology but which don't have the same geometric background.

coin paradox

Consider two round coins of equal size. Imagine holding one still and then rolling the other coin around it, making sure that it doesn't slip and that the rims are touching at all times. How many times will the moving coin have rotated after it has completed one revolution of the stationary coin? Most people believe that the answer will be once and are therefore surprised to discover that the truth is in fact *twice.*

coincidence

What an amazing coincidence! Well, not really. Coincidences are bound to happen. In a world where there are a great many potential coincidences each with a small probability of happening, someone, somewhere is going to see one—and be amazed by it. The fact that there are countless numbers of noncoincidences and many people who don't see a significant coincidence in the same period of time is overlooked. Also, we tend to underestimate the probabilities of coincidences in certain situations and are therefore more surprised than we should be when coincidences happen. A classic example of this is the **birthday paradox**.

Obviously some things are extraordinarily unlikely. What are the chances, for example, of a meteorite hitting your car? Next to nothing, but not quite nothing. There are a lot of cars and there are dozens of meteorites that strike Earth every day. Sooner or later, it's bound to happen. In fact, it did happen to Michelle Knapp's Chevy Malibu parked outside her home in Peekskill, New York, on the evening of October 9, 1992. A 12-kilogram space rock smashed through the car's trunk and ended up on the driveway below.

Does coincidence completely explain away all events that might otherwise be put down to precognition? On April 15, 1912, the SS *Titanic* sunk on her maiden voyage, having been holed by an iceberg, and over 1,500 people died. Fourteen years earlier a novel had been published by Morgan Robertson that seemed to foretell the disaster. The book described a ship the same size as *Titanic* that struck an iceberg on its maiden voyage on a misty April night. The name of Robertson's fictional ship was the *Titan*. Mere happenstance or evidence of something deeper? Numerologists often spot matchups that would go unnoticed by the rest of us. Is it so strange that there are almost exactly 500 million inches in the pole-to-pole diameter of Earth? Not if you work in centimeters. And should we make such a fuss over the fact that the speed of light is within 0.1% of 300,000 kilometers per second when we give no attention to the miles-per-second value of 186,282? Yet, surely, there can be no doubt that Shakespeare wrote the Bible. The King James Version was published in 1611, when Shakespeare was 46 years old. Look up Psalm 46. Count 46 words from the beginning of the Psalm. You will find the word "Shake." Count 46 words from the end of the Psalm. You will find the word "Spear." To some, an obvious coded message. See also **thirteen**.

Collatz problem
A problem first posed by the German mathematician Lothar Collatz (1910–1990) in 1937, that is also known variously as the *3n + 1 problem, Kakutani's problem,* the *Syracuse problem, Thwaites' conjecture,* and *Ulam's conjecture.* It runs as follows. Let *n* be any integer. (1) If *n* is odd, put *n* equal to $3n + 1$; otherwise, put *n* equal to $n/2$. (2) If $n = 1$, stop; otherwise go back to step 1. Does this process always terminate (i.e., end in 1) for any value of *n*? To date, this question remains unanswered, though the process has been found to stop for all *n* up to 5.6×10^{13}. British mathematician Bryan Thwaites (1996) has offered a £1,000 reward for a resolution of the problem. However, John **Conway** has shown that Collatz-type problems can be formally undecidable, so it not known if a solution is even possible. The members of sequences produced by the Collatz problem are sometimes known as **hailstone sequences**.[144]

combination
A set of objects selected without reference to the order in which they are arranged. Compare with **permutation**. See also **binomial coefficient**.

combinatorics
The study of the ways of choosing and arranging objects from given collections and the study of other kinds of problems relating to counting the number of ways to do something.

commensurable
Two lines or distances are commensurable if the ratio of their lengths is a **rational number**. If the ratio is an **irrational number**, they are called *incommensurable.*

common fraction
A **fraction** that consists of the **quotient** of two **integers**.

communication theory
See **information theory**.

commutative
Two numbers, *x* and *z*, are said to be *commutative under addition* if

$$x + y = y + x$$

and to be *commutative under multiplication* if

$$x \times y = y \times y.$$

In general, two elements *a* and *b* of a **set** *S* are commutative under the binary operation (an operation that works on two elements at a time) * if

$$a * b = b * b.$$

Compare with **associative** and **distributive**.

complement
That which is needed to complete something. For instance, the complement of a number is what needs to be added to it to make a specified value; the complement of an angle is the angle required to turn it into a right angle. The complement of a **set** is composed of all the elements that are not members of that set.

complete
Describes a **formal system** in which all statements can be proved as being true or false. Most interesting formal systems are not complete, as demonstrated by **Gödel's incompleteness theorem**.

complete graph
A **connected graph** in which exactly one **edge** connects each pair of vertices (see **vertex**). A complete graph with *n* vertices, denoted K_n, has $n(n - 1)/2$ edges (i.e., the *n*th **triangular number**), $(n - 1)!$ **Hamilton circuits**, and a **chromatic number** of *n*. Every vertex in K_n has **degree** $n - 1$; therefore K_n has an **Euler circuit** if and only if *n* is odd. In a *weighted complete graph,* each edge has a number called a *weight* attached to it. Each path then has a total weight, which is the sum of the weights of the edges in the path. See also **traveling salesman problem**.

complex adaptive system (CAS)

A nonlinear, interactive, **complex system** with the ability to adapt to a changing environment. CASs evolve by random mutation, **self-organization**, the transformation of their internal models of the environment, and natural selection. Examples include living organisms, the nervous system, the immune system, the economy, corporations, and societies. In a CAS, semiautonomous agents interact according to certain rules of interaction, evolving to maximize some measure like fitness. The agents are diverse in form and capability and they adapt by changing their rules and, hence, behavior, as they gain experience. CASs evolve historically—their experience determines their future trajectory. Their adaptability can either be increased or decreased by the rules shaping their interaction. Moreover, unanticipated, emergent structures can play a determining role in the evolution of such systems, which is why they are highly unpredictable. On the other hand, CASs have the potential of a great deal of creativity that was not programmed into them from the beginning.

complex analysis

The study of **function**s of a complex variable. Often, the most natural proofs for statements in real **analysis** or even **number theory** use techniques from complex analysis. Unlike real functions, which are commonly represented as two-dimensional graphs, complex functions have four-dimensional graphs and may usefully be illustrated by color-coding a three-dimensional graph to suggest four dimensions.

complex number

A **real number** plus a real number times the square root of -1; in other words, a number of the form $z = a + ib$, where a and b are real and $i = \sqrt{-1}$. The term ib is known as an **imaginary number** or the *imaginary part* of the complex number $a + ib$; a is called the real part. The names "complex," "real," and "imaginary," which came about historically, are totally misleading because complex numbers are not particularly complex and imaginary numbers are no less real than real numbers! Another way to represent a complex number is as an **ordered pair** of real numbers (a, b) together with the operations: $(a, b) + (c, d) = (a + c, b + d)$ and $(a, b) \times (c, d) = (ac - bd, bc + ad)$. Alternatively, complex numbers can be shown as points on an **Argand diagram** (a representation of the *complex plane*) in which the horizontal axis is the real number line and the vertical axis represents all possible purely imaginary numbers. Any point that appears on the complex plane off-axis has both real and imaginary parts. On an Argand diagram a complex number can also be shown as a **vector**, or directed line segment (a line of a certain length with an arrow), extending from the origin $(0 + 0i)$ to the number

$(a + bi)$. The *absolute value* or *magnitude* of a complex number z, thought of as a point on a plane, is its Euclidean distance from the origin, and is denoted $|z|$; this is always a nonnegative real number. Algebraically, if $z = a + ib$, we can define $|z| = \sqrt{(a^2 + b^2)}$. If the complex number z is written in polar coordinates $z = r\, e^{i\phi}$, then $|z| = r$.

Complex numbers are a natural extension of real numbers and form what is called an *algebraically closed field*. Because of this, mathematicians sometimes consider the complex numbers to be more "natural" than the real numbers: all **polynomial** equations have solutions among the complex numbers, which is not true for the real numbers. Complex numbers are used in electrical engineering and other branches of physics as a convenient description for periodically varying signals. In an expression $z = r e^{i\phi}$ one may think of r as the amplitude and ϕ as the phase of a sine wave of given frequency. In special and general **relativity theory**, some formulas for the metric on **space-time** become simpler if the time variable is taken to be imaginary.

complex plane

See **Argand diagram**.

complex system

A collection of many simple nonlinear units that operate in parallel and interact locally with each other so as to produce emergent (see **emergence**) behavior.

complexity

A phenomenon that has two distinct and almost opposite meanings. The first, and probably the oldest mathematically, goes back to Andrei **Kolmogorov**'s attempt to give an **algorithm**ic foundation to notions of **random**ness and probability and to Claude **Shannon**'s study of communication channels via his notion of information. In both cases, complexity is synonymous with *disorder* and a lack of structure. The more random a process, the greater its complexity. An ideal gas, for example, with its numerous molecules bouncing around in complete disarray, is complex as far as Kolmogorov and Shannon are concerned. Thus, in this sense, complexity equates to the degree of complication.

The second, and more recent notion of complexity refers instead to how structured, intricate, hierarchical, and sophisticated a natural process is. In particular, it's a property associated with **dynamical system**s in which new, unpredictable behavior arises on scales above the level of the constituent components. The distinction between these two meanings can be revealed by answering a simple question about a system: Is it complex or is it merely complicated? Measures of complexity include **algorithmic complexity**, **fractal dimension**ality; **Lyapunov fractals**, and **logical depth**.

complexity theory

A part of the theory of computation that has to do with the resources needed to solve a given problem. The most common resources are *time* (how many steps it takes to solve a problem) and *space* (how much memory it takes to solve a problem). Complexity theory differs from **computability theory**, which deals with whether a problem can be solved at all, regardless of the resources required.

composite number

A positive integer that can be factored into smaller positive integers, neither of which is one. If a positive integer is not composite (4, 6, 8, 9, 10, 12, . . .) or one, then it is a **prime number** (2, 3, 5, 7, 11, 13, 17, . . .). As Karl **Gauss** put it in his *Disquisitiones Arithmeticae* (1801): "The problem of distinguishing prime numbers from composite numbers and of resolving the latter into their prime factors is known to be one of the most important and useful in arithmetic." One reason for its importance today is that many secret codes and much of the security of the Internet depends in part on the relative difficulty of factoring large numbers. But more basic to a mathematician is that this problem has always been central to **number theory**. Numbers that, for their size, have a lot of factors are sometimes referred to as *highly composite numbers*. Examples include 12, 24, 36, 48, 60, and 120.

compound polyhedron

An assemblage of two or more polyhedra, usually interpenetrating and having a common center. There are two types: a combination of a solid with its **dual** and an interpenetrating set of several copies of the same **polyhedron**. The simplest example of a compound polyhedron is the compound of two tetrahedra, known as the *stella octangula* and first described by Johannes **Kepler**. This shape is unique in that it falls under both of the above classes, because the **tetrahedron** is the only self-dual **uniform polyhedron**; the edges of the two tetrahedra form the diagonals of the faces of a **cube** in which the stella octangula can be inscribed.

compound polyhedron A compound of duals: the cube and the octahedron. *Robert Webb, www.software3d.com; created using Webb's Stella program*

compound polyhedron   A compound polyhedron of three cubes (left), such as that used in Escher's picture "Waterfall." The compound of four cubes (right) is also known as Bakos's compound. *Robert Webb, www.software3d.com; created using Webb's Stella program*

Another example of a compound follows from an important Platonic relationship: a cube can be inscribed within a **dodecahedron**. There are five different positions for a cube within a dodecahedron; superimposing all five gives the compound known as the *rhombic triacontahedron*.

compressible
Having a description that is smaller than itself; not **random**; possessing regularity.

computability theory
The part of the theory of computation that deals with problems that are solvable by **algorithms** or—what amounts to the same thing—by **Turing machine**s. Computability theory is concerned with four main questions: What problems can Turing machines solve? What other systems are equivalent to Turing machines? What problems require more powerful machines? What problems can

be solved by less powerful machines? Not all problems can be solved computationally. An *undecidable problem* is one that can't be solved by any algorithm, no matter how much time, processing speed, or memory is available. Many examples are known, one of the most famous of which is the **Halting problem**. See also **cellular automaton**.

computable number
A **real number** for which there is an **algorithm** that, given n, calculates the nth digit. Alan **Turing** was the first to define a computable number and the first to prove that almost all numbers are uncomputable. An example of a number that, even though well-defined, is uncomputable is **Chaitin's constant**.

concave
Curved inward, like the inner surface of a sphere; the word comes from the Latin *concavus* for "hollow." A figure, such as a **polygon** or **polyhedron**, is said to be concave if a line

segment joining any points inside the figure goes outside the figure. Similarly, a **set** is concave if it doesn't contain all the line segments connecting any pair of its points.

conchoid

A shell-shaped curve. Given a point A and a curve C, if we pick a point Q on C and draw a line L through A and Q and mark points P and P' on L at some fixed distance in either direction from Q, then the locus of P and P' as Q moves on C is a conchoid. The *conchoid of Nichomedes* is a conchoid in which the given line is a straight line; that is, given a line C and a point A we pick a point Q on C, draw a line L through A and Q, and mark P and P' on L at some fixed distance from Q. The conchoid of Nicomedes is the locus of P and P' as Q moves along C. It has the polar equation $R = a \sec\theta + k$. The *conchoid of de Sluze* is the curve with the Cartesian equation $a(x - a)(x^2 + y^2) = k^2 x^2$.

cone

A shape (its name comes from the Greek *konos* for pinecone) that has a circular or elliptical base and a **vertex**, also known as an *apex,* lying outside the plane of the base and that is formed from all the line segments joining points on the edge of the base to the vertex. If the base is a circle, the shape is a *circular cone;* if the line, or *axis,* from the center of the base to the vertex is perpendicular to the base, then it is a *right cone* (an ice-cream cone is a right circular cone); otherwise it's an *oblique cone.* The curved lateral surface of the cone is called a *nappe.* If the cone is extended in both directions from the vertex, the result is a *double cone* or *bicone.* A section through a double cone that has been extended indefinitely in both directions to form a *conic surface* is known as a **conic section.** Another way to think of a cone is as a **surface of revolution** generated by a line that rotates around a fixed point, at a fixed angle from another line (the axis), both lines passing through that fixed point. The volume of a cone, of perpendicular height h and circular base of radius r, is $\frac{1}{3}\pi r^2 h$.

Take a solid cylinder of radius r and height $2r$. Remove the right double cone that passes through the center of the cylinder and extends to meet the circular disks on the cylinder's top and bottom. Interestingly, the volume of the remaining object and the volume of a sphere of radius r are the same.

PUZZLE

The *Cone Puzzle* (no. 202) from Henry **Dudeney**'s *Amusements in Mathematics*[88] runs as follows: "I have a wooden cone. How am I to cut out of it the greatest possible cylinder?"

Solutions begin on page 369.

conformal mapping

A map from the plane to itself that preserves angles. Conformal mapping results in the angle between any two curves being the same as the angle between their images. The Mercator map is a conformal map of Earth's surface.

congruent

In the case of geometric figures, having exactly the same shape and size.

congruum problem

Find a **square** number x^2 such that, when a given number h is added or subtracted, new square numbers are obtained, so that $x^2 + h = a^2$ and $x^2 - h = b^2$. This problem was posed by the mathematicians Théodore and Jean de Palerma in a mathematical tournament organized by Frederick II in Pisa in 1225. The solution is $x = m^2 + n^2$ and $h = 4mn(m^2 - n^2)$, where m and n are integers.

conic section

An important, familiar, and ubiquitous family of curves obtained by slicing a right circular double **cone**, extended indefinitely in both directions, with a plane. Depending on the angle of the slice to the axis of the cone, the resulting curve may be a **circle**, an **ellipse**, a **parabola**, or a **hyperbola**. The circle is a limiting case of the ellipse, when the slice is made at right angles to the axis, while the parabola is the limiting case of both the ellipse and the hyperbola, when the slice is made parallel to the side of the cone. The name *conic sections* comes from the eight-volume work *Conics* (Κωνικα) by **Apollonius**, who also gave us the names *ellipse, parabola,* and *hyperbola.*

Another geometric way to define the conics is as the **locus** of all points in the plane whose distances, r, from a fixed point called the *focus,* and a, from a given straight line called the **directrix**, have a constant ratio. This ratio, r/a, is known as the *eccentricity, e*. The circle has an eccentricity of zero. As the eccentricity increases from near zero, corresponding to a nearly circular ellipse, the ellipse stretches until the right-hand side of it disappears to infinity, e becomes 1, and the ellipse turns into a parabola, with just one open branch. Like the circle, the parabola has only one shape, though it may look different depending on how much it is enlarged or diminished. As the eccentricity increases beyond 1, the "lost" right-hand end of the ellipse reappears from the other side of infinity, so to speak, and turns into the left-hand branch of a hyperbola.

Because a hyperbola is effectively an ellipse split in two by infinity, it comes as no surprise that these curves are related in an inverse way. An ellipse consists of all points whose distances from two foci have a constant sum, while a hyperbola is made from all points whose

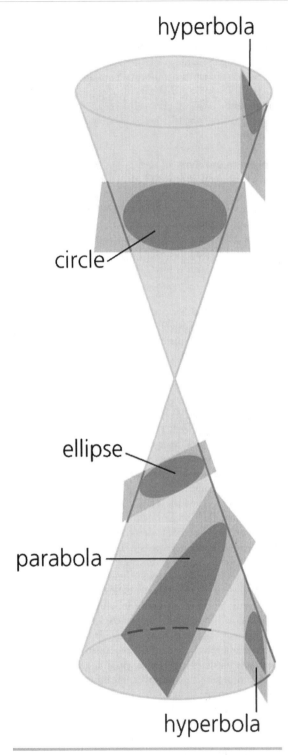

hyperbola

circle

ellipse

parabola

hyperbola

conic section The circle, ellipse, parabola, and hyperbola, obtained by slicing a right double cone in various ways.

distances from two foci have a constant difference. These definitions also apply to the circle and the parabola, if the two foci are considered to coincide in the case of the circle and to be separated by an infinite distance in the case of the parabola.

In terms of algebra, the family of conics represents all the possible **real number** solutions to the general **quadratic** equation $ax^2 + bxy + cy^2 + dx + ey + f = 0$. In other words, the graph of any quadratic with real solutions is always a conic section. The key quantity is the difference $b^2 - 4ac$. If this is less than zero, the graph is an ellipse, a circle, a point, or no curve. If $b^2 - 4ac = 0$, the graph is a parabola, two parallel lines, one line, or no curve; if it is greater than zero, the graph is a hyperbola or two intersecting lines.

conical helix
See **helix**.

conjecture
A mathematical statement that has been put forward as a true statement, but that no one has yet been able to prove or disprove; in mathematics, a conjecture and a hypothesis are essentially the same thing. When a conjecture has been proven to be true, it becomes known as a **theorem**. Famous conjectures include the **Riemann hypothesis**, the **Poincaré conjecture**, the **Goldbach conjecture**, and the **twin primes** conjecture. Just to show how terminology can be used inconsistently, however, the most famous of all conjectures, for centuries before its proof in 1995, was always known as **Fermat's last theorem**!

conjugate
(1) *Conjugate angles* add up to 360°. (2) The *complex conjugate* of a **complex number** $a + bi$ is $a - bi$. (3) *Conjugate lines* of a **conic section** have the property that each contains the pole point of the other, while *conjugate points* of a conic have the property that each lies on the polar line of the other. In general, conjugate indicates that there is a symmetrical relationship between two objects A and B; in other words, there is an operation that will turn A into B and B into A.

connected
A space S is said to be connected if any two points in S can be connected by a curve lying wholly within S. Two spaces can be added by what is called a *connected sum*. Roughly speaking, this involves pulling out a disk from each surface, creating **holes**, and then sewing the two surfaces together along the boundaries of the holes. In this way, a one-holed **torus** can be added to a two-holed torus to give a three-holed torus; alternatively, a **projective**

plane can be added to a projective plane to give a **Klein bottle**. The operation is **commutative** and **associative** and there is even an **identity** element: for example, adding a sphere to any surface simply returns the same surface. See also **simply connected**.

connected graph

A **graph** in which a path exists between all pairs of vertices (see **vertex**). If the graph is also a **directed graph**, and there exists a path from each vertex to every other vertex, then it is a *strongly connected graph*. If a connected graph is such that exactly one edge connects each pair of vertices, then it is said to be a **complete graph**. See also **Euler path** and **Hamilton path**.

connectionism

A computational approach to modeling the brain that relies on the interconnection of many simple units to produce complex behavior.

connectivity

The amount of interaction in a system, the structure of the weights in a **neural network**, or the relative number of edges in a **graph**.

consistency

An **axiom**atic theory is said to be consistent if it's impossible (within the confines of the theory) to prove simultaneously a statement and its negation. **Godel's incompleteness theorem** states that any (sufficiently powerful) consistent axiomatic theory is incomplete.

constructible

In classical geometry, a figure or length that can be drawn using only an unmarked straightedge and a compass. The Greeks were adept at constructing **polygons**, but the question of proving which **regular polygons** are constructible and which are not had to wait for the genius of Carl **Gauss**. At the age of only 19, Gauss found that a regular polygon with *n* sides is constructible if and only if *n* is a prime **Fermat number**. The only known such primes are 3, 5, 17, 257, 65,537. It is also possible to construct certain numbers, known as *constructible numbers,* that correspond to line segments, including **rational numbers** and some **irrational numbers**, but no **transcendental numbers**. It turns out that all constructions possible with a compass and straightedge can be done with a compass alone, as long as a line is considered constructed when its two endpoints are located. The reverse is also true, since Jakob **Steiner** showed that all constructions possible with straightedge and compass can be done using only a straightedge, as long as a fixed circle and its center (or two intersecting circles without their

centers, or three nonintersecting circles) have been drawn beforehand. Such a construction is known as a *Steiner construction.* The Greeks were unable to achieve certain constructions, such as **squaring the circle, duplicating the cube,** and **trisecting an angle**, despite numerous attempts, but it wasn't until hundreds of years later that the problems were proved to be actually impossible under the limitations imposed.

continued fraction

A representation of a **real number** in the form

$$x = a_0 + \cfrac{1}{a_1 + \cfrac{1}{a_2 + \cfrac{1}{a_3 + \ldots}}}$$

which, mercifully for typesetters, can be written in compact notation as

$$x = [a_0; a_1, a_2, a_3, \ldots],$$

where the integers a_i are called *partial quotients.* Although rarely encountered in school and even college math courses, continued fractions (CFs) provide one of the most powerful and revealing forms of numerical expression. Numbers whose decimal expansions look unremarkable turn out, when unfolded as CFs, to have extraordinary symmetries and patterns. CFs also offer a way of constructing rational approximations to **irrational numbers** and of discovering the most irrational numbers.

CFs first appeared in the sixth century in the works of the Indian mathematician Aryabhata, who used them to solve linear equations. They surfaced in Europe in the fifteenth and sixteenth centuries and **Fibonacci** attempted to define them in a general way. The term "continued fraction" first appeared in 1653 in an edition of *Arithmetica Infinitorum* by John **Wallis**. Their properties were also studied by one of Wallis's English contemporaries, William Brouncker, who, along with Wallis, was one of the founders of the Royal Society. At about the same time, in Holland, Christiaan **Huygens** made practical use of CFs in his designs of scientific instruments. Later, in the eighteenth and early nineteenth centuries, Carl **Gauss** and Leonhard **Euler** delved into many of their deeper properties.

CFs can be finite or infinite in length. Finite CFs can be evaluated level by level (starting at the bottom) and will always reduce to a rational fraction; for example, the

CF [1; 3, 2, 4] = 40/31. By contrast, infinitely long CFs produce representations of irrational numbers. Here are the leading terms from a few notable examples of infinite CFs:

$$e = [2; 1, 2, 1, 1, 4, 1, 1, 6, 1, 1, 8, 1, 1, 10, \ldots]$$
$$\sqrt{2} = [1; 2, 2, 2, 2, 2, 2, 2, 2, 2, 2, 2, 2, 2, 2, 2, \ldots]$$
$$\sqrt{2} = [1; 2, 1, 2, 1, 2, 1, 2, 1, 2, 1, 2, 1, 2, 1, \ldots]$$
$$\pi = [3; 7, 15, 1, 292, 1, 1, 1, 2, 1, 3, 1, 14, 2, 1, 1, 2, 2,$$
$$2, 2, 1, 84, 2, \ldots]$$

Each of these expansions has a simple pattern except that for π (see **pi**), which has no obvious pattern at all. There's also a preference for the quotients to be small numbers.

If an infinite CF is truncated after a finite number of steps, the result is a *rational approximation* to the original irrational. In the case of π, chopping the CF at [3; 7] gives the familiar approximation for π of 22/7 = 3.1428571 .... Keeping two more terms leads to [3; 7, 15, 1] = 353/113 = 3.1415929 ..., which is an even better approximation to the true value of π (3.14159265 ...). The more terms retained in the CF, the better the rational approximation becomes. In fact, the CF gives yields the best possible rational approximations to a general irrational number. Notice also that if a large number occurs in the expansion of quotients, then truncating the CF after that will produce an especially good rational approximation. Most CF quotients are small numbers (1 or 2), so the appearance in the CF of π of a number as large as 292 so early in the expansion is unusual. It also leads to an extremely good rational approximation to π = [3; 7, 15, 1, 292] = 103,993/33,102.

continuity

A mathematical property that has to do with how smooth or "well-behaved" a **function** or curve is. If two adjacent points on a graph, for example, are not connected or are separated by a jump, this marks a breakdown of continuity. At such a discontinuity it is impossible to obtain a **derivative**, or slope, of the curve. Usually if a curve does misbehave like this, it is only at one or two isolated places; elsewhere the curve is likely to be both continuous and differentiable. However, it is possible to construct a continuous function that has "problem points" everywhere and, therefore, is nowhere differentiable! The first example was found by Karl **Weierstrass** in 1872 and came as a total surprise. It is defined as an infinite series

$$f(x) = \sum_{n=0}^{\infty} B^n \cos(A^n \pi x),$$

where A and B can be any numbers such that B is between 0 and 1, and $A \times B$ is bigger than $1 + (3\pi/2)$.

continuum

Any **set** that can be brought into **one-to-one** correspondence with the set of **real numbers**. Examples include a finite line segment, a square, a circle, and a disk.

continuum hypothesis

In 1874 Georg **Cantor** discovered that there is more than one level of **infinity**. The lowest level is called *countable infinity;* higher levels are known as *uncountable infinities.* The **natural numbers** are an example of a countably infinite set and the **real numbers** are an example of an uncountably infinite set. The continuum hypothesis, put forward by Cantor in 1877, says that the number of real numbers is the *next* level of infinity above countable infinity. It is called the continuum hypothesis (CH) because the real numbers are used to represent a linear continuum. Let c be the cardinality of (i.e., number of points in) a continuum, **aleph**-null (χ_0) be the cardinality of any countably infinite set, and χ_1 be the next level of infinity above χ_0. CH is equivalent to saying that there is no **cardinal number** between χ_0 and c, and that $c = \chi_1$. CH has been, and continues to be, one of the most hotly pursued problems in mathematics.

convergence

A property of some **sequences**. A sequence u_i is said to be convergent if there exists a value u with the property that by choosing a large enough value of i, we can make u_i as close as we wish to u.

convex

Curved outward, like the exterior surface of a sphere; the word comes from the Latin *convexus* for "vaulted." A figure, such as a **polygon** or a **polyhedron**, is said to be convex if every line segment that joins two interior points remains inside the figure. Similarly, a **set** is convex if it contains all the line segments connecting any pair of its points.

Conway, John Horton (1937–)

A British-born (Liverpool) mathematician, who studied and taught at Cambridge University and is now a professor at Princeton University. Conway has been an extraordinarily fertile source of new ideas in mathematics and of mathematical games. His most significant contribution was the discovery of **surreal numbers**, to which he was led after watching the British **Go** champion play at Cambridge. In 1967, he found a cluster of three new sporadic **groups**, now sometimes called *Conway's constellation,* building on an earlier discovery by John Leech of an extremely dense **packing** of unit spheres in a space of 24 dimensions. He has also been active in the field of **knots** and in **coding theory**. Among amateur mathematicians, Conway is best known as the inventor of the games of **Life**, **Sprouts**, and **Phutball**,

Conway, John Horton *Princeton University*

as well as for his detailed analyses of many other games and puzzles, such as the **Soma Cube**.

Conway's chained-arrow notation

One of various methods that have been devised recently for representing extremely **large numbers**. Developed by John Conway, it is based on **Knuth's up-arrow notation** but is even more powerful. The two systems are related thus:

$$a \to b \to 1 = a \uparrow b$$
$$a \to b \to 2 = a \uparrow\uparrow b$$
$$a \to b \to 3 = a \uparrow\uparrow\uparrow b$$
$$a \to b \to c = a \uparrow\uparrow \ldots \uparrow\uparrow b \ (c \text{ up arrows})$$

Longer chains are evaluated by the following general rules:

$$a \to \ldots \to b \to c \to 1 = a \to \ldots \to b \to c$$
$$a \to \ldots \to b \to 1 \to d + 1 = a \to \ldots \to b$$
$$\text{and} \quad a \to \ldots \to b \to c + 1 \to d + 1$$
$$= a \to \ldots \to b \to (a \to \ldots \to b \to c \to d) \to d$$

It's important to recognize that the Conway arrow isn't an ordinary *dyadic* operator. Where three or more numbers are joined by arrows, the arrows don't act separately but rather the whole chain has to be considered as a unit. The chain might be thought of as a **function** with a variable number of arguments, or as a function whose single argument is an ordered list or **vector**. The **Ackermann function** is equivalent to a three-element chain: $A(m, n) = (2 \to (n + 3) \to (m - 2)) - 3$. It can also be shown that **Graham's number** is bigger than $3 \to 3 \to 64 \to 2$ and smaller than $3 \to 3 \to 65 \to 2$.

coordinate

One of a set of variables that specifies the location of a point in space. If the coordinates are distances measured along perpendicular axes, they are known as **Cartesian coordinates**. See also **polar coordinates**.

coordinate geometry

See **analytical geometry**.

coprime

Two or more numbers are coprime if they have no **factors** in common other than 1.

cosine

See **trigonometric function**.

countable set

A **set** that is either finite or *countably infinite*. A countably infinite set is one that can be put in **one-to-one** correspondence with the **natural numbers** and thus has a **cardinal number** ("size") of **aleph**-null ($\aleph_0$). Examples of countable sets include the set of all people on Earth and the set of all fractions. See also **infinity**.

counterfeit coin problem

Among n coins, identical in size, shape, and appearance, one is a counterfeit and has a slightly different weight than the others. Using only a two-pan balance, what is the smallest number of weighings that would guarantee finding the fake coin? The problem of the counterfeit coin (or some other object), especially involving 8, 10, 12, or 13 coins, has cropped up in many guises over the years. Typically, the problem also involves finding whether the counterfeit coin is lighter or heavier than the rest. The answer depends on the specific problem and can involve quite a number of steps.

covariance

The tendency of two **random** variables to move in tandem. This is important in applications such as survey-taking and sociology, as well as in many branches of science, because if two things tend to vary together, there is a good chance they may be causally linked. See also **causality**.

Coxeter, Harold Scott MacDonald (1907–2003)

A British-born, Cambridge-educated mathematician who spent most of his career (from 1936 on) at the University of Toronto and was regarded as the greatest classical geometer of his generation. Always known as "Donald," he is best known for his work on hyperdimensional geometries and regular **polytopes**.

In 1926, at the age of 19, Coxeter discovered a new regular polyhedron, having six hexagonal faces at each vertex. He went on to study the mathematics of kaleidoscopes and, by 1933, had enumerated the *n*-dimensional kaleidoscopes. His algebraic equations expressing how many images of an object may be seen in a kaleidoscope are now known as *Coxeter groups*. His research on icosahedral symmetries played an important role in the discovery by scientists at Rice University, Texas, of the carbon-60 molecule (see **buckyball**), for which they won the 1996 Nobel Prize in Chemistry.

Coxeter was a close friend of the artist M. C. **Escher**, whom he met in 1954, and also of Buckminster Fuller, who used Coxeter's ideas in his architecture. Indeed Coxeter's work was motivated by a strong artistic temperament and a sense of what is beautiful. He had originally intended to be a composer but fascination for symmetry took him toward mathematics and a career about which he said "I am extremely fortunate for being paid for what I would have done anyway."

Several of Coxeter's books are considered classics, including *The Real Projective Plane* (1955), *Introduction to Geometry* (1961),[74] *Regular Polytopes* (1963),[75] *Non-Euclidean Geometry* (1965)[72] and, written jointly with S. L. Greitzer, *Geometry Revisited* (1967). In 1938, he revised and updated Rouse **Ball**'s *Mathematical Recreations and Essays*.[24]

cross

A shape that consists in its most basic form of an upright section and a transverse section. The *Latin cross* has the shape of an irregular **dodecahedron** with a single (vertical) line of symmetry, and can be folded up to make a cube. The *Greek cross* has the shape of a plus sign, has four lines of symmetry, and is used as the emblem of the Red Cross organization. A version of the Greek cross that has flared ends is also known as the *crux immissa* or *cross patée*. A cross of Saint Andrew is an ordinary Greek cross rotated through 45°, and is also called the *crux decussata;* it served as the basis for the multiplication sign. A cross of Saint Anthony takes the form of a capital T. The *Maltese cross* is an irregular dodecahedron whose cross pieces flange out from the center.

crunode

A point where a curve intersects itself so that two branches of the curve have distinct **tangent** lines.

cryptarithm

A number puzzle in which a group of arithmetical operations has some or all of its digits replaced by letters or symbols, and where the original digits must be found. In such a puzzle, each letter or symbol represents a unique digit. The first example appeared in *American Agriculturist* in 1864. Specific types of cryptarithm include the **alphametic**, the **digimetic**, and the **skeletal division**.

cross From left to right: a Latin cross; a *crux immissa* (a Greek cross with flared ends), also sometimes called a Latin cross; and a Maltese cross.

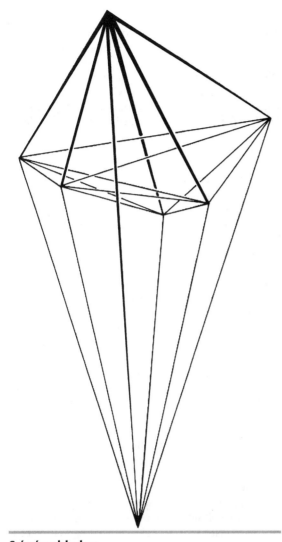

Csász ár polyhedron

first possible polyhedron beyond the tetrahedron has exactly one hole; this is the Császár polyhedron, which is thus topologically equivalent to a **torus** (donut). The Császár polyhedron has 7 vertices, 14 faces, and 21 edges, and is the **dual** of the **Szilassi polyhedron**. It isn't known if there are any other polyhedra in which every pair of vertices is joined by an edge. The next possible figure would have 12 faces, 66 edges, 44 vertices, and 6 holes, but this seems an unlikely configuration–as, indeed, to an even greater extent, does any more complex member of this curious family.

cube

(1) The **Platonic solid** that has a square for every one of its 6 faces; it also has 12 edges and 8 vertices (corners). The $60' \times 30' \times 30'$ Double Cube Room of Wilton House (the seat of the Earl of Pembroke), near Salisbury, is considered, together with the Single Cube Room of the same domicile, among the finest surviving rooms in England from the mid-seventeenth century. A favorite with filmmakers, it has provided locations for *Barry Lyndon* by Stanley Kubrick, *The Madness of King George,* and *Sense and Sensibility.* See also **Atomium, the**. (2) To cube something is to raise it to the power of three. The result of cubing is a *cube number:* $1^3 = 1$, $2^3 = 8$, $3^3 = 27$, and so on. To take the *cube root* is the reverse process; thus, 4 cubed (4^3) is 64 and the cube root of 64 ($\sqrt[3]{64}$) is 4. For cube **dissection** problems, see **Hadwiger problem**, **Slothouber-Graatsma puzzle** and **Soma cube**. See also **tesseract** and **Prince Rupert's problem**.

cubic curve

An **algebraic curve** described by a **polynomial** equation of the general form

$$ax^3 + bx^2y + cxy^2 + dy^3 + ex^2 + fxy + gy^2 + hx + iy + j = 0,$$

where $a, b, c, d, e, f, g, h, i,$ and j are constants, such that at least one of $a, b, c,$ and d is nonzero, and x and y are variables. One of Isaac **Newton**'s many accomplishments was the classification of the cubic curves. Newton found 72 different species of curve; later investigators found six more, and it is now known that there are precisely 78 different types of cubic curves. Interesting examples include the **folium** of Decartes and the Witch of **Agnesi**.

cubic equation

A **polynomial** equation of the third degree, the general form of which is

$$ax^3 + bx^2 + cx + d = 0,$$

where $a, b, c,$ and d are constants. There was a great controversy in sixteenth-century Italy between Girolamo **Cardano** and Niccoló **Tartaglia** about who should get credit for solving the cubic. At this time symbolic algebra

cryptography

The science and mathematics of encoding and decoding information. See also **cipher** and **cryptarithm**.

Császár polyhedron

A **polyhedron**, first described in 1949 by the Hungarian mathematician Ákos Császár,[79] that is a solution to an interesting problem, namely: How many polyhedra exist such that every pair of vertices is joined by an edge? The first clear example is the well known **tetrahedron** (triangular pyramid). Some simple **combinatorics** specify how many vertices, edges, faces, and holes such polyhedra must have. It turns out that, other than the tetrahedron, any such polyhedron must have at least one hole. The

hadn't been developed, so all the equations were written in words instead of symbols. Early studies of cubics helped legitimize **negative numbers**, give a deeper insight into equations in general, and stimulate work that eventually led to the discovery and acceptance of **complex numbers**. Cardano, in his *Ars Magna*, found negative solutions to equations, but called them "fictitious." He also noted an important fact connecting solutions of a cubic equation to its **coefficients**, namely, that the sum of the solutions is the negation of b, the coefficient of the x^2 term. At one other point, he mentions that the problem of dividing 10 into two parts so that their product is 40 would have to be $5 + v(-15)$ and $5 - v(-15)$. Cardano didn't go further than this observation of what later came to be called complex numbers, but a few years later Rafael Bombelli (1526–1672) gave several examples that involved these strange new mathematical beasts.

cubit

A measure of length used in the ancient world. It is approximately equal to the length of a person's forearm, that is, the part of the arm from the elbow to the fingers. The Romans used a cubit equal to 17.4 modern inches; the Egyptians used one of 20.64 inches.

cuboctahedron

A **polygon** obtained by cutting the corners off a cube or an **octahedron**. It has eight faces that are equilateral triangles and six faces that are squares.

cuboid

Also called a *rectangular prism*, a **hexahedron** of which all of the faces are rectangles and all of the opposite faces are identical. It is not known whether a *perfect cuboid*, whose sides, face diagonals, and space diagonals are all integers, exists. The general suspicion is that it doesn't, although several near misses have been found, including one in which $a = 240$, $b = 117$, $c = 44$, $dab = 267$, $dac = 244$, and $dbc = 125$. If there is a perfect cuboid, it has been shown that the smallest side must be at least $2^{32} = 4,294,967,296$.

Cullen number

A number of the form $(n \times 2^n) + 1$, denoted C_n, and named after the Reverend James Cullen (1867–1933), an Irish Jesuit priest and schoolmaster. Cullen noticed that the first, $C_1 = 3$, was a **prime number**, but with the possible exception of the fifty-third, the next 99 were all composite. Soon afterward, Cunningham discovered that 5,591 divides C_{53}, and noted that all the Cullen numbers are **composite numbers** for n in the range $2 \leq n \leq 200$, with the possible exception of 141. Five decades later Robinson showed that C_{141} is a prime. Currently, the only known Cullen primes are those with $n = 1$, 141,

4,713, 5,795, 6,611, 18,496, 32,292, 32,469, 59,656, 90,825, 262,419, 361,275, and 481,899. Although the vast majority of Cullen numbers are composite, it has been conjectured that there are infinitely many Cullen primes. Whether n and C_n can simultaneously be prime isn't known. Sometimes, the name "Cullen number" is extended to include the Woodall numbers, $W_n = (n \times 2^n) - 1$. Finally, a few authors have defined a number of the form $(n \times b^n) + 1$, with $n + 2 > b$, to be a *generalized Cullen number*.

Cunningham chain

A sequence of **prime numbers** in which each member is twice the previous one plus one. For example, {2, 5, 11, 23, 47} is the first Cunningham chain of length 5 and {89, 179, 359, 719, 1,439, 2,879} is the first of length 6. In general, a *Cunningham chain of length k of the first kind* is a sequence of k prime numbers, each of which is twice the preceding one plus one. A *Cunningham chain of length k of the second kind* is a sequence of k primes, each of which is twice the preceding one minus one. For example, {2, 3, 5} is a Cunningham chain of length 3 of the second kind and {1,531, 3,061, 6,121, 12,241, 24,481} is a Cunningham chain of length 5 of the second kind. Prime chains of both these forms are said to be complete if they can't be extended by adding either the next larger or the next smaller terms. See also **Sophie Germain prime**.

cup

The symbol ∪, which is used to denote the **union** of two sets.

curvature

A measure of the amount by which a curve, a surface, or any other **manifold** deviates from a straight line, a plane, or a hyperplane (the multidimensional equivalent of a plane). For a plane curve, the curvature at a given point has a magnitude equal to one over the radius of an *osculating* circle (a circle that "kisses," or just touches, the curve at the given point) and is a **vector** pointing in the direction of that circle's center. The smaller the radius r of the osculating circle, the greater the magnitude of the curvature $(1/r)$ will be. A straight line has zero curvature everywhere; a circle of radius r has a curvature of magnitude $1/r$ everywhere.

For a two-dimensional surface, there are two kinds of curvature: a *Gaussian* (or *scalar*) *curvature* and a *mean curvature*. To compute these at a given point, consider the intersection of the surface with a plane containing a fixed normal vector (an arrow sticking out perpendicularly) at the point. This intersection is a plane and has a curvature; if the plane is varied, this curvature also changes, and there are two extreme values—the maximal and the minimal curvature—which are known as the *main curva-*

tures, $1/R_1$ and $1/R_2$. (By convention, a curvature is taken to be positive if its vector points in the same direction as the surface's chosen normal, otherwise it is negative.) The Gaussian curvature is equal to the product $1/R_1R_2$. It is everywhere positive for a **sphere**, everywhere negative for a **hyperboloid** and **pseudosphere**, and everywhere zero for a plane. It determines whether a surface has elliptic (when it is positive) or hyperbolic (when it is negative) geometry at a point. The integral of the Gaussian curvature over the whole surface is closely related to the surface's **Euler characteristic**. The mean curvature is equal to the sum of the main curvatures, $1/R_1 + 1/R_2$.

A **minimal surface**, like that of a soap film, has a mean curvature of zero. In the case of higher-dimensional manifolds, curvature is defined in terms of a curvature **tensor**, which describes what happens to a vector that is transported around a small loop of the manifold.

curve

A continuous mapping from a one-dimensional space to an *n*-dimensional space. The most familiar mathematical curves are two- and three-dimensional **graph**s. A curve, such as a circle, that lies entirely in a plane is called a *plane curve;* by contrast, a curve that may pass through any region of three-dimensional space is called a *space curve.* See also **space-filling curve**.

curve of constant width

A curve that, when rotated in a square, makes continuous contact with all four sides. It may seem, at first sight, as if there is only one such curve–a circle. But, in fact, there are infinitely many different curves of constant width. The circle is the one with the largest area. The simplest noncircular one, and the one with the smallest area, is the **Reuleaux triangle**. Others can be constructed starting with equilateral (but not necessarily equiangular) stars. Every curve of constant width is **convex**. Moreover, *Barbier's theorem* states that every curve of constant width *w* has the same perimeter, πw. (The width of a convex figure is defined as the distance between parallel lines–known as *supporting lines*–that bound it.) A curve of constant width can be used in a special drill chuck to cut square holes. A generalization gives solids of constant width. These do not have the same surface area for a given width, but their shadows are curves of constant width with the *same* width.

cusp

In mathematics, a point on a curve where two branches, coming from different directions, meet and have a common **tangent**. If the two branches of the curve approach the tangent from opposite sides the cusp is called a *keratoid* (from the Greek *kera* for "horn") or *first-order cusp.* This is the case, for example, with the curve given by the equation

$y^2 = x^2y + x^5$. If the two branches of the curve approach the tangent from the same side the result is a *ramphoid* or *second-order cusp.* "Cusp" derives from the Latin *cuspis* for "sharp." Outside of mathematics, the points of a crescent moon are called cusps and the sharp pointed premolar teeth of children are known as bicuspids.

cute number

A number *n* such that a square can be cut into *n* squares of, at most, two different sizes. For example, 4 and 10 are cute numbers.

Cutler, William (Bill)

An Australian puzzle maker and solver who, in 1977, became the first to completely analyze, using a computer, six-piece burrs used to make solid six-piece **burr puzzles**. Martin Gardner devoted his January 1978 "Mathematical Games" column in *Scientific American* to this and other of Cutler's discoveries. In 2003, Cutler used a computer to enumerate all solutions of the **Loculus of Archimedes**.

cybernetics

The theoretical study of communication and control processes in biological, mechanical, and electronic systems, especially the comparison of these processes in biological and artificial systems. It was pioneered by Norbert **Wiener**.

cyclic number

A number with *n* digits, which, when multiplied by 1, 2, 3, . . . , *n* produces the same digits in a different order. For example, 142,857 is a cyclic number: $142,857 \times 2 = 285,714$; $142,857 \times 3 = 428,571$; $142,857 \times 4 = 571,428$; $142,857 \times 5 = 714,285$; $142,857 \times 6 = 857,142$, and so on. It has been conjectured, but not yet proven, that an infinite number of cyclic numbers exist.

cyclic polygon

A **polygon** with vertices (see **vertex**) that all lie on the same circle. All triangles are cyclic (but not all of any other kind of polygon) because any set of three points, not lying on a single line, can have a circle drawn through it.

cycloid

The shape defined by a fixed point on a wheel as it rolls; more precisely, it is the **locus** of a point on the rim of a circle rolling along a perfectly straight line. The cycloid was named by Galileo in 1599. It is the solution to both the **tautochrone problem** and the **brachistochrone problem**. In 1634, the French mathematician Gilles de Roberval (1610–1675) showed that the area under a cycloid is three times the area of its generating circle. In 1658, the English architect Christopher Wren showed

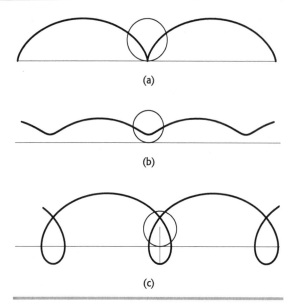

(a)

(b)

(c)

cycloid An ordinary cycloid is traced out by a point on a wheel as it rolls along a flat surface (a). A curtate cycloid is traveled by a point on the wheel that is inside the circumference (b). If the point lies outside the circumference of the wheel, the result is a prolate cycloid (c).

that the length of a cycloid is four times the diameter of its generating circle. But there was a lot of bickering and a lack of public sharing of information around this time that led to much duplication of effort, particularly over questions related to the cycloid. In fact, the confusion

was so bad that the curve was nicknamed the Helen of Geometers, and Jean Montucla referred to it as "*la pomme de discorde*" (the apple of discord).

As well as the ordinary cycloid there is the *curtate cycloid*, which is the path traced out by a point on the inside of a rolling circle, and the *prolate cycloid*, which is followed by a point on the outside of the circle. A prolate cycloid is traced out, for example, by points on the flange of the wheels of a locomotive, which extends below the top of the tracks. This leads to the surprising conclusion that even as the locomotive is moving forward there are always parts of its wheels that are going backward for a moment before moving forward again. See also **epicycloid** and **hypocycloid**.

cylinder

A three-dimensional surface described by the Cartesian equation $(x/a)^2 + (y/b)^2 = 1$. If $a = b$ then the surface is a *circular cylinder*, otherwise it is an *elliptic cylinder*. The cylinder is a *degenerate quadric* because at least one of the coordinates (in this case z) doesn't appear in the equation, though by some definitions the cylinder isn't considered to be a quadric at all. In common usage, a *cylinder* is taken to mean a finite section of a right circular cylinder with its ends closed to form two circular surfaces. If the cylinder has a radius r and a length h, then its volume is $V = \pi r^2 h$ and its surface area is $A = 2\pi r^2 + 2\pi rh$. For a given volume, the cylinder with the smallest surface area has $h = 2r$. For a given surface area, the cylinder with the largest volume has $h = 2r$. More unusual types of cylinder include the *imaginary elliptic cylinder*: $(x/a)^2 + (y/b)^2 = -1$, the *hyperbolic cylinder*: $(x/a)^2 - (y/b)^2 = 1$, and the *parabolic cylinder*: $x^2 + 2y = 0$.

d'Alembert, Jean Le Rond (1717–1783)

A French mathematician named for the church of St. Jean Baptiste de Rond upon whose steps he was abandoned as a baby, the illegitimate son of a Parisian society hostess. He clarified the concept of a **limit** in calculus, discovered the *Cauchy-Riemann equations* decades before Augustin **Cauchy** or Bernhard **Riemann**, was the first to find and solve the wave equation, and recast **Newton's** third law in a new and powerful form through what has become known as *d'Alembert's principle*.

Dandelin spheres

If a **cone** is sliced through by a plane, the two spheres that just fit inside the cone, one on each side of the plane and both tangent to it and touching the cone, are known as Dandelin spheres. They are named after the Belgian mathematician and military engineer Germinal Pierre Dandelin (1794–1847) who gave an elegant proof that the two spheres touch the **conic section** at its foci. In 1826, Dandelin showed that the same result applies to the plane sections of a **hyperboloid** of revolution.

dart

Also known as an *arrowhead*, a special kind of quadrilateral that has one **reflex angle**. See also **Penrose tiling**.

de L'Hôpital, Guillaume François Antoine, Marquis de (1661–1704)

A French mathematician who wrote the first textbook on differential calculus, *Analyse des infiniment petits pour l'intelligence des lignes courbes* (1696). This contains the rule, now known as *L'Hôpital's rule,* for finding the limit of a rational function whose numerator and denominator tend to zero at a point. Along with Isaac **Newton**, Gottfried **Leibniz**, and Jacob Bernoulli (see **Bernoulli family**), de L'Hôpital was among the first to solve the **brachistochrone problem**.

de L'Hôpital's cubic

See **Tschirnhaus's cubic**.

de Malves's theorem

Given a **tetrahedron** in which the edges meeting at one **vertex**, *X,* form three right angles (i.e., the tetrahedron is the result of chopping off the corner of a **cuboid**), the square of the face opposite *X* is equal to the sum of the squares of the other three faces.

de Méré's problem

A question posed in the mid-seventeenth century to Blaise **Pascal** by a French nobleman and inveterate gambler, the Chevalier de Méré, which marked the birth of **probability theory**. One of de Méré's favorite bets was that at least one six would appear during a total of four rolls of a die. From past experience, he knew that this gamble paid off more often than not. Then, for a change, he started betting that he would get a double-six on 24 rolls of two dice. However, he soon realized that his old approach to the game was more profitable. He asked his friend Pascal why. Pascal showed that the probability of getting at least one six in four rolls of a die is $1 - (\frac{5}{6})^4 \approx 0.5177$, which is slightly higher than the probability of at least one double-six in 24 throws of two dice, $1 - (\frac{35}{36})^{24} \approx 0.4914$. This problem and others posed by de Méré are thought to have been the original inspiration for a fruitful exchange of letters on probability between Pascal and Pierre de **Fermat**. To tackle these problems, Fermat used combinatorial analysis (finding the number of possible outcomes in ideal games of chance by computing permutation and combination numbers), while Pascal reasoned by recursion (an iterative process that determines the result of the next case by the present case). Their combined work laid the foundations for probability theory as we know it today.

de Moivre, Abraham (1667–1754)

A French-British mathematician who founded analytical trigonometry and stated what has become known as **de Moivre's theorem**. He also worked on probability theory and the normal distribution, and was a good friend of Isaac **Newton**. In 1698 he wrote that the theorem had been known to Newton as early as 1676.

de Moivre's theorem

A theorem, named after Abraham **de Moivre**, that links **complex numbers** and trigonometry. It states that for any real number *x* and any integer *n*,

$$(\cos x + i\sin x)^n = \cos(nx) + i\sin(nx).$$

By expanding the left-hand side and then comparing real and imaginary parts, it is possible to derive useful expressions for $\cos(nx)$ and $\sin(nx)$ in terms of $\sin(x)$ and $\cos(x)$. Furthermore, the formula can be used to find explicit expressions for the nth **root of unity**: complex numbers z such that $z^n = 1$. It can be derived from (but historically preceded) **Euler's formula** $e^{ix} = \cos x + i \sin x$ and the exponential law $(e^{ix})^n = e^{inx}$.

de Morgan, Augustus (1806–1871)

A British mathematician, born in India, who was an important innovator in the field of mathematical logic. The system he devised to express such notions as the contradictory, the converse, and the transitivity of a relation, as well as the union of two relations, laid some of the groundwork for his friend George **Boole**. De Morgan lost the sight of his right eye shortly after birth, entered Trinity College, Cambridge, at the age of 16, and received his B.A. However, he objected to a theological test required for the M.A. and returned to London to study for the bar. In 1827, he applied for the chair of mathematics in the newly founded University College, London and, despite having no mathematical publications, he was appointed. In 1831, he resigned on principle (after another professor was fired without explanation) but regained his job five years later when his replacement died in an accident. He resigned again in 1861.

His most important published work, *Formal Logic,* included the concept of the *quantification of the predicate,* an idea that solved problems that were impossible under the classic Aristotelian logic. De Morgan coined the phrase **"universe of discourse,"** was the first person to define and name mathematical **induction**, and developed a set of rules to determine the **convergence** of a mathematical series. In addition, he devised a decimal coinage system, an almanac of all full moons from 2000 B.C. to A.D. 2000, and a theory on the probability of life events that is still used by insurance companies. De Morgan was also deeply interested in the history of mathematics. In *Arithmetical Books* (1847) he describes the work of over fifteen hundred mathematicians and discusses subjects such as the history of the length of a foot, while in *A Budget of Paradoxes* he gives a marvelous compendium of eccentric mathematics including the poem

Great fleas have little fleas upon their backs to bite
'em,
And little fleas have lesser fleas, and so ad infinitum,
And the great fleas themselves, in turn, have greater fleas to go on,
While these again have greater still, and greater still, and so on.

The first lines of this poem paraphrase a similar rhyme by Jonathan Swift.

PUZZLE
On one occasion, when asked his age, de Morgan replied: "I was x years old in the year x^2." How old must he have been at the time?
Solutions begin on page 369.

decagon

A **polygon** with 10 sides.

decimal

The commonly used **number system**, also known as *denary,* in which each place has a value 10 times the value of the place at its right. For example, 4,327 in the decimal (base 10) system is shorthand for $(4 \times 10^3) + (3 \times 10^2) + (2 \times 10^1) + (7 \times 10^0)$, where $10^0 = 1$. "Decimal" comes from the Latin *decimus* for "tenth." The verb *decimare,* literally "to take a tenth of," was used to describe a form of punishment applied to mutinous units in the Roman army. The men were lined up and every tenth soldier was killed as a lesson to the rest. From this custom comes our word *decimate,* which we use more loosely—in fact, incorrectly—to indicate near-total destruction. The Latin *decimare* was also used in a less ferocious sense to mean "to tax to the amount of one tenth." However, the usual word describing a one-tenth tax in English is *tithe,* which comes from the Old English *teogotha,* a form of *tenth.*

decimal fraction

A number consisting of an **integer** part, which may be zero, and a decimal part less than unity that follows the decimal marker (which may be a point or a comma). A *finite* or *terminating decimal fraction* has a sequence of decimals with a definite break-off point after which all the places are zeros. Other fractions produce endless sequences of decimals that are *periodic nonterminating.*

Dedekind, (Julius Wilhelm) Richard (1831–1916)

A German mathematician whose most important contribution was the discovery of what became known as the *Dedekind cut.* He realized that every **real number** r divides the **rational number**s into two subsets: those greater than r and those less than r. Dedekind's brilliant idea was to represent the real numbers by such divisions of the rationals. He also provided important support for Georg **Cantor**'s set theory, which was highly controversial at the time.

Dee, John (1527–1609)

A notable English alchemist, mathematician, and astronomer, sometimes referred to as the "last magician" because of his astrological services to Queen Elizabeth I; Dee may also have influenced the writings of Shakespeare. He enrolled at St. Johns College, Cambridge, at age 15, but found the atmosphere there stifling and later went to the Continent to study and lecture. Upon his return to England, Dee cast the horoscope for Queen Mary and later visited Mary's half-sister Elizabeth in jail to determine when Mary would die. Accused of black magic, he was jailed and then released in 1555, three years before Mary's death. When Elizabeth came to the throne she consulted Dee on many matters, including the geography of newly discovered lands, and paid him well. Some of his income he spent on extensive traveling, which may have involved some spying on behalf of his sponsor.

Dee had a large library of books on witchcraft, the occult, and magic, and he wrote 79 manuscripts, only a few of which were published. He married three times and fathered eight children. He also struck up an uneasy partnership with Edward Kelly, a bad-tempered Irishman who claimed to have discovered the alchemical secret of transmuting base metal into gold but had lost his ears for forgery. In 1585 Dee and Kelly went on a four-year trek across the Continent conducting astrological readings for nobility and royalty. But Dee and Kelly had many arguments and eventually parted company. Back in England Dee found his house ransacked and many of his possessions stolen or destroyed. Elizabeth helped pay for the damage and made him warden of Christ's College in Manchester in 1595. However, Elizabeth died in 1603 and her successor James I opposed magic. Dee was forced to retire, his life ending in poverty.

deficient number
See **abundant number**.

degree
(1) The unit of measurement for angles; one degree is $1/360$ of a circle. (2) The **exponent** of a variable. For example, the degree of $7x^5$ is 5. See also **degree of freedom**.

degree of freedom
A positive integer that gives the number of pieces of data that are independent.

deletable prime
See **truncatable prime**.

delta curve
A curve that can be turned inside an equilateral **triangle** while continuously making contact with all three sides. There are an infinite number of delta curves, but the simplest are the **circle** and lens-shaped delta-biangle. All the delta curves of height h have the same perimeter $2\pi h/3$. See also **Reuleaux triangle** and **rotor**.

deltahedron
A **polyhedron** whose faces consist of equilateral triangles that are all the same size. Although there are an infinite number of different deltahedra, only eight of them are convex, as O. Rausenberger first showed in 1915. Among this group of eight, faces made of coplanar equilateral triangles sharing an edge (such as the rhombic dodecahedron) aren't allowed. The eight convex deltahedra have 4, 6, 8, 10, 12, 14, 16, and 20 faces.

deltoid
A **hypocycloid** with three cusps, also known as a *tricuspoid* or *Steiner's hypocycloid* after the Swiss mathematician Jakob **Steiner** who investigated the curve in 1856. The deltoid, so-named because it looks like an uppercase Greek delta, Δ, is formed by a point on the circumference of a circle rolling inside another circle with a radius three times as large. While working on a problem in optics in 1745, Leonhard **Euler** was among the first to study its properties. The parametric equations of the cycloid with inner circle of radius r are:

$$x(t) = 2r\cos t + r\cos 2t$$
$$y(t) = 2r\sin t - r\sin 2t$$

The length of the path of the deltoid is $16r/3$, and the area inside the deltoid is $2\pi r^2$. If a tangent is drawn to the deltoid at some point, P, and the points where the tangent crosses the deltoids other two branches are called points A and B, then the length of AB equals $4r$. If the deltoid's tangents are drawn at points A and B, they will

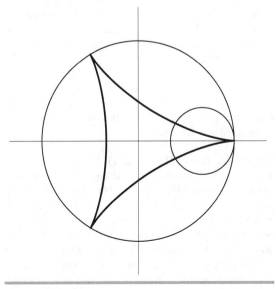

deltoid The deltoid curve. © Jan Wassenaar, www.2dcurves.com

be perpendicular, and they will intersect at a point inside the deltoid that is the 180° rotation of point P about the center of the fixed circle.

denominator

In a **rational number**, the number below the fraction bar; it indicates into how many parts the whole is divided.

derivative

The result of differentiating a **function**; that is, the infinitesimal change in a function caused by an infinitesimal change in the variable(s) upon which it depends. The derivative gives the rate of change of a function (the slope of its curve) at a particular point. Second and third derivatives give the rate at which the rate of change is changing and the rate at which the rate of rate of change is changing, respectively. In an article in 1996, Hugo Rossi wrote: "In the fall of 1972 President Nixon announced that the rate of increase of inflation was decreasing. This was the first time a sitting president used the third derivative to advance his case for reelection."[269]

Here is a fallacious "proof" that $x = 2x$ based on derivatives. Consider the function $f(x) = x^2$, the derivative of which is $2x$. What is wrong with the following?

$$x^2 = x + x + \ldots + x \text{ (repeated } x \text{ times)}$$

Taking the derivative of both sides gives

$$(x^2)' = 1 + 1 + \ldots + 1 = x.$$

But we have already said that the derivative of x^2 is $2x$. Therefore, $x = 2x$. The error stems from taking the derivative of x different x's. Each of the terms depends not only on x, which was accounted for in taking the derivative, but also on the *number* of terms (which could be fractional) which depends on x, too, and this was not accounted for. Put another way, the derivative measures the rate of change of x^2 as x changes, but as x changes, the number of terms on the right, as well as the terms themselves, increases. For positive x, the correct answer must be larger than x–as indeed it is.

Desargues, Girard (1591–1661)

A French mathematician who is regarded as the chief founder of perspective geometry. His 12-page treatise *La perspective* (1636) consists of a single worked example in which Desargues sets out a method for constructing a perspective image without using any point lying outside the picture field. He considers the representation in the picture of a plane of lines that meet at a point and also of lines that are parallel to each another. In the last paragraph of the work he considers the problem of finding the perspective image of a conic section. Three years later, he wrote his treatise on projective geometry *Brouillon project d'une atteinte aux evenemens des rencontres du cone avec un plan* (Rough draft for an essay on the results of taking plane sections of a cone). The first part of this deals with the properties of sets of straight lines meeting at a point and of ranges of points lying on a straight line. In the second part, the properties of conics are investigated in terms of properties of ranges of points on straight lines and the modern term "point at infinity" appears for the first time. Desargues shows that he has completely grasped the connection between conics and perspective; in fact he treats the fact that any conic can be projected into any other conic as obvious. Given such innovative work it may seem surprising that the subject didn't develop rapidly in the following years. That may be partly due to mathematicians failing to recognize the power of what had been put forward. On the other hand, the algebraic approach to geometry put forward by René **Descartes** at almost exactly the same time (1637) may have diverted attention from Desargues's projective methods.

Descartes, René (1596–1650)

If you would be a real seeker after truth, it is necessary that at least once in your life you doubt, as far as possible, all things.

A hugely influential philosopher and mathematician, born in La Haye (now named Descartes after its most famous son), Indre-et-Loire, France, who is often referred to as the father of modern philosophy and one of the founders of modern mathematics. He studied law at the University of Poitiers but never practiced it, served in the military for a while, and then lived in Holland for 20 years where he did the bulk of his great work. In his *Meditations on First Philosophy*, he tried to establish what can be known as true beyond doubt. His tool was methodological skepticism: the assumption that any idea that can be doubted is false. He gives the example of dreaming: in a dream, one senses things that seem to be real, but that don't actually exist. Thus, the data of the senses can't be fully trusted. Then again, he mused, perhaps there is an "evil genius"–a supremely powerful and devious being who sets out to prevent anyone from knowing the true nature of reality. Given these possibilities, what is it that one can know for certain? Descartes argues that if "I" am being deceived, then surely "I" must exist–the statement famously referred to as *cogito ergo sum* ("I think, therefore I am"), though these words don't actually appear anywhere in the *Meditations*. Descartes concludes that he can be certain that he exists. But in what form? If the senses are unreliable, Descartes reasons, all he can say for sure is that he is a *thinking thing*. He then proceeds to build a system of knowledge, discarding perception as unreliable and instead admitting only deduction as a method. Halfway through the *Meditations* he also claims to prove the existence of a benevolent **God** who has provided him with a working mind and sensory system, and who cannot desire to deceive him, and thus, finally, he establishes the possibility of acquiring knowledge about the world based on deduction *and* perception.

In mathematics, Descartes is important for his discovery of **analytical geometry**. Up to Descartes's time, **geometry**, dealing with lines and shapes, and **algebra**, dealing with numbers, were regarded as completely independent aspects of mathematics. Descartes showed how almost all problems in geometry can be translated into problems in algebra, by regarding them as questions asking for the length of a line segment, and using a coordinate system to describe the problem. Descartes's theory provided the basis for **calculus**, developed by Isaac **Newton** and Gottfried **Leibniz**, and thus for much of modern mathematics. This is particularly amazing when you consider that Descartes intended it merely as an example to support his *Discours de la méthode pour bien conduire sa raison, et chercher la verité dans les sciences* (Discourse on the method to rightly conduct the reason and search for the truth in sciences), better known under the shortened title *Discours de la méthode*.

Descartes died of pneumonia in Stockholm, where he had been invited to serve as tutor to the energetic 19-year-old Queen Christina of Sweden. Accustomed to working in a warm bed till noon, he was shocked into a rapid decline by having to teach philosophy at 5 A.M. in a freezing library. Seventeen years after his death, the Roman Catholic Church placed his works on the Index of Prohibited Books.

Descartes's circle theorem
See **Soddy's formula**.

determinant
A quantity obtained from a square ($n \times n$) array of numbers that can be useful, among other things, in solving systems of **linear** equations (equations in which the unknowns are raised to at most the first power). More generally, a determinant transforms a square **matrix** into a **scalar**—an operation that has many important properties. Two-by-two determinants were considered by Girolamo **Cardano** at the end of the sixteenth century and ones of arbitrary size by Gottfried **Leibniz** about a century later. Determinants are so named because, when applied to systems of linear equations, they "determine" if the systems are singular—that is, have multiple solutions. They also have important geometric applications, because they describe the area of a **parallelogram** and, more generally, the volume of a **parallelepiped**. A three-rowed determinant is defined by:

$$\begin{vmatrix} a_{11} & a_{12} & a_{13} \\ a_{21} & a_{22} & a_{23} \\ a_{31} & a_{32} & a_{33} \end{vmatrix} = \begin{matrix} a_{11} \times a_{22} \times a_{33} + a_{12} \times a_{23} \times a_{31} \\ + a_{13} \times a_{21} \times a_{32} - a_{13} \times a_{22} \times a_{31} \\ - a_{12} \times a_{21} \times a_{33} - a_{11} \times a_{23} \times a_{32}. \end{matrix}$$

deterministic system
A system in which the later states of the system follow from, or are determined by, the earlier ones. Such a system contrasts with a *stochastic* or *random system* in which future states are not determined from previous ones. An example of a stochastic system would be the sequence of heads or tails of an unbiased coin, or radioactive decay.

If a system is deterministic, this doesn't necessarily imply that later states of the system are predictable from a knowledge of the earlier ones. In this way, **chaos** is similar to a random system. Chaos has been termed "deterministic chaos" since, although it is determined by simple rules, its property of sensitive dependence on initial conditions makes a chaotic system, in practice, largely unpredictable.

devil's curve
Also known as the *devil on two sticks,* a curve with the Cartesian equation

$$y^4 - a^2 y^2 = x^4 - b^2 x^2$$

and the polar equation

$$r^2 (\sin^2\Theta - \cos^2\Theta) = a^2 \sin^2\Theta - b^2\cos^2\Theta).$$

Early studies of it were carried out in 1750 by the Swiss mathematician Gabriel Cramer (1704–1752), who is

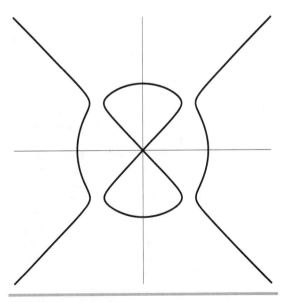

devil's curve   © Jan Wassenaar, www.2dcurves.com

most famous for his work on **determinant**s, and in 1810 by Lacroix. For $a = {}^{25}/24$, the curve is called the *electric motor curve*.

Dewdney, Alexander Keewatin (1941–)

A Canadian computer scientist and mathematician at the University of Western Ontario, Canada, best known for his popular books and articles, most notably *The Planiverse: Computer Contact with a Two-dimensional World*, first published in 1984.[81] For several years, Dewdney wrote the "Mathematical Recreations" column for *Scientific American*.

diagonal

A line that joins any two vertices of a **polygon**, if the vertices are not next to each other; or a line that joins two vertices of a **polyhedron** that are not on the same face.

diagonal matrix

A **matrix** that has zero entries along all nondiagonal entries, that is, only the main diagonal may have nonzero values.

diameter

The distance across a **circle** through the center.

dice

Small polyhedra (see **polyhedron**), usually cubes, whose faces are numbered from one to six by patterns of dots, with opposite sides totaling seven. They are thrown, singly or in groups, from the hand or from a cup, onto a flat surface, to provide **random** numbers for gambling and other games. The face of each die that is uppermost when it comes to rest provides the value of the throw. Typical of their use today is the game of craps, in which two dice are thrown together, and bets placed on the total face-up value. Dice probably evolved from knucklebones, which are approximately tetrahedral. Even today, dice are sometimes colloquially referred to as "bones." Ivory, bone, wood, metal, and stone materials have been commonly used to make dice, though the use of plastics is now nearly universal.

Dice found in ancient tombs in the Orient point to an Asiatic origin and dicing is mentioned as an Indian game in the Rig-veda. In its primitive form, knucklebones was essentially a game of skill, played by women and children; gradually, a derivative form evolved for gambling in which four sides of the bones received different values and were counted like dice. Gambling with three, sometimes two, dice was a popular form of amusement in Greece, especially with the upper classes, and was an almost invariable accompaniment to the symposium, or drinking banquet. The Romans were passionate gamblers, and dicing was a favorite form, though it was for-

bidden except during the festival of Saturnalia (December 17). Throwing dice for money led to many special laws in Rome, one of which decreed that no suit could be brought by a person who allowed gambling in his house, even if he'd been cheated or assaulted! Professional gamblers were common, and some of their loaded dice are preserved in museums.

The Roman historian Tacitus states that the Germans also were passionately fond of dicing—so much so, that, having lost everything, they would even stake their personal liberty. Centuries later, in medieval times, dicing became the favorite pastime of knights, and both dicing schools and guilds of dicers flourished.

Dice are frequently used to randomize allowable moves in board games such as **backgammon**. Loaded dice can be made in many ways to cheat at such games. Weights can be added, or some edges made round while others are sharp, or some faces made slightly off-square, to make some outcomes more likely than would be predicted by pure chance. Dice with non-cubical shapes were once almost exclusively used by fortune-tellers and in other occult practices, but they have become popular lately among players of role-playing and war-games.

difference equation

An equation that describes how something changes in discrete time steps. Numerical solutions to **integral**s are usually realized as difference equations.

differential

A term such as dx used in an expression such as $ydx - xdy$ to denote first-order small changes in the variable. *Differentiation* is the method by which a differential is found.

differential equation

A description of how something continuously changes over time (see **continuity**). Some differential equations have an exact *analytical solution* such that all future states can be known without simulating the time evolution of the system. However, most have a *numerical solution* with only limited accuracy. A differential equation involves the first or higher derivatives of the **function** to be solved for. If the equation only involves first derivatives, it is known as an equation of *order* one, and so on. If only nth powers of the derivatives are involved, the equation is said to have *degree n*. Equations of degree one are called *linear*. Equations in only one variable are called **ordinary differential equation**s to distinguish them from **partial differential equation**s, which have two or more.

differential geometry

The study of geometry using **calculus**; it has many applications in physics, especially in **relativity theory**. The objects studied by differential geometry are known as *Riemannian manifold*s. These are geometrical objects, such as surfaces, that locally look like **Euclidean space** and therefore allow the definition of analytical concepts such as tangent **vectors** and tangent space, differentiability (see **differential**), and vector and **tensor** fields. Riemannian manifolds have a **metric**, which opens the door to measurement because it allows distances and angles to be evaluated locally and concepts such as **geodesics**, **curvature**, and torsion to be defined.

differential topology

A branch of **topology** concerned with those properties of **differential geometry** that are preserved by continuous transformation.

differentiation

The method by which the **derivative** of a **function** is found.

digimetic

A **cryptarithm** in which digits are used to represent other digits.

digit

A symbol or numeral that is used to represent an **integer** in a positional **number system**. Examples of digits include the decimal characters 0 through 9, the binary characters 0 or 1, and the hexadecimal digits $0 \ldots 9$, $A \ldots F$. The word comes from the Latin *digitus* for "finger" or "toe," and retains this meaning, reminding us of the origins of our base 10 number system. The earlier Indo-European root *deik* is related to many other words that hark back to the use of the hands and fingers to "point" out objects, including index, indicate, token, and teach.

digital root

Take a number, *n*, add its digits, then add the digits of numbers derived from it, and so on, until the remaining number has only one digit. This single digit result is called the digital root of *n*. For example, in the case of 5381: $5 + 3 + 8 + 1 = 19$; $1 + 9 = 10$; $1 + 0 = 1$; thus, the digital root of 5381 is 1. See also **casting out nines**.

digraph

A **graph** in which each edge has a direction associated with it.

dihedral angle

The angle defined by two given faces meeting at an edge; for example, all the dihedral angles of a cube are 90°. An almost-spherical **polyhedron** (with many faces) has small dihedral angles.

dimension

An extension in some unique direction or sense; the word comes from the Latin *dimetiri* for "measured out." The most common way to think of a dimension is as one of the three spatial dimensions (up-down, left-right, back-forth) in which we live. Mathematicians and science fiction writers alike have long imagined what it would be like in a world with a different number of spatial dimensions. Speculation has particularly focused on **two-dimensional worlds** and, to an even greater extent, on the **fourth dimension**. **Time** is also thought of as a dimension; indeed, in **relativity theory** and as a component of **space-time**, it is treated almost exactly the same as a dimension of space. The universe may have additional spatial dimensions—a total of 10, 11, or 26 are especially favored—according to some theories of the subatomic world (see **string theory** and **Kaluza-Klein theory**), though the additional ones are "curled up" incredibly small and only become important at scales far smaller than those that can be experimentally probed today.

In mathematics, the term *dimension* is used in many different ways. Some of these correspond to the everyday idea of an extension in physical space or to some of the more esoteric meanings in physics. Others are purely abstract and exist only in certain types of theoretical, mathematical space. There are, for example, *Hamel dimensions, Lebesgue covering dimensions,* and **Hilbert spaces**. So-called **Hausdorff dimension**s are used to characterize **fractals**—mathematical objects that have fractional dimensions—by giving a precise meaning to the idea of how well something, such as an extremely "wriggly" curve or surface, fills up the space in which it is embedded.

dinner party problem
See **Ramsey theory**.

Diocles (c. 240–c. 180 B.C.)

A Greek mathematician and contemporary of **Apollonius** who studied the **cissoid** as part of an attempt to duplicate the cube and also was the first to prove the focal property of a parabolic mirror.

Diophantine approximation

The approximation of a **real number** by a **rational number**.

Diophantine equation

An equation that has integer **coefficients** and for which integer solutions are required. Such equations are named after **Diophantus**. The best known examples are those from **Pythagoras's theorem**, $a^2 = b^2 + c^2$, when *a*, *b*, and *c* are all required to be whole numbers—a **Pythagorean triplet**. Despite their simple appearance Diophantine equations can be fantastically difficult to solve. A notorious example comes from **Fermat's last theorem** (recently solved), $a^n = b^n + c^n$ for $n > 2$. To give a specific example, suppose we want to find integer values for x and y such that

$$x^2 = 1620y^2 + 1.$$

A trial-and-error approach using a computer would quickly find the solution: $y = 4$, $x = 161$. However, just a slight change to the equation to make it

$$x^2 = 1621y^2 + 1$$

would leave the trial-and-error method floundering, even with the resources of the most powerful computers on Earth. The smallest integer solution to this innocent looking formula involves a y-value that is on the order of a thousand trillion trillion trillion trillion trillion trillion! One of the challenges (the tenth one) that David **Hilbert** threw down to twentieth-century mathematicians in his famous list was to find a general method for solving equations of this type. In 1970, however, the Russian mathematician Yu Matijasevic showed that there is no general algorithm for determining whether a particular Diophantine equation is soluble: the problem is undecidable.[217, 218]

Diophantus of Alexandria (A.D. c. 200–c. 284)

A Greek mathematician who developed his own algebraic notation and is sometimes called "the father of algebra." His works were preserved by the Arabs and translated into Latin in the sixteenth century, when they served to inspire momentous new advances. **Diophantine equations** are named in his honor. It was in the margin of a French translation of Diophantus's work *Aritmetike* from c. 250 that Pierre de **Fermat** scribbled his famous comment that became known as **Fermat's last theorem**.

Diophantus's riddle

One of the oldest known age problems (see **age puzzles and tricks**). It comes from the *Greek Anthology*, a collection of puzzles compiled by Metrodorus in about A.D. 500, and purports to tell how long **Diophantus** lived in the form of a riddle engraved on his tombstone:

> God vouchsafed that he should be a boy for the sixth part of his life; when a twelfth was added, his cheeks acquired a beard; He kindled for him the light of marriage after a seventh, and in the fifth year after his marriage He granted him a son. Alas! late-begotten and miserable child, when he had reached the measure of half his father's life, the chill grave took him. After consoling his grief by this science of numbers for four years, he reached the end of his life.

If d and s are the ages of Diophantus and his son when they died, then the epitaph boils down to these two equations,

$$d = (\tfrac{1}{6} + \tfrac{1}{12} + \tfrac{1}{7})d + 5 + s + 4$$
$$s = \tfrac{1}{2}\,d,$$

which can be solved simultaneously to give $s = 42$ years and $d = 84$ years.

Dirac, Paul Adrien Maurice (1902–1984)

A British theoretical physicist who played a major role in the development of **quantum mechanics** and predicted the existence of antiparticles. He made his first great breakthrough at Cambridge University in 1928, when he found a wave equation for the electron. This explained aspects of the electron that had previously been observed but not understood, and, incidentally, is the only equation to appear in Westminster Abbey, where it is engraved on Dirac's commemorative plaque. Dirac's electron equation also made the remarkable prediction that there exists a previously unseen type of matter—a particle like the electron, but with the opposite charge. This was startling at the time because only two subatomic particles, the electron and the proton, were known, and there was no suspicion that others might be waiting in the wings. The prediction was fulfilled four years later when the *positron*, as it is now called, was first seen. A central theme of Dirac's work was his belief that **beauty and mathematics** go hand in hand. When a journalist once asked him to explain the concept of mathematical beauty, Dirac asked the journalist "Do you know mathematics?" and when the journalist replied "No," Dirac said, "Then you can't understand the concept of mathematical beauty." A shy, retiring person, Dirac is not as famous as his achievements warrant.

Dirac string trick

Take a cardboard square and tie the four corners to another larger square by loose string. Rotate the small square by 360° about a vertical axis, that is, in a horizontal plane. The strings will become somewhat tangled, and it is not possible to untangle them without rotating the square. Turn the square through another 360°, for a total of 720°. Contrary to all expectations, it is now possible to untangle the string, without further rotation of the square, simply by allowing enough space for the strings to be looped over the top of the square!

Another version of the Dirac string trick has been called the Philippine wineglass trick. A glass of water held in the hand can be rotated continuously through 720° without spilling any water. Surprisingly, these geometrical demonstrations are related to the physical fact that an

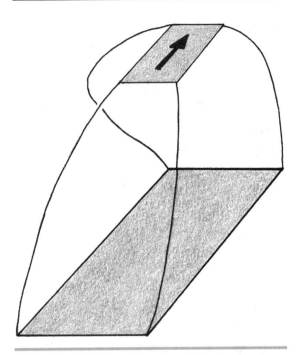

Dirac string trick The equipment needed to simulate an electron's spin property.

electron has spin ½. A particle with spin ½ is something like a ball attached to its surroundings with string. Its amplitude changes under a 360° (2π) rotation and is restored on rotation to 720° (4π).

direct proportion
The relationship two quantities have if the graph of one against the other is a straight line through the origin; so if one doubles then the other doubles, and so forth.

directed graph
Also known as a *digraph,* a **graph** in which each edge is replaced by a directed edge, indicated by an arrow. A directed graph having no multiple edges or loops is called a *simple directed graph.* A **complete graph** in which each edge is bidirected is called a complete directed graph. A directed graph having no symmetric pair of directed edges (i.e., no bidirected edges) is known as an *oriented graph.* A complete oriented graph (i.e., a directed graph in which each pair of vertices is joined by a single edge having a unique direction) is called a *tournament.*

directrix
The line that, together with a point called the **focus,** serves to define a **conic section** as the **locus** of points

whose distance from the focus is proportional to the horizontal distance from the directrix. If the ratio $r = 1$, the conic is a **parabola,** if $r < 1$, it is an **ellipse,** and if $r > 1$, it is a **hyperbola.**

Dirichlet, Peter Gustav Lejeune (1805–1859)
A German mathematician who made significant contributions to **number theory, analysis,** and mechanics, and who is credited with the modern formal definition of a **function.** He taught at the universities of Breslau (1827) and Berlin (1828–1855) and in 1855 succeeded Carl **Gauss** at the University of Göttingen but died of a heart attack only three years later. Dirichlet continued Gauss's great work on number theory, publishing on **Diophantine equations** of the form $x^5 + y^5 = kz^5$. His book *Lectures on Number Theory* (published posthumously in 1863) is similar in stature to Gauss's earlier *Disquisitiones* and founded modern **algebraic number theory.** In 1829 he gave the conditions sufficient for a **Fourier series** to converge (though the conditions *necessary* for it to converge are still undiscovered).

Dirichlet's theorem
For any two positive **coprime** integers, a and b, there are infinitely many **prime number**s of the form $an + b$, where $n > 0$. This theorem was first conjectured by Karl **Gauss** and proved by Peter **Dirichlet** in 1835.

discontinuity
Also called a *jump,* a point at which a **function** is not **continuous.**

discrete
Taking only non**continuous** values, for example, **Boolean** or **natural number**s.

discriminant
A quantity that gives valuable information about the solutions of an equation. In the case of the **quadratic** equation $ax^2 + bx + c = 0$, the discriminant is given by $d = b^2 - 4ac$. If $d > 0$, the roots of the equation are two different **real number**s; if $d = 0$, the roots are real and equal; if $d < 0$, the roots are **complex number**s. The concept of discriminant can also be applied in the case of **polynomial**s, **elliptic curve**s, and **metric**s.

disk
Roughly speaking, the "filling" of a **circle.** A flat (two-dimensional) disk of radius r consists of all the points that are at a distance $\leq r$ (*closed disk*) or $< r$ (*open disk*) from a fixed point in the plane. More generally, an n dimensional disk of radius r is the set of all points at a distance $\leq r$ (*closed*) or $< r$ (*open*) from a fixed point in

Euclidean *n*-space. A disk is the two-dimensional analog of a **ball**.

disme

An old word for "tenth." A notation for **decimal** fractions was introduced for the first time in 1585 in a pamphlet called *La Disme* by Simon Stevin of Holland.

dissection

Cutting apart one or more figures and rearranging the pieces to make another figure. Dissection puzzles have been around for thousands of years. The problem of dissecting two equal squares to form one larger square using four pieces dates back to at least the time of Plato (427–347 B.C.). In the tenth century, Arabian mathematicians described several dissections in their commentaries on Euclid's *Elements*. The eighteenth-century Chinese scholar Tai Chen presented an elegant dissection for approximating the value of **pi**. Others worked out dissection proofs of the **Pythagoras's theorem**. In the nineteenth century, dissection puzzles by Sam **Loyd**, Henry **Dudeney**, and others became tremendously popular in magazine and newspaper columns. A classic example is the **Haberdasher's puzzle**. Dissections can get quite elaborate: an eight-piece octahedron becomes a hexagon, a nine-piece five-pointed star becomes a pentagon, and so on. See also **tangrams** and **Loculus of Archimedes**.[201, 202]

dissipative system

A **dynamical system** that contains internal friction that deforms the structure of its **attractor**. Dissipative systems often have internal structure despite being far from **equilibrium**, like a whirlpool that preserves its basic form despite being in the midst of constant change.

distortion illusion

An illusion that distorts an image's shape and/or size. Famous examples include **Poggendorff illusion**, **Zöllner illusion**, **Titchener illusion**, **irradiation illusion**, **Fraser spiral**, **Müller-Lyer illusion**, **Orbison's illusion**, **vertical-horizontal illusion**, and **Ames room**.

distributive

Three numbers *x, y,* and *z* are said to be distributive over the operation + if they obey the identity

$$x(y + z) = xy + xz.$$

Compare with **associative** and **commutative**.

diverge

If a **sequence** doesn't converge it is said to diverge (see **convergence**). This can be if it goes to infinity, or if it simply cycles between two or more values without ever staying on one of them. For example, the sequences: 1, 2, 4, 8, 16, 32, . . . and 1, 0, 1, 0, 1, 0, . . . are both divergent.

division

A counterpart to **multiplication** defined so that if

$$a \times b = c$$

where b is nonzero, then

$$a = c/b.$$

In this equation, *a* is the *quotient,* b is the *divisor,* and c is the *dividend.* A **skeletal division** is a long division in which most or all of the digits are replaced by a symbol (usually asterisks) to form a **cryptarithm**.

dodecagon

A **polygon** with 12 sides.

dodecahedron

A **polyhedron** with 12 faces. A *regular dodecahedron* is made from faces that are identical regular **pentagon**s and is one of the **Platonic solid**s.

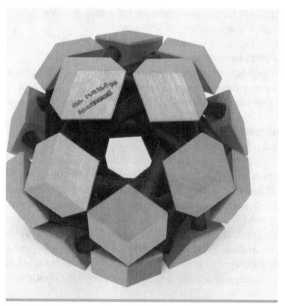

dodecahedron   A mechanical puzzle in the form of a dodecahedron. *Mr. Puzzle Australia, www.mrpuzzle.com.au*

Dodgson, Charles Lutwidge

See **Carroll, Lewis**.

dollar

The elaborate designs on the various denominations of American dollar bills can be used for some amusing games of the "Can you find . . . ?" variety. On a $1 bill, there is an owl in the upper left-hand corner of the "1" encased in the shield, and a spider hidden in the front upper right-hand corner. There are also at least nine occurrences of **thirteen** things: 13 steps on the pyramid, 13 letters in the Latin above the pyramid, 13 letters in "E Pluribus Unum," 13 stars above the eagle, 13 plumes of feathers on each span of the eagle's wing, 13 bars on the shield, 13 leaves on the olive branch, 13 fruits, and 13 arrows. On the back of a $5 bill the number 172 can be found in the bushes at the base of the Lincoln Memorial, while on the back of the old $10 bill are four cars and eleven light posts. On the new $100 dollar bill the time on the clock tower of Independence Hall reads 4:10.

domain

The set of numbers x for which the **function** $f(x)$ is defined. See also **codomain** and **range**.

domino

A small rectangular tile, marked with spots, that is used to play games. There are many varieties of dominoes and games based on them. Most domino tiles, however, have roughly 2-to-1 proportions and each half of each domino has spots arranged as on a six-sided die; the set generally contains tiles with all possible combinations of two numbers. A tile is identified by the number of spots on each half: for example, "1-6" or "3-3." English/American dominoes include blank sides; Chinese dominoes don't but do duplicate some whole tiles. English/American dominoes can also be bought in larger sets with numbers of spots up to nine or twelve per side. Other than games of strategy, there are many mathematical puzzles that involve dominoes. Some of these puzzles involve tiling variations on the standard 8 × 8 chessboard.

PUZZLE

A standard chessboard can easily be tiled by using four dominos in each row. But what if two squares are removed, one each from diagonally opposite corners of the chessboard? Can this reduced board be completely tiled by nonoverlapping dominos?

Solutions begin on page 369.

Another common pastime using domino tiles is to stand them on edge in long lines, then topple the first tile, which falls on and topples the second, and so forth, resulting in all of the tiles falling. Arrangements of thousands of tiles have been made that take several minutes to fall. By analogy, phenomena of chains of small events each causing similar events leading to eventual catastrophe are called *domino effects*. The word *domino* was first used to refer to the hooded black cape worn by priests, and later to black masks (of the Lone Ranger type) worn at masquerade balls. The domino is the simplest form of **polyomino**.

domino problem

Is there an algorithm (a set of instructions) that, when given a particular shape as an input, decides if the shape can be used to tile the entire plane? The solution to this unresolved problem is tied up with **Heesch numbers**. The domino problem in turn has a deep connection with the **Einstein problem**.

dozen

See **twelve**.

dragon curve

A classic example of a recursively generated **fractal** shape. Benoit **Mandelbrot** called it the "Harter-Heighway" dragon curve and it formed the subject of one of Martin

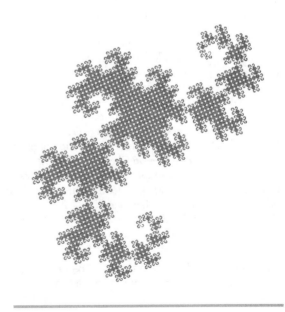

dragon curve   *Jos Leys, www.josleys.com*

Gardner's *Mathematical Games* columns in *Scientific American* in 1967.[116] The dragon curve fills out an "island" of positive area with a fractal boundary.

dual

(1) The *dual of a solid* is formed by joining the centers of adjacent faces with straight lines. In the resulting dual solid, each **vertex** of the dual corresponds to a face on the original, each face on the dual to an original vertex, while the edges match, one for one. (2) The *dual of a tessellation* is obtained by replacing each tile with a point at its center, and each edge between tiles with an edge joining vertices. The dual of a regular tessellation is a regular tessellation; the dual of a semi-regular tessellation is not semi-regular.

Dudeney, Henry Ernest (1857–1930)

An English writer and puzzle-maker who became one of the greatest exponents of recreational mathematics of his time. **Chess** and chess problems captivated him from an early age and he was only 9 when he started contributing puzzles to a local newspaper. His education was limited and he started work as a clerk in the civil service at the age of 13. However, he kept up his interest in math and chess, wrote articles for magazines under the pseudonym "Sphinx," and joined a literary circle that included Arthur Conan Doyle. In 1893 he struck up a correspondence with the American puzzle-maker Sam **Loyd**, the other leading mathematical recreationist of the day, and the two shared many ideas. However, a rift developed after Dudeney accused Loyd of publishing many of Dudeney's puzzles under his own name. One of Dudeney's daughters "recalled her father raging and seething with anger to such an extent that she was very frightened and, thereafter, equated Sam Loyd with the devil." Dudeney was a columnist for the *Strand Magazine* for over 30 years and wrote six books. The first of these, *The Canterbury Puzzles*,[87] published in 1907, purports to include a collection of problems posed by the characters in Chaucer's *The Canterbury Tales*. The answer to the so-called **Haberdasher's puzzle** is Dudeney's best known geometrical discovery. His other books include *Amusements in Mathematics* (1917)[88] and *The World's Best Word Puzzles* (1929). See also **spider-and-fly problem** and **polyomino**.[233]

Dunsany, Lord Edward Plunkett (1878–1957)

An Irish writer who was one of the founders of the fantasy genre of literature. Edward John Moreton Drax Plunkett was born in London to a family whose roots in Ireland predate the Norman invasion. He inherited his father's title in 1899, fought in the Boer War, and

Dunsany, Lord Edward Plunkett *The Dunsany Estate*

returned to the ancestral home, Dunsany Castle, in 1901. Lord Dunsany was a keen marksman and hunter, a fine player of cricket (Dunsany had its own cricket ground near the village), tennis (there is a court beside the Castle), and **chess** (he was an amateur champion and once drew with Grand Master Capablanca. He also wrote chess puzzles for the *Times* over many years and invented his own variant of the game. His first of many books, *The Gods of Pegana*, was published in 1905. In writings that spanned fantasy, drama, poetry, and science fiction, he was an early explorer of such ideas as chess-playing computers (in "The Three Sailors' Gambit" from *The Last Book of Wonder* and, again, in his 1951 novel *The Last Revolution*) and paradoxes in **time travel** (e.g., in "Lost" from *The Fourth Book of Jorkens* and "The King That Was Not" from *Time and the Gods*).

Dupin cyclide

The envelope of all **spheres** touching three given fixed spheres. (Each of the fixed spheres is to be touched in an assigned manner, either externally or internally.) Equivalently, the envelope of all spheres whose centers lie on a given **conic section** and which touch a given sphere. Also equivalently, the inverse of a **torus**.

duplicating the cube

A classical mathematical challenge of antiquity: using only a straightedge and compass, construct a **cube** whose volume is exactly twice that of a given cube. It is often called the "Delian" problem because of a legend that surrounds its origin. The citizens of Athens were being devastated by a plague so that, in 430 B.C., they sought advice from the oracle at Delos on how to rid their community of this pestilence. The oracle replied that the altar of Apollo, which was in the form of a cube, had to be doubled in size. Thoughtless builders merely doubled the edges of the cube, failing to appreciate that this increased the volume of the altar eightfold. The oracle said the gods had been angered; the plague grew worse. Other delegations consulted Plato. When informed of the oracle's admonition, Plato told the citizens "the god has given this oracle, not because he wanted an altar of double the size, but because he wished in setting this task before them, to reproach the Greeks for their neglect of mathematics and their contempt of geometry." Many Greek mathematicians attacked the problem. All failed, because the so-called *Delian constant*, $\sqrt[3]{2}$ (the required ratio of sides of the original cube and that to be constructed), needed for the duplication can't be constructed as prescribed. Cube duplication is possible, however, using a **Neusis construction**. See also **cissoid** of Diocles.

Dürer, Albrecht (1471–1528)

Sane judgment abhors nothing so much as a picture perpetrated with no technical knowledge, although with plenty of care and diligence. Now the sole reason why painters of this sort are not aware of their own error is that they have not learnt Geometry, without which no one can either be or become an absolute artist.

—from *The Art of Measurement*, 1525

A German printmaker who, through applying mathematics to art, brought important ideas to mathematics itself especially in the area of perspective geometry. Dürer was born in Nüremberg, one of 18 children, and showed an early talent for art. After a four-year apprenticeship in painting and woodcutting, he began traveling Europe, especially Italy, in search of new styles and ideas. Back in Nüremberg, he began a serious study of mathematics, absorbing *Elements* by **Euclid** and *De architectura* by the great Roman architect Vitrivius, and studying the work of Leone Alberti (1404–1472) and Luca **Pacioli** on mathematics and art, in particular their work on proportion. His mastery of perspective is clear in woodcuts *Life of a Virgin*

(1502–1505). In about 1508, Dürer began to collect material for a major work on mathematics and its applications to the arts. This work was never finished but Dürer did use parts of the material in later published work. One of his most famous engravings *Melancholia*, produced in 1514, contains the first **magic square** to be seen in Europe, cleverly including the date 1514 as two entries in the middle of the bottom row. Also of mathematical interest in *Melancholia* is the **polyhedron** in the picture, the faces of which appear to consist of two equilateral triangles and six somewhat irregular pentagons. In 1825 Dürer published a four-volume treatise, *Underweysung der Messung* (available in English translation as *Painter's Manual*), which dealt with, among other things, the construction of various curves, polygons, and other solid bodies. One of the first books to teach the methods of perspective, it was highly regarded throughout the sixteenth century and presents the earliest known examples of polyhedral **net**s, that is, polyhedra unfolded to lie flat for printing.

Dürer traveled to Italy to learn about perspective and was keen to publish the methods so they weren't kept secret among a few artists. Who he learned from is not known, but Luca Pacioli is a likely possibility. Some of

Dürer, Albrecht The famous woodcut *Melancholia,* by Albrecht Dürer, features a magic square and an unusual polyhedron.

the techniques and illustrations also follow closely the work of Piero della Francesca.

Dürer's final work, his *Treatise on Proportion,* was published posthumously and laid the groundwork for descriptive geometry and its rigorous mathematical treatment by Gaspard **Monge**.

Dürer's shell curve

Given a **parabola** and a line that is **tangent** to the parabola, the **glissette** of a point on a line sliding between the parabola and the tangent. It has the equation $(x^2 + xy + ax - b^2) = (b^2 - x^2)(x - y + a)^2$.

dynamical system

A **nonlinear**, interactive system that evolves over time, showing transformations of behavior and an increase in **complexity**. Key to this evolution is the presence and emergence of **attractor**s, most notably **chaotic attractor**s.

The changes in the system's organization and behavior are known as **bifurcation**s. Dynamical systems are **deterministic system**s, although they can be influenced by random events. Times series data of dynamical systems can be graphed as phase portraits in **phase space** in order to indicate the qualitative or topological properties of the system and its attractors. For example, various physiological systems, such as the heart, can be conceptualized as dynamical systems. Seeing physiological systems as dynamical systems opens up the possibility of studying various attractor regimes. Moreover, certain diseases can be understood now as "dynamical diseases," meaning that their temporal phasing can be a key to understanding pathological conditions.

dynamics

Pertaining to the change in behavior of a system over time.

e

Pi goes on and on and on . . .
And e is just as cursed.
I wonder: Which is larger
When their digits are reversed?

—Martin Gardner

Possibly the most important number in mathematics. Although **pi** is more familiar to the layperson, *e* is far more significant and ubiquitous in the higher reaches of the subject. One way to think of *e* is as the number of dollars you would have in the bank at the end of a year if you invested $1 at the start of the year and the bank paid an annual interest rate of 100% compounded *continuously*. Compound interest doesn't behave in quite the way intuition suggests. Because more frequent compounding causes the principal to grow faster, it might seem that continuous compounding would make the investor very rich in short order. But the effect tails off. At the end of one year, the $1 would have grown to a mere $2.72, rounded to the nearest cent. To a better approximation, *e* is 2.718281828459045 . . . , its decimal expansion stretching out forever, never repeating in any permanent pattern, because *e* is a **transcendental number**. It is the base of natural **logarithms**, which is equivalent to the fact that the area under the curve (the integral of) $y = 1/x$ between $x = 1$ and $x = e$ is exactly equal to one unit. It also features in the **exponential** function $y = e^x$, which is unique in that its value (y) is exactly equal to its growth rate (dy/dx in calculus notation) at every point. As well as showing up in problems involving growth or decay (including compound interest) or in calculus, whenever logarithmic or exponential functions are involved, *e* is at the heart of the statistical **bell curve**; the shape of a hanging cable, known as a **catenary**; the study of the distribution of **prime numbers**; and Stirling's formula for approximating **factorial**s.

Like π, *e* pops up as the limit of many **continued fractions** and **infinite series**. Leonhard **Euler**, who was the first to study and to use the symbol *e* (in 1727), found it could be expressed as the curious fraction:

$$1 + \cfrac{1}{0 + \cfrac{1}{1 + \cfrac{1}{1 + \cfrac{1}{2 + \cfrac{1}{1 + \cfrac{1}{1 + \cfrac{1}{4 + \cfrac{1}{1 + \cfrac{1}{1 + \cfrac{1}{6 + \dots}}}}}}}}}}$$

No less remarkable is this infinite series of which *e* is the sum:

$$1 + 1/1! + 1/2! + 1/3! + 1/4! + \dots$$

But of all the places that *e* appears in mathematics none is more extraordinary than **Euler's formula** from which comes the most profound relationship in mathematics: $e^{i\pi} + 1 = 0$, linking *e* and π with **complex numbers**.[215]

Earthshapes

A series of 12 hypothetical Earths as conceived by American airman Joseph Portney in 1968 during a flight to the North Pole onboard a U.S. Air Force KC-135. As the North Pole was reached, Portney looked on the icy terrain below and asked himself, "What if the Earth were . . . ?" The hypothetical Earths, cylindrical, conic, donut-shaped, and so forth, were sketched and captioned by Portney and given to the Litton Guidance & Control Systems graphic arts group to create models. These models were then photographed and became the theme of a Litton publication entitled *Pilots and*

Navigators Calendar for 1969. Each month was introduced with a different one of the 12 hypothetical Earths. The result was an international sensation, attracting awards and heavy fan mail.

eccentricity
See **conic section**.

economical number
A number that has no more digits than there are digits in its prime factorization (including powers). If a number has fewer digits than are in its prime factorization it is known as a *frugal number.* The smallest frugal is 125, which has three digits, but can be written as 5^3, which has only two. The next few frugals are 128 (2^7), 243 (3^5), 256 (2^8), 343 (7^3), 512 (2^9), 625 (5^4), and 729 (3^6). An *equidigital number* is an economical number that has the same number of digits as make up its prime factorization. The smallest equidigitals are 1, 2, 3, 5, 7, and 10 ($= 2 \times 5$). All **prime number**s are equidigital. An *extravagant number* is one that has fewer digits than are in its prime factorization. The smallest extravagant number is 4 ($= 2^2$), followed by 6, 8, and 9. There are infinitely many of each of these kinds of numbers. Are there also arbitrarily long sequences of consecutive ones? Seven-member strings of consecutive economical numbers start at each of 157; 108,749; 109,997; 121,981; and 143,421. On the other hand, the longest string of consecutive frugal numbers up to 1 million is just two (for example, 4374 and 4375). Even so, it has been proved that if a certain conjecture about prime numbers known as *Dickson's conjecture* is true, then there are arbitrarily long strings of frugals.

Eddington number
"I believe there are 15,747,724,136,275,002,577,605,653, 961,181,555,468,044,717,914,527,116,709,366,231,425, 076,185,631,031,296 protons in the universe and the same number of electrons." So wrote the English astrophysicist Sir Arthur Eddington (1882–1944) in his book *Mathematical Theory of Relativity* (1923). Eddington arrived at this outrageous conclusion after a series of convoluted (and wrong!) calculations in which he first "proved" that the value of the so-called fine-structure constant was exactly $1/136$. This value appears as a factor in his prescription for the number of particles (protons + electrons; neutrons were not discovered until 1930) in the universe: $2 \times 136 \times 2^{256} = 17 \times 2^{260} \cong 3.149544 \ldots \times 10^{79}$ (double the number written out in full in the quote above). This is the Eddington number, notable for being the largest *specific* integer (as opposed to an estimate or

approximation) ever thought to have a unique and tangible relationship to the physical world. Unfortunately, experimental data gave a slightly lower value for the fine-structure constant, closer to $1/137$. Unfazed, Eddington simply amended his "proof" to show that the value had to be exactly $1/137$, prompting the satirical magazine *Punch* to dub him "Sir Arthur Adding-One." See also **large number**.

edge
A line segment where two faces meet. A **cube**, for example, has 12 edges.

edge coloring theorem
See **Tait's conjecture**.

edge of chaos
The hypothesis that many natural systems tend toward dynamical behavior that borders static patterns and the chaotic regime. See also **chaos**.

egg
Specifically, a chicken's egg and its mathematical equivalent. Eggs are often described as being **oval** in shape, which is effectively tautological since "oval" comes from the Latin *ovus* for "egg." Strictly speaking, an oval is a flat two-dimensional curve, so it is more accurate to say that an egg is shaped like the **surface of revolution** of an oval. In real life, eggs, like ovals, come in a variety of forms all of which can be loosely described as "like an **ellipsoid** but with one end more pointed than the other." Because eggs vary in shape, so too do their mathematical descriptions. Having said this, there are a variety of ways to approximate the shape of a hen's egg by modifying the equation of an ellipsoid, $x^2/a^2 + y^2/b^2 + z^2/c^2 = 1$, so as to

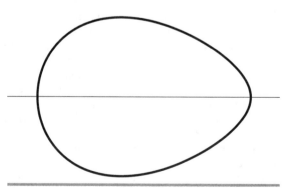

egg A good egg shape is obtained by drawing four circle arcs of different radii. © *Jan Wassenaar, www.2dcurves.com*

introduce an asymmetry about the long (say, z-) axis. These involve multiplying z^2/c^2 by a suitable term, so that y becomes larger on the right side of the y-axis and smaller on the left side. For example, $x^2/a^2 + y^2/b^2 + z^2/c^2$ $(1 - kx) = 1$ gives a good egg. Other useful egg approximations come from surfaces of revolution of **Cartesian ovals**, **Cassinian ovals**, and sections through cones and cylinders. In France, where tennis first became popular, a zero on the scoreboard looked like an egg and was called *l'oeuf*, which is French for "egg." When tennis was introduced in the United States, Americans pronounced it "love."

Why is a hen's, or other bird's, egg shaped as it is? Because it gives strength even though the eggshell is thin enough to allow the young bird to peck its way out when ready. To demonstrate this strength, try balancing a pile of books on four half egg shells. It is even possible for a person's weight to be supported in this way.

Another trick with eggs is to distinguish between a raw egg and a hardboiled one without cracking them open to see which is which. Lay both eggs on their sides on a table, and spin them as you would a spinning top. With a bit of practice, the cooked egg will be made to rise up for a few seconds, while the raw one will remain on its side. The physics of this odd behavior was finally cracked by two mathematicians, Keith Moffat of Cambridge University and Yutaka Shimomura of Keio University, who reported their findings in 2002. They concluded that friction between the egg and the surface produces a gyroscopic effect, which causes some of the kinetic energy of the object to be translated into potential energy, raising its center of gravity (see also **Tippee Top**). As the hardboiled egg spins, its curved surface causes it to touch the tabletop at only one point. The contact point changes and traces out a little circle. If the texture of the tabletop is just right (neither too slippery nor too sticky) the egg will slide a bit as it spins. This sliding slows the spin a bit and causes a wobble. This in turn tilts the egg, lifting one end off the table more than the other, at which point the gyroscopic effect kicks in and swaps some of the kinetic energy of the spinning egg into potential energy and raises its center of gravity in a seemingly paradoxical way. This effect is heightened by the fact that as the end of the egg rises, the egg draws in closer to the axis of spin, causing it to spin more quickly—just as figure-skaters can make themselves pirouette faster by raising their arms above their heads. Why doesn't the effect occur with a raw egg? Because the inside of the egg is runny and it lags behind the shell. This lag serves as a drag, which reduces the spin rate and dissipates the egg's kinetic energy. This in turn reduces the friction between the egg and tabletop, and

means that not enough energy is available to be turned into potential energy to raise the egg's center of gravity. As well as solving the mystery of the balancing egg, Moffat also found time to write a limerick to commemorate the event:

Place a hard-boiled egg on a table,
And spin it as fast as you're able;
It will stand on one end
With vectorial blend
Of precession and spin that's quite stable.

See also **superegg**.

Egyptian fraction
A unit **fraction**; in other words, a fraction in which the numerator (the number on top) is one. This type of fraction was the only kind used by the ancient Egyptians and appears extensively in the **Rhind papyrus**. Other fractions can be obtained by adding Egyptian fractions together; for example, $5/7 = \frac{1}{2} + \frac{1}{6} + \frac{1}{21}$. In 1201 **Fibonacci** proved that every **rational number** can be written as a sum of Egyptian fractions.

eigenvalue
A **complex number**, λ, that satisfies the equation $Ax = \lambda x$, where A is an $n \times n$ **matrix** and x is some **vector**. In this case, x is called an *eigenvector*.

eight
The second smallest cube number (after 1^3): $8 = 2^3 = 2 \times 2 \times 2$. A queen or king in **chess** can move in eight different directions, in the same way that a compass has eight principal points: north, northeast, east, southeast, south, southwest, west, and northwest. In three dimensions, there are eight diagonal ways to move, corresponding to the eight *octants* into which three-dimensional space is divided by three mutually perpendicular planes. Add a **fourth dimension** and movement becomes possible back and forth along four directions at right angles to each other: up and down, left and right, forward and back, and to one other! The Spanish dollar was a gold coin with a value of eight reales, and was sometimes actually cut into eight wedge-shaped pieces—"pieces of eight"—to make change.

eight curve
A curve, also known as the *lemniscate of Gerono,* that has the Cartesian equation

$$x^4 = a^2(x^2 - y^2)$$

and the appearance of a figure eight lying on its side.

Einstein problem

(1) Is there a single shape that will tile a plane aperiodically (see **aperiodic tiling**)? An answer of "no" would imply the existence of a decision method for the **domino problem**. This problem, which is named because of the German translation (*ein* = "one," *stein* = "stone") and was not an invention of the famous scientist, remains unsolved. (2) A logic problem, invented by Albert Einstein, who claimed that 98% of the people in the world couldn't solve it.

PUZZLES

1. There are 5 houses (along a street) in 5 different colors: blue, green, red, white, and yellow.
2. In each house lives a person of a different nationality: Briton, Dane, German, Norwegian, and Swede.
3. These 5 owners drink a certain beverage: beer, coffee, milk, tea, or water; smoke a certain brand of cigar: Blue Master, Dunhill, Pall Mall, Prince, or blend; and keep a certain type of pet: cat, bird, dog, fish, or horse.
4. No owners have the same pet, smoke the same brand of cigar, or drink the same beverage.
5. The Briton lives in a red house. The Swede keeps dogs as pets. The Dane drinks tea. The green house is on the left of the white house (next to it). The green house owner drinks coffee. The person who smokes Pall Mall rears birds. The owner of the yellow house smokes Dunhill. The man living in the house right in the center drinks milk. The Norwegian lives in the first house. The man who smokes blend lives next to the one who keeps cats. The man who keeps horses lives next to the man who smokes Dunhill. The owner who smokes Blue Master drinks beer. The German smokes Prince. The Norwegian lives next to the blue house. The man who smokes blend has a neighbor who drinks water.

The question is: Who keeps the fish?

Solutions begin on page 369.

---

elementary function

Any real-value algebraic function or transcendental function (trigonometric, hyperbolic, exponential, logarithmic).

eleven

A **palindromic number**, the smallest integer that is not a **Harshad number**, a **prime number** that is a member of a **twin prime** (11 and 13), and the largest integer that is not the sum of two or more distinct primes. There are 11

players on a soccer team and on a cricket team. Strange but true: the youngest pope was 11 years old.

ellipse

A shape that looks like a squashed circle. It is one of the **conic sections** and can be defined as the **locus** of all points in a plane that have the same sum of distances from two given fixed points known as *foci*. If the two foci coincide then the ellipse is a **circle**. The line segment connecting the foci is called the *major axis* of the ellipse; half this is the *semimajor axis, a*. The line passing through the center of the ellipse (the midpoint of the foci) at right angles to the major axis is called the *minor axis*, half of which is the *semiminor axis, b*. An ellipse centered at the origin of an *x-y* coordinate system with its major axis along the *x*-axis is defined by the equation

$$x^2/a^2 + y^2/b^2 = 1.$$

The shape of an ellipse is expressed by a number called the *eccentricity, e*, which is related to *a* and *b* by the formula $b^2 = a^2(1 - e^2)$. The eccentricity is a positive number less than 1, or 0 in the case of a circle. The greater the eccentricity, the larger the ratio of *a* to *b*, and therefore the more elongated the ellipse. The distance between the foci is $2ae$. The area enclosed by an ellipse is πab. The circumference of an ellipse is $4aE(e)$, where the function E is the complete elliptical integral of the second kind.

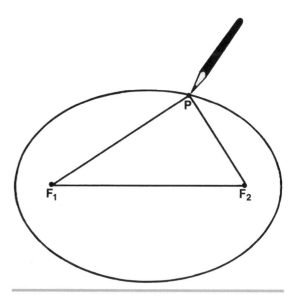

ellipse   Nails mark the foci, F_1 and F_2, of an ellipse that is drawn by pencil whose moving tip, P, keeps the string threaded around the nails taut.

ellipsoid

A **quadratic** surface that is the three-dimensional analog of an **ellipse**. The general equation of an ellipsoid in Cartesian coordinates is

$$x^2/a^2 + y^2/b^2 + z^2/c^2 = 1,$$

where a, b, and c are positive real numbers determining the shape. If two of these numbers are equal, the ellipsoid is a **spheroid**; if all three are equal, it is a **sphere**. The intersection of an ellipsoid with a plane is a single point or an ellipse. Ellipsoids can also be defined in higher dimensions.

elliptic curve

The set of solutions to a type of **cubic equation** whose solutions lie on a **torus** (a donut-shaped surface). The particular type of cubic equation whose solutions lead to elliptic curves takes the form

$$y^2 + axy + by = x^3 + cx^2 + dx + e.$$

Elliptic curves, which are said to have a **genus** of 1, have an unusually rich theory and structure, and their study is linked to many other important areas of mathematics and their applications. For example, it was work done on elliptic curves by Andrew **Wiles** that finally led to a proof of **Fermat's last theorem**.

elliptic function

In **complex analysis**, a **function** defined on the complex plane that is **periodic** in two directions. The elliptic functions can be thought of as analogs of the **trigonometric functions** (which have only a single period). Leading eighteenth-century mathematicians, including Leonhard **Euler** and Joseph **Lagrange**, had studied elliptic integrals, such as the integral that gives the arc length of an ellipse; however, these cannot be expressed in terms of the elementary functions (polynomials, exponentials, and trigonometric functions). It was the insight of Karl **Jacobi**, and also of Karl **Gauss** and Niels **Abel**, that the *inverse* functions of elliptic integrals are much easier to study. They turn out to be doubly periodic functions of a complex variable. While a singly periodic function like sine has a number a (specifically $a = 2\pi$) so that $\sin(x + a) = \sin(x)$, a doubly periodic function f has the property that there are two numbers a, b, not rational multiples of each other, so that $f(x + a) = f(x + b) = f(x)$. As Jacobi proved in 1834, the ratio a/b is necessarily an imaginary number.

elliptical geometry

One of the two most important types of **non-Euclidean geometry**: the other is **hyperbolic geometry**. In ellipti-cal geometry, Euclid's **parallel postulate** is broken because no line is parallel to any other line. The original form of elliptical geometry, known as *spherical geometry* or *Riemannian geometry*, was pioneered by Bernhard **Riemann** and Ludwig **Schläfli** and treats lines as **great circles** on the surface of a sphere. The most familiar example of such circles, which are **geodesics** (shortest routes) on a spherical surface, are the lines of longitude on Earth. In spherical geometry any two great circles always intersect at exactly two points. Two lines of longitude, for example, meet at the North and South Poles. Working in spherical geometry produces some surprising, nonintuitive results. For instance, it turns out that the shortest flying distance from Florida to the Philippine Islands is a path across Alaska—even though the Philippines are at a more southerly latitude than Florida! The reason is that Florida, Alaska, and the Philippines lie on the same great circle and so are collinear in spherical geometry. Another odd property of spherical geometry is that the sum of the angles of a triangle is greater than 180°. This is always the case on a surface that bulges out or, in mathematical parlance, has positive **curvature**. It was Felix **Klein** who first saw clearly how to rid spherical geometry of its one blemish: the fact that two lines have not one but two common points. He redefined the notion of a point as a **set** of antipodal points. With this definition, any two points determine a unique line so that the traditional form of Euclid's first postulate is restored. Thus modified, spherical geometry became what Klein called elliptical geometry.

embedding

Putting one mathematical object inside another, such as a subgroup within a group or one topological space inside another, while preserving all topological properties.

emergence

The arising of new, unexpected structures, patterns, or processes in a self-organizing system (see **self-organization**). These *emergents* have their own rules, laws, and possibilities, and can be understood as existing on a higher level than that of the components from which they came. The term was first used by the nineteenth-century philosopher G. H. Lewes and came into greater currency in the scientific and philosophical movement known as *emergent evolutionism* in the 1920s and 1930s.

emirp

A **prime number** that becomes a different prime number when its digits are reversed ("emirp" is "prime" spelled backward). The first twenty emirps are 13, 17, 31, 37, 71, 73, 79, 97, 107, 113, 149, 157, 167, 179, 199, 311, 337, 347,

359, and 389. Compare with a *palindromic prime* (see **palindromic number**), which gives the same prime when reversed.

empty set

The **set**, denoted by $\varnothing$ or {}, that has no members; also known as the *null set*. This is not the same as **zero**, which is the number of members of $\varnothing$. Nor is $\varnothing$ the same as **nothing** because a set with nothing in it is still a set, and a set is something. The empty set, for example, is the set of all triangles with four sides, the set of all numbers that are bigger than nine but smaller than eight, and the set of all opening moves in chess that involve a king. Applying the concept of the empty set helps distinguish between the different ways that "nothing" is used in everyday language. In his book *What Is the Name of This Book?* (1978), Raymond **Smullyan** wrote:[304] "Which is better, eternal happiness or a ham sandwich? It would appear that eternal happiness is better, but this is really not so! After all, nothing is better than eternal happiness, and a ham sandwich is certainly better than nothing. Therefore a ham sandwich is better than eternal happiness."

What is wrong with this declaration? The first statement is equivalent to "The set of things that are better than eternal happiness is $\varnothing$." The second statement is equivalent to "The set {ham sandwich} is better than the set $\varnothing$." The confusion arises because the first is comparing individual things, while the second is comparing sets of things, and $\varnothing$ plays a different role in each.

enantiomorph

The mirror image of a given **chiral** polyhedron or other figure.

enormous theorem

The largest theorem in mathematics; it concerns the classification of finite **simple group**s and encapsulates the work of hundreds of mathematicians over many years.

entropy

A measure of a system's degree of **random**ness or disorder.

envelope

A curve or a surface that touches every member of a family of lines, curves, planes, or surfaces.

epicycloid

The path traced out by a point on the circumference of a circle of radius b rolling on the outside of a **circle** of radius a. It is described by the parametric equations:

$$x = (a + b) \cos(t) - b \cos((a/b + 1)t)$$
$$y = (a + b) \sin(t) - b \sin((a/b + 1)t).$$

An epicycloid is like a **cycloid** on the circumference of a circle and is closely related to the **epitrochoid**, **hypocycloid**, and **hypotrochoid**. An epicycloid with one cusp is called a **cardioid**, one with two cusps is called a **nephroid**, and one with five cusps is called a **ranunculoid** (after the buttercup genus *Ranunculus*).

Epimenides paradox

See **liar paradox**.

epitrochoid

A curve traced out by a point that is a distance c from the center of a circle of radius b, where $c < b$, that is rolling around the outside of another circle of radius a. It is described by the parametric equations

$$x = (a + b) \cos(t) - c \cos((a/b - 1)t)$$
$$y = (a + b) \sin(t) - c \sin((a/b + 1)t).$$

Closely related to the epitrochoid are the **epicycloid**, **hypocycloid**, and the **hypotrochoid**. An example of an epitrochoid appears in Albrecht **Dürer**'s work *Instruction in Measurement with Compasses and Straightedge* (1525).

EPORN

An equal product of reversible numbers; defined by the Indian recreational mathematician Shyam Sunder Gupta as a number that can be expressed as the product of two reversible numbers (numbers whose digits are reversed) in two different ways. For example: $4,030 = 130 \times 031 = 310 \times 013$ and $144,648 = 861 \times 168 = 492 \times 294$. The smallest EPORN, $2,520 = 120 \times 021 = 210 \times 012$, is also the least common multiple of all single digit natural numbers in decimal system. The **digital root**, i.e. the ultimate sum of digits, of all EPORNs is always 1, 4, 7, or 9. For example, $2,520 = 2 + 5 + 2 + 0 = 9$; $4,030 = 4 + 0 + 3 + 0 = 7$; $9,949,716 = 9 + 9 + 4 + 9 + 7 + 1 + 6 = 36$ and $3 + 6 = 9$.

equichordal point

A point inside a closed **convex** curve in the plane, all the **chord**s through which have the same length.

equilateral

Having sides of equal length, as in the case of an equilateral **polygon**. The equilateral **triangle**, with its three equal angles of 60°, is widely found in historic buildings and structures across Europe. See **Triangular Lodge**.

equilibrium

A term indicating a rest state of a system, for example, when a **dynamical system** is under the sway of a **fixed-point attractor** or **periodic attractor**. The concept originated in ancient Greece when **Archimedes** experimented with levers in balance, literally "equilibrium." The idea was elaborated through the Middle Ages, the Renaissance, and the birth of modern mathematics and physics in the seventeenth and eighteenth centuries. "Equilibrium" has come to mean pretty much the same thing as *stability*, that is, a system that is largely unaffected by internal or external changes since it easily returns to its original condition after being perturbed.

equivalent numbers

Numbers such that the sums of their **aliquot parts** (proper divisors) are the same. For example, 159, 559, and 703 are equivalent numbers because their aliquot parts all sum to 57.

Eratosthenes of Cyrene (c. 276–194 B.C.)

A Greek mathematician, astronomer, and geographer who was born in Cyrene, a Greek colony to the west of Egypt. He studied at Plato's school in Athens and eventually became the chief librarian of the great Library at Alexandria. He wrote works on geography, philosophy, history, astronomy, mathematics, and literary criticism. One of Eratosthenes' contributions to mathematics was his measurement of Earth's circumference, which he calculated to be about 252,000 stadii, or 24,700 miles (about one-tenth the actual value, but still a big improvement on earlier estimates). Eratosthenes is also known in number theory for his **sieve of Eratosthenes**, which finds all **prime numbers** less than a given integer n.

Eratosthenes's sieve

See **sieve of Eratosthenes**.

Erdös, Paul (1913–1996)

A Hungarian mathematician (his name is pronounced "AIR-dosh"), one of the greatest mathematicians of the twentieth century and, in terms of the number of papers published (more than 1,500), the most prolific in history—beating out even Leonhard **Euler** and inspiring the term "Erdös number." A mathematician has an Erdös number of 1 if he or she has published a paper with Erdös, of 2 if he or she has published with someone who published a paper with Erdös, and so on. Erdös worked almost nonstop, 19 hours a day, 7 days a week. "A mathematician," he quipped, "is a machine for turning coffee into theorems." At age 20, Erdös discov-

ered an elegant proof of a famous theorem in number theory, known as *Chebyshev's theorem*, which says that for each number greater than one, there is always at least one **prime number** between it and its double. **Number theory** remained one of his chief interests, though his work spread across many fields, and he became renowned for posing and solving problems that were often simple to state but notoriously difficult to solve. He did groundbreaking work in a branch of mathematics known as **Ramsey theory** long before it became fashionable in the late 1950s. Bent and slight, often wearing sandals, Erdös had no time for the material side of life. "Property is nuisance," he said. Focused totally on mathematics, Erdös traveled from meeting to meeting, carrying a half-empty suitcase and staying with mathematicians wherever he went. His colleagues took care of him, lent him money, fed him, bought him clothes, and even did his taxes. In return, he showered them with ideas and challenges—with problems to be solved and brilliant ways of attacking them. Ernst Straus, who worked with both Albert Einstein and Erdös, wrote a tribute to Erdös shortly before his own death in 1983. He said of Erdös: "In our century, in which mathematics is so strongly dominated by 'theory doctors,' he has remained the prince of problem solvers and the absolute monarch of problem posers."[170]

ergodic

The property of a **dynamical system** such that all regions of a **phase space** are visited with similar frequency and that all regions will be revisited (within a small proximity) if given enough time.

Escher, Maurits Cornelius (1898–1972)

My work is a game, a very serious game.

A Dutch artist whose graphic explorations of **tiling**, figure-ground ambiguities, **impossible figures**, and regression has attracted the interest of mathematicians and scientists. His experiences of Moorish art (see **Alhambra**) and his contact with mathematicians, most notably Harold **Coxeter**, led him to explore the way repetitive shapes can be used to tile the plane, and from this to ideas about duality and transformation. Escher's preoccupation with dualities is a constant presence in his work in the form of foreground/background, light and dark, flatness and dimensionality, representation and decoration, frame and scene, large and small, viewpoint and vanishing point, form and negative space, positive and negative, observer and observed, as well as the metaphysical aspects of good and evil. Self-referential images (see **self-referential sentence**) resonate throughout Escher's works—reflections of the artist, hands that draw themselves, the visitor in a picture gallery who looks at a print that contains him. It was for this reason that Douglas **Hofstadter** wove Escher, along with Kurt Gödel and Bach, into the "eternal golden braid" at the heart of his Pulitzer Prize–winning book.[172] Having seen some of Escher's art, Roger **Penrose** was inspired to devise impossible figures, including the **Penrose triangle**, which Escher then incorporated into several of his later works. After the artist's death Penrose regretted that Escher had not lived long enough to take advantage of the discovery of **Penrose tiling**.

escribed circle

A circle that is **tangent** to one side of a triangle and to the extensions of the other sides.

Eternity Puzzle

An enormously difficult jigsaw consisting of 209 pieces, each one different and each made from a unique configuration of equilateral triangles and half-triangles with the same total area as six triangles. The puzzle was to fit them together into an almost-regular 12-sided figure aligned to a triangular grid. The puzzle's inventor, Christopher Monckton, announced a prize of $1 million when the puzzle was released commercially in June 1999, for the first correct solution submitted, assuming there was one, when all the solutions were opened in September 2000. Monckton had run computer searches on much smaller versions of the puzzle, which had convinced him that the sheer size of Eternity would make it intractable. However, the prize was won by two British mathematicians, Alex Selby and Oliver Riordan, with the help of a couple of computers, who sent in a correct tiling on May 15, six weeks ahead of the only other puzzler known to have found a correct solution. Early on, Selby and Riordan made a surprising discovery. As the number of pieces in an Eternity-like puzzle increased, so did the difficulty—but only up to a point. The critical size is about 70 pieces, which would be almost impossible to solve. For larger puzzles, however, the number of possible correct solutions increases. In the case of Eternity itself, with its 209 pieces, there are thought to be at least 10^{95} solutions—far more than the number of subatomic particles in the universe but far, far less than the number of nonsolutions. The puzzle itself is much too large to solve by an exhaustive search but not, as it turns out, by more savvy methods that take into account what shaped regions are easiest to tile and what shaped pieces are easiest to fit. By steadily refining their search algorithm, Selby and Riordan were able to prune out the vast majority of nonsolutions and, with a bit of

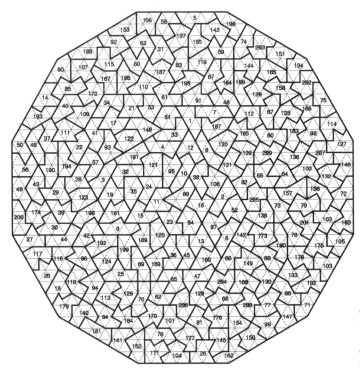

Eternity Puzzle The solution to the Eternity Puzzle that was awarded a $1 million prize. *Eternity pieces are copyright © 1999 by Christopher Monckton*

good fortune, to hit upon a correct solution and claim the prize.

Euclid of Alexandria (c. 330–270 B.C.)

A Greek mathematician who compiled and systematically arranged the geometry and number theory of his day into the famous text *Elements*. This text, used in schools for about 2,000 years, earned him the name "the father of geometry." Even today, the geometries that don't satisfy the fifth of Euclid's "common notions" (now called axioms or postulates) are called non-Euclidean geometries. When, according to the Greek philosopher Proclus, the Egyptian ruler Ptolemy asked if there was a shorter way to the study of geometry than the *Elements,* Euclid told the pharaoh that "there is no royal road to geometry." Little is known of Euclid's life. Proclus wrote (c. A.D. 350) that Euclid lived during the reign of Ptolemy and founded the first school of mathematics in Alexandria–the site of the most impressive library of ancient times with perhaps as many as 700,000 volumes. He wrote books on other subjects such as optics and conic sections, but most of them are now lost. See also **Euclidean geometry.**

Euclidean geometry

Geometry of the type described originally by **Euclid** in his book *Elements* and based on five axioms (see **Euclid's postulates**), one of which is the controversial **parallel postulate.** Various forms of **non-Euclidean geometry** began to emerge in the nineteenth century, with enormous implications for science and philosophy. See also **Euclidean space.**

Euclidean space

Any *n*-dimensional mathematical space that is a generalization of the familiar two- and three-dimensional spaces described by the axioms of **Euclidean geometry.** The term "*n*-dimensional Euclidean space" (where *n* is any positive whole number) is usually abbreviated to "Euclidean *n*-space", or even just "*n*-space". Formally, Euclidean *n*-space is the set R^n (where R is the set of **real numbers**) together with the *distance function,* which is obtained by defining the distance between two points $(x_1, \ldots, x_n)$ and $(y_1, \ldots, y_n)$ to be the square root of $\Sigma(x_i - y_i)^2$, where the sum is over $i = 1, \ldots, n$. This distance function is based on **Pythagoras's theorem** and is called the *Euclidean metric.*

Euclid's postulates

The five postulates, which together with 23 definitions and five "common notions," form the basis of **Euclid's** great work on geometry, *Elements.* The postulates are:

1. A straight line may be drawn from any one point to any other point.

2. A finite straight line may be produced to any length in a straight line.

3. A circle may be described with any center at any distance from that center.

4. All right angles are equal.

5. If a straight line meets two other lines, so as to make the two interior angles on one side of it together less than two right angles, the other straight lines will meet if produced on that side on which the angles are less than two right angles.

The last postulate is not as obvious as the other four, and Euclid himself was reluctant to use it. Later mathematicians, finding the fifth postulate to be complicated, thought it might be possible to derive it from the other four. However, they only succeeded in replacing it with equivalent statements. The most common of these is the **parallel postulate.**

Eudoxus of Cnidus (c. 408–c. 355 B.C.)

A Greek astronomer, mathematician, and physician whose work on ratios formed the basis for Book V of **Euclid's** *Elements* and anticipated some aspects of algebra, such as cross multiplying, which is otherwise absent from ancient Greek mathematics. Eudoxus constructed many geometric proofs, found formulas for measuring pyramids, cones, and cylinders, and developed the **method of exhaustion,** a forerunner of integration, later extended by Archimedes. He also studied the kampyle curve, often known as the **kampyle of Eudoxus,** in connection with the classical problem of **duplicating the cube.**

Euler, Leonhard (1707–1783)

A great Swiss mathematician; the second most prolific mathematician in history, after Paul **Erdös.** His greatest contributions were to **number theory,** but Euler also did important work in **calculus,** geometry, algebra, probability, acoustics, optics, mechanics, astronomy, artillery, navigation, and finance. He had a knack for coming up with important results by intuition, he cast calculus and trigonometry in their modern forms, and he showed the importance of the number *e.* Even the amusing puzzles he invented and, in some cases, solved have opened up new mathematical fields. The **bridges of Königsberg** problem, for example, heralded the beginning of **graph** theory and **topology,** while his **thirty-six officers problem** stimulated important work in **combinatorics.** Euler also worked on **magic squares** and the problem of the **knight's tour.** Having learned some math from his father, a Calvinist preacher, Euler studied at the University of

Basle where he became close friends with members of the **Bernoulli family**. In 1727, he moved to St. Petersburg, to the court of Catherine the Great, becoming professor of physics (1730) and of mathematics (1733). While in Russia, Euler, a devout Christian, met the encyclopedist and philosopher René Diderot, a notorious atheist. When Diderot heard that Euler had a mathematical proof of the existence of **God**, he asked for it and was quoted the equation now often referred to as **Euler's formula**.

Upon losing the use of his right eye, Euler said "Now I will have less distraction." Indeed, the quantity of his output seemed to be inversely proportional to the quality of his sight, because his rate of publication increased after he became almost totally blind in 1766. Euler died moments after calculating the orbit of Uranus on September 18, 1783.

Euler characteristic

An important kind of number, known as a *topological invariant,* that describes a closed surface. In the case of polyhedra, the Euler characteristic is the number of vertices and faces minus the number of edges (see **Euler's formula for polyhedra**).

Euler circuit

A **connected graph** such that starting at a **vertex** *a*, one can traverse every **edge** of the graph once to each of the other vertices and return to vertex *a*. In other words a Euler circuit is a **Euler path** that is a circuit. Thus, using the properties of odd and even degree vertices given in the definition of a Euler path, a Euler circuit exists if and only if every vertex of the graph has an even degree. See also **mazes**.

Euler line

A line that connects the **centroid** and the **circumcenter** of a triangle.

Euler path

A path along a **connected graph** that connects all the vertices (see **vertex**) and that traverses every **edge** of the graph only once. Note that a vertex with an odd degree allows one to travel through it and return by another path at least once, while a vertex with an even degree only allows a number of traversals through, but one cannot end a Euler path at a vertex with even degree. Thus, a connected graph has a Euler path which is a circuit (a **Euler circuit**) if all of its vertices have even degree. A connected graph has a Euler path which is non-circuitous if it has exactly two vertices with odd degree. See also **Hamilton path**.

Euler square

A square array made by combining *n* objects of two types such that the first and second elements form a **Latin square**. Euler squares are also known as *Graeco-Latin squares, Graeco-Roman squares,* or *Latin-Graeco squares.* For many years, Euler squares were known to exist for *n* = 3, 4, and for every odd *n* except *n* = 3*k*. *Euler's Graeco-Roman squares conjecture* maintained that there are no Euler squares of order *n* = 4*k* + 2 for *k* = 1, 2, ... However, such squares were found to exist in 1959 by Bose and Shrikande, refuting the conjecture.

Euler-Mascheroni constant (γ)

Also known as *Euler's constant* or *Mascheroni's constant,* the limit (as *n* goes to infinity) of

$$1 + \tfrac{1}{2} + \tfrac{1}{3} + \tfrac{1}{4} + \tfrac{1}{5} + \ldots + 1/n - \log n$$

It is often denoted by a lowercase gamma, γ, and is approximately 0.5772156649 . . . . Even though over one million digits of this number have been calculated, it isn't yet known if it is a rational number (the ratio of two integers *a/b*). If it is rational, the denominator (*b*) must have more than 244,663 digits. The constant γ crops up in many places in **number theory**. For example, in 1898, the French mathematician Charles de la Vallée Poussin (who proved the **prime number** theorem) proved the following: Take any positive integer *n* and divide it by each positive integer *m* less than *n*. Calculate the average (mean) fraction by which the quotient *n/m* falls short of the next integer. The larger *n* gets, the closer the average gets to gamma.

Euler's conjecture

It always takes *n* terms to sum to an *n*th power: two squares, three cubes, four fourth powers, and so on. This hypothesis is now known to be wrong. In 1966, L. J. Lander and T. R. Parkin found the first counterexample: four fifth powers that sum to a fifth power. They showed that $27^5 + 84^5 + 110^5 + 133^5 = 144^5$. In 1988, Noam Elkies of Harvard University found a counterexample for fourth powers: $2{,}682{,}440^4 + 15{,}365{,}639^4 + 187{,}960^4 = 20{,}615{,}673^4$. Subsequently, Roger Frye of Thinking Machines Corporation did a computer search to find the smallest example: $95{,}800^4 + 217{,}519^4 + 414{,}560^4 = 422{,}481^4$.

Euler's constant

See **Euler-Mascheroni constant**.

Euler's formula

For any **real number** *x*, Euler's formula is

$$e^{ix} = \cos x + i \sin x$$

where e is a fundamental constant (the base of natural logarithms) and $i = \sqrt{-1}$. If we now put $x = \pi$, we get

$$e^{i\pi} = \cos \pi + i \sin \pi,$$

and since $\cos(\pi) = -1$ and $\sin(\pi) = 0$, this reduces to

$$e^{i\pi} = -1$$

so that

$$e^{i\pi} + 1 = 0.$$

This most extraordinary equation first emerged in Leonhard **Euler**'s *Introductio,* published in 1748. It is remarkable because it links the most important mathematical constants, e and **π**, the imaginary unit i, and the basic numbers used in counting, 0 and 1. In describing the equation to students, the Harvard mathematician Benjamin Peirce said: "Gentlemen, that is surely true, it is absolutely paradoxical; we cannot understand it, and we don't know what it means, but we have proved it, and therefore, we know it must be the truth."

Euler's formula for polyhedra

The earliest known equation in **topology**. If F is the number of faces of a **polygon**, E the number of edges, and V the number of vertices, Euler's formula can be written as

$$F - E + V = 2$$

where F – E + V is known as the **Euler characteristic**. For example, the surface of a cube has six (square) faces, twelve edges, and eight vertices and, sure enough, $6 - 12 + 8 = 2$.

even function

A function $f(x)$ such that $f(x) = f(-x)$ for all x.

evolute

The **locus** of the centers of curvature (the envelope) of a plane curve's normals. The original curve is then said to be the **involute** of its evolute. For example, the evolute of an **ellipse** is a **Lamé curve** and the evolute of a **tractrix** is a **catenary**.

excluded middle law

A law in (two-valued) logic which states that there is no third alternative to truth or falsehood. In other words, for any statement A, either A or not-A must be true and the other must be false. This law no longer holds in three-valued logic, in which "undecided" is a valid state, nor does it hold in **fuzzy logic**.

existence

A term that has several different meanings within **mathematics**. In the broadest sense there is the question of what it means for certain concepts, such as **pi**, to exist. Was π, for example, invented or discovered? In other words, does π exist only as an intellectual construct or was it somehow already "out there" waiting for people to find it. If it does exist independently of the human mind, when did its existence start? Does π predate the physical universe? Such ontological questions become even more difficult when applied to more complex or abstract mathematical concepts such as the **Mandelbrot set**, **surreal numbers**, or **infinity**. A narrower and more technical type of "existence" in math is implied by an *existence theorem*. Such a theorem is used to prove that a number or other object with particular properties definitely exists, but does not necessarily give a specific example. Finally, there is existence in the sense of particular solutions to problems. If at least one solution can be determined for a given problem, a solution to that problem is said to exist. Something of the flavor of all three types of mathematical existence mentioned here are captured in the following anecdote:

An engineer, a chemist, and a mathematician are staying in three adjoining cabins at an old motel. First the engineer's coffee-maker catches fire. He smells the smoke, wakes up, unplugs the coffee maker, throws it out the window, and goes back to sleep. Later that night the chemist smells smoke, too. He wakes up and sees that a cigarette butt has set the trash can on fire. He thinks to himself, "How does one put out a fire? One can reduce the temperature of the fuel below the flash point, isolate the burning material from oxygen, or both. This could be accomplished by applying water." So he picks up the trash can, puts it in the shower stall, turns on the water, and, when the fire is out, goes back to sleep. The mathematician has been watching all this out the window. So later, when he finds that his pipe ashes have set the bed sheet on fire, he's not in the least taken aback. "Aha!" he says, "A solution exists!" and goes back to sleep.

exponent

A number that gives the **power** to which a **base** is raised. For example, in 3^2 the base is 3 and the exponent is 2.

exponential

Who has not been amazed to learn that the function
y = e^x, like a phoenix rising again from its own
ashes, is its own derivative?

—Francois le Lionnais

Anything that grows at a rate proportional to its size is said to grow exponentially. The simplest form of the exponential function is just $y = e^x$, where e is about 2.712 . . . The exponential function to **base** a can be written as $f(x) = a^x$.

extrapolate

See **interpolate**.

extravagant number

See **economical number**.

F

face

A **polygon** bounding a **polyhedron**. A **cube**, for example, has six square faces. The plane angle formed by adjacent edges of a polygonal angle in space is called a *face angle*.

factor

Also known as a *divisor,* a number or variable that divides evenly into another number or algebraic expression. For example, the factors of 28 are 1, 2, 4, 7, 14, and 28. Although it is true that 28 is also divisible by the negative of each of these, "factors" is usually taken to mean only the positive divisors. *Factorization,* or *factoring,* is the decomposition of an object into a product of factors. For example, the number 15 factorizes into **prime numbers** as 3×5; and the **polynomial** $x^2 - 4$ factorizes as $(x - 2)(x + 2)$. The aim of factoring is usually to reduce something to basic building blocks, such as numbers to prime numbers, or polynomials to linear expressions. Factoring integers is covered by the **fundamental theorem of arithmetic** and factoring polynomials by the **fundamental theorem of algebra**. *Integer factorization* for large integers appears to be a difficult problem; there are no known methods for solving it quickly, and, for this reason, it has formed the basis of some public key cryptography algorithms.

factorial

The function, denoted $n!,$ that is the product of the positive integers less than or equal to n. For example, $1! = 1$; $5! = 5 \times 4 \times 3 \times 2 \times 1 = 120$; $10! = 10 \times 9 \times 8 \times 7 \times 6 \times 5 \times 4 \times 3 \times 2 \times 1 = 3,628,800$. $0!$ is defined to be 1, by working the relationship $n! = n \times (n - 1)!$ backward. An interesting equality is $1!\ 10!\ 22!\ 1! = 11!\ 0!\ 2!\ 21!$ in which the same digits are broken up two different ways into factorials. This may be the smallest such example. Factorials are important in **combinatorics** because there are $n!$ different ways (permutations) of arranging n distinct objects in a sequence. They also turn up in formulas in calculus, for instance in *Taylor's theorem,* because the nth derivative of the function x^n is $n!$.

factorion

A natural number that equals of the sums of the **factorials** of its digits in a given **base**. The only known **decimal** factorions are $1 = 1!, 2 = 2!, 145 = 1! + 4! + 5!,$ and $40,585 = 4! + 0! + 5! + 8! + 5!$.

Fadiman, Clifton (1904–1999)

An American essayist, literary critic, and noted intellectual who, among many other works, edited *Fantasia Mathematica*[96] and *The Mathematical Magpie*.[97] He became well known for the encyclopedic knowledge he displayed on the *Information Please* radio programs in the 1930s and '40s.

Fagnano's problem

In a given acute triangle $ABC,$ find the inscribed triangle whose perimeter is as small as possible. The answer is the *orthic triangle* of $ABC,$ that is, the triangle whose vertices are endpoints of the altitudes from each of the vertices of ABC. The problem was proposed and solved using calculus by Giovanni Fagnano (1715–1797) in 1775. Once the answer became known, several purely geometric solutions were also discovered.

fair division

See **cake-cutting**.

Farey sequence

A sequence of numbers named after the English geologist John Farey (1766–1826) who wrote about such sequences in an article called "On a curious property of vulgar fractions" in the *Philosophical Magazine* in 1816. Farey says that he noted the "curious property" while examining the tables of *Complete decimal quotients* produced by Henry Goodwin. To obtain the Farey sequence for a fixed number n, consider all **rational numbers** between 0 and 1 which, when expressed in their lowest terms, have a denominator (the number on the bottom of a fraction) not exceeding n. Write the sequence in ascending order of magnitude beginning with the smallest. The "curious property" is that each member of the sequence is equal to the rational number whose numerator (the number on top of a fraction) is the sum of the numerators of the fractions on either side, and whose denominator is the sum of the denominators of the fractions on either side. For example, the Farey sequence for $n = 5$ is ($^0\!/_1$, $^1\!/_5$, $^1\!/_4$, $^1\!/_3$, $^2\!/_5$, $^1\!/_2$, $^3\!/_5$, $^2\!/_3$, $^3\!/_4$, $^4\!/_5$, $^1\!/_1$), from which it can be seen that $^2\!/_5 = (1 + 1)/(3 + 2)$, $^1\!/_3 = (1 + 2)/(4 + 5)$, $^1\!/_2 = (2 + 3)/(5 + 5)$, $^2\!/_3 = (3 + 3)/(5 + 4)$, and so forth. Farey wrote: "I am not acquainted whether this curious property of vulgar fractions has been before pointed out?; or whether it may admit of some easy or general demonstration?; which are points on which I should be glad to learn the sentiments of some of your mathematical readers."

One "mathematical reader" was Augustin **Cauchy**, who gave the necessary proof in his *Exercises de mathématique,* published in the same year as Farey's article. Farey was not the first to notice the property. C. Haros, in 1802, wrote a paper on the approximation of decimal fractions by common fractions. He explains how to construct what is in fact the Farey sequence for $n = 99$ and Farey's "curious property" is built into his construction.

Fechner, Gustav Theodor (1801–1887)

A German physicist and psychologist who studied aesthetic aspects of the **golden ratio** and published his findings in *Vorschule der Aesthetik* (Introduction to aesthetics) (1876), arguing that this ratio turns up commonly in human-made rectangular objects and is judged by people to be the most pleasing to the eye (though some later researchers have called his results into question).

Federov, E. S. (1853–1919)

A Russian geologist and crystallographer who helped lay the theoretical foundations for modern crystallography. In his famous two-part paper "Symmetry of Regular Systems of Figures" published in 1891, he proved that there are exactly 17 distinct symmetries in the **wallpaper group**.

feedback

The mutually **reciprocal** effect of one system or subsystem on another. *Negative feedback* is when two subsystems act to dampen the output of the other. For example, the relation of predators and prey can be described by a negative feedback loop since more predators lead to a decline in the population of prey, but when prey decrease too much so does the population of predators since they don't have enough food. *Positive feedback* means that two subsystems are amplifying each other's outputs, e.g., the screech heard in a public address system when the mike is too close to the speaker. The microphone amplifies the sound from the speaker which in turn amplifies the signal from the microphone, and so on. Feedback is a way of talking about the nonlinear interaction among the elements or components in a system and can be modeled by nonlinear differential or difference equations as well as by the activity of cells in a **cellular automaton** array.

Feigenbaum's constant

A universal constant, denoted δ, that governs the behavior of systems that are approaching **chaos**; it was discovered by the American mathematical physicist Mitchell Feigenbaum (1944–) in 1975 and has the value δ =

4.6692 . . . . All one-dimensional chaotic systems have a behavior, as they approach instability, known as period doubling. The Feigenbaum constant gives the rate at which the period of the system doubles.

Fermat, Pierre de (1601–1665)

A French lawyer, magistrate, and gentleman scholar, often called the "Prince of Amateurs," who is best known for the **conjecture**, now proved, known as **Fermat's last theorem**. Although employed as a senior government official, Fermat somehow managed to find time to do an astonishing amount of math, for which he sought little acclaim or acknowledgment. In fact, he published only one important manuscript in his entire lifetime and even then used fake initials. When his fellow French mathematician Gilles Roberval offered to edit and publish some of his works, Fermat replied, "Whatever of my works is judged worthy of publication, I do not want my name to appear there." Most of his results are known through letters to friends, notes in book margins, and challenges to other mathematicians to find proofs for theorems he had devised.

Fermat was one of the founders, with René **Decartes**, of **analytical geometry** and, with Blaise **Pascal**, of **probability theory**. His work on the maxima and minima of curves and tangents to them was seen, by Isaac **Newton**, as a starting point for **calculus**. Yet his greatest love was for **number theory**. In 1640, while studying **perfect numbers**, Fermat wrote to Mersenne that if p is prime, then $2p$ divides $2^p - 2$. Shortly after he expanded this into what is now called **Fermat's little theorem**. As usual, Fermat stated "I would send you a proof, if I did not fear its being too long." His most famous statement of this form accompanied his hasty notes on the "last theorem." See also **Fermat number**.

Fermat number

A number defined by the formula $F_n = 2^{2^n} + 1$ and named after Pierre **Fermat** who conjectured, wrongly, that all such numbers would be **prime**. The first five Fermat numbers, $F_0 = 3$, $F_1 = 5$, $F_2 = 17$, $F_3 = 257$, and $F_4 = 65{,}537$, are prime. However, in 1732, Leonhard **Euler** discovered that 641 divides F_5. It takes only two trial divisions to find this factor because Euler showed that every factor of a Fermat number F_n with n greater than 2 has the form $k \times 2^{n+2} + 1$. In the case of F_5 this is $128k + 1$, so we would try 257 and 641 (129, 385, and 513 are not prime). It is likely that there are only finitely many Fermat primes. Gauss proved that a **regular polygon** of n sides can be inscribed in a circle with Euclidean methods (e.g., by straightedge and compass) if and only if n is a power of two times a product of distinct Fermat primes.

Fermat, Pierre de

Fermat's last theorem

A challenge for many long ages
Had baffled the savants and sages.
Yet at last came the light:
Seems old Fermat was right—
To the margin add 200 pages.

—Paul Chernoff

A conjecture put forward by Pierre de **Fermat** in 1637 in the form of a note scribbled in the margin of his copy of the ancient Greek text *Arithmetica* by **Diophantus**. The note was found after his death, and the original is now lost. However, a copy was included in the appendix to a book published by Fermat's son. Fermat's note read: "It is impossible to write a cube as a sum of two cubes, a fourth power as a sum of fourth powers, and, in general, any power beyond the second as a sum of two similar powers. For this, I have found a truly wonderful proof, but the margin is too small to contain it."

Fermat claimed that the **Diophantine equation** $x^n + y^n = z^n$ has no integer solutions for $n > 2$. It turns out he was right. But the proof had to wait 350 years and involved such advanced techniques, virtually none of which existed in the seventeenth century, that is seems very unlikely that Fermat really had found an elementary proof. Fermat's last theorem—now truly a theorem—was finally proved correct by Andrew **Wiles** in 1994.[353] In order to reach that dizzy height, however, Wiles had to draw on and extend several ideas at the core of modern mathematics. In particular, he tackled the *Shimura-Taniyama-Weil conjecture,* which provides links between the branches of mathematics known as **algebraic geometry** and **complex analysis**. This conjecture dates back to 1955, when it was published in Japanese as a research problem by the late Yutaka Taniyama. Goro Shimura of Princeton and Andre Weil of the Institute for Advanced Study provided key insights in formulating the conjecture, which proposes a special kind of equivalence between the mathematics of objects called elliptic curves and the mathematics of certain motions in space. Interestingly, the Wiles proof of Fermat's last theorem was a byproduct of his deep inroads into proving the Shimura-Taniyama-Weil conjecture. Now, the Wiles effort could help point the way to a general theory of three variable Diophantine equations. Historically, mathematicians have always had to state and solve such problems on a case-by-case basis. An overarching theory would represent a tremendous advance. See also **ABC conjecture**.

Fermat's little theorem

If P is a **prime number** then for any number a, $(a^P - a)$ must be divisible by P. This theorem is useful for testing if a number is *not* prime, though it can't tell if a number is prime. As usual, Pierre de **Fermat** didn't provide a proof (this time saying "I would send you the demonstration, if I did not fear its being too long"). Leonhard **Euler** first published a proof in 1736, but Gottfried **Leibniz** left virtually the same proof in an unpublished manuscript from sometime before 1683.

Fermat's spiral

A parabolic **spiral**.

Fibonacci (c. 1175–1250)

The pen name of Leonardo of Pisa, one of the greatest mathematicians of the Middle Ages. The son of a Pisan merchant who also served as a customs officer in North Africa, he traveled widely in Barbary (Algeria) and was later sent on business trips to Egypt, Syria, Greece, Sicily, and Provence. In 1200 he returned to Pisa and used the knowledge gained on his travels to write *Liber Abaci* (The book of the abacus), published in 1202, which introduced to western Europe the Hindu-Arabic numerals and **decimal** number system that remain in use today. The first chapter of Part 1 begins: "These are the nine figures of the Indians: 9 8 7 6 5 4 3 2 1. With these nine figures, and with this sign *0* which in Arabic is called zephirum, any number can be written, as will be demonstrated."

Fibonacci also showed he was capable of some amazing feats of calculation. For example, he found the positive solution of the **cubic** equation $x^3 + 2x^2 + 10x = 20$ using the Babylonian number system with base 60 (a strange choice, in view of his public advocacy of the decimal system!). He gave the result as 1, 22, 7, 42, 33, 4, 40 which is equivalent to

$$1 + \frac{22}{60} + \frac{7}{60^2} + \frac{42}{60^3} + \frac{33}{60^4} + \frac{4}{60^5} + \frac{40}{60^6}.$$

How on Earth he obtained this, nobody knows; it was 300 years before anyone else could obtain such accurate results. As well as serious mathematics, *Liber Abaci* contains many playful passages and it is for one of these, concerning a problem about counting the offspring of a pair of rabbits, that Fibonacci became best known after Edouard **Lucas** called the sequence of numbers discussed by the rabbit problem the **Fibonacci sequence**.

Fibonacci sequence

The sequence that arises in answer to this problem posed in **Fibonacci**'s great work *Liber Abaci:* "A certain man put a pair of rabbits in a place surrounded on all sides by a wall. How many pairs of rabbits can be produced from that pair in a year if it is supposed that every month each pair begins a new pair which from the second month on becomes productive?"

The number of pairs of rabbits in the nth month begins 1, 1, 2, 3, 5, 8, 13, 21, 34, 55, 89, . . . , where each term is the sum of the two terms preceding it. This sequence can be defined recursively as follows: $F(1) = F(2) = 1$, $F(n + 1) = F(n) + F(n - 1)$ for $n > 2$, where $F(n)$ is the nth Fibonacci number. Johannes **Kepler** was the first to point out that the growth rate of the Fibonacci numbers, that is, $F(n + 1) / F(n)$, converges to the **golden ratio**, ϕ (phi).

In the nineteenth century Fibonacci numbers were discovered in many natural forms. For example, many types of flower have a Fibonacci number of petals: certain types of daisies tend to have 34 or 55 petals, while sunflowers have 89 or, in some cases, 144. The seeds of sunflowers spiral outward both to the left and the right in a Fibonacci

number of spirals. Similarly, the whorls on a pinecone, the numbers of rings on the trunks of palm trees, the patterns of snail shells, and the genealogy of the male bee all follow a sequence of Fibonacci numbers. The arrangement of plant leaves, or phyllotaxis, unfolds to the same pattern because this results in an optimal solution in terms of the spacing of the leaves or the amount of light that can reach them. A familiar spiral form, known as the **logarithmic spiral**, emerges when seeds on a plant grow and space themselves according the Fibonacci sequence. The logarithmic spiral is approximated by the rule: start at the origin of the Cartesian coordinate system, move $F(1)$ units to the right, move $F(2)$ units up, move $F(3)$ units to the left, move $F(4)$ units down, move $F(5)$ units to the right, and so on. By growing in this way, on structures such as sunflowers, pinecones, and pineapples, seeds are able to pack themselves together most efficiently.

Fibonacci numbers have so many interesting mathematical properties that an entire journal, *The Fibonacci Quarterly*, is devoted to them. The sequence of final digits in Fibonacci numbers repeats in cycles of 60. The last two digits repeat in 300, the last three in 1,500, the last four in 15,000, etc. The product of any four consecutive Fibonacci numbers is the area of a **Pythagorean triangle**. The shallow (least steep) diagonals of **Pascal's triangle** sum to Fibonacci numbers. Let m and n be positive integers, then

$F(n)$ divides $F(mn)$

$\gcd(F(n), F(m)) = F(\gcd(m, n))$, where "gcd" stands for "greatest common divisor."

$(F(n))^2 - F(n + 1)F(n - 1) = (-1)^{n-1}$.

$F(1) + F(3) + F(5) + \ldots + F(2n - 1) = F(2n)$.

For every n, there are n consecutive composite Fibonacci numbers.

An interesting use of the Fibonacci sequence is for converting miles to kilometers. For instance, if you want to know about how many kilometers 5 miles is, take the Fibonacci number (5) and look at the next one (8) (5 miles is about 8 kilometers). This works because it happens that the conversion factor between miles and kilometers is roughly equal to the Golden Ratio.

The first few Fibonacci numbers that are also **prime numbers** are 3; 5; 13; 89; 233; 1,597; 28,657; 514,229; . . . . It seems likely that there are infinitely many Fibonacci primes, but this has yet to be proven. However, it is relatively easy to show that for $n \geq 4$, $u_n + 1$ is never prime. The Fibonacci sequence is a special case of the **Lucas sequence**.

The *tribonacci series* is made by adding the last two digits: 1, 1, 2, 4, 7, 13, 24, 44, 81, . . . and from this the quadbonacci series, the pentbonacci series, and the

Fibonacci sequence   The number of spirals of seeds on a sunflower is always a Fibonacci number—an arrangement that keeps the seeds uniformly packed no matter how large the seed head. *Thomas Stromberg*

hexbonacci series, all the way up to the *n*-bonacci series. Each ratio of successive terms forms a special constant, analogous to φ.

field

A number system in which addition, subtraction, multiplication, and division (except by zero) are always defined, and the **associative** and **distributive** laws are valid. For example, the set of **rational numbers** is a field, whereas the set of **integers** is not a field, because the result of dividing one integer by another is not necessarily an integer. The **real numbers** also constitute a field, as do the **complex numbers**. Compare with **ring**.

Fields Medal

By convention, the most prestigious award for research in mathematics. It is awarded every four years to between two and four mathematicians under the age of 40.

Fifteen Puzzle

A sliding-tile puzzle invented by Sam **Loyd** in the 1870s that became a worldwide obsession, much as **Rubik's cube** did a century later. Fifteen little tiles, numbered 1 to 15, were placed in a four by four frame in serial order except for tiles 14 and 15, which were swapped around; the lower right-hand square was left empty. The object of the puzzle was to get all the tiles in the correct order; the only allowed moves were sliding counters into the empty square. Everyone it seemed was caught up with the craze—playing the game in horse-drawn trams, during their lunch breaks, or when they were supposed to be working. The game even made its way into the solemn halls of the German parliament. "I can still visualize quite clearly the gray-haired people in the Reichstag intent on a square small box in their hands," recalled the geographer and mathematician Sigmund Gunter who was a deputy during the puzzle epidemic. "In Paris the

Fifteen Puzzle   A version of the Fifteen Puzzle produced in England by Fairylite. *Sue & Brian Young/Mr. Puzzle Australia, www.mrpuzzle.com.au*

puzzle flourished in the open air, in the boulevards, and proliferated speedily from the capital all over the provinces," wrote a contemporary French author. "There was hardly one country cottage where this spider hadn't made its nest lying in wait for a victim to flounder in its web." Loyd offered a $1,000 reward for the first correct solution. But, although many claimed it, none were able to reproduce a winning series of moves under close scrutiny. There is a simple reason for this, which is also the reason that Loyd was unable to obtain a U.S. patent for his invention. According to regulations, Loyd had to submit a working model so that a prototype batch could be manufactured from it. Having shown the game to a patent official, he was asked if it were solvable. "No," he replied. "It's mathematically impossible." Upon which the official reasoned there could be no working model and thus no patent!

The puzzle's theory reveals that the more than 20 bil-lion possible starting arrangements of the tiles fall into just two groups: one in which all the tiles can be maneu-vered into ascending numerical order (call this group I), and one in which tiles 14 and 15 will be inverted (group II). It's impossible to combine arrangements from these two groups and impossible to turn a group I arrangement into a group II, or vice versa, using the normal rules of the game. Given a random arrangement of tiles, can we know in advance if we have the unsolvable kind? Very easily. Simply count how many instances there are of a tile numbered n appearing after the tile numbered $n + 1$. If there are an even number of such inversions, the puz-zle is solvable, otherwise you are wasting your time!

figure number

A number sequence found by creating consecutive geo-metrical figures from arrangements of equally spaced points. Here is an example:

```
1   3   5   7
*   *   *   *
    **  *   *
        *** *
            ****
```

The points can be arranged in one, two, three, or more dimensions. There are many different kinds of figurate numbers, such as **polygonal number**s and **tetrahedral numbers**.

films and plays involving mathematics

Mathematicians and mathematics rarely make an entrance on the silver screen or the stage, unless as parodies in the form of mad professors and meaningless scrawled equations. A notable exception is *A Beautiful Mind* (2001), directed by Ron Howard and starring Russell Crowe as the brilliant but mentally troubled mathematician John **Nash**. Although a fine love story and a well-crafted film, which won four Oscars, *A Beautiful Mind* is weak on math and inaccurate in many of its details of Nash's life and his battle with schizophrenia. *Rain Man* (1988), also based on a true story, costars Dustin Hoffman as an autistic savant with a photographic memory and a genius for mental arithmetic.

Good Will Hunting (1997), written by and starring Matt Damon and Ben Affleck, and also starring Robin Williams, is about a young man who has led a troubled life but has an amazing talent for mathematics. His abilities are discovered when he comes into conflict with the law, and he soon has to decide if he should pursue his mathematical future and leave his family and friends behind. In Darren Aronofsky's disturbing independent film *Pi* (1998), the main character is a mathematician obsessed with his search for patterns within **pi**'s infinite decimal places. He believes they can be used to predict chaotic behaviors, including that of the stock market. Throughout the film he is pursued by ruthless stock market players and by rabbis trying to find a mathematical way to communicate with God. In the science fiction film *Cube* (1997), six people awake to find themselves trapped in a deadly maze, and one of the characters uses mathematical skills to solve the puzzle and find a way to escape.

Lesser known films with strong mathematical themes include Mario Martone's *Death of a Neapolitan Mathematician;* Peter Greenaway's *Drowning by Numbers;* George Csicsery's *N Is a Number;* and *Moebius,* made by students and faculty at the Universidad del Cine of Buenos Aires. Mathematics has also found its way onto the stage. The musical *Fermat's Last Tango* (2000), a fictionalized account of Andrew **Wiles**'s struggle to prove **Fermat's last theorem**, was performed in New York by the York Theatre company. It followed the Pulitzer prize–winning play *Proof* by David Aubern, about the death of a brilliant mathematician and the repercussions for his daughters and his student.

finite

Limited in extent or scope. In mathematics, a *finite set* is such that the number of elements it contains can be described by a **natural number**. For instance, the set of integers between −18 and 5 is finite, because it has a natural number (17) of elements. The set of all **prime numbers**, on the other hand, is not finite. In physics, "finite" is used to mean both "not infinite" and "nonzero."

finite-state automaton (FSA)

The simplest computing device. Although it is not nearly powerful enough to perform universal computation, it can recognize **regular expression**s. FSAs are defined by a state transition table that specifies how the FSA moves from one state to another when presented with a particular input. FSAs can be drawn as **graph**s.

Fisher, Adrian

A British professional designer and constructor of **maze**s; his company, Adrian Fisher Maze Design, has built a huge variety of mazes in Britain, continental Europe, the United States, and elsewhere. These include the formal hedge maze at Leeds Castle and the largest brick pavement maze in the world at Kentwell Hall in Long Melford. The latter is based on a Tudor rose and has 15 sepals used as locations for a board game in which live players take part in Tudor costume.[101]

Fitchneal

An Irish version of the Viking game **Hnefa-Tafl**; played on a 7×7 board (as was the Scottish equivalent, known as *Ard-Ri,* "High King"), it is mentioned in the *Mabinogion* and *Cormac's Glossary* of the ninth century.

five

The length of the hypotenuse of the smallest **Pythagorean triangle** (a right triangle having integral sides). Five is the only **prime number** that is a member of two pairs of **twin primes.** Every integer is the sum of five positive or negative cubes in an infinite number of ways. Five is the smallest degree of a polynomial equation for which there is no general formula for the solutions (see **quintic**).

fixed-point attractor

An **attractor** that is represented by a particular point in **phase space**, sometimes called an *equilibrium point.* As a point it corresponds to a very limited range of possible behaviors of the system. For example, in the case of a pendulum, the fixed-point attractor represents the pendulum when the bob is at rest. This state of rest attracts the system because of gravity and friction.

Flatland: A Romance of Many Dimensions

A satirical novel by Edwin A. **Abbott**,[1] first published in 1884, that portrays a **two-dimensional world**, like the surface of a map, over which its inhabitants move. Flatlanders have no concept of up and down, and appear to each other as mere points or lines. From our three-dimensional perspective we can look down on Flatland and see that its people are "really" a variety of shapes, including straight lines (females), narrow isosceles triangles (soldiers and workmen), equilateral triangles (lower middle-class men), squares and pentagons (professional men, including the pseudonymous author of the tale, A. Square), hexagons and other regular polygons with still more sides (the nobility), and circles (priests). Abbott uses these geometrical distinctions, especially the appearance of Flatland females and the working class, as a commentary on the discrimination against women, the rigid class stratification, and the lack of tolerance for "irregularity" that was prevalent in Victorian Britain.

In a dream, A. Square visits the one-dimensional world of Lineland where he tries, unsuccessfully, to persuade the king that there is such a thing as a second dimension. In turn, the narrator is told of three-dimensional space by a sphere who moves slowly through the plane of Flatland, growing and shrinking as his cross-section changes in size. (If a **hypersphere** were to move through our three-dimensional world, we would see a sphere appear, grow to a maximum size, and then shrink again before disappearing.) Abbott is aware that he cheats a little in his description of what the inhabitants of Flatland actually see. In his preface to the second edition, he gives a lengthy but not-too-convincing reply to the objection, raised by some readers, that a Flatlander, "seeing a Line, sees something that must be *thick* to the eye as well as *long* to the eye (otherwise it would not be visible . . .)." The curious and often-neglected fact is that we are just as unable to imagine what it would truly be like to see in two dimensions as we are to conceive of four dimensions! No matter how hard we try we cannot imagine being able to see a line of *zero* thickness.

flexagon

A flat model constructed from a folded strip of paper, which, when flexed, can be made to reveal a number of hidden faces. Flexagons are amusing toys but they have also caught the interest of mathematicians. They are usually square or rectangular (tetraflexagons) or hexagonal (hexaflexagons). A prefix can be added to the name to indicate the number of faces that the model can display,

including the two faces (back and front) that are visible before flexing. For example, a hexaflexagon with a total of six faces is called a hexahexaflexagon. The discovery of the first flexagon, a trihexaflexagon, is credited to the British student Arthur H. Stone who was studying at

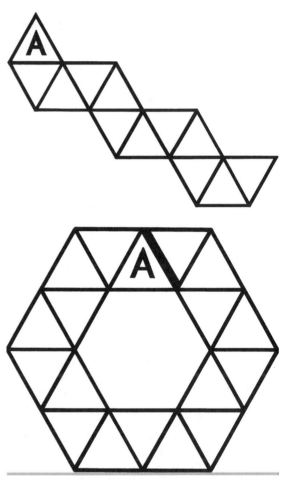

flexagon Two nets for folding into hexaflexagons: the 4-flexagon (top) and the 6-flexagon (bottom). To use the nets, photocopy and enlarge them, and label each side with these numbers:

4-flexagon $\begin{pmatrix} 2231 \ 2231 \ 2231 \\ 3144 \ 3144 \ 3144 \end{pmatrix}$,

6-flexagon $\begin{pmatrix} 662554 \ 662554 \ 662554 \\ 231431 \ 231431 \ 231431 \end{pmatrix}$.

Start at A using the top row of numbers. Number the other side of the net with the bottom row of numbers so that the top and bottom numbers appear on either side of the first triangle, and so on. *Jill Russell*

Princeton University in 1939. Stone's colleagues Bryant Tuckermann, Richard P. Feynman, and John W. Tukey became interested in the idea. Tuckerman worked out a topological method, called the *Tuckerman traverse,* for revealing all the faces of a flexagon. Tukey and Feynman developed a complete mathematical theory that has not been published. Flexagons were introduced to the general public by Martin **Gardner** writing in *Scientific American.*[110]

floor function
The greatest integer in x, that is, the largest integer less than or equal to x.

Flower of Life
One of the beautiful arrangements of circles found at the Temple of Osiris at Abydos, Egypt. The pattern also appears in Phoenician art from the ninth century B.C. The circles are placed with six-fold symmetry, forming a mesmerizing pattern of circles and lenses. A related design from the same temple, which recurs in Italian art from the thirteen century, is called the *Seed of Life.*

fly between trains problem
Two trains are approaching each another and a fly is buzzing back and forth between the two trains. Given the (constant) speed of the trains and their initial separation distance, and the (constant) speed of the fly, calculate how far the fly will travel before the trains collide. This problem appears to have been first posed by Charles Ange Laisant (1840–1921) in his *Initiation Mathématique.* There is a long-winded method of getting the answer and a much shorter way. Suppose the trains start out 200 miles apart and are each traveling at 50 miles per hour, and the fly–a speedster of its kind–is moving at 75 miles per hour. The long method involves considering the length of the back-and-forth path that the fly takes and evaluating this as the sum of an infinite series. The quick solution is to notice that the trains will collide in 2 hours and that in this time the fly will travel $2 \times 75 = 150$ miles! When this problem was put to John **von Neumann**, he immediately gave the correct answer. The poser, assuming he had spotted the shortcut, said: "It is very strange, but nearly everyone tries to sum the infinite series." Von Neumann replied: "What do you mean, strange? That's how I did it!"

focal chord
A **chord** of a **conic section** that passes through a **focus**.

focal radius
A line segment from the **focus** of an **ellipse** to a point on the **perimeter** of the ellipse.

focus
A defining point in the construction of a **conic section**. The word comes from the Latin for hearth or fireplace and appears to have been first used in mathematics, in describing an **ellipse**, by Johannes **Kepler**.

foliation
A decoration of a **manifold** in which the manifold is partitioned into sheets of some lower dimension, and the sheets are locally parallel. More technically, the foliated manifold is locally **homeomorphic** to a **vector space** decorated by cosets of a subspace.

folium
A curve, first described by Johannes **Kepler** in 1609, that corresponds to the general equation

$$(x^2 + y^2)(y^2 + x(x + b)) = 4axy^2,$$ in Cartesian form, or
$$r = -b \cos\theta + 4a \cos\theta \sin^2\theta,$$ in polar coordinates.

The Latin *folium* means "leaf-shaped." Three types, known as the simple folium, the bifolium (or double folium), and the trifolium, correspond to the cases when $b = 4a, b = 0,$ and $b = a$, respectively. The *folium of Descartes* is given by the Cartesian equation $x^3 + y^3 = 3axy$ and was first discussed by René **Descartes** in 1638. Although he found the correct shape of the curve in the positive quadrant, he wrongly thought that this leaf shape was repeated in each quadrant like the four petals of a flower. The problem to determine the tangent to the curve was proposed to Gilles de Roberval who, having made the same incorrect assumption, called the curve *fleur de jasmin* after the four-petal jasmine bloom, a name that was later dropped. The folium of Descartes has an asymptote $x + y + a = 0$.

formal system
A mathematical formalism in which statements can be constructed and manipulated with logical rules. Some formal systems, such as **Euclidean geometry**, are built around a few basic **axiom**s and can be expanded with theorems that can be deduced through proofs.

formalism
A mathematical school of thought that was headed by the German mathematician David **Hilbert**. Formalists argue that mathematics must be developed through **axiom**atic systems. Formalists agree with Platonism on the principles of mathematical proof, but Hilbert's followers don't recognize an external world of mathematics. Formalists argue that mathematical objects don't exist until we define them. Humans create the **real number** system, for example, by establishing axioms to describe it. All that mathematics needs are inference rules to

progress from one step to the next. Formalists tried to prove that within the framework of established axioms, theorems, and definitions, a mathematical system is consistent and, in the mid-twentieth century, formalism became the predominant philosophical attitude in math textbooks. However, it was undermined by **Gödel's incompleteness theorem** and also the general recognition that results can be usefully applied without having to be proved or derived axiomatically.

Fortune's conjecture

A conjecture about **prime numbers** made by the New Zealander social anthropologist Reo Fortune (1903–1979), who had a reputation for unstable behavior bordering on the psychotic. He once attempted to settle an academic dispute with a colleague, Thomas McIlwraith, at the University of Ontario, by challenging him to a duel with any weapon of his choice from the collections of the Royal Ontario Museum. Fortune proposed that if q is the smallest prime greater than $P + 1$, where P is the product of the first n primes, then $q - P$ is prime. For example, if n is 3, then P is $2 \times 3 \times 5 = 30$, $q = 37$, and $q - P$ is the prime 7. These numbers, $q - P$, are now known as *Fortunate numbers*. The conjecture remains unproven but is generally thought to be true. The sequence of Fortunate numbers begins

$$3, 5, 7, 13, 23, 17, 19, 23, 37, 61, 67, \ldots$$

four

The smallest **composite number**, the second smallest **square number**, the first non-Fibonacci number (see **Fibonacci sequence**), the smallest **Smith number**, and the smallest number that can be written as the sum of two **prime numbers**. Four is the number of dimensions that make up **space-time** (three of space and one of time). It is the most number of colors needed to color any map so that no two neighboring areas are the same color (see **four-color map problem**). There are four cardinal points on the compass, four Riders of the Apocalypse, and four Gospels.

four-color map problem

A long-standing problem that dates back to 1852 when Francis Guthrie, while trying to color a map of the counties of England noticed that four colors were enough to ensure that no adjacent counties were colored the same. He asked his brother Frederick if it was true that *any* map could be colored using four colors in such a way that adjacent regions (i.e., those sharing a common boundary segment, not just a point) receive different colors. Frederick Guthrie then passed on the conjecture to Augustus **de Morgan**. The first printed reference is due to Arthur **Cayley** in 1878. A year later the false "proof," by the English

barrister Alfred Kempe, appeared; *its* incorrectness was pointed out by Percy Heawood 11 years later. Another failed proof is due to Peter **Tait** in 1880, a gap in *his* argument being pointed out by Julius Petersen in 1891. Both false proofs did have some value, though. Kempe discovered what became known as *Kempe chains,* and Tait found an equivalent formulation of the four-color theorem in terms of three-edge coloring.

The next major contribution came from George Birkhoff whose work allowed Philip Franklin in 1922 to prove that the four-color conjecture is true for maps with at most 25 regions. It was also used by other mathematicians to make various forms of progress on the four-color problem. In the 1970s, the German mathematician Heinrich Heesch developed the two main ingredients needed for the ultimate proof—reducibility and discharging. While the concept of reducibility was studied by other researchers as well, it seems that the idea of discharging, crucial for the unavoidability part of the proof, is due to Heesch, and that it was he who conjectured that a suitable development of this method would solve the four-color problem. This was confirmed by Kenneth Appel and Wolfgang Haken of the University of Illinois in 1977, when they published their proof of the four-color theorem.[12] Their controversial proof challenges the basic assumptions of what mathematical proof is. They used more than 1,200 hours of supercomputer time to analyze 1,478 different configurations that in turn can produce every possible map on a plane. Not everyone was happy with the method of the breakthrough, as Appel himself pointed out:

> For almost a century and a half, a Holy Grail of graph theory has been a simple incisive proof of the Four Color Theorem. It has troubled our profession that a problem that can be understood by a school child has yet to be solved in a way that better illuminates the reason that only four colors are needed for planar maps. The feelings of many mathematicians were summed up for me by Herb Wilf's response to being told that it appeared that one could prove the theorem by a long reducibility argument which used computers to test the reducibility of a large number of configurations. He simply said, "God would not allow such a beautiful theorem to have such an ugly proof."

Martin **Gardner** commented, "Whether a simple, elegant proof not requiring a computer will ever be found, is still an open question." It's interesting that such a simple, intuitive puzzle can be so difficult to settle! The four-color theorem is true for maps on a plane or on a sphere. The answer is different for geographic maps on a **torus**: in this case, seven colors are necessary and sufficient.[277]

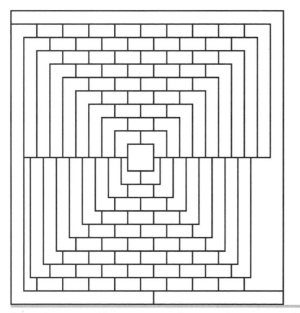 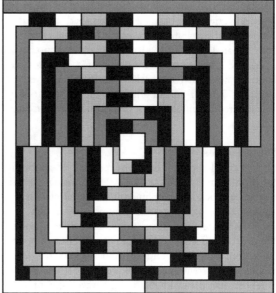

four-color map problem   Martin Gardner's spoof counterexample to the four-color theorem (left) and a solution using four colors (right). The different colors are represented here in black, white, and gray.

four coins problem

Given three coins of possibly different sizes, which are arranged so that each is tangent to the other two, find the coin that is tangent to the other three coins. The solution is the inner **Soddy circle**.

four fours problem

Using arithmetic combinations of four 4's, express all the numbers from 1 to 100. For example, 1 = 44/44 and 2 = (4 × 4)/(4 + 4). The problem was first presented in *The Schoolmaster's Assistant: Being a Compendium of Arithmetic Both Practical and Theoretical* (first edition c. 1744), a popular textbook by the English schoolteacher and cleric Thomas Dilworth (d. 1780). Operations and symbols that are allowed include the four arithmetic operations (+, ×, −, /), concatenation (e.g., the use of 44), decimal points (e.g., 4.4), powers (e.g., 4^4), square roots, factorials (e.g., 4!), and overbars for repeating digits (e.g., .4 with an overbar to express $^4/_9$). Ordinary use of parentheses are allowed. One of the trickiest numbers to represent in this way is 73, which calls for something as contorted as

$$\sqrt{(\sqrt{(\sqrt{(4^4)})})} + 4 \, / \, .4'$$

(where .4′ is shorthand for .444 . . .). Of course, the problem can be extended to represent integers greater than 100. The highest value achievable in the four four's puzzle is $10^{8.0723047260281 \times 10153} = 4^{4^{4^4}}$.

four knights puzzle

On a 3 × 3 chessboard are two white knights at the top left-hand and top right-hand squares and two black knights at the bottom left-hand and bottom right-hand squares. The problem is to exchange the black knights with the white knights in the minimum possible number of moves. One move is a normal knight's move on any vacant cell of the board, which renders the center square inaccessible.

Fourier, (Jean Baptiste) Joseph, Baron (1768–1830)

A French mathematician known chiefly for his contribution to the mathematical analysis of heat flow. Although he trained for the priesthood, Fourier didn't take his vows but instead turned toward mathematics. He first studied and later taught mathematics at the newly created Ecole Normale. In 1798 he joined Napoleon's army in its invasion of Egypt as scientific advisor, to help establish educational facilities there and to carry out archaeological explorations. After his return to France in 1801 he was appointed prefect of the department of Isere. Fourier became famous for his *Theorie analytique de la Chaleur* (1822), a mathematical treatment of how heat conducts in solid bodies. He established the **partial differential equation** governing heat flow and solved it by using an infinite series of trigonometric functions, now known as **Fourier series**. Though these series had been used before, Fourier investigated them in much greater detail and prepared the way for later work on trigonometric series and the theory of functions of a real variable. Fourier's belief that his health would be improved by wrapping himself up in blankets proved fatal: thus encumbered he tripped down the stairs of his house and died.

Fourier series

Named after Joseph **Fourier**, the expansion of a **periodic** function as an infinite sum of sines and cosines of various frequencies and amplitudes. This is similar to the approximation of an irrational number by a sum of a series of rational numbers (or a decimal expansion). Human ears effectively produce Fourier series automatically from complex sounds. Tiny hairs, known as *cilia*, vibrate at different specific frequencies. When a wave enters the ear, the cilia vibrate if the wave function contains any component of the corresponding frequency. This enables the hearer to distinguish sounds of various pitches. Fourier series are used a great deal in science and engineering to find solutions to **partial differential equations**, such as those in problems involving heat flow. They can also be used to construct some pathological functions such as ones that are continuous but nowhere differentiable. The study and computation of Fourier series is known as **harmonic analysis**.

fourth dimension

"Do you think that there are things which you cannot understand, and yet which are; that some people see things that others cannot?" said Dr. Van Helsing in Bram Stoker's *Dracula*. Instead of vampires, he may just as easily have been talking about the fourth dimension—an extension at right-angles to the three familiar directions of up-down, forward-backward, and side-to-side. In physics, especially **relativity theory**, **time** is often regarded as the fourth dimension of the **space-time** continuum in which we live. But what meaning can be attached to a fourth *spatial* dimension? The mathematics of the fourth dimension (4-d) can be approached through a simple extension of either the algebra or the geometry of one, two, and three dimensions.

Algebraically, each point in a multidimensional space can be represented by a unique sequence of **real numbers**. One-dimensional space is just the **number line** of real numbers. Two-dimensional space, the plane, corresponds to the set of all **ordered pairs** (x, y) of real numbers, and three-dimensional space to the set of all ordered triplets (x, y, z). By extrapolation, four-dimensional space corresponds to the set of all ordered quadruplets $(x, y,$ $z, w)$. Linked to this concept is that of **quaternions**, which can also be viewed as points in the fourth dimension.

Geometric facts about the fourth dimension are just as easy to state. The fourth dimension can be thought of as a direction perpendicular to every direction in three-dimensional space; in other words, it stretches out along an axis, say the w-axis, that is mutually perpendicular to the familiar x-, y-, and z-axes. Analogous to the cube is a hypercube or **tesseract**, and to the sphere is a 4-d **hypersphere**. Just as there are five regular polygons, known as the **Platonic solids**, so there are six four-dimensional regular **polytopes**. They are: the 4-simplex (constructed from five tetrahedra, with three tetrahedra meeting at an edge); the tesseract (made from eight cubes, meeting three per edge); the 16-cell (made from 16 tetrahedra, meeting four per edge); the 24-cell (made from 24 octahedra, meeting three per edge); the 120-cell (made from 120 dodecahedra, meeting three per edge); and the monstrous 600-cell (made from 600 tetrahedra, meeting five per edge).

Geometers have no difficulty in analyzing, describing, and cataloging the properties of all sorts of 4-d figures. The problem starts when we try to *visualize* the fourth dimension. This is a bit like trying to form a mental picture of a color different from any of those in the known rainbow from red to violet, or a "lost chord," different from any that has ever been played. The best that most of us can hope for is to understand by analogy. For example, just as a sketch of a cube is a 2-d perspective of a real cube, so a real cube can be thought of as a perspective of a tesseract. At a movie, a 2-d picture represents a 3-d world, whereas if you were to watch the action live, in three-dimensions, this would be like a screen projection in four dimensions.

Many books have been written and schemes devised to nudge our imaginations into thinking four-dimensionally. One of the oldest and best is Edwin **Abbott**'s *Flatland* [1] written more than a century ago, around the time that mathematical discussion of **higher dimensions** was becoming popular. H. G. Wells also dabbled in the fourth dimension, most notably in *The Time Machine* (1895), but also in *The Invisible Man* (1897), in which the central character drinks a potion "involving four dimensions," and in "The Plattner Story" (1896), in which the hero of the tale, Gottfried Plattner, is hurled into a four-spatial dimension by a school chemistry experiment that goes wrong and comes back with all his internal organs switched around from right to left.[285] The most extraordinary and protracted attack on the problem, however, came from Charles **Hinton**, who believed that, through appropriate mental practice involving a complicated set of colored blocks, a higher reality would reveal itself, "bring[ing] forward a complete system of four-dimensional thought [in] mechanics, science, and art."

Victorian-age spiritualists and mystics also latched on to the idea of the fourth dimension as a home for the spirits of the departed. This would explain, they argued, how ghosts could pass through walls, disappear and reappear at will, and see what was invisible to mere three-dimensional mortals. Some distinguished scientists lent their weight to these spiritualist claims, often after being duped by clever conjuring tricks. One such unfortunate was the astronomer Karl Friedrich Zöllner who wrote about the four-dimensional spirit world in his *Transcendental Physics* (1881) after attending séances by Henry Slade, the fraudulent American medium.

Art, too, became enraptured with the fourth dimension in the early twentieth century. When the Cubist painter and theorist Albert Gleizes said, "Beyond the three dimensions of Euclid we have added another, the fourth dimension, which is to say, the figuration of space, the measure of the infinite," he united math and art and brought together two major characteristics of the fourth dimension in early Modern Art theory—the geometric orientation as a higher spatial dimension and the metaphorical association with infinity.[159] See also **Klein bottle**.[25, 156, 163, 212, 213, 231, 254, 267, 271, 273, 340]

Fox and Geese

An English board game that dates back to the Middle Ages and is unusual in that the two sides are unequal, thus making this an example of a **Tafl game**. The lone fox attempts to capture 13 (or, in later versions, 17) geese, while the geese try to hem the fox in so that it can't move. The geese start out by filling up all the points on one side of the cross-shaped grid on the board. The fox—the one counter of a different color—begins on any vacant point remaining. The fox moves first. Each side, in its turn, may move one counter. Both fox and geese can move along a line, forwards, backwards, or sideways, to the next contiguous point. The fox may move along a line or jump over a goose to an empty point, capturing the goose and removing it from the board. Two or more geese may be captured by the fox in one turn, provided that he is able to jump to an empty point after each one. The fox wins if he depletes the gaggle of geese to a number that makes it impossible for them to trap him. The geese can't jump over the fox or capture the fox but instead must try to mob him and trap him in a corner. The geese win if they make it impossible for the fox to move. A modification of this game spread with the British to India, where during the Great Mutiny the game became known as "Officers and Sepoys." In this variant, two officers in a fort attempt to hold off 24 sepoys, who must storm the fort.

fractal

A geometric shape that can be subdivided at any scale into parts that are, at least approximately, reduced-size copies of the whole. The name "fractal," from the Latin *fractus* meaning a broken surface, was coined by Benoit **Mandelbrot** in 1975. The key property of fractals is **self-similarity**, which means that zooming in or zooming out of a fractal produces no overall change in appearance.

One of several technical definitions of a fractal is "a set of points whose topological dimension is less than its **Hausdorff dimension**." The topological dimension is an object's ordinary dimensionality—one in the case of a curve, two in the case of a surface, and so forth—and is always a whole number. The Hausdorff dimension, on the other hand, measures how much space an object fills, and can take non-integer values if the object is very complex and twisty.

Some fractals show a strong regularity and rigid self-similarity and are produced by the repeated application of a set of rules that may be quite simple. Among the best known of these "iterated function" systems are the **Koch snowflake**, the **Peano curves**, the **Sierpinski carpet**, and the **Sierpinski gasket**. Other fractals, defined by a recurrence relation at each point in space, are among the most complex, beautiful, and beguiling mathematical structures known. They include the well-known **Mandelbrot set** and **Lyapunov fractals**. Finally, there are random fractals generated by stochastic rather than deterministic processes, for example fractal landscapes. Random fractals have the greatest practical use, and can be used to describe many highly irregular real-world objects, including clouds, mountains, coastlines, and trees. See also **fractal dimension**.

fractal dimension

A non-integer measure of the irregularity or **complexity** of a system; it is an extension of the notion of dimension found in **Euclidean geometry**. Knowing the fractal dimension helps one determine the degree of irregularity and pinpoint the number of variables that are key to determining the dynamics of the system.

fraction

A number that represents a part, or several equal parts, of a whole; examples include one-half, two-thirds, and three-fifths. The word comes from the Latin *frangere*, meaning "to break." A *simple, common,* or *vulgar fraction* is of the form *a/b*, where *a* may be any integer and *b* may be any integer greater than 0. If *a* < *b*, the fraction is said to be *proper* ("bottom heavy"); otherwise it is *improper* ("top heavy"). A **decimal fraction** has a denominator (number on the bottom) of 10, 100, 1000, and so forth. See also **continued fraction**.

fractal A deep zoom of part of the Mandelbrot set. *Christopher Rowley*

Fraser spiral

A **distortion illusion** in which overlapping black arc segments appear to form a spiral but are in reality a series of overlapping concentric circles. This is easily demonstrated by following one of the curves with your finger. The illusion is named after the British psychologist James Fraser (1863–1936) who first published it in 1908.[104]

Fredholm, Erik Ivar (1866–1927)

A Swedish mathematician who founded the modern theory of **integral equations**. This became a major research topic in the first quarter of the twentieth century and underpinned important theoretical developments in physics; David **Hilbert**, in particular, extended Fredholm's work to arrive at the concept of **Hilbert space**. Fredholm also devoted time to actuarial science and made a particularly important contribution by proposing an elegant formula to determine the surrender value of a life insurance policy. He earned his Ph.D. from the University of Uppsala but then spent the rest of his academic career at the University of Stockholm.

Freemish crate

See **impossible figure**.

Freeth's nephroid

See **nephroid**.

Frege, Friedrich Ludwig Gottlob (1848–1925)

A German mathematician and philosopher who virtually founded the modern discipline of mathematical logic. In *Die Grundlagen der Arithmetik* (The foundations of arithmetic, 1884), he used **set theory** to define the **cardinal number** of a given class as the class of all classes that are similar (i.e. can be placed in a **one-to-one** correspondence) to the given class. In *Grundgesetze der Arithmetik* (The basic laws of arithmetic, 2 vols., 1893 and 1903), Frege began attempting to build up mathematics from arithmetic and symbolic logic on a rigorous and contradiction-free basis. When the second volume was in the process of being printed, Bernard **Russell** pointed out a paradox in Frege's work. The paradox, which became known as **Russell's paradox**, stems from the question: "Is the class of all classes that are not members of itself a member of itself or not?" The question leads to a contradiction and cannot be resolved. Frege was thus forced to admit that the foundation of his reasoning was worthless. As he stated at the end of his work, "A scientist can hardly encounter anything more undesirable than to have the foundation collapse just as the work is finished. I was put in this position by a letter from Mr. Bertrand Russell when the work was almost through the press."

Frénicle de Bessy, Bernard (1602–1675)

An eminent French amateur mathematician who extensively researched **magic squares**; his *Des quassez ou tables magiques,* published posthumously in 1693, first identified all 880 magic squares of the fourth order. Frénicle also corresponded with **Descartes**, **Fermat**, **Huygens**, and **Mersenne**, mostly on **number theory**, the work for which he is best known.

frequency

The number of times a value occurs in some time interval.

friendly number

See **amicable numbers**.

Frogs and Toads

A puzzle in which three counters or pegs representing frogs are placed on three successive positions on the left of a string of seven squares, and three different tokens representing toads are placed on the three rightmost squares. Frogs only move to the right, toads only to the left. Every move is either a slide to the adjacent square or a jump over one position, which is allowed only if the latter is occupied by a member of the other species. No two animals are ever allowed on the same square. The goal is to move the toads into the three leftmost positions and the frogs into three rightmost positions in the fewest possible moves. Many different versions of this puzzle have appeared over the centuries and it may be Arabic in origin. The number of pieces on each side may vary, as may the number of empty starting places in the middle; other names for the puzzle have included *Sheep and Goats* and *Sphinxes and Pyramids.*

frugal number

See **economical number**.

frustum

Part of a solid cut off between two parallel planes; in particular, for a **cone** or a **pyramid**, a frustum is determined by the plane of the base and a plane parallel to the base. *Frustum* is Latin for "a piece broken off."

function

Old mathematicians never die; they just lose some of their functions.

–Anonymous

A way of expressing the dependence of one quantity on another quantity or quantities. Traditionally, functions were specified as explicit rules or formulas that converted some input value (or values) into an output value. If f is the name of the function and x is a name for an input value, then $f(x)$ denotes the output value corresponding to x under the rule f. An input value is also called an *argument* of the function, and an output value is called a *value* of the function. The **graph** of the function f is the collection of all pairs $(x, f(x))$, where x is an argument of f. For example, the circumference C of a circle depends on its diameter d according to the formula $C = \pi d$; therefore, one can say that the circumference is a function of the diameter, and the functional relationship is given by $C(d) = \pi d$. Equally well, the diameter can be considered a function of the circumference, with the relationship given by $d(C) = d/\pi$. In modern mathematics, the insistence on specifying an explicit effective rule has been abandoned; all that is required is that a function f associate with every element of some **set** X a unique element of some set Y. This makes it possible to prove the existence of a function without necessarily being able to calculate its values explicitly. Also, it enables general properties of functions to be proved independently of their form. The set X of all admissible arguments is called the *domain* of f; the set Y of all admissible values is called the *codomain* of f. We write $f: X \rightarrow Y$.

fundamental group

A group of a **topological space** X that is constructed by looking at how closed paths in X can be combined to get new paths. Under a suitable way of identifying paths (known as **homotopy**) one can get a group structure on the set which gives an algebraic invariant of the space X.

fundamental theorem of algebra

The result that any **polynomial** with real or complex coefficients has a root in the complex plane.

fundamental theorem of arithmetic

Every positive **integer** greater than 1 is a **prime number** or can be expressed as a unique product of primes and powers of primes.

fuzzy logic

A departure from classical two-valued logic in which something is either true or false, to allow a continuous range of truth values. Fuzzy logic was introduced by Lotfi Zadeh of the University of California at Berkeley in the 1960s as a means to model the uncertainty of natural language.

Gabriel's horn

The **surface of revolution** of

$$y = 1/x$$

for x greater than 1. Surprisingly this has a finite volume of **pi** cubic units but an *infinitely large* surface area! Gabriel's horn is also known as *Torricelli's trumpet* because it was investigated by the Italian Evangelista Torricelli (1608–1647). As a young man Torricelli studied in Galileo's home at Arcetri, near Florence, and then, upon Galileo's death, succeeded his teacher as mathematician and philosopher for their good friend and patron, the grand duke of Tuscany. Torricelli was amazed by the strange property of his mathematical trumpet and tried various ways to avoid the conclusion that a finite volume could be enclosed by a vessel with an infinite surface area. Unfortunately, he lived before **calculus** came along to explain the apparent paradox in terms of **infinitesimal**s.

Galois, Évariste (1811–1832)

A French mathematician who led a short, dramatic life and is often credited with founding modern **group** theory, though the Italian Paolo Ruffini (1765–1822) came up with many of the ideas first. Galois's work wasn't widely acknowledged by his contemporaries, partly because he didn't present his material very well and partly because he held unpopular political views. In fact, he was a republican revolutionary who was twice imprisoned because of his activities. During his second incarceration he fell in love with the daughter of the prison physician, Stephanie-Felice du Motel, and after being released, was killed in a duel over her with Perscheux d'Herbinville. His death started republican riots and rallies which lasted for several days. See also **Galois theory**.

Galois theory

The study of certain **groups**, known as *Galois groups*, that can be associated with **polynomial** equations. Whether

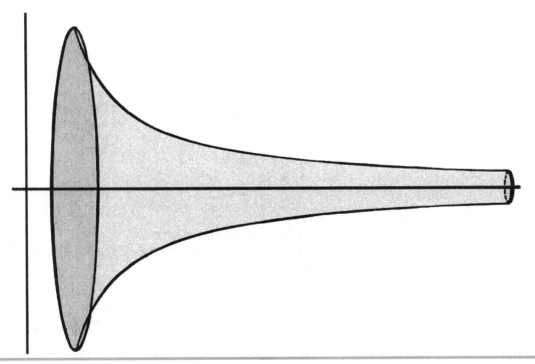

Gabriel's horn   The horn for x values between 1 and 10.

or not the solutions to an equation can be written down using rational functions and square roots, cube roots, and so forth depends on certain group-theoretic properties of *Galois groups*.

game

A conflict, with formal rules and a finite number of choices of what to do at each stage, between two or more players. The study of games belongs to a branch of mathematics and logic known as **game theory**. If a game is simple enough, it can be solved for all possible outcomes. This is the case, for example, with **tic-tac-toe** and **Nim**. By harnessing the power of computers to check through vast numbers of moves, even more complicated games are succumbing to a complete analysis. In the 1990s, **nine men's morris** was shown, by searching through billions of possible endgames, to be a certain draw if both players work to an optimal strategy. Checkers may be the next to be fully determined: its roughly 500 million trillion possible positions may soon be within reach of the most powerful supercomputers. See also **blackjack, chess, Frogs and Toads, Ovid's game, TacTix,** and **Wythoff's game**.

game theory

A mathematical formalism used to study human **game**s, economics, military conflicts, and biology. The goal of game theory is to find the optimal **strategy** for one player to use when his opponent also plays optimally. A strategy may incorporate randomness, in which case it is referred to as a **mixed strategy**.

Early ideas of game theory can be found in writings throughout history as diverse as the Bible and works by René **Descartes**, Sun Tzu (author of the 2,400-year-old *The Art of War*), and Charles Darwin. The basis of modern game theory is an outgrowth of several books that deal with related subjects such as economics and probability. These include Augustin Cournot's *Researches into the Mathematical Principles of the Theory of Wealth* (1838), which gives an intuitive explanation of what would eventually be formalized by John **Nash** as *Nash equilibrium;* Francis Edgeworth's *Mathematical Psychics,* which explored the notion of competitive equilibria in a two-type (or two-person) economy; and Emile **Borel**'s *Algebre et calcul des probabilites* (1927), which gave the first insight into so-called mixed strategies.[49] Game theory finally came of age through the efforts of two European immigrants to the United States working at the Institute of Advanced Studies in Princeton. Around 1940, the idea of the **utility function** was taken up by John **von Neumann**, who had been forced to flee his native Hungary when the Nazis invaded, and the economist Oskar Morgenstern (1902–1976), who had left Austria because he loathed the National Socialists. In Princeton the two immigrants worked together on what they initially thought would be a short paper on the theory of games, but that kept growing until it finally appeared in 1944 as an opus of 600 pages with the title *Theory of Games and Economic Behavior.*[230]

GLOSSARY OF GAME THEORY

categorical game	A game in which a tie is impossible.
finite game	A game in which each player has a finite number of moves and a finite number of choices at each move.
futile game	A game that allows a tie when played properly by both players.
impartial game	A game in which the possible moves are the same for each player in any position.
mixed strategy	A collection of moves together with a corresponding set of weights which are followed probabilistically in the playing of a game.
partisan game	A game for which each player has a different set of moves in any position.
payoff matrix	An $m \times n$ matrix that gives the possible outcome of a two-person zero-sum game when player A has m possible moves and player B has n moves.
strategy	A set of moves that a player plans to follow while playing a game.
zero-sum game	A game in which players make payments only to each other. One player's loss is the other player's gain, so the total amount of "money" available is constant.

gamma

See **Euler-Mascheroni constant**.

gamma function

A generalization of the **factorial** function to the real line and to the complex plane. It is defined by:

$$\Gamma(n+1) = \int_0^\infty x^n\, e^{-x}\, dx$$

If n is an integer, then $\Gamma(n+1) = n!$ See also **beta function**.

Gardner, Martin (1914–)

An American recreational mathematician best known for his "Mathematical Games" column, which ran in *Scien-*

tific American for 25 years. Through this column he introduced many subjects, including **flexagons**, **polyominos**, Piet Hein's **Soma Cube**, and John Conway's Game of **Life**, to a wider audience. He is also an accomplished amateur magician and an active member of the skeptical movement associated with James Randi. Gardner is the author of more than 60 books, including various collections of his *Scientific American* columns, *The Ambidextrous Universe*, and *The Annotated Alice*.[108–131]

gauge theory

A force field in nature, or an analogous **vector** field in mathematics with an enormous amount of **symmetry** that expresses the redundancy or ambiguity of many parameters. The simplest example is the *electromagnetic field*.

Gauss, Carl Friedrich (1777–1855)

A German mathematician, often called the "Prince of Mathematics," whose stature and range of interests rivaled those of **Aristotle** and Isaac **Newton**. Some inkling of what was in store came when, as a 3-year-old, he corrected a mistake in one of his father's lengthy payroll calculations. In school, at age 10, when his teacher gave the class the task of adding all the integers from 1 to 100, Gauss immediately wrote down the correct answer, 5050, on his slate. He had spotted that the numbers can be paired off as $(100 + 1)$, $(99 + 2)$, $(98 + 3)$, . . . , $(51 + 50)$ so that the problem reduces to multiplying 101 by 50. At age 19, Gauss found a way to construct a heptadecagon (a **regular polygon** with 17 sides) using only a straightedge and compass—a feat that had eluded the Greeks. Then Gauss entered the mathematical stratosphere of his time by proving what is now called the *fundamental theorem of algebra*, namely, that every **polynomial** has at last one root that is a **complex number**; in fact, he gave four different proofs, the first of which appeared in his dissertation. In 1801, he proved the *fundamental theorem of arithmetic* (that every natural number can be represented as the product of **prime numbers** in only one way); published a brilliant tour de force on the properties of integers in his *Disquisitiones Arithmeticae*, which systematized the study of **number theory**; and showed that every number is the sum of at most three **triangular numbers**. In the same year, he also developed the method of least squares fitting and, though he didn't publish it, used it to calculate the orbit of the asteroid Ceres, that had recently been discovered by Piazzi, from only three observations. Gauss published his monumental treatise on celestial mechanics *Theoria Motus* in 1806. He became interested in the compass through surveying, and developed the magnetometer, an instrument with which, together with Wilhelm Weber, he measured the intensity of magnetic forces. With Weber, he also built the first successful telegraph.

Unfortunately for mathematics, Gauss reworked and improved papers incessantly, and, in keeping with his motto "*pauca sed matura*" (few but ripe), he published only a fraction of his work. Many of his results were subsequently repeated by and attributed to others, since his terse diary remained unpublished for years after his death. Only 19 pages long, this diary later confirmed his priority on many breakthroughs, including work on an alternative to the **parallel postulate**, which really makes him the earliest pioneer of **non-Euclidean geometry** despite the fact that János **Bólyai** and Nikolai Lobachevsky are normally given this accolade. Gauss did, however, publish his seminal treatment on **differential geometry** in *Disquisitiones circa superticies curvas*, and Gaussian **curvature** is named for him. Gauss wanted a heptadecagon placed on his gravestone, but the carver refused, saying it would be indistinguishable from a circle. The heptadecagon appears in the shape of a pedestal with a statue erected in his honor in his hometown of Braunschweig.

Gaussian

Normally distributed (with a bell-shaped curve) and having a **mean** at the center of the curve with tail widths proportional to the **standard deviation** of the data about the mean.

Gelfond's theorem

Also known as the *Gelfond-Schneider theorem*: a^b is a **transcendental number** if (1) a is an **algebraic number** and not equal to either 0 or 1, and (2) b is algebraic and also an **irrational number**. Gelfond's theorem enables the seventh of David **Hilbert**'s famous problems to be solved.

general relativity

See **relativity theory**.

general topology

See **point-set topology**.

genetic algorithm

A type of evolving computer program, developed by the computer scientist John Holland, whose strategy of arriving at solutions is based on principles taken from genetics. Basically, the genetic **algorithm** utilizes the mixing of genetic information in sexual reproduction, random mutations, and natural selection at arriving at solutions.

genus

In **topology**, roughly speaking, the number of **holes** in a surface. Spheres, bowling balls (the finger holes aren't true holes because they don't go all the way through), and wine glasses have a genus of 0 and can be represented by **quadratic** equations. Bagels, inner tubes, and teacups have a genus of 1 and can be described by **cubic** equations. Humans are more difficult to specify. However, you will certainly increase your genus by one if you have your ear pierced! Different definitions of genus apply to other types of mathematical objects such as a **curve**, a **knot**, or a **set**.

geoboard

A device commonly used in elementary schools to aid in the teaching of basic geometric concepts. A simple geoboard can be made from a square piece of wood and 25 nails arranged in an evenly spaced grid of 5 vertical lines and 5 horizontal lines. These nails represent the **lattice** points in the plane. Figures are made on the geoboard by stretching rubber bands from one nail to another until the desired shape is formed.

geodesic

A path on a given **surface** that is as straight as possible; in other words, a path that doesn't deviate either to the left or to the right, and only bends when forced to do so by the **curvature** (if any) of the surface. If the surface is an ordinary **plane**, the geodesics are straight lines; on a sphere the geodesics are **great circles**.

geometric magic square

A square array of $n \times n$ cells each occupied by a distinct geometrical figure (or piece or tile), such that the n pieces contained in every row, column, and diagonal can be fitted together to produce (i.e., tile or pack) a constant shape known as the target. The figures may be of any dimension, but are normally planar (topological disks). In tessellating the target, which may be of any shape, planar pieces are allowed to be flipped. Pieces of three or more dimensions are considered distinct from their mirror images. Geometric **magic squares** using one-dimensional entries have been known for centuries; they are the traditional magic squares in which straight lines pave a constant length, as usually represented by numbers adding to a constant total. The properties of generalized geometric magic squares were first investigated by Lee **Sallows**.

geometric mean

The geometric mean of n numbers is the nth root of the product of the numbers.

geometric sequence

Also known as a *geometric progression,* a finite **sequence** of at least three numbers, or an infinite sequence, whose terms differ by a constant multiple, known as the *common ratio*. For example, starting with 3 and using a common ratio of 2 leads to the finite geometric sequence: 3, 6, 12, 24, 48, and also to the infinite sequence 3, 6, 12, 24, 48, . . . , (3×2^n). . . . In general, the terms of a geometric sequence have the form $a_n = ar^n$ ($n = 0, 1, 2, \ldots$) for fixed numbers a and r. If the terms of a geometric sequence are added together the result is a *geometric series*. If it is a finite series, then we add its terms to get the series sum, $S_n = a + ar + ar^2 + \ldots + ar^n = (a - ar^{n+1})/(1 - r)$. In the case of an **infinite series**, if $|r| < 1$, the sum is $a/(1 - r)$. If $|r| \geq 1$, however, the series diverges and thus has no sum. See also **arithmetic sequence**.

geometry

The study of the properties of shapes and of spaces. See also **Euclidean geometry** and **non-Euclidean geometry**.

geometry puzzles

One of the attractions of puzzles involving shapes, especially **dissection** problems, is that they appeal to the eye and very often don't call for much ability in solving equations and the like. Anyone can try to assemble the pieces of a jigsaw, whether it be of a picture or of a geometric shape, so a mathematical game such as **tangrams** or the **Soma cube** is within everyone's reach. On the other hand, some geometric puzzles call for a basic knowledge of more abstract fields such as algebra and calculus. They may also exploit our sometimes faulty intuition about how different quantities vary in one, two, and three dimensions and about how much information is needed to solve a problem.

As an example of faulty intuition, imagine that Earth, taken to be a perfect sphere with a radius r of 6,378 km, is completely covered by a thin membrane. Now suppose that 1 square meter is added to the area of this membrane to form a larger sphere. By how much does the radius and the volume of this membrane increase? This can be worked out from the formulae for the volume of a sphere ($V = (4/3)\pi r^3$) and the area of a sphere ($A = 4\pi r^2$), respectively. It turns out that if the area of the cover is increased by 1 square meter, then the volume it contains is increased by about 3.25 million cubic meters. This seems like a huge amount. However, the new cover wouldn't be very high above the surface of the planet—only about 6 nanometers! As an example of a problem that is both counterintuitive *and* seems to lack sufficient data for its solution see **hole-through-a-sphere problem**.

Gergonne point

In a triangle, the point at which the lines from the vertices (see **vertex**) to the points of contact of the opposite sides with the **inscribed circle** meet.

Germain, Sophie (1776–1831)

A French mathematician who made notable contributions to **number theory** and to mathematical physics despite a lack of formal training and the social prejudices of her day. She taught herself, against her parents' wishes, often at night, during the Reign of Terror following the French Revolution, with books from her father's library. When deprived of heat and light, she would wrap herself in quilts and use candles. Finally her parents acquiesced to her "incurable" passion for mathematics and let her study. Through Joseph **Lagrange**, to whom she had originally submitted work under a pseudonym, she gained access to a circle of distinguished mathematicians, including Carl **Gauss**. Among her most important work was an analysis of Ernst **Chladni**'s studies of vibrating surfaces, and her proof that if x, y, and z are integers and if $x^5 + y^5 = z^5$ then at least one of x, y, or z must be divisible by 5 (a result now known as *Germain's theorem*); this was an important early step towards proving **Fermat's last theorem**.

Get Off the Earth

A famous **vanishment puzzle** by Sam **Loyd**. The picture is made from a rectangular background topped with a circular card, representing the world, that can be rotated. Parts of a number of Chinese men are on each piece. With the world orientated so that the large arrow on it points to

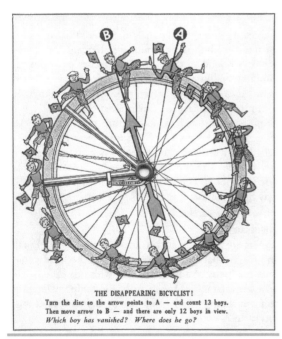

THE DISAPPEARING BICYCLIST!
Turn the disc so the arrow points to A — and count 13 boys.
Then move arrow to B — and there are only 12 boys in view.
Which boy has vanished? Where does he go?

Get Off the Earth   An unusual variant of Loyd's Get Off the Earth puzzle called "The Disappearing Bicyclist." *From the collection of William Waite*

the N.E. point on the background, 13 Chinamen can be counted. But when the earth is turned slightly, so that the arrow points N.W., there are only 12 characters. Where did the thirteenth Chinaman go? The cleverness of the puzzle is that there are many bits of Chinamen–arms, legs, bodies, heads, and swords–and each has tiny slivers missing. When the earth is turned, these pieces get slightly rearranged. In particular, each of the 12 Chinamen gains a sliver of a Chinaman from his neighbor.

Gettier problem

A thought experiment in philosophy that throws into question the long-held supposition that to know something is equivalent to holding a belief about something that is both true and for which there is justification. Consider a case in which a lecturer has two students in her class called Mr. Havenot and Mr. Havegot. Mr. Havenot claims to own a Ferrari, drives one around, and has papers that state that the car is his. But in fact he does not actually own the car. Mr. Havegot, on the other hand, who shows no sign of Ferrari ownership, secretly has one of these rare cars. On the basis of the evidence, the teacher concludes that one of her students owns a Ferrari–and is correct in this belief. However, there is something wrong. Despite the combination of truth, justification, and belief, it seems that there is no real knowledge. The first examples of such problems were published in 1963 by the American philosopher Edmund Gettier (1927–).

Giant's Causeway

A natural structure that occurs on the coast in County Antrim, Ireland; it is one of the few places in the world where volcanic basalt has cooled in a columnar formation. The columns approximately form a **hexagon**al tessellation (see **tiling**) and tend to break off to produce a pavement with this pattern. The full length of the columns can't be seen, but it is estimated that they may be 20 feet (about 6 meters) high before merging into the underlying irregular basaltic mass. About 99% of the columns are believed to be hexagonal and only one triangular column is known. Though many of the hexagons are fairly regular, some have a side twice as long as their smallest side. Side lengths vary from 8 to 18 inches (20 to 46 cm) and the pillars break up into sections 6 to 36 inches (15 to 90 cm) long, with a concavo-convex junction rather than a plane junction. Other examples of such formations occur at Kirkjubaejarklaustri, in Iceland, and the Devil's Postpile, in California.

Gilbreath's conjecture

A strange hypothesis concerning **prime numbers** that was first suggested in 1958 by the American mathematician and amateur magician Norman L. Gilbreath following

Giant's causeway Hexagonal paving on the Giant's Causeway. *Martin Melaugh, University of Ulster*

some doodlings on a napkin. Gilbreath started by writing down the first few primes:

2, 3, 5, 7, 11, 13, 17, 19, 23, 29, 31, . . .

Under these he put their differences:

1, 2, 2, 4, 2, 4, 2, 4, 6, 2, . . .

Under these he put the unsigned difference of the differences:

1, 0, 2, 2, 2, 2, 2, 2, 4, . . .

And he continued this process of finding iterated differences:

1, 2, 0, 0, 0, 0, 0, 2, . . .
1, 2, 0, 0, 0, 0, 2, . . .
1, 2, 0, 0, 0, 2, . . .
1, 2, 0, 0, 2, . . .
1, 2, 0, 2, . . .
1, 2, 2, . . .
1, 0, . . .
1, . . .

Gilbreath's conjecture is that, after the initial two rows, the numbers in the first column are all one. No exception has been found to date, despite searches out to several hundred billion rows, and the conjecture is generally assumed to be true. However, it may have nothing to do with primes as such. The English mathematician Hallard Croft has suggested the conjecture may apply to *any* sequence that begins with 2 and is followed by odd numbers that increase at a "reasonable" rate and with gaps of "reasonable" size. If this is the case, Gilbreath's conjecture may not be as mysterious as it first seems, though it may be very difficult to prove.

glissette
If there are two fixed curves, and a curve S of fixed shape and length that slides with its ends on the fixed curves, then the **locus** of a point moving with S is called a glissette. An example is the locus of the midpoint of a line segment sliding with its ends on two perpendicular lines; this locus is a circle.

gnomon magic square
A 3×3 array in which the elements in each 2×2 corner have the same sum. See also **magic square**.

Go

A two-player board game that originated in China in about 2000 B.C. It is often compared and contrasted with **chess**. Go is played on a board marked with 19×19 lines. Round, lens-shaped pieces called stones are placed, one per move, on the intersections of this grid, of which there are 361. As in chess, the pieces are colored black and white, but in Go black plays first. The board starts blank and pieces once played are not thereafter moved except to be taken off as prisoners. Pieces are captured singly or en masse by being surrounded so that they are not connected to any adjacent open intersection. The player with highest score at the end of the game, or following resignation or time expiry, wins. Go is considered to be a deeply strategic game, unlike chess, which is largely tactical. There are 32,940 opening moves, after symmetry is taken into account, 992 of which are deemed strong. Estimates of the number of possible board configurations vary but are typically on the order of 10^{174}.

God

I'm still an atheist, thank God.
—Luis Buñuel (Spanish film director, 1900–1983)

Mathematicians, logicians, and scientists have long debated the nature, existence, and dice-playing ability of a Higher Power. The pre-Renaissance French philosopher Jean Buridan (c. 1295–1358) used a version of the *liar paradox* to "prove" the existence of God. He wrote these two sentences:

God exists.

None of the sentences in this pair is true.

The only consistent way to have these two sentences be either true or false is for "God exists" to be true. (However, there is nothing to say that such consistency is necessary.) Blaise **Pascal** gave a more persuasive argument, not for the existence of God but for why we should believe in that existence: "If I believe in God and life after death and you do not, and if there is no God, we both lose when we die. However, if there is a God, you still lose and I gain everything." Pierre **Laplace**, on the other hand, replying to **Napoleon Bonaparte**, who had asked why his celestial mechanics made no mention of God, said: "Sir, I have no need of this hypothesis." The German mathematician **Kronecker** thought that "God made the Integers, all the rest is the work of man." In *The City of God*, however, Saint Augustine seems to imply that the integers were independent of God. He wrote: "**Six** is a number perfect in itself, and not because God created the world in six days; rather the contrary is true. God created the world

in six days because this number is perfect, and it would remain perfect, even if the work of the six days did not exist." Augustine's statement can be taken to suggest that six would be a perfect number not only if the universe didn't exist, but even if God didn't exist. As to God's mathematical specialty, Plato said, "God ever geometrizes" while Charles Jacobi insisted that "God ever arithmetizes." James Jeans thought, "The Great Architect of the Universe now begins to appear as a pure mathematician," and Einstein ("God does not play dice") was sure He wasn't a probabilist.

Gödel, Kurt (1906–1978)

An Austrian-American mathematician and logician who, in 1931, proved that within a **formal system** questions exist that are neither provable nor disprovable on the basis of the **axiom**s that define the system. This is known as Gödel's undecidability theorem. He also showed that in a sufficiently rich formal system in which decidability of all questions is required, there will be contradictory statements. This is called **Gödel's incompleteness theorem**. In establishing these theorems Gödel showed that

Gödel, Kurt

there are problems that can't be solved by any set of rules or procedures; instead, for these problems one must always extend the set of axioms. This disproved a common belief at the time that the different branches of mathematics could be integrated and placed on a single logical foundation. Gödel was a close friend of Albert Einstein at Princeton and contributed to his general **relativity theory** and cosmology. The so-called *Gödel universe* is a rotating model of the universe in which it is theoretically possible to travel into the past (see **time travel**).

Gödel's incompleteness theorem

In a nutshell: All consistent axiomatic systems contain undecidable propositions. What does this mean? An axiomatic system consists of some undefined terms, a number of **axiom**s that refer to those terms and partially describe their properties, and a rule or rules for deriving new propositions from already existing propositions. Axiomatic systems are powerful because they reduce large bodies of math to a simple description. Also, because they're very abstract, they allow all, and only, the results that follow from things having the formal properties specified by the axioms to be derived. An axiomatic system is *consistent* if, given the axioms and the derivation rules, it doesn't lead to any contradictions. One of the first modern axiomatic systems was a formalization of simple arithmetic (adding and multiplying whole numbers) by Giuseppe **Peano** and now known as *Peano arithmetic*. Kurt **Gödel** showed that every syntactically correct proposition in Peano arithmetic can be represented by a unique integer, called its *Gödel number*. The trick is to replace each symbol in the proposition, *including numerals*, by a different string of digits. If we represent "1" by 01, "2" by 02, "+" by 10, and "=" by 11, then the Gödel number of "1 + 1 = 2" is 0110011102. This allowed Gödel to write down, unambiguously, propositions about propositions. In particular, he was able to write down self-referential (see **self-referential sentence**) propositions—ones that include their own Gödel number. Gödel was then able to prove that, *either* the system of Peano arithmetic is inconsistent, *or* there are true propositions that can't be reached from the axioms by applying the derivation rules. The system is thus *incomplete*, and the truth of those propositions is *undecidable* (within that system). Such undecidable propositions are known as *Gödel propositions* or *Gödel sentences*. Nobody knows what the Gödel sentences for Peano arithmetic are, though people have their suspicions about the **Goldbach conjecture** (every even number is the sum of two prime numbers).

The results of an axiomatic system pertain to more than just Peano arithmetic, they apply to all kinds of things that satisfy the axioms. There are an immense number of other axiomatic systems, which either include Peano numbers among their basic entities or can be constructed from them. It follows that these systems, too, contain undecidable propositions, and are incomplete.

A common misconception is that Gödel's theorem imposes some profound limitation on knowledge, science, and mathematics. In the case of science, this ignores that Gödel's theorem applies to *deduction from axioms*, which is only one source of knowledge and not even a very common mode of reasoning in science. More generally, Gödel's incompleteness result doesn't touch directly on the most important sense of completeness and incompleteness, namely, descriptive completeness and incompleteness—the sense in which an axiom system describes a given field. In particular, the result represents no threat to the notion of truth.

Goldbach conjecture

One of the oldest and easiest-to-understand hypotheses in mathematics that remains unproven. In its original form, now known as the *weak Goldbach conjecture*, it was put forward by the Prussian amateur mathematician and historian Christian Goldbach (1690–1764) in a letter dated June 7, 1742, to Leonhard **Euler**. In this guise it says that every whole number greater than five is the sum of three **prime number**s. Euler restated this, in an equivalent form, as what is now called the *strong Goldbach conjecture* or, simply, the Goldbach conjecture: every even number greater than two is the sum of two primes. Thus, $4 = 2 + 2$, $6 = 3 + 3$, $8 = 3 + 5$, $10 = 3 + 7, \ldots$, $100 = 53 + 47, \ldots$. In fact René **Descartes** knew about the two-prime version of Goldbach's conjecture before either Goldbach or Euler did. So, is it misnamed? Paul **Erdös** said, "It is better that the conjecture be named after Goldbach because, mathematically speaking, Descartes was infinitely rich and Goldbach was very poor." In any event, there is a much more important question, namely, is the conjecture true? The general assumption is that it is, but no one knows for sure. The most significant step toward a proof came in 1966 when the Chinese mathematician Chen Jing-Run showed that every sufficiently large even integer is the sum of a prime and a number that has at most two prime factors. Using powerful computers, the Goldbach conjecture has been checked out to about 400 trillion. But there is no great optimism among mathematicians that a final breakthrough is on the horizon. Even a reward of $1 million dollars for a proof offered by the publishing house Faber & Faber in 2000, to help publicize the novel *Uncle Petros and Goldbach's Conjecture* by the Greek mathematician and author Apostolos Doxiadis, went unclaimed.[85]

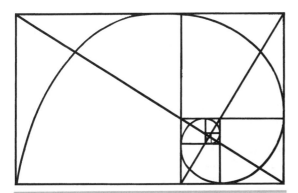

golden ratio A golden rectangle and a logarithmic spiral emerge from a whirling pattern of squares built up from two small squares of equal size at the spiral's center.

golden ratio (phi, φ)

A remarkable number that, like **pi** and *e*, pops up all over the place in mathematics but, in some ways, has a more "human" connection, in that it seems to be linked to aesthetics. Its name, which is also given as the *golden mean,* the *golden section,* the *golden number,* and the *divine proportion,* reflects this sense of a harmonious or pleasing ideal. The golden ratio is an **irrational number** of the type known as an **algebraic number** (in contrast with π and *e*, which are transcendental) and is represented by the Greek letter ϕ (phi). It can be defined in various ways. For example, it is the only number equal to its own reciprocal plus 1, that is, $\phi = (1/\phi) + 1$, so that $\phi^2 = \phi + 1$. From this comes the quadratic equation $\phi^2 - \phi - 1 = 0$ of which the golden ratio is the positive solution, $(1 + \sqrt{5}) / 2 \approx 1.6180339887 \ldots$. The golden ratio is also approximated by the ratio of successive terms in the **Fibonacci sequence**; in fact, $F(n + 1) / F(n)$ gets closer and closer to ϕ as *n* tends to infinity. Because $1/(1 - \phi) = \phi$, the **continued fraction** representation of ϕ is

$$\phi = 1 + 1/(1 + 1/(1 + 1/(1 + 1/(1 + 1/ \ldots$$
$$= [1;1,1,1,1, \ldots].$$

Two quantities are said to be in the golden ratio, if the ratio of the larger one, *a*, to the smaller one, *b*, is the same as the ratio of the smaller one to their difference, that is, $a/b = b/(a - b)$. The so-called *golden rectangle* is one whose sides *a* and *b* stand in the golden ratio. It is famously said to have great aesthetic appeal and is closely approximated by the dimensions of the front of the Parthenon in Rome. Leonardo da Vinci's masterpiece the *Mona Lisa* is said to have a face that is framed by a golden rectangle; what is certain is that Leonardo was a close personal friend of Luca **Pacioli**, who published a three-volume

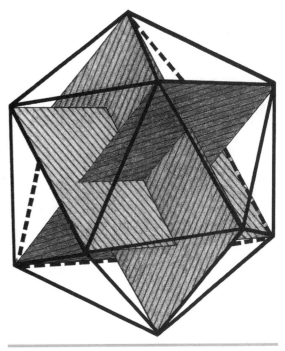

golden ratio The corners of an icosahedron meet the corners of three orthogonal golden rectangles.

treatise on the golden ratio, *Divina Proportione,* in 1509. The Swiss-French architect and painter Le Corbusier designed an entire proportional system called the "Modulor," that was based on the golden ratio. The Modulor was supposed to provide a standardized system that would automatically confer harmonious proportions to everything, from door handles to high-rise buildings. Another artist who deliberately used the golden ratio is the surrealist Salvador Dali. The ratio of the dimensions of Dali's *Sacrament of the Last Supper* is equal to the golden ratio. Dali also incorporated in the painting a huge dodecahedron (a twelve-faced Platonic solid in which each side is a pentagon) engulfing the supper table. The dodecahedron, which according to Plato is the solid "which the god used for embroidering the constellations on the whole heaven," is intimately related to the golden ratio—both the surface area and the volume of a dodecahedron of unit edge length are simple functions of the golden ratio. In fact, ϕ turns up frequently in figures that have pentagonal symmetry. For instance the ratio of a regular pentagon's side and diagonal is equal to ϕ, and the vertices of a regular **icosahedron** are located on three orthogonal golden rectangles. The golden ratio is also related to **Penrose tiling** and to the **plastic number**.[205]

Golomb, Solomon W. (1932–)

A mathematician and electrical engineer at the University of Southern California who is best known for his seminal studies of **polyominos**. His article "Checker Boards and Polyominos," published in the *American Mathematical Monthly* in 1954 when Golomb was a 22-year-old graduate student at Harvard, defined a polyomino as a simply connected set of squares (i.e., a set of squares joined along their edges). Golomb also originated the idea of **graceful graph**s.[137]

golygon

A series of straight-line segments that have lengths of one, two, three, and so on, up to some finite number of units, with the property that every segment connects at a right angle to the segment that is one unit larger, except the longest segment, which meets the shortest segment at a right angle. The name "golygon" was invented by Lee **Sallows**. Golygons have inspired some interesting puzzles as well as some intriguing research problems.

googol

A named coined in 1938 by Milton Sirotta, the 9-year-old nephew of the mathematician Edward **Kasner**, when the child was asked by his uncle to come up with a name for a very **large number**; at the same time, *googolplex* was suggested for a still larger number. Kasner defined these numbers as follows:

1 googol = 10^{100} (i.e., 1 followed by 100 zeros);
1 googolplex = $10^{googol} = 10^{10^{100}}$ (i.e., 1 followed by a googol number of zeros)

A googol is very roughly the number of years thought to be needed for all the black holes in the universe to evaporate by a process known as Hawking radiation. It is much larger than the number of protons and neutrons in the known universe (about 10^{80}), but much smaller than the number of protons and neutrons needed to pack every cubic centimeter of the known universe (about 10^{128}). The googolplex is the largest number with a proper name of which many people have heard. It is dwarfed, however, by such esoterica as **Graham's number**.

Gordian knot

The earliest reference to a **string puzzle**. In Greek mythology, a Phrygian peasant called Gordius, the father of Minos (see **maze**), became king because he was first to arrive in town after an oracle commanded the Phrygians to select as ruler the first person to drive into the public square in a wagon. In gratitude, Gordius dedicated his wagon to Zeus and placed it in the temple grove, tying the wagon pole to the yoke with a rope of bark. The knot was so intricately entwined that no one could undo it. A saying developed that whoever succeeded in untying the knot would become ruler of all Asia. Many tried, but all failed. According to legend, even Alexander the Great was unable to untie the Gordian knot, so he drew his sword and cut it through with a stroke. The expression "to cut the Gordian knot" is used to refer to a situation in which a difficult problem is solved by a quick and decisive action.

graceful graph

A **graph** of points and connecting lines that can be numbered in a certain way. Say the graph has p points and e lines ("e" for edges) connecting them. Each of the points is assigned an integer; the lowest integer (by convention) is taken to be 0, and no two integers may be alike. Each of the lines is labeled with the difference between the two integers of the points that it connects. Then, if the numbers corresponding with the lines run from 0 through e, the graph is said to be graceful. Graceful graphs were originally defined and developed by Solomon **Golomb**.

gradient

A **vector** of partial **derivative**s of a **function** that operates on vectors. Intuitively, the gradient represents the slope of a high-dimensional surface.

Graham, Ronald L. (1936–)

An American mathematician and leading combinatorialist after whom **Graham's number** is named. Graham is also one of the country's best jugglers and former president of the International Juggler's Association. In his youth, he and two friends were professional trampolinists who performed with a circus as the Bouncing Baers. His office ceiling is covered with a large net that he can lower and attach to his waist so that when he practices juggling with six or seven balls, any that are dropped will roll back to him. Graham is a professor in the department of computer science and engineering at the University of California at San Diego.

Graham's number

A stupendously **large number** that found its way in to the *Guinness Book of Records* as the biggest number ever obtained as part of a mathematical proof; it is named after its discoverer, Ronald **Graham**. Graham's number is the upper bound solution to a very exotic problem in **Ramsey theory**, namely: What is the smallest dimension n of a **hypercube** such that if the lines joining all pairs of corners are two-colored, a planar complete graph K_4 of one color will be forced? This is exactly equivalent to a problem that can be stated in plain language: Take any number of people, list every possible committee that can be formed from them, and consider every possible pair of committees. How many people must be in the original group so that no graph **139**

matter how the assignments are made, there will be four committees in which all the pairs fall in the same group, and all the people belong to an even number of committees. Graham's number is the *greatest* value that the answer could take. It is so large that it can only be written using special big-number notation, such as **Knuth's up-arrow notation**. Even then, it must be built up in stages. First, construct the number $G_1 = 3 \uparrow\uparrow \ldots \uparrow\uparrow 3$, where there are $3\uparrow\uparrow\uparrow\uparrow3$ up-arrows. This, in itself, is a number far, far beyond anyone's ability to even remotely comprehend. Next, construct $G_2 = 3 \uparrow\uparrow \ldots \uparrow\uparrow 3$, where there are G_1 up-arrows; then construct $G_3 = 3 \uparrow\uparrow \ldots \uparrow\uparrow 3$, where there are G_2 up-arrows; then continue this pattern until the number G_{62} has been made. Graham's number $G = 3 \uparrow\uparrow \ldots \uparrow\uparrow 3$, where there are G_{62} up-arrows. While the unthinkably large upper boundary to the problem described earlier is given by Graham's number, nobody, including Graham himself, believes the solution is nearly so large. In fact, it is thought that the actual answer is probably 6![115]

grandfather paradox

One of the most powerful and commonly used arguments against **time travel**. It points out that if you were able to travel into the past you could (if you were so inclined) kill your grandfather when he was very young and thus render your own birth impossible. A simpler version is that you could kill a younger version of yourself so that you would not be alive in the future to travel back in time. The grandfather paradox shows how one form of time travel could violate **causality** by eliminating the cause of a phenomenon that has already taken place in the present.

The most bizarre adaptation of the grandfather paradox is found in Robert Heinlein's classic short story "All You Zombies." A baby girl is mysteriously left at an orphanage in Cleveland in 1945. "Jane" grows up lonely and dejected, not knowing who her parents are, until one day in 1963 she is strangely attracted to a drifter. She falls in love with him. But just when things are finally looking up for Jane, a series of disasters strike. First, she becomes pregnant by the drifter, who then disappears. Second, during the complicated delivery, doctors find that Jane has both sets of sex organs, and to save her life, they are forced to surgically convert "her" to a "him." Finally, a mysterious stranger kidnaps her baby from the delivery room. Reeling from these disasters, rejected by society, scorned by fate, "he" becomes a drunkard and drifter. Not only has Jane lost her parents and her lover, but he has lost his only child as well. Years later, in 1970, he stumbles into a lonely bar, called Pop's Place, and spills out his pathetic story to an elderly bartender. The bartender offers the drifter the chance to avenge the stranger who left her pregnant and abandoned, on the condition that he (Jane) join the "time travelers corps." Both of

them enter a time machine, and the bartender drops off the drifter in 1963. The drifter is strangely attracted to a young orphan woman, who subsequently becomes pregnant. The bartender then goes forward nine months, kidnaps the baby girl from the hospital, and drops off the baby in an orphanage back in 1945. Then the bartender drops off the thoroughly confused drifter in 1985, to enlist in the time travelers corps. The drifter eventually gets his life together, becomes a respected and elderly member of the time travelers corps, and then disguises himself as a bartender and has his most difficult mission: a date with destiny, meeting a certain drifter at Pop's Place in 1970. The question is: who is Jane's mother, father, grandfather, grandmother, son, daughter, granddaughter, and grandson? The girl, the drifter, and the bartender, of course, are all the same person. As an exercise (on the road to insanity) try drawing Jane's family tree. You will find that not only is she her own mother and father, she is an entire family tree unto herself!

graph

I'll do algebra, I'll do trig, and I'll even do statistics, but graphing is where I draw the line!
—Anonymous

(1) In common usage, a plot of x values (the domain) against y values (the codomain) for a given **function**, $y = f(x)$. Such a graph is also known as a *function graph* or the *graph of a function*. (2) In strict mathematical usage, any set of dots, known as *nodes* or *vertices,* in which at least some pairs are joined by lines known as *edges* or *arcs.* What follows applies only to this second definition.

Often the lines on a graph are used to represent relationships between objects (represented by dots). Depending on the application, edges may or may not have a direction, as indicated by an arrow (see **directed graph**); edges joining a node to itself may or may not be allowed, and nodes and/or edges may be assigned weights. A *path* is a series of nodes such that each node is adjacent to both the preceding and succeeding node. A path is considered *simple* if none of the nodes in the path is repeated. The *length* of a path is the number of edges that the path uses, counting multiple edges multiple times. If it's possible to establish a path from any node to any other node of a graph, the graph is said to be a **connected graph**. A *circuit* or *cycle* is a path that begins and ends with the same node and has a length of at least two. A *tree* is a connected acyclic graph, that is, a graph without any circuits. A **complete graph** is one in which every node is adjacent to every other node. An **Euler path** in a graph is a path that uses each edge precisely once. If such a path exists, the graph is said to be *traversable.* An **Euler circuit** is a path that traverses each edge precisely once. A **Hamilton path**

Great Monad

in a graph is a path that visits each node once and only once; a **Hamilton circuit** is a **circuit** that visits each node once and only once. Well known problems whose solution involves graphs and graph theory include the **four-color problem** and the **traveling salesman problem**.

graph theory
The study of **graph**s, either for their own sake, or as models of such diverse things as **groups** (in pure mathematics) or computer networks.

great circle
A circle that goes all the way around a **sphere** and is centered at the center of the sphere. The shortest route between two points on a sphere, such as Earth, is along the great circle that connects these points. A great circle is a **geodesic**.

Great Monad
Also known as *T'ai-Chi*, an important and ubiquitous symbol in traditional Chinese philosophy and cosmology. It represents the underlying harmony of the universe when its opposites or dualities—male and female (*yang* and *yin*), hard and soft, sun and moon, and so forth—are in balance. It occurs everywhere in Chinese art: in books, on walls, porcelain, tablets, and stitched into brocade.

greatest common divisor
The largest integer that divides each of a sequence of integers exactly. Also called the *greatest common factor*.

greatest lower bound
The largest **real number** that is smaller than each of the numbers in a **set** of real numbers.

Green, George (1793–1841)
An English mathematician who published work in the fields of hydrodynamics, electricity, and magnetism, but is best known for his theorem (see **Green's theorem**), which is the basis of **potential theory**. Green took over a bakery and adjoining windmill after the death of his father, but studied mathematics in his spare time. In 1828, he wrote his most important paper, "An Essay on the Application of Mathematical Analysis to the Theories of Electricity and Magnetism," which, though generally overlooked at the time, is now regarded as the beginning of mathematical physics in England.

Green's theorem
A connection between path integrals over a well-connected region in the plane and the area of the region bounded in the plane. Green's theorem is a form of the *fundamental theorem of calculus,* and is used today in almost all computer codes that solve **partial differential equation**s.

Grelling's paradox
An equivalent, from the world of words and grammar, of **Russell's paradox**. Grelling's paradox involves dividing all adjectives into two sets: self-applicable and not self-applicable. Words like "English," "written," and "short" are self-applicable, while "Russian," "spoken," and "long" are not self-applicable. Now, define the adjective *heterological* to mean "not self-applicable." To which set of adjectives does "heterological" belong? This strange quandry was devised by the logician and philosopher Kurt Grelling (1886–1941/2), who was persecuted by the Nazis; it is not certain whether he died with his wife in the Auschwitz concentration camp in 1942, or whether he was killed in 1941 in the Pyrenees while trying to escape into Spain.

gross
A group of 144 items. The word comes from the Latin *grossus,* for "thick" or "large," via the Old French *gross douzaine* or "large dozen" (12 dozen), though this grouping may have started out in Germany. "Grocer" has the same origins as "gross" because a grocer is someone who deals in large quantities of food. A *great gross,* or a dozen gross, is 1728. See also **twelve**.

group
Wherever groups disclosed themselves, or could be introduced, simplicity crystallized out of comparative chaos.

—Eric Temple Bell

An abstract and crucially important way of representing **symmetry** and one of the most fundamental concepts in modern **algebra**. Groups were brought into mathematics in the early nineteenth century by the radical young French student Evariste **Galois** as a tool to help solve one of the outstanding problems of his day: to find a formula for solving **polynomial** equations of order five—**quintic**s—and higher. Galois showed, in notes scribbled down the night before he died in a duel, that no such formula exists. The reason for this is that the possible symmetries, or permutations, of the **roots** of fifth-degree polynomial equations are more complex than are the symmetries that can be represented by arithmetical formulas. This fact emerged from the development of the idea of a *permutation group* by Galois and, independently at about the same time, by Niels **Abel**. Half a century later, another Norwegian, Sophus **Lie**, showed how important groups are to the whole of mathematics. The theory of what became known as **Lie group**s links the discrete structure of permutations with the continuous variation of **differential equation**s. Not surprisingly, because group theory forms a common underpinning to algebra and to geometric features such as rotation, reflection, and symmetry, it crops up routinely in modern physics, from the classification of elementary particles to crystallography.

A group is a **set** whose elements are defined by a single operation. The group is called *additive* if the symbol for the operation is "+" and is called *multiplicative* if the symbol is "·" for multiplication. But any other symbol can be substituted for these. There is always a unique element (1, for multiplicative, and 0, for additive, groups) that leaves elements unchanged under the defined operation, like $a + 0 = a$. Also, for every element a there exists a unique inverse b such that, for example, in the case of the additive symbol, $a + b = 0$ and $b + a = 0$. Most often, however, the inverse is denoted as a^{-1}. Lastly, the group operation must be **associative** as in $a \cdot (b \cdot c) = (a \cdot b) \cdot c$. A group is *commutative* or *Abelian* if its operation is symmetric, as in $a + b = b + a$.

Groups come in two types: finite and infinite. The symmetry group of the roots of a polynomial equation is a finite group, because there is only a limited number of permutations possible among the roots of a given polynomial. In contrast, the Lie groups that represent symmetries of solutions of differential equations are infinite because they represent continuous transformations, and continuity carries the potential of an infinite number of changes. Finite groups can be built up from combinations of smaller groups by a process analogous to multiplication. In the same way that a whole number can be written as a product of **prime number**s, a finite group can be expressed as a combination of certain factors known as *simple groups*. Most simple groups belong to one of three families: the *cyclic groups,* the *alternating groups,* or the *groups of Lie type.* Cyclic groups consist of cyclic permutations of a prime number of objects. Alternating groups consist of even permutations—those formed by interchanging the positions of two objects an even number of times. Sixteen subfamilies make up the simple groups of Lie type, each associated with a family of infinite Lie groups. (Confusingly, a Lie group is not a group of Lie type, since the former is infinite and the latter is finite!) Altogether, there are 18 specific families of finite simple groups. There are also 26 simple groups, known as *sporadic groups,* that are highly irregular and fall outside these families. Five sporadic groups were found in the nineteenth century by Emile Mathieu. Then came a hiatus until the 1960s, when suddenly a rush of new sporadics came to light. The most remarkable of these is the so-called **monster group**, which appears to be intimately related to the structure of the universe at the subatomic level.

Grundy's game
See **Nim**.

Guy, Richard Kenneth (1916–)
A British-born mathematician who is professor emeritus of mathematics at the University of Calgary, Canada, and an expert in **combinatorics** and in **number theory**. Guy is the author of more than 250 papers and 10 books, including (as a coauthor) the game theory classic *Winning Ways*. He has been an editor of the "Problems" section of the *American Mathematical Monthly* since 1971.

Haberdasher's puzzle

The greatest mathematical discovery of Henry **Dudeney**, it was first published in the *Weekly Dispatch in* 1902 and then as problem no. 26 in his *The Canterbury Puzzles* (1907).[87] One must decide how to cut an equilateral triangle into four pieces that can be rearranged to make a square. The accompanying diagram shows the solution, which Dudeney describes as follows:

> Bisect AB in D and BC in E; produce the line AE to F making EF equal to EB; bisect AF in G and describe arc AHF; produce EB to H, and EH is the length of the side of the required square; from E with distance EH, describe the arc HJ, and make JK equal to BE; now from the points D and K drop perpendiculars on EJ at L and M.

A remarkable feature of the solution is that each of the pieces can be hinged at one vertex, forming a chain that can be folded into the square or the original triangle. Two of the hinges bisect sides of the triangle, while the third hinge and the corner of the large piece on the base cut

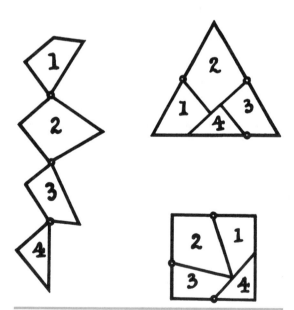

Haberdasher's puzzle The puzzle and its solution as illustrated by Henry Dudeney.

the base in the approximate ratio 0.982:2:1.018. Dudeney showed just such a model of the solution, made of polished mahogany with brass hinges, at a meeting of the Royal Society on May 17, 1905.

Hadwiger problem

In d dimensions, define $L(d)$ to be the largest integer n for which a cube cannot be cut into n cubes (not necessarily different). The Hadwiger problem is to find $L(d)$. Definite solutions are only known in two and three dimensions: $L(2) = 5$ and $L(3) = 47$. However, it is known that $L(4) \leq 853$ and $L(5) \leq 1,890$, and it is considered likely that $L(d)$ is odd for all values of d. See also **dissection**.

hailstone sequence

A sequence of numbers produced by the rules of the **Collatz problem**; in other words, a sequence formed in the following way: Start with any positive integer n. (1) If n is even, divide it by 2; if n is odd, multiply it by 3 and add 1. (2) If the result is not 1, repeat step (1) with the new number. For $n = 5$, this produces the sequence 5, 16, 8, 4, 2, 1, 4, 2, 1, ... For $n = 11$, the resulting sequence is 11, 34, 17, 52, 26, 13, 40, 20, 10, 5, 16, 8, 4, 2, 1, 4, 2, 1, ... The name "hailstone" comes from the fact that the numbers in these sequences rise and fall like hailstones in a cloud before finally falling to Earth. It seems from experiment that such a sequence will always eventually end in the repeating cycle 4, 2, 1, 4, 2, 1, ..., but some values for n generate many values before the repeating cycle begins. An unsolved mystery is whether all such sequences eventually hit 1 (and then 4, 2, 1, 4, 2, 1, ...) or whether there are some sequences that never settle down to a repeating cycle.

hairy ball theorem

If a sphere is covered with hair or fur, like a tennis ball, the hair cannot be brushed so that it lies flat at every point. In mathematical terms: any continuous tangent vector field on the sphere must have a point where the vector is zero. This theorem also means that somewhere on Earth's surface there has to be a point where the horizontal wind speed is zero, even if it's windy everywhere else. Does the same apply to a **torus**? Is there a hairy donut theorem? No! The number of "problem points," where the hair would stick up on a surface, is related to a quantity called the **Euler characteristic** of that surface. Basically, every point on a surface has an index that describes how many times the vector field rotates in a neighborhood of the problem point. The sum of the indices of all the vector fields is the Euler characteristic. Since the torus has Euler number 0, it is possible to have a covering of hair–a vector field–on it that lies flat at every point.

half-line
A **ray**.

half-plane
The part of a **plane** that lies on one side of a given line.

halting problem
Given a program and inputs for it, decide whether it will run forever or will eventually stop. This is not the same thing as actually running a given program and seeing what happens. The halting problem asks whether there is any *general prescription* for deciding how long to run an arbitrary program so that its halting or non-halting will be revealed. In a celebrated 1936 paper,[337] Alan **Turing** proved that the halting problem is undecidable: there's no way to construct an **algorithm** that is *always* able to determine whether another algorithm halts or not. From this it follows that there can't be an algorithm that decides whether a given statement about **natural numbers** is true or not. The undecidability of the halting problem provides an alternative proof of **Gödel's incompleteness theorem**. This is because if there were a complete and consistent axiomatization of all true statements about natural numbers, then we would be able to create a set of rules that decides whether such a statement is true or not. Another amazing consequence of the undecidability of the halting problem is *Rice's theorem,* which states that the truth of *any* nontrivial statement about the **function** that is defined by an algorithm is undecidable. So, for example, the decision problem "will this algorithm halt for the empty string" is already undecidable. Note that this theorem holds for the *function defined by the algorithm* and not the algorithm itself. It is, for example, quite possible to decide if an algorithm will halt within 100 steps, but this isn't a statement about the function that is defined by the algorithm. Many problems can be shown to be undecidable by reducing them to the halting problem. However, Gregory **Chaitin** has given an undecidable problem in algorithmic information theory that doesn't depend on the halting problem.

While Turing's proof shows that there can be no general method or algorithm to determine whether algorithms halt, individual instances of that problem may very well be susceptible to attack. Given a specific algorithm, one can often show that it must halt, and in fact computer scientists often do just that as part of a correctness proof.

But every such proof requires new arguments: there is no *mechanical, general way* to determine whether algorithms halt. And there's another caveat. The undecidability of the halting problem relies on the fact that computers are assumed to have a memory of potentially infinite size. If the memory and external storage of a machine is limited, as it is for any real computer, then the halting problem for programs running on that machine can be solved with a general algorithm (albeit an extremely inefficient one).

ham sandwich theorem
Given a sandwich in which bread, ham, and cheese (three finite volumes) are mixed up, in any way at all, there is always a flat slice of a knife (a plane) that bisects each of the ham, bread, and cheese. In other words, however messed up the sandwich–even if it's been in a blender–you can always slice through it in such a way that the two halves have exactly equal amounts, by volume, of the three ingredients. This theorem generalizes to higher-dimensional ham sandwiches, when it essentially becomes the **Borsuk-Ulam theorem**: in n-dimensional space in which there are n globs of positive volume, there is always a hyperplane that cuts all the globs exactly in half.

Hamilton, William Rowan (1805–1865)
An Irish mathematician who, among other things, invented **quaternions** and a new theory of dynamics. Having excelled in Greek and mathematical physics at Trinity College, Cambridge, Hamilton was appointed Astronomer Royal of Ireland; in this position he served from 1827 to his death and, during all that time, lived in Dunsink Observatory, Dunsink Lane, to the northwest of Dublin. However, he quickly lost interest in staying up at nights to make observations–he hired three of his sisters to help run the place–and preferred instead to write poetry (badly). He was friends with Samuel Coleridge, who introduced him to the philosophy of Kant, which had a great influence on him, and with William Wordsworth, who advised him against writing any more poems.

Hamilton did early work on **caustic** curves and was led from this to his discovery of the *law of least action,* which enabled many physical problems to be expressed more elegantly. One of his greatest triumphs was his treatment of **complex number**s as pairs of real numbers, an approach that finally exorcised long-standing suspicions about the reality of **imaginary number**s, and helped clear the way for other algebras. From this he was led to consider ordered quartets of numbers, which he called quaternions. The idea for quaternions came to Hamilton suddenly on October 16, 1843, while he was standing on Brougham ("Broom") Bridge, where Broombridge Street crosses the Royal Canal, Dublin. A commemorative plaque under the bridge, on the towpath, was unveiled by the Taoiseach (head of the Irish parliament), Eamon De Valera, on November 13, 1958. Of his invention, Hamilton wrote:

> The quaternion was born, as a curious offspring of a quaternion of parents, say of geometry, algebra, metaphysics, and poetry. . . . I have never been able to give a clearer statement of their nature and their aim than I have done in two lines of a sonnet addressed to Sir John Herschel:
> "And how the One of Time, of Space the Three
> Might in the Chain of Symbols girdled be."

Hamilton's interest in complex numbers was stimulated by his friend and compatriot John Graves, who pointed Hamilton in the direction of John Warren's *A Treatise on the Geometrical Representation of the Square Root of Negative Quantities*. This book explained the concept of the complex plane, which Hamilton turned from geometry into algebra. One of Hamilton's last inventions was a curiosity called the *icosian calculus,* which was another outcome of his friendship with Graves. After a visit to the latter's house, Hamilton wrote: "Conceive me shut up and revelling for a fortnight in John Graves' Paradise of Books! of which he has really an astonishingly extensive collection, especially in the curious and mathematical kinds. Such new works from the Continent he has picked up! and such rare old ones too!" Graves posed some puzzles to Hamilton, and either Graves or his books got Hamilton to thinking about regular polyhedra. When Hamilton returned to Dublin he thought about the symmetry group of the **icosahedron**, and used it to invent an algebra he called the "icosians" and also a game called the **Icosian game**. The only complete example of this game, inscribed to Graves, is now in the keeping of the Royal Irish Academy, of which Hamilton was the president from 1837 to 1847. (In early 1996, a second example of the Icosian game came to light but only included the board.)

In some ways, Hamilton was too far ahead of his time. The operator now referred to as the *Hamiltonian* and the so-called *Hamilton-Jacobi equation* that relates waves and particles only became important when **quantum mechanics** came along, and Felix **Klein** introduced Wernher Schrödinger, the father of wave mechanics, to Hamilton's work.

Hamilton's personal life was not always happy. He fell deeply in love with a woman named Catherine Disney, who was forced by her parents to marry a wealthy man 15 years older than her. Hamilton remained hopelessly in love with her the rest of his life, though he eventually married someone else. He became an alcoholic, then foreswore drink, then relapsed. Many years after their early romance, Catherine began a secret correspondence with Hamilton. Her husband became suspicious, and she

attempted suicide by taking laudanum. Five years later, she became seriously ill. Hamilton visited her and gave her a copy of his *Lectures on Quaternions*. They kissed at last, and she died two weeks later. He carried her picture with him ever afterward and talked about her to anyone who would listen.[149]

Hamilton circuit
A **Hamilton path** that starts and ends at the same vertex. See also **traveling salesman problem**.

Hamilton path
Named after William **Hamilton**, a path that traverses every vertex of a **connected graph** once and only once. The problem of the **knight's tour** is equivalent to finding a Hamilton path (or, in the case of a reentrant tour, a **Hamilton circuit**) that corresponds to the legal moves of the knight. Compare with **Euler path**.

Hankel matrix
A **matrix** in which all the elements are the same along any **diagonal** that slopes from northeast to southwest.

happy number
If you iterate the process of summing the squares of the decimal digits of a number and if this process terminates in 1, then the original number is called a happy number. For example $7 \rightarrow (7^2)\ 49 \rightarrow (4^2 + 9^2)\ 97 \rightarrow (9^2 + 7^2)\ 130 \rightarrow (1^2 + 3^2)\ 10 \rightarrow 1$. See also **amicable number**.

Hardy, Godfrey Harold (1877–1947)
One of the most prominent English mathematicians of the twentieth century; his legendary collaboration with John Littlewood lasted 35 years and produced nearly 100 papers. Hardy was a precocious child, whose tricks included factorizing hymn numbers during sermons. In 1919, he became Savilian Professor of Geometry at Oxford but returned to Cambridge in 1931 as professor of pure mathematics. His work was mainly in **analysis** and **number theory**.

Hardy had only one other passion in his life—the game of cricket. His daily routine would begin with reading *The Times* and studying the cricket scores over breakfast. Then he would do mathematical research from 9 o'clock till 1 o'clock. After a light lunch, he would walk down to the university cricket ground to watch a game. In the late afternoon he would walk slowly back to his rooms at the college, and take dinner followed by a glass of wine. Hardy was known for his eccentricities. He couldn't stand having his photo taken and only five snapshots of him are known to exist. He also hated mirrors and his first action on entering any hotel room was to cover any mirror with a towel.

Hardy's book *A Mathematician's Apology* (1940)[151] is one of the most vivid descriptions of how a mathematician thinks and the pleasure of mathematics. But the book is more, as C. P. Snow writes:

> *A Mathematician's Apology* is . . . a book of haunting sadness. Yes, it is witty and sharp with intellectual high spirits: yes, the crystalline clarity and candor are still there: yes, it is the testament of a creative artist. But it is also, in an understated stoical fashion, a passionate lament for creative powers that used to be and that will never come again. I know nothing like it in the language: partly because most people with the literary gift to express such a lament don't come to feel it: it is very rare for a writer to realise, with the finality of truth, that he is absolutely finished.

See also **Ramanujan**.

harmonic analysis

The method of expressing **periodic** functions as sums of sines and cosines.

harmonic division

The division of a line segment by two points such that it is divided externally and internally in the same ratio.

harmonic mean

The harmonic mean of two numbers a and b is $2ab/(a+b)$.

harmonic sequence

The sequence: $1, \frac{1}{2}, \frac{1}{3}, \frac{1}{4}, \frac{1}{5} \ldots$ Added together, these become the terms of the *harmonic series:* $1 + \frac{1}{2} + \frac{1}{3} + \frac{1}{4} + \frac{1}{5} + \ldots$. This series diverges (has no finite sum), though very slowly—a result first proved by the French philosopher and theologian, Nichole d' Oresme (c. 1325–1382). In fact, it still diverges if you take away every other term, and even if you take away nine out of every ten terms. However, if you take the sum of reciprocals of all natural numbers that do not contain the number nine (when written in decimal expansion) the series converges! To show this, group the terms based on the number of digits in their denominator. There are eight terms in $(\frac{1}{1} + \ldots + \frac{1}{8})$, each of which is no larger than 1. Consider the next group $(\frac{1}{10} + \ldots + \frac{1}{88})$. The number of terms is *at most* the number of ways to choose two ordered digits out of the digits $0 \ldots 8$, and each such term is clearly no larger than $\frac{1}{10}$. So this group's sum is no larger than $9^2/10$. Similarly, the sum of the terms in $(\frac{1}{100} + \ldots + \frac{1}{999})$ is at most $9^3/10^2$, etc. So the entire sum is no larger than

$$9 \times 1 + 9 \times (\tfrac{9}{10}) + 9 \times (9^2/10^2) + \ldots + 9 \times (9^n/10^n) + \ldots$$

This is a geometric series that converges. Thus by the comparison test, the original sum (which is smaller term-by-term) must converge.

Harshad number

A number that is divisible by the sum of its own digits; also known as a Niven number. For example, 1,729 is a Harshad number because $1 + 7 + 2 + 9 = 19$ and $1,729 = 19 \times 91$. A *Harshad amicable pair* is an amicable pair (m, n) such that both m and n are Harshad numbers (see **amicable numbers**). For example, 2,620 and 2,924 are a Harshad amicable pair because 2,620 is divisible by $2 + 6 + 2 + 0 = 10$ and 2,924 is divisible by $2 + 9 + 2 + 4 = 17$ (2,924/17 = 172). There are 192 Harshad amicable pairs in the first 5,000 amicable pairs.

hat problem

A team of three contestants, Alice, Bob, and Cedric, enter a room and a hat is placed on each one's head so that he or she can't see it. The color of each hat is based on a coin toss—blue (B) for heads, red (R) for tails. After all the contestants enter the room, they look at the colors of one anothers' hats and, based on this information, they guess the color of their own hat. Each can guess red or blue, or, if she can't make up her mind, she can pass. No communication is allowed during the competition, but the players are allowed to agree on a strategy before play begins. The team wins if at least one of them guesses correctly, and none of them guesses incorrectly. What is the team's best strategy? At first sight, it may seem as if no effective strategy is possible beyond each contestant guessing his or her own hat color. In fact, this is the very worst approach since, to succeed, it requires that everyone guess correctly and the probability of this is only $\frac{1}{2} \times \frac{1}{2} \times \frac{1}{2} = \frac{1}{8}$. A far better plan is for the contestants to agree that two of them will pass while the third takes a stab at the color of his own hat. Then the odds improve to one in two. Beyond this it's hard to see any way that the probability of success could be increased. Yet there is an even better strategy. The key is to realize that there are only two cases (RRR and BBB) where everyone's hat is the same color but *six* cases where two hats are the same color and the other hat is a different color (RRB, RBR, BRR, BBR, BRB, and RBB). This suggests the following strategy for members of the team: *if you see two hats of opposite colors, pass. If you see two hats of one color, guess that your hat is the other color.* If everyone's hat is the same color, all players on the team will guess wrong and the team will lose. But the chance of this happening is only $\frac{2}{8}$ $(= \frac{1}{4})$. In every other possible case, the odd person out will guess correctly and their teammates will pass, so the team will win. This strategy wins $\frac{6}{8}$ $(= \frac{3}{4})$ of the time, and can't be improved upon. Since half of each player's guesses will be wrong, it's impossible to do better than a strategy in which each player in turn guesses correctly alone three times out of four, and the fourth time all guess wrong.

What if there are more players in the team? Say the number of players is n. By the reasoning described previously, it's clear that the team can't hope to win more than $n/(n + 1)$ of the time. Yet it isn't obvious that it can do this well. Having more people, it seems, might make it harder for them to synchronize their wrong guesses. However, it turns out that, if the number of people in the team is *one less than a power of 2,* this best-possible value can be achieved. For example, with a team of 7, the team can win $7/8$ of the time, and with a team of 15, they can win $15/16$ of the time. The strategy involved is complicated but is closely linked to *Hamming codes* (see **coding theory**), which are a method of encoding and transmitting information so that even if a small number of errors occur during transmission, the original information can be entirely recovered. For teams of other sizes, such as 9, 10, or 13, mathematicians have yet to find an optimal strategy or establish what proportion of the time the team can be expected to win.

Hausdorff, Felix (1868–1942)

A German mathematician who is considered to be one of the founders of modern **topology** and who also did significant work in **set theory** and functional analysis. Among several concepts named after him is the **Hausdorff dimension**, which gives a way of assigning a fractional dimension to a curve or shape. Hausdorff also published philosophical and literary works under the pseudonym "Paul Mongré." He studied at Leipzig and taught mathematics there until 1910, when he became professor of mathematics at Bonn. When the Nazis came to power, Hausdorff, a Jew, felt that as a respected university professor he would be safe from persecution. However, his abstract mathematics was denounced as useless and "un-German" and he lost his position in 1935. He sent his daughter to Britain but stayed with his wife in Germany. In 1942, when he could no longer avoid being sent to a concentration camp, he committed suicide together with his wife and sister-in-law.

Hausdorff dimension

A way to accurately measure the **dimension** of complicated **sets** such as **fractals**. The Hausdorff dimension, named after Felix **Hausdorff**, coincides with the more familiar notion of dimension in the case of well-behaved sets. For example, a straight line or an ordinary curve, such as a circle, has a Hausdorff dimension of 1; any **countable set** has a Hausdorff dimension of 0; and an n-dimensional **Euclidean space** has a Hausdorff dimension of n. But a Hausdorff dimension is not always a natural number. Think about a line that twists in such a complicated way that it starts to fill up the plane. Its Hausdorff dimension increases beyond 1 and takes on values that get closer and closer to 2. The same idea of ascribing a fractional dimension applies to a plane that contorts more and more in the third dimension: its Hausdorff dimension gets closer and closer to 3. As a specific example, the fractal known as the **Sierpinski carpet** has a Hausdorff dimension of just over 1.89.

Heesch number

The maximum number of times that a closed plane figure–a tile–can be completely surrounded by copies of itself. The Heesch number of a triangle, quadrilateral, regular hexagon, or any other single shape that can completely tile the plane (see **tiling**), is infinity. *Heesch's problem* is to find the largest possible *finite* Heesch number, or, more generally, what values other than zero and infinity can Heesch numbers take. In considering this problem, it's helpful to define the Heesch number more precisely. In a tiling, the *first corona* of a tile is the set of all tiles that have a common boundary point with the tile, including the original tile itself. The *second corona* is the set of tiles that share a point with anything in the first corona; and so on. The Heesch number is the maximum value of coronas (k) that can surround a shape. For a long time the record holder for the largest finite value of k was a shape found by the American computer scientist

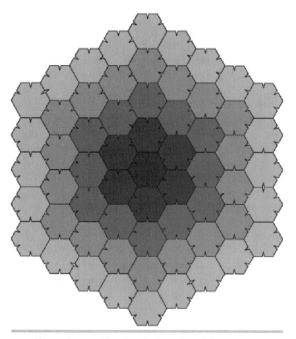

Heesch number A tiling in which copies of the same shape are used out to the fourth surrounding layer, or corona. *David Eppstein*

Robert Ammann, which consists of a regular hexagon with small bumps on two sides and matching notches on three sides. This was thought to have a Heesch number of three; however, in 2000, Alex Day argued that the Ammann hexagon actually has a Heesch number of four, though it isn't clear whether the difference has to do with a definition of tiling. In any event, it has since been shown by Casey Mann, of the University of Arkansas that there exists an infinite family of tiles (consisting of indented and outdented pentahex) with Heesch number five (or six by Day's reckoning)–the largest finite value currently known. Are there any polygons that have higher Heesch numbers? The answer is unknown but Mann thinks that more rounded **polyomino**s than the long skinny ones he's been using may have a better chance of giving unbounded Heesch numbers.

The Heesch number question is connected to two other famous unsolved tiling problems: the **domino problem** and the **Einstein problem**. Aperiodic tiling seems to act as a barrier to the existence of tiling algorithms, so it isn't expected that both of these problems have the same answer. On the other hand, if there's a maximum finite Heesch number k, then it seems that this could be used as the basis of an algorithm to test whether a shape tiles: simply attempt to fill out a tiling to the $(k + 1)$st corona; if successful, the shape must tile the plane, and if not, the shape will not tile. Similar questions can be asked about Heesch numbers for tilings in higher dimensions.

Hein, Piet (1905-1996)
An extraordinarily creative Danish mathematician, scientist, inventor, and poet who often wrote under the Old Norse pseudonym Kumbel, meaning "tombstone." A direct descendant of the Dutch naval hero of the sixteenth century who had the same name, Piet Hein was born in Copenhagen and studied at the Institute for Theoretical Physics of the University of Copenhagen (later the Niels Bohr Institute), the Technical University of Denmark, and the Royal Swedish Academy of Fine Art. He was later awarded an honorary doctorate by Yale University. A good friend of Albert Einstein, he is famed for his many mathematical games, including **Hex**, **Tangloids**, Polytaire, **TacTix**, and the **Soma cube**. These games were featured in numerous columns of Martin **Gardner**'s "Mathematical Recreations" column in *Scientific American* and often achieved worldwide attention in this way. As an artist and constructor, Hein gave form, in the 1950s and 1960s, to elegant pieces of furniture that helped "Scandinavian design" attract international recognition. These pieces, including a dining-room table created in cooperation with the Swedish designer Bruno Mathsson, were based on the **superellipse** curve–a shape

that Hein also brought to bear in applications as varied as city planning (it's the basis for Sergel's Square in the center of Stockholm) and toy making (see **superegg**). Hein was a prolific and excellent writer of light verse, producing thousands of short, aphoristic poems known as Grooks. For him there was no unbridgeable gap between the subjectivity of fine art and the objective world of science. "Art," he said, "is a solution to problems which cannot be formulated clearly before they have been solved." His philosophy of life was summed up by his aphorism "co-existence or no existence."

helicoid
The second oldest known **minimal surface**; it was discovered by Jean-Baptiste Meusnier in 1776, thirty years after the **catenoid**. It is the only minimal surface, apart from the simple plane, that is also a **ruled surface**. The helicoid is the surface swept out by a line that always intersects a fixed axis at right angles and that rotates uniformly as its point of intersection moves uniformly along the axis. This line intersects any cylinder concentric with the axis in a **helix**. The helicoid has a wide variety of shapes and is a familiar sight in everyday life, taking the form of many things from spiraling parking ramps to screw threads.

helix
A curve in three dimensions, the tangent to which makes a constant angle with a fixed line. A *circular helix* is formed by winding a line around a **cylinder** so the radius is always the same. A *conical helix* is formed by winding a line around a **cone**, so that, consequently, its radius constantly changes. Springs often take the form of various kinds of helices. In nature, the DNA molecule is in the shape of a double helix.

helicoid *Richard Palais*

Henon, Michele

An astronomer at the Nice Observatory in southern France. For a number of years, particularly during the 1960s, he studied the dynamics of stars moving within galaxies, using computers as a way to understand the stability of their motions. His work was very much in the spirit of Henri **Poincaré**'s approach to the classical **three-body problem**: What important geometric structures govern their behavior? The main property of these systems is that the energy of their motion is constant to a very good approximation. Consequently, their chaotic dynamics are not described by **simple attractor**s, but by objects that are markedly more difficult to analyze and visualize, existing on energy "surfaces" in three and higher dimensions. During the 1970s, Henon discovered a very simple iterated mapping that showed a **chaotic attractor**, now called *Henon's attractor*, that allowed him to make a direct connection between deterministic data and **fractals**. The Henon attractor is self-similar (see **self-similarity**): if you zoom in on the attractor in its state space you find more and more layers, much like filo dough or a croissant.

Henstock integration

See **integration**.

heptagon

A **polygon** with 7 sides.

Hermann grid illusion

An illusion first described by the German physiologist Ludimar Hermann (1838–1914) in 1870. While reading a book on sound by the Irish physicist John Tyndall, Hermann saw gray spots in the intersections of spaces among the figures that Tyndall had arranged in a matrix. Despite the fact that the same intensity of light is reflected all the way along the white spaces in the Hermann Grid, the intersections appear gray. To explain this, consider two regions of the retina. One region views an intersection of a white horizontal and vertical band, while the other views a white band between two intersections (the region going away from the intersection). Although the two regions themselves receive the same amount of light, the situation in their neighboring regions is different. At the intersection, light comes in from all four sides, but the white band that lies between the two intersections is surrounded by two dark sides. This leads to an effect called *lateral inhibition,* which causes a bright surround to an area appear darker and, conversely, a dark surround to an area appear lighter. A similar but more powerful illusion, known as the *Lingelbach illusion* or the *Scintillating grid illusion,* was discovered in 1994 by Elke Lingelbach, the wife of a German mathematics professor, and has not

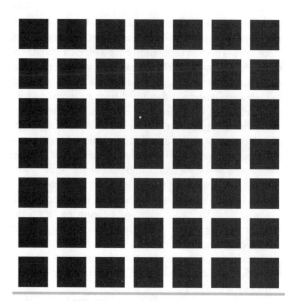

Hermann grid illusion The original illusion, in which the viewer sees gray spots at the intersections.

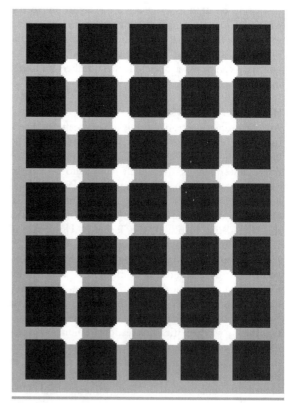

Hermann grid illusion The more striking and recently discovered "scintillating" version of the illusion.

yet been fully explained.[286] Curiously, the effect of the scintillation is lessened by tilting the head through 45°!

Hermite, Charles (1822–1901)

A French mathematician whose work in the theory of functions includes the application of **elliptic functions** to provide the first solution to the general equation of the fifth degree, the **quintic** equation. He also showed that e is a **transcendental number**, studied a class of **differential equation**s now known as *Hermite polynomials,* which later proved to be of importance in some applications of quantum mechanics, and discovered the properties of *Hermitian matrices.*

Heron of Alexandria (c. A.D. 60)

Also called *Hero,* a Greek geometer and inventor whose writings have helped preserve a knowledge of the mathematics and engineering of Babylonia, ancient Egypt, and the Greco-Roman world. His most important geometric work, *Metrica,* was lost until a fragment was discovered in 1894, followed by a complete copy in 1896. It is a compendium, in three books, of geometric rules and formulas, the best known of which is Proposition 1.8, now known as **Heron's formula**. He invented many devices operated by water, steam, or compressed air, including a fountain, a fire engine, siphons, and an engine in which the recoil of steam revolves a ball or a wheel.

Heron's formula

An important formula in plane geometry that allows the area of any triangle to be calculated without knowing the altitude (perpendicular height) of any of its sides. Let a, b, and c be the side lengths of a triangle and A its area. Heron's formula states that

$$A^2 = s(s - a)(s - b)(s - c),$$

where $s = (a + b + c)/2$. The origin of this formula is historically obscure. A medieval Arab source, for example, ascribes it to **Archimedes**. However, the first definite reference we have to it is by **Heron of Alexandria**. His proof is extremely convoluted, and it seems clear that it must have been determined by an entirely different thought process, and then dressed up in the usual synthetic form that the classical Greeks preferred for their presentations. Heron's formula contains **Pythagoras's theorem** as a degenerate case. A *Heronian triangle* is one with integer sides and integer area.

Herring illusion

A **distortion illusion** first published in 1861 by the German psychologist Ewald Herring (1834–1918), and now named after him. As in the case of the **Zöllner illusion** and others, it shows how geometrical relationships can

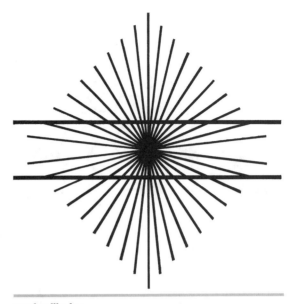

Herring illusion

seem to be distorted by their background (a lined background can make circles, squares, and triangles seem distorted, too). The straight horizontal lines in the illusion appear to bow out in the center. This can be explained if the brain interprets the radiating lines in terms of depth, making the central spot in the Herring diagram, and thus also the heavy black lines near the center of the diagram, appear to be farther away than the edges. Because the heavy black lines are the same thickness at the center as at the edges but are assumed to be farther away, the brain thinks they must be more widely spaced at the center.

heuristic argument

An educated guess: something that helps in finding the solution to a problem but is otherwise unjustified or incapable of justification.

Hex

A board game played by two players on a hexagonal grid, usually in the shape of an 11×11 rhombus. It was invented by Piet **Hein** in 1942 and independently by John **Nash** in 1948. Hein said that the game occurred to him while contemplating the **four-color problem** and it soon became popular in Denmark under the name *Polygon.* Nash's version was played by math students at Princeton and a number of other American campuses. Players use differently colored pieces–say, red and blue. They take alternate turns placing a piece of their color inside a hexagon, filling in that hexagon with their color. Red's goal is

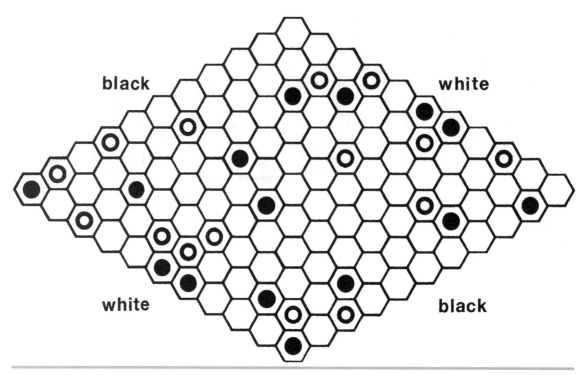

Hex A position in a game of Hex.

to form a red path connecting the top and bottom sides of the parallelogram; Blue's goal is to form a path connecting the left and right sides. The game can never end in a tie, a fact found by Nash. The only way to prevent your opponent from forming a connecting path is to form a path yourself. When the sides of the grid are equal, the game favors the first player and the first player has a winning strategy. There are two ways to make the game fairer. One is to make the second player's sides closer together, playing on a parallelogram rather than a rhombus; however, this has been proven to result in a win for the second player, so it theoretically doesn't improve matters. A better way is to allow the second player to choose his color after the first player makes the first move or to make the first three moves, which encourages the first player to intentionally even out the game.

hexa-

The Greek prefix meaning "six." A *hexagon* is a **six**-sided **polygon**. A *hexahedron* is a six-sided **polyhedron**, otherwise known as a **cube** if it is regular. *Hexadecimal* is the number system with **base** 16 (i.e., six more than the decimal system) and is used mostly in computing (because

four binary digits can represent 16 different numbers). *Hexagonal numbers* are **figurate number**s (numbers that can be represented by a regular geometric arrangement of equally spaced points) of the form $n(2n-1)$; the first few are 1, 6, 15, 28, 45, ... For *hexaflexagon* see **flexagon**. For *hexomino* see **polyomino**. See also **Giant's Causeway**.

higher dimensions

Dimensions beyond the familiar three spatial dimensions (up-down, left-right, back-forth) of which we are aware in every-day life. Intense speculation, both scientific and fictional, has naturally been directed toward the possibility of a **fourth dimension**. One way to think of points in four-dimensional space is as ordered sets of four numbers. Clearly, this algebraic representation can be extended to many arbitrary dimensions: *n*-dimensional space is defined as the set of the set of points $(a_1, a_2, . . . , a_n)$ where a_1 to a_n can take any **real number** value. There has been much conjecture that the universe in which we live contains many more than three spatial dimensions. This speculation began with the **Kaluza-Klein theory** but is now firmly embedded in modern **string theory**.[25, 55, 81]

Hilbert, David (1862–1943)

One can measure the importance of a scientific work by the number of earlier publications rendered superfluous by it.

A great German mathematician who was one of the colossi in the field in the twentieth century. His most important discovery was of what is now called **Hilbert space**. He was also a master of mathematical organization. During the early phase of his career, Hilbert reorganized **number theory**, crystallizing his conclusions in the classic book *Der Zahlbericht* (The theory of algebraic number fields, 1897). He then moved into geometry and performed a similar service by setting forth the first rigorous set of geometrical axioms in his *Grundlagen der Geometrie* (Foundations of geometry, 1899). He invented a simple **space-filling curve** now known as the *Hilbert curve* and also proved **Waring's conjecture**. At the Paris International Congress of 1900, Hilbert proposed 23 outstanding problems in mathematics to whose solutions he believed twentieth-century mathematicians should devote themselves. These problems have come to be known as *Hilbert's problems,* and a number still remain unsolved today. Hilbert's mathematical philosophy is partly re-vealed by a couple of remarks, one of which he made after learning that a student in his class had dropped the subject in order to become a poet. "Good," he said. "He did not have enough imagination to become a mathematician." Whether he really believed the second is open to question: "Mathematics is a game played according to certain simple rules with meaningless marks on paper."

Hilbert space

A space of infinite dimensions, named after David **Hilbert**, in which distance is preserved by making the sum of squares of coordinates a convergent sequence; it is of crucial importance in the mathematical formulation of **quantum mechanics**. See also **Fredholm, Erik Ivar**.

Hinton, Charles Howard (1853–1907)

An English-born mathematician best known for his writings and inventions aimed at helping to visualize the **fourth dimension**; he may also have coined the name *tesseract* for the four-dimensional analogue of a cube. Hinton matriculated at Oxford and continued to study there, earning a B.A. (1877) and an M.A. (1886), while he also taught, first at Cheltenham Ladies' School and then, from 1880 to 1886, at Uppingham School. At this time, another teacher at Uppingham was Howard Candler, who was a friend of Edwin **Abbott** and thus provides a possible link between these two explorers of other dimensions. In the early 1880s, Hinton published a series of pamphlets starting with "What Is the Fourth Dimension?" and "A Plane World" (a contemporary of Abbott's *Flatland: A Romance of Many Dimensions*), which were reprinted in the two-volume *Scientific Romances* (1884). Hinton's descriptions owed much to the mathematical models of William **Clifford**, whose theories about four-dimensional spaces were then in vogue. But Hinton went much further in his attempts to break free of three-dimensional thought. He devised an elaborate set of small colored cubes to represent the various cross sections of a tesseract and then memorized the cubes and their many possible orientations in order to gain a window on the fourth dimension.

At the time he was teaching in England, Hinton married Mary Everest Boole, the eldest daughter of George **Boole**, the founder of mathematical logic. Regrettably, he also married a Maud Wheldon and was tried at the Old Bailey in London for bigamy. After serving a day in prison for the offence, he fled with his (first) family to Japan, where he taught for some years, before taking up a post at Princeton University. There, in 1897, he designed a species of baseball gun which, with the help of gunpowder charges, would shoot out balls at speeds of 40 to 70 miles per hour. It was used by the Princeton team for several seasons before being abandoned by the players in fear of their lives.

After a brief spell at the University of Minnesota, Hinton joined the Naval Observatory in Washington, D.C. At the same time, he more rigorously developed his ideas on the fourth dimension and presented his results before the Washington Philosophical Society in 1902. Hinton asked: What would prove the existence of a real fourth spatial dimension? He offered three possibilities, two of which involved a specific molecular structure and a particular case of electrical induction, and have since been explained by science in more mundane ways. However, Hinton's other case, pertaining to right- and left-handedness remains open because there are instances of right- and left-handedness in nature, such as the spin of elementary particles, to which his example could be applied. In any event, Hinton's final assessment that we can only regard a four-dimensional space as possible if three-dimensional mechanics fails to explain known physical phenomena still rings true today.[166, 167, 272] See also **Boole (Stott), Alicia**.

Hippias of Elis (c. 5th century B.C.)

An itinerant Greek philosopher who contributed significantly to mathematics by discovering the quadratrix, a special curve he may have used to trisect an angle (see **quadratrix of Hippias**). Hippias is one of the first mathematicians about whom a good deal is known. He came from a state in the northwest corner of Peloponnesia that was the home of the Olympic games. According to Plato, Hippias boasted, during one of his visits to the Olympics, that everything he wore—his clothing, sandals, ring, and oil flask—he'd made himself. Later, in Athens, Hippias became one of the first to teach for money, a practice forbidden by the Pythagoreans and scorned by Plato. He and other paid teachers became known as "sophists," which was a derogatory term at the time but has since come to mean "wise man."

hippopede

A **quartic** curve described by the equation

$$(x^2 + y^2)^2 + 4b(b-a)(x^2 + y^2) - 4b^2x^2 = 0,$$

where a and b are positive constants. Hippopede means "foot of a horse." It is often known as the *hippopede of Proclus,* after **Proclus** who was the first to study it (together with **Eudoxus**, who used it in his theory of how the planets move), and also the *horse fetter* and the *curve of Booth* because of work done on it by J. Booth (1810–1878). Any hippopede is the intersection of a **torus** (donut) with one of its **tangent** planes—that is, a plane parallel to its axis of rotational symmetry. The curve takes any of a variety of forms depending on where the donut is sliced. It may be a simple oval; an indented oval or *elliptical lemniscate of Booth* ($0 < b < a$); two isolated circles; a figure-eight curve or **lemniscate of Bernoulli** (the only hippopede that is also a **Cassinian oval**); or a *hyperbolic lemniscate of Booth* ($0 < a < b$).

Hi-Q

See **peg solitaire**.

Hnefa-Tafl

The Viking equivalent of chess and a particular form of a **Tafl game**. It effectively models the kind of internal conflict familiar to the Vikings and recounted in *Njal's Saga,* the best known of the Icelandic sagas. The king or chieftain sits in his hall, surrounded by his thanes. His enemies gather in secret, and in numbers sufficient to overwhelm the king's standing forces in a lightning raid. They gather around the king's hall and set it alight, forcing the defenders to fight in the open or burn in the hall. If the king can, by some desperate stratagem, break free and escape the trap, he can rally his people and strike back at his enemies. If not, he dies. Hnefa-Tafl reflects this mode of contest. The board has 19×19 grid lines (though sometimes has as few as 7×7 grid lines), and the pieces are placed at the points of the intersections (curiously analogous to the oriental game of **Go**, which also uses 19×19 lines and where play occurs at the intersections). The opposing forces are unequal in size, and have different objectives: the attackers attempt to trap the king in his hall, while the defenders try to open an escape route for him.

Hoffmann, Louis "Professor" (1839–1919)

The pseudonym of Angelo John Lewis, an English barrister who was the leading writer on magic, cards, and "parlor amusements" at the turn of the twentieth century. His *Puzzles Old and New*[171] (1893) is a major source of information about mathematical recreations.

Hofstadter, Douglas R. (1945–)

A physicist and philosopher best known for his 1980 Pulitzer Prize–winning book *Gödel, Escher, Bach: An Eternal Golden Braid*.[172] He is currently a professor of cognitive science and computer science at Indiana University, Bloomington, and has particular interests in themes of the mind, consciousness, self-reference, translation, and mathematical games. He is the son of the Nobel Prize–winning physicist Robert Hofstadter.

Hofstadter's law

It always takes longer than you think, even when you take Hofstadter's law into account.

Hogben, Lancelot Thomas (1895–1975)

An English zoologist and geneticist famed for his best-selling *Mathematics for the Million* (1933)[173] of which Albert Einstein said, "It makes alive the contents of the elements of mathematics" and H. G. Wells said, "A great book, a book of first class importance." Hogben was born in Southsea, Hampshire, and studied at Cambridge and London. Imprisoned in 1916 as a conscientious objector during World War I, he was released only when his health went into serious decline. He held various academic posts in Britain, Canada, and South Africa, becoming professor of social biology at London University in 1930. During World War II he was put in charge of the medical statistics records for the British Army. After the war he became professor of medical statistics at the University of Birmingham, where he remained until his retirement in 1961. Hogben first began to apply mathematical principles to the study of genetics in the 1930s, focusing on the study of generations of the fruit fly *Drosophila* and how it related to research on heredity in humans. In addition to *Mathematics for the Million,* he authored half a dozen other books, including the popular *Science for the Citizen.* Though trained as a scientist, Hogben was passionately interested in linguistics. In *The Loom of Language,* which he edited, he set out the principles of his own invented language, "Interglossa," based on Greek and Latin roots but with a syntax resembling that of Chinese.

hole

A topological structure (see **topology**) that prevents any object in which it occurs from being continuously shrunk to a point. A **sphere** has no holes; a **torus** and a teacup each have one hole. See also **genus**.

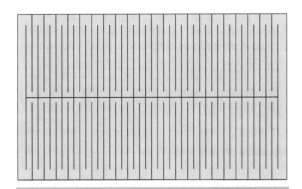

hole-in-a-postcard problem Photocopy the diagram, cut along the lines, and step through the postcard!

hole-in-a-postcard problem

With a pair of scissors, make cuts in a regular-sized postcard to create a hole large enough for a person to step through it. The solution is shown in the diagram. The number of lines determines the size of the resulting aperture. With enough cuts you could literally drive a horse and cart through the card!

hole-through-a-sphere problem

A number of **geometry puzzles** hinge on a surprising fact about spheres that have had holes bored through them. Imagine you have a round bead that is 1 inch (about 2.5 cm) in diameter and that you drill a hole exactly through the middle of this so that the remaining part of the sphere is only half an inch thick. Now imagine that an enormously large drill has been used to bore a hole though Earth so large that the part of Earth that is left behind is only half an inch thick. Amazingly, the residual volumes of these two holey spheres, the drilled bead and the drilled Earth, are exactly the same! It just happens that even though Earth is vastly larger than the bead, the drill has to take out proportionately more in order to make the thickness of the hole the same, so that the volume left doesn't depend separately on the initial size of the sphere or of the hole, but only on their *relation*, which is forced by requiring the hole to be half an inch long. This fact enables the following poem-problem to have a solution even though its seems as if not enough information has been provided:

> Old Boniface he took his cheer,
> Then he bored a hole through a solid sphere,
> Clear through the center, straight and strong,
> And the hole was just six inches long.
> Now tell me, when the end was gained,
> What volume in the sphere remained?

Sounds like I haven't told enough,
But I have, and the answer isn't tough.

Having already learned the secret that the volume that remains of a drilled sphere doesn't depend on the initial size of the sphere, we can cheat and give a kind of meta-argument that is much shorter than the geometric proof. The volume left behind of any sphere with a 6-inch-long hole through it must be the same as the volume left behind of a 6-inch-diameter sphere with a hole of 0 diameter drilled through it. This is equal to $\frac{4}{3}(\pi 6^3)$, or approximately 905 cubic inches.

hole-through-the-earth problem

I wonder if I shall fall right through *the earth!*
—Alice in *Alice in Wonderland*

Imagine there is a hole going from one point on Earth's surface, all the way through the center of Earth, to the exact opposite (antipodal) point on the other side. What would happen if you dropped something into this hole? The Greek historian Plutarch considered the problem some 2,000 years ago. In 1624, van Etten argued that a millstone dropped down such a hole at 1 mile per minute would take more than $2\frac{1}{2}$ days to reach the center, where "it would hang in the air." The first correct answer was given by Galileo is his *Dialogue Concerning the Two Chief World Systems* (1632). Galileo realized a dropped object would accelerate until it reached the center of Earth, travel through to the other side, then oscillate back and forth.

holism

The idea that the whole is greater than the sum of the parts. Holism is credible on the basis of **emergence** alone, since **reductionism** and bottom-up descriptions of nature often fail to predict complex higher-level patterns.

homeomorphic

In **topology**, two objects are said to be homeomorphic if they can be smoothly deformed into each other. See also **homeomorphism**.

homeomorphism

A **one-to-one** continuous transformation that preserves **open** and **closed** sets.

homology

A way of attaching **Abelian group**s, or more elaborate algebraic objects, to a **topological space** so as to obtain algebraic invariants. In a sense, it detects the presence of "holes" of various dimensions in the space. The methods developed to handle this led to what is now called *homological algebra*, a subject in which homological invariants are calculated for many purely algebraic structures.

homomorphism

A **function** that preserves the operators associated with the specified structure.

homotopy

A continuous transformation from one path in a **topological space** to another, or more generally, of one **function** to another (see **continuity**). Paths connected by a homotopy are called *homotopic* and are said to be in the same *homotopy class*. Properties left unchanged by such homotopies are known as *homotopy invariants*. Homotopy classes of paths can be composed to form the **fundamental group**, or *first homotopy group*. Other maps can be used to form *higher homotopy groups*.

Hordern, L. Edward (d. 2000)

An English puzzlist and leading authority and writer on **sliding-piece puzzles**.[175]

hundred

The smallest three-digit number in the decimal system and the smallest square of a two-digit number (10). A hundred today means 100 but, over the years and in different places, it has stood for different values including 112, 120, 124, and 132. The remnants of these old measures still persist in the *hundredweight* of some countries representing 112 or 120 pounds. A hundred is also a measure of land area, frequently used in colonial America and in England to signify a division of a county or a shire having its own court. A strange custom is invoked if a member of the English Parliament's lower house, the House of Commons, wishes to resign his seat (something that is technically illegal). That member accepts stewardship of the "Chiltern Hundreds," an area of chalk hills near Oxford and Buckingham, which effects his release from Parliament. The "Hundred Years' War" between the English and the French actually lasted 116 years.

hundred fowls problem

A Chinese puzzle found in the sixth-century work of the mathematician Zhang Qiujian. Similar problems involving two constraints and three unknowns are found in early European and Arabic mathematics from about the eighth century A.D. on.

PUZZLE

If a rooster is worth five coins, a hen three coins, and three chicks together are worth one coin, how many roosters, hens, and chicks totaling 100 can be bought for 100 coins? It turns out that there are three different solutions. These can be found the long way, by trial and error, or by using algebra. Call the number of roosters R, the number of hens H, and the number of

---

chicks C. The problem gives two constraints. First, the total number of fowl must be 100, so $R + H + C = 100$. Second, the total cost of the fowl must be 100. The cost of roosters is $5R$, the cost of hens is $3H$, and the cost of chicks is $(\frac{1}{3})C$, so $5R + 3H + (\frac{1}{3})C = 100$. These two equations can be used to get rid of one of the unknowns; then it's a question of guess and check.

Solutions begin on page 369.

Hunter, James Alston Hope

An American mathematician and puzzlist who has written numerous articles and several books on recreational mathematics (two in partnership with Joseph **Madachy**), and was the author of a syndicated puzzle column read throughout the United States and Canada. In 1955, he coined the name "**alphametic**" and is probably the most prolific producer of **cryptarithm**s.

Huygens, Christiaan (1629–1695)

A Dutch scientist and mathematician who solved the **tautochrone problem**, proposed a new wave theory of light, designed a new pendulum clock, discovered Saturn's largest moon (Titan), and sketched the first feature on the surface of another planet (Syrtis Major on Mars). In his final years Huygens composed one of the earliest discussions of extraterrestrial life, published after his death as the *Cosmotheoros* (1698).

Hypatia of Alexandria (c. A.D. 370–417)

The first woman known to have made a significant contribution to mathematics. Although there is no evidence that Hypatia did any original research, she assisted her father, Theon of Alexandria, in writing his 11-part commentary on Ptolemy's great work on astronomy and mathematics, the *Almagest*. It's thought that she also helped in producing a new version of **Euclid**'s *Elements*, which formed the basis for all later editions of Euclid. Hypatia became head of the Platonist school at Alexandria in about A.D. 400 and, as a pagan, represented a menace to some Christian sects who felt threatened by her learning and depth of scientific knowledge. In the end, although the exact circumstances are unclear, she was murdered by a mob. The event served as a trigger for the departure of many scholars and the beginning of the decline of Alexandria as a major academic center.

hyperbola

One of the **conic section** family of curves, which also includes the **circle**, the **ellipse**, and the **parabola**; it is obtained if a double cone is cut by a plane inclined to the axis of the cone such that it meets both branches of the cone. Of the four conic curves, the hyperbola is the one least encountered in everyday life. A rare chance to see the complete shape is when a lamp with a cylindrical or conical shade throws shadows on a nearby wall. Part of a hyperbola is produced by the liquid that climbs by capillary action between two microscope slides held vertically and almost touching.

A hyperbola is the path followed by a smaller object that is traveling fast enough to escape completely from the gravitational pull of a larger object. Some comets, for example, have hyperbolic or "open" orbits so that, after one swing around the Sun, they head off into interstellar space never to return. It can be difficult to tell, in some cases, whether a comet's orbit is hyperbolic or is highly elliptical and, therefore, closed. In fact, one way to think of a hyperbola is as a kind of ellipse that is split in half by infinity. Not surprisingly, the hyperbola and the ellipse share many inverse relationships. For example, whereas the eclipse is the **locus** of all points whose distances from two fixed points, called *foci*, have a constant sum, the hyperbola is the locus of all points whose distances, r_1 and r_2, from two fixed points, F_1 and F_2, is a constant difference, $r_2 - r_1 = k$. If a is the distance from the origin to either of the x intercepts of the hyperbola, then $k = 2a$. Also, let the distance between the foci, $F_2 - F_1 = 2c$. Then the *eccentricity*, a measure of the flatness of the hyperbola, is given by $e = c/a$. For all hyperbolas, $e > 1$; the larger the value of e, the more the hyperbola resembles two parallel lines. Just as the circle (for which $e = 0$) is the limiting case of the ellipse (for which $0 < e < 1$), so the parabola ($e = 1$) is the limiting case of both the ellipse and the hyperbola.

A hyperbola has two **asymptotes**: the never-quite-attainable limits of the curve's branches as they run away to infinity. The *transverse axis* of the hyperbola is the line on which both foci lie and that also intersects both vertices (turning points); the *conjugate axis* goes through the center and is perpendicular to the transverse axis.

A *rectangular hyperbola* has an eccentricity of $\sqrt{2}$, asymptotes that are mutually perpendicular, and the property that when stretched along one or both of its asymptotes remains unchanged. This special case of the hyperbola was first studied by **Menaechmus**. **Euclid** and Aristaeus wrote about the general hyperbola but only studied one branch of it, while **Apollonius** was the first to study the two branches of the curve of the hyperbola and is generally thought to have given it its present name.

The **pedal curve** of a hyperbola with one focus as the pedal point is a circle. The pedal of a rectangular hyperbola with its center as pedal point is a **lemniscate of Bernoulli**. The **evolute** of a hyperbola is a **Lamé curve**. If the center of a rectangular hyperbola is taken as the center of inversion, the rectangular hyperbola inverts to a lemniscate. If the vertex is used as the center of inversion, the rectangular hyperbola inverts to a right **strophoid**. If the focus of a hyperbola is taken as the center of inversion, the hyperbola inverts to a limaçon (see **limaçon of Pascal**). In this last case if the asymptotes of the hyperbola make an angle of π/3 with the axis that cuts the hyperbola, then it inverts to the Maclaurin **trisectrix**. See also **hyperboloid**.

hyperbolic geometry

One of the two main types of **non-Euclidean geometry** and the first to be discovered. It is concerned with saddle-surfaces, which have negative **curvature** and on which the **geodesic**s are **hyperbolas**. In hyperbolic geometry, contrary to the **parallel postulate**, there exists a line *m* and a point *p* not on *m* such that at least two distinct lines parallel to *m* pass through *p*. As a result, the sum of the angles of a triangle is less than 180° and, for a right triangle, the square of the hypotenuse is greater than the sum of the squares of the other two sides. See also **elliptical geometry**.

hyperbolic spiral

The curve whose equation in **polar coordinates** is $r\theta = a$.

hyperboloid

A **quadratic** surface of which there are two basic forms: a *hyperboloid of one sheet*, generated by spinning a **hyperbola** around its conjugate axis, and a *hyperboloid of two sheets* produced by rotating a hyperbola about its transverse axis. The hyperboloid of one sheet, first described by **Archimedes**, has some particularly remarkable properties. In 1669 Christopher Wren, the architect who de-

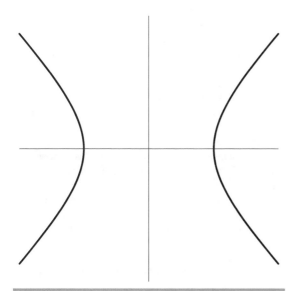

hyperbola © Jan Wassenaar, www.2dcurves.com

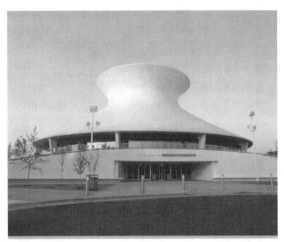

hyperboloid The McDonnell Planetarium in St. Louis is an example of a hyperboloid. *Courtesy of the St. Louis Science Center*

signed St. Paul's Cathedral in London, showed that this kind if hyperboloid is what mathematicians now call a **ruled surface**—a surface composed of infinitely many straight lines. This fact enables a close approximation to a hyperboloid to be made in the form of a string model. Two circular disks, of the same size, are held parallel, one exactly above the other, by a framework. Strings are then run through holes near the circumference of one circle to corresponding holes in the other circle that are a fixed distance farther around the circumference. Each string is perfectly straight but the surface that emerges takes the curved form of a hyperboloid. For the same reason, a cube spun rapidly on one of its corners will appear to describe a hyperbolic curve when viewed side-on.

Prominent examples of hyperboloids can be seen in the form of cooling towers at power stations and, most strikingly, in the shape of the McDonnell Planetarium in St. Louis, Missouri. The designer of this building, Gyo Obata, chose the design because the hyperbolic paths of some comets suggest "the drama and excitement of space exploration."

hypercube

A higher dimensional analog of a cube. A four-dimensional hypercube is known as a **tesseract**.

hyperellipse

See **superellipse**.

hyperfactorial

A number such as 108, which is equal to $3^3 \times 2^2 \times 1^1$. In general, the nth hyperfactorial $H(n)$ is given by

$$H(n) = n^n \, (n-1)^{n-1} \ldots 3^3 \, 2^2 \, 1^1.$$

The first eight hyperfactorials are 1, 4, 108, 27,648, 86,400,000, 4,031,078,400,000, 3,319,766,398,771,200,000, and 55,696,437,941,726,556,979,200,000. See also **large numbers** and **superfactorials**.

hypergeometric function

The sum of the *hypergeometric series:*

$$F(a; b; c; x) = 1 + \frac{ab}{1.c}\,x + \frac{a(a+1)b(b+1)}{1.2.c(c+1)}\,x^2 + \ldots$$

Many common functions can be written as hypergeometric functions.

hyperreal number

Any of a colossal set of numbers, also known as *nonstandard reals,* that includes not only all the **real numbers** but also certain classes of infinitely large (see **infinity**) and **infinitesimal** numbers as well. Hyperreals emerged in the 1960s from the work of Abraham **Robinson** who showed how infinitely large and infinitesimal numbers can be rigorously defined and developed in what is called **nonstandard analysis**. Because hyperreals represent an extension of the real numbers, R, they are usually denoted by *R.

Hyperreals include all the reals (in the technical sense that they form an ordered field containing the reals as a subfield) and they also contain infinitely many other numbers that are either infinitely large (numbers whose absolute value is greater than any positive real number) or infinitely small (numbers whose absolute value is less than any positive real number). No infinitely large number exists in the real number system and the only real infinitesimal is zero. But in the hyperreal system, it turns out that each real number is surrounded by a cloud of hyperreals that are infinitely close to it; the cloud around zero consists of the infinitesimals themselves. Conversely, every (finite) hyperreal number x is infinitely close to exactly one real number, which is called its *standard part,* st(x). In other words, there exists one and only one real number st(x) such that $x - $ st(x) is infinitesimal.

hypersphere

A four-dimensional analog of a **sphere**; also known as a *4-sphere.* Just as the shadow cast by a sphere is a circle, the shadow cast by a hypersphere is a sphere, and just as the intersection of a sphere with a plane is a circle, the intersection of a hypersphere with a hyperplane is a sphere. These analogies are reflected in the underlying mathematics.

$x^2 + y^2 = r^2$ is the Cartesian equation of a circle of radius r;

$x^2 + y^2 + z^2 = r^2$ is the corresponding equation of a sphere;

hyperboloid   A hyperboloidal sculpture at the Fermi National Accelerator Laboratory. *FNAL*

$x^2 + y^2 + z^2 + w^2 = r^2$ is the equation of a hypersphere, where w is measured along a **fourth dimension** at right angles to the x-, y-, and z-axes.

The hypersphere has a *hypervolume* (analogous to the volume of a sphere) of $\pi^2 r^4/2$, and a *surface volume* (analogous to the sphere's surface area) of $2\pi^2 r^3$. A solid angle of a hypersphere is measured in *hypersteradians,* of which the hypersphere contains a total of $2\pi^2$. The apparent pattern of 2π radians in a circle and 4π steradians in a sphere does not continue with 8π hypersteradians because the *n-volume, n-area,* and number of *n-radians* of an *n-sphere* are all related to the **gamma function** and the way it can cancel out powers of π halfway between integers. In general, the term "hypersphere" may be used to refer to any *n-sphere.*

hypocycloid
A curve formed by the path of a point attached to a circle of radius b that rolls around the inside of a larger circle of radius a. The parametric equations for a hypocycloid can be written as:

$$x = (a - b)\cos(t) + b\cos((a/b - 1)t)$$
$$y = (a - b)\sin(t) - b\sin((a/b - 1)t)$$

The type of hypocloid depends on where the point whose path is being traced is located on the rolling circle. If it lies on the circumference of the circle, the curve generated is an ordinary hypocycloid. If it lies elsewhere the result is a **hypotrochoid**. A hypocycloid has a closed form–that is, the moving point eventually retraces its steps–when the ratio of the rolling circle and the larger circle, fixed, is equal to a rational number. When this ratio is in its simplest form, the numerator is the number of revolutions covered inside the fixed circle before the curve closes. In the same family of curves as the hypocycloid (and hypotrochoid) are the **epicycloid** and **epitrochoid**.

hypoellipse
See **superellipse**.

hypotenuse
The longest side of a **right triangle**–the one opposite the right angle. The word comes from the Greek roots *hypo* meaning "under" (which also appears, for example, in hypodermic, "under the skin") and *tein* or *ten,* for stretch. Thus hypotenuse is the line segment "stretched under" the right angle.

hypothesis
See **conjecture**.

hypotrochoid
A curve formed by the path of a point attached to a point c, which is not on the circumference, of a circle of radius b that rolls around the inside of a larger circle of radius a. The parametric equations for a hypotrochoid can be written as:

$$x = (a - b)\cos(t) + c\cos((a/b - 1)t)$$
$$y = (a - b)\sin(t) - c\sin((a/b - 1)t)$$

The hypotrochoid is a generalized **hypocycloid** and comes in two varieties: a *prolate hypocycloid* if the starting point is outside the circumference of the rolling circle and a *curtate hypocycloid* if the starting point is inside the rolling circle. In the same family of curves as the hypotrochoid (and hypocycloid) are the **epicycloid** and **epitrochoid**.

I

i

The square root of minus one; *i* is also known as the *imaginary unit number*, $\sqrt{-1}$. The product of *i* and a real number is known as an *imaginary number*. For details, see **complex number**.

I-Ching

A system of Chinese divination based on 64 hexagrams, each consisting of six horizontal lines each of which may be solid (representing the male principle, or yang) or have a break in the middle (representing the female principle, or yin). The first three lines of the hexagram, from the bottom up, constitute the lower trigram and symbolize the inner world. The fourth, fifth, and sixth lines constitute the upper trigram and symbolize the outer world. One of the first Europeans to see the hexagram structure when it was brought out of China in the early 1700s was Gottfried **Leibniz**, the first mathematician to work with base two, or **binary**, arithmetic.

icosahedron

A **polyhedron** with 20 faces. A *regular icosahedron* has faces that are all equilateral **triangle**s, and is one of the five **Platonic solid**s. The length from **vertex** to opposing vertex of a regular icosahedron is $5^{1/4} \times \phi^{1/2} \times d$ where ϕ is the **golden ratio** and *d* is the length of the side of one of the triangular faces. Chopping off each vertex (corner) of a regular icosahedron reveals the 12 pentagonal and 20 hexagonal faces of the *truncated icosahedron,* which is one of the 13 **Archimedean solid**s (shapes made from truncating Platonic solids in certain ways). See also **buckyball**.

Icosian game

A game devised by William **Hamilton** and first described by him in 1857 at a meeting of the British Association for the Advancement of Science in Dublin. The object of the game is to find a way around the edges of a **dodecahedron** so that every vertex (corner) is visited once and only once. A path such as this became known as a **Hamilton circuit**, though the task of finding a circuit that passes just once through every vertex of a shape seems to have arisen first in connection with Leonhard **Euler**'s study of the **knight's tour**. Two years before Hamilton introduced his game, Thomas **Kirkman** posed the problem explicitly in a paper that he submitted to the Royal Society: Given

a graph of a **polyhedron**, does there exist a cycle passing through every vertex?

The Icosian game stemmed from Hamilton's invention of a curious kind of algebra that he called *icosians,* based on the symmetry properties of the **icosahedron**. Hamilton connected the mathematics of his icosians with the problem of traveling along the edges of a dodecahedron, hitting each vertex just once, and coming back to the starting point. His friend and fellow Irishman John Graves (1806–1870) suggested turning the problem into a commercial game and put Hamilton in contact with the London company of John Jacques and Sons, toy-makers and manufacturer of high quality chess sets. Jacques bought the rights to the game for £25 and marketed two versions of it, under the name Around the World. One version, for the parlor, was played on a flat board; another, for the "traveler," consisted of an actual dodecahedron. In both cases, nails at each vertex stood for a major city of the world and the player wrapped a piece of string around these nails as they went. The game was a

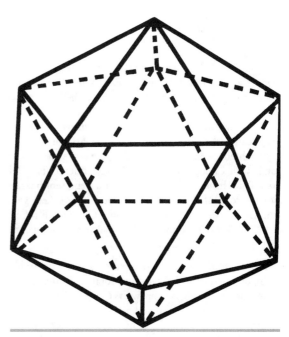

icosahedron   The most multifaceted of the Platonic solids.

complete sales flop, mainly because it was too easy, even for children—but not for Hamilton himself who always used the icosian calculus to figure out his moves, instead of just trying different paths like everyone else! See also **traveling salesman problem**.

idempotent

The element x in some algebraic structures is called idempotent if $x \times x = x$.

identity

A statement that two expressions are always equal, for any values of their variables.

imaginary number

A number whose square is negative. Every imaginary number can be written in the form ib where b is a **real number** and i the *imaginary unit* with the property that $i^2 = -1$. Imaginary numbers are **complex number**s in which the real part is zero. The term "imaginary" is unfortunate because it suggests something that has less reality than a "real" number, which isn't the case.

impossibilities in mathematics

> *Alice laughed: "There's no use trying," she said; "one can't believe impossible things." "I daresay you haven't had much practice," said the Queen. "When I was younger, I always did it for half an hour a day. Why, sometimes I've believed as many as six impossible things before breakfast."*
> —Lewis Carroll, *Alice in Wonderland*

Mathematicians are used to believing things that most people would consider impossible or, at least, too outrageous to contemplate, such as the **Banach-Tarski paradox**. However, there are some genuine impossibilities, even in mathematics, including **trisecting an angle**, **duplicating the cube**, and **squaring the circle** using only a straightedge and compass; finding the center of a given circle with a straightedge alone; deriving Euclid's **parallel postulate** from the other four; and representing the **square root of 2** as a rational fraction a/b. Less well known is this little gem from Gustave Flaubert (1821–1880), who sounds as if he had seen too much of this type of problem in school:

> Since you are now studying geometry and trigonometry, I will give you a problem. A ship sails the ocean. It left Boston with a cargo of wool. It grosses 200 tons. It is bound for Le Havre. The mainmast is broken, the cabin boy is on deck, there are 12 passengers aboard, the wind is blowing East-North-East, the clock points to a quarter past three

in the afternoon. It is the month of May. How old is the captain?

impossible figure

An image in two dimensions of an object that, because of spatial inconsistencies, is impossible to realize fully in three dimensions. Among the best known impossible figures are the **Penrose triangle**, the **Penrose stairway**, the **Impossible trident**, and the Freemish crate. Notable pioneers of this peculiar form of representation have been Ocar **Reutersvärd**, Roger **Penrose** (and his father), and M. C. **Escher**. See also **optical illusion**.

impossible tribar

See **Penrose triangle**.

impossible trident

One of the most notorious **impossible figure**s. It was first seen by many when it appeared on the cover of the March 1965 issue of *Mad* magazine. The two halves of the figure seem perfectly reasonable in themselves. When the top part is covered, the bottom part is taken to be three separate cylinders or tubes. With the bottom part hidden, the foreground figure is interpreted as being built of flat faces making two rectangular prongs. The trouble is that these two aspects of the figure are totally incompatible. Somewhere in the middle, the foreground and background swap places and give rise to an irreconcilable paradox. Over the years, countless adaptations of the trident have appeared with names such as the devil's fork, the three-stick clevis, the blivit, the impossible columnade, the trichotometric indicator support, and, most extravagantly, the triple encabulator tuned manifold. Swedish artist Oscar **Reutersvärd**'s mastery of such figures has led him to draw thousands of variations on the theme. When the figure is drawn long, it is easy to perceive locally as a three-dimensional object and to overlook its inherent inconsistency, because the contradictory clues are too well separated. When the figure is of medium length, the figure is easily interpreted as a three-dimensional object, and its impossibility is quickly perceived. If the prongs are very short, the two different interpretations vie for acceptance within the same local area; thus there is no consistent interpretation and the illusion breaks down. Some early writers commented that the impossible trident couldn't be built in any form in three dimensions. However, this has been shown to be false. In 1985, the Japanese artist Shigeo Fukuda made a three-dimensional model of the trident in the form of classical columns in which the illusion works—from one critical angle.

The origins of the figure are uncertain. It turns out that *Mad* magazine bought the illustration rights from a con-

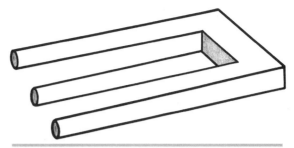

impossible trident

tributor who claimed that it was original; however, the magazine's management soon found out to their embarrassment that the figure had been previously published. It began to surface in several popular engineering, aviation, and science-fiction periodicals in May and June of 1964. D. H. Schuster published an article that same year in the *American Journal of Psychology,* which first brought the figure to the attention of the psychological community.

incircle
The **inscribed circle** of a triangle.

incomputable number
A **real number** with an infinite decimal (or binary) expansion that cannot be enumerated by any **universal computer**.

induction
A method of reasoning by which one infers a generalization from a series of instances. Say there is a hypothesis H that contains the variable n, which is a whole number. To prove by induction that H is true for every value of n is a two-step process: (1) prove that H is true for $n = 1$; (2) prove that H being true for $n = k$ implies that H is true for $n = k + 1$. This is sufficient because (1) and (2) together imply that H is true for $n = 2$, which, from (2), then implies H is true for $n = 3$, which implies H is true for $n = 4$, and so on. H is called an *inductive hypothesis.* Some philosophers don't accept this kind of proof, because it may take infinitely many steps to prove something; however, most mathematicians are happy to use it.

inequality
The statement that one quantity is less than or greater than another.

infinite dimensions
In mathematics, the concept of an infinite-dimensional space considered literally. It is a **vector space** with an infinite basis or a space with infinitely many coordinates.

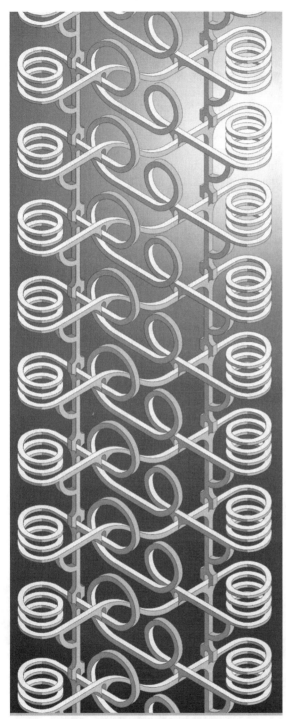

impossible figure "The Rollercoaster," a rendering in two dimensions of a structure impossible to build in three dimensions. *Jos Leys, www.josleys.com*

infinite series

An infinite sum of the form

$$a_1 + a_2 + a_3 + \ldots = \sum_{k=1}^{\infty} a_k.$$

Such series crop up in many areas of modern mathematics. Their development began in the seventeenth century and was continued by Leonhard **Euler,** who, in the process, solved many important problems.

infinitesimal

A number that is greater than zero yet smaller than any positive **real number**. In a sense, infinitesimals are to small what **infinity** is to large. They were first introduced by Isaac **Newton** and Gottfried **Leibniz** in their early versions of **calculus**; however, the lack of a rigorous definition for them stood in the way of calculus being fully accepted. As Bertrand **Russell** later put it: "Calculus required continuity, and continuity was supposed to require the infinitely little; but nobody could discover what the infinitely little might be." In the 1800s, calculus was put on a firmer footing by Augustin **Cauchy**, Karl **Weierstrass**, and others, who clarified and redefined the notion of a **limit** without reference to infinitesimals. When a function $f(x)$ can be made as close as desired to L by taking x close enough to a, then L is the limit of $f(x)$ as x approaches a. This is the classical or *epsilon-delta* formulation of calculus, named for the common use of δ for $|x - a|$ and ε for $|f(x) - L|$. For a long time, this was the only rigorous foundation for calculus, and it is still the one taught in most calculus classes. But in 1960, Abraham **Robinson** discovered **nonstandard analysis**, which provides a rigorous formulation of infinitesimals, confers on them a new significance, and brings them closer to the vision of Newton and Leibniz.

infinity

Mystery has its own mysteries, and there are gods above gods. We have ours, they have theirs. That is what's known as infinity.

—Jean Cocteau (1889–1963),
French author and filmmaker

A concept that has always fascinated philosophers and theologians, linked as it is to the notions of unending distance or space, eternity, and God, but that was avoided or met with open hostility throughout most of the history of mathematics. Only within the past century or so have mathematicians dealt with it head on and accepted infinity as a number—albeit the strangest one we know.

An early glimpse of the perils of the infinite came to Zeno of Elae through his paradoxes, the best known of which pits Achilles in a race against a tortoise (see **Zeno's paradoxes**). Confident of victory, Achilles gives the tortoise a head start. But then how can he ever overtake the sluggish reptile? First he must catch up to the point where it began, by which time the tortoise will have moved on. When he makes up the new distance that separated them, he finds his adversary has advanced again. And so it goes on, indefinitely. No matter how many times Achilles reaches the point where his competitor was, the tortoise has progressed a bit farther. So perplexed was Zeno by this problem that he decided not only was it best to avoid thinking about the infinite but also that motion was impossible!

A similar shock lay in store for **Pythagoras** and his followers who were convinced that everything in the universe could ultimately be understood in terms of whole numbers (even common fractions being just one whole number divided by another). The **square root of 2**–the length of the diagonal of a right-angled triangle whose shorter sides are both one unit long–refused to fit into this neat cosmic scheme. It was an **irrational number**, inexpressible as the ratio of two integers. Put another way, its decimal expansion goes on forever without ever settling into a recurring pattern.

These two examples highlight the basic problem in coming to grips with infinity. Our imaginations can cope with something that hasn't yet reached an end: we can always picture taking another step, adding one more to a total, or visualizing another term in a long series. But infinity, taken as a whole, boggles the mind. For mathematicians this was a particularly serious problem because mathematics deals with precise quantities and meticulously well-defined concepts. How could they work with things that clearly existed and went on indefinitely–a number like $\sqrt{2}$ or a curve that approached a line ever more closely–while avoiding a confrontation with infinity itself? **Aristotle** provided the key by arguing that there were two kinds of infinity. *Actual infinity,* or *completed infinity,* which he believed could not exist, is endlessness fully realized at some point in time. *Potential infinity,* which Aristotle insisted was manifest in nature–for example, in the unending cycle of the seasons or the indefinite divisibility of a piece of gold–is infinitude spread over unlimited time. This fundamental distinction persisted in mathematics for more than 2,000 years. In 1831, no less a figure than Karl **Gauss** expressed his "horror of the actual infinitude," saying: "I protest against the use of infinite magnitude as something completed, which is never permissible in mathematics. Infinity is merely a way of speaking, the true meaning being a limit which certain ratios approach indefinitely close, while others are permitted to increase without restriction."

By confining their attention to potential infinity, mathematicians were able to address and develop crucial concepts such as those of **infinite series**, **limit**, and **infinitesimals**, and so arrive at **calculus**, without having to grant that infinity itself was a mathematical object. Yet as early as the Middle Ages certain paradoxes and puzzles arose, which suggested that actual infinity was not an issue to be easily dismissed. These puzzles stem from the principle that it is possible to pair off, or put in *one-to-one correspondence,* all the members of one collection of objects with all those of another of equal size. Applied to indefinitely large collections, however, this principle seemed to flout a commonsense idea first expressed by **Euclid**: the whole is always greater than any of its parts. For instance, it appeared possible to pair off *all* the positive integers with *only those that are even:* 1 with 2, 2 with 4, 3 with 6, and so on, despite the fact that positive integers also include odd numbers. Galileo, in considering such a problem, was the first to show a more enlightened attitude toward the infinite when he proposed that "infinity should obey a different arithmetic than finite numbers." Much later, David **Hilbert** offered a striking illustration of how weird the arithmetic of the endless can get.

Imagine, said Hilbert, a hotel with an infinite number of rooms. In the usual kind of hotel, with finite accommodation, no more guests can be squeezed in once all the rooms are full. But "Hilbert's Grand Hotel" is dramatically different. If the guest occupying room 1 moves to room 2, the occupant of room 2 moves to room 3, and so on, all the way down the line, a newcomer can be placed in room 1. In fact, space can be made for an infinite number of new clients by moving the occupants of rooms 1, 2, 3, etc, to rooms 2, 4, 6, etc, thus freeing up all the odd-numbered rooms. Even if an infinite number of coaches were to arrive each carrying an infinite number of passengers, no one would have to be turned away: first the odd-numbered rooms would be emptied as above, then the first coach's load would be put in rooms $3n$ for $n = 1, 2, 3, \ldots$, the second coach's load in rooms $5n$ for $n = 1, 2, \ldots$, and so on; in general, the people aboard coach number i would empty into rooms p_n where p is the $(i + 1)$th prime number.

Such is the looking-glass world that opens up once the reality of **sets** of numbers with infinitely many elements is accepted. That was a crucial issue facing mathematicians in the late nineteenth century: Were they prepared to embrace actual infinity as a number? Most were still aligned with Aristotle and Gauss in opposing the idea. But a few, including Richard **Dedekind** and, above all, Georg **Cantor**, realized that the time had come to put the concept of infinite sets on a firm logical foundation.

Cantor accepted that the well-known pairing-off principle, used to determine if two finite sets are equal, is just as applicable to infinite sets. It followed that there really

are just as many even positive integers as there are positive integers altogether. This was no paradox, he realized, but the *defining property* of infinite sets: the whole is no bigger than some of its parts. He went on to show that the set of all positive integers, 1, 2, 3, . . . , contains precisely as many members—that is, has the same **cardinal number** or cardinality—as the set of all rational numbers (numbers that can be written in the form p/q, where p and q are integers). He called this infinite cardinal number **aleph**-null, $\aleph_0$ ("aleph" being the first letter of the Hebrew alphabet). He then demonstrated, using what has become known as *Cantor's theorem*, that there is a hierarchy of infinities of which $\aleph_0$ is the smallest. Essentially, he proved that the cardinal number of all the subsets—the different ways of arranging the elements—of a set of size $\aleph_0$ is a bigger form of infinity, which he called $\aleph_1$. Similarly, the cardinality of the set of subsets of $\aleph_1$ is a still bigger infinity, known as $\aleph_2$. And so on, indefinitely, leading to an infinite number of different infinities.

Cantor believed that $\aleph_1$ was identical with the total number of mathematical points on a line, which, astonishingly, he found was the same as the number of points on a plane or in any higher n-dimensional space. This infinity of spatial points, known as the *power of the continuum, c,* is the set of all **real numbers** (all rational numbers plus all irrational numbers). Cantor's **continuum hypothesis** asserts that $c = \aleph_1$, which is equivalent to saying that there is no infinite set with a cardinality between that of the integers and the reals. Yet, despite much effort, Cantor was never able to prove or disprove his continuum hypothesis. We now know why—and it strikes to the very foundations of mathematics.

In the 1930s, Kurt **Gödel** showed that it is impossible to *disprove* the continuum hypothesis from the standard **axiom**s of set theory. Three decades later, Paul Cohen showed that it *cannot be proven* from those same axioms either. Such a situation had been in the cards ever since the emergence of **Gödel's incompleteness theorem**. But the independence of the continuum hypothesis was still unsettling because it was the first concrete example of an important question that could not be proven either way from the universally accepted system of axioms on which most of mathematics is built.

Currently, the preference among mathematicians is to regard the continuum hypothesis as being false, simply because of the usefulness of the results that can be derived this way. As for the nature of the various types of infinities and the very existence of infinite sets, these depend crucially on what number theory is being used. Different axioms and rules lead to different answers to the question *what lies beyond all the integers?* This can make it difficult or even meaningless to compare the various types of infinities that arise and to determine their relative size, although within any given number system the infinities can usually be put into a clear order. Certain extended number systems, such as the **surreal numbers**, incorporate both the ordinary (finite) numbers and a diversity of infinite numbers. However, whatever number system is chosen, there will inevitably be inaccessible infinities—infinities that are larger than any of those the system is capable of producing.[198, 214, 275]

inflection
A point of inflection of a plane curve is a point where the curve has a stationary **tangent**, at which the tangent is changing from rotating in one direction to rotating in the opposite direction.

information theory
A mathematical theory of information born in 1948 with the publication of Claude **Shannon**'s landmark paper, "A Mathematical Theory of Communication." Its main goal is to discover the laws governing systems designed to communicate or manipulate information, and to set up quantitative measures of information and of the capacity of various systems to transmit, store, and otherwise process information. Among the problems it treats are finding the best methods of using various communication systems and the best methods for separating the wanted information, or signal, from the noise. Another of its concerns is setting upper bounds on what it is possible to achieve with a given information-carrying medium (often called an information channel). The theory overlaps heavily with communication theory but is more oriented toward the fundamental limitations on the processing and communication of information and less oriented toward the detailed operation of the devices used.

injection
A **one-to-one** mapping.

inscribed angle
The angle formed by two **chord**s of a curve that meet at the same point on the curve.

inscribed circle
A circle inside a triangle or other **polygon** whose edges are **tangent**s to the circle. Such a polygon is said to be *circumscribed* to the circle. The (unique) circle inscribed to a triangle is called the *incircle*.

instability
The condition of a system when it is easily disturbed by internal or external forces or events, in contrast to a *stable*

system, which will return to its previous condition when disturbed. A pencil resting vertically on its eraser or a coin standing on edge are examples of systems that have the property of instability since they easily fall over at the slightest breeze or movement of the surface they are resting on. An unstable system is one whose **attractor**s can change; thus, instability is a characteristic of a system that is far from equilibrium or at **bifurcation**.

integer
Any positive or negative whole number or zero: . . ., –3, –2, –1, 0, 1, 2, 3, ... *Integer* is Latin for "whole" or "intact." The **set** of all integers is denoted by **Z**, which stands for *zahlen* (German for "numbers"). The integers are an extension of the **natural number**s to include **negative number**s and so make possible the solution of all equations of the form $a + x = b$, where a and b are natural numbers. Integers can be added and subtracted, multiplied, and compared. Like the natural numbers, the integers form a countably infinite set. However, the integers don't form a **field** since, for instance, there is no integer x such that $2x = 1$; the smallest field containing the integers is that of the **rational number**s. An important property of the integers is *division with remainder:* given two integers a and b with $b \neq 0$, it is always possible to find integers q and r such that $a = bq + r$, and such that $0 \leq r < |b|$. q is called the *quotient* and r is called the *remainder* resulting from division of a by b. The numbers q and r are uniquely determined by a and b. From this follows the *fundamental theorem of arithmetic,* which states that integers can be written as products of prime numbers in an essentially unique way.

integral
The area, or a generalization of area, under any section of a graph that is described by a **function**; in other words, the continuous cumulative sum of a function (see **continuity**). Not all functions have an exact formula that allows an integral to be found. In such cases, *numerical integration* has to be used, in which the area is found using approximate numerical techniques. Integrals, together with **derivative**s, are the fundamental objects of **calculus**.

integral equation
An equation that involves an **integral** of the **function** which is to be solved for.

integration
An operation that corresponds to the informal idea of finding the area under the graph of a **function**. The first theory of integration was developed by **Archimedes** with

his method of *quadratures,* but this could be applied only in circumstances where there was a high degree of geometric symmetry. In the seventeenth century, Isaac **Newton** and Gottfried **Leibniz** independently discovered the idea that integration was a sort of opposite of **differentiation** (which they had just invented); this allowed mathematicians to calculate a broad class of integrals for the first time. However, unlike Archimedes's method, which was based on **Euclidean geometry**, Newton's and Leibniz's integral calculus lacked a secure foundation.

In the nineteenth century, Augustin **Cauchy** finally developed a rigorous theory of **limits**, and Bernhard **Riemann** followed this up by formalizing what is now called the **Riemann integral**. To define this integral, one fills the area under the graph with smaller and smaller rectangles and takes the limit of the sums of the areas of the rectangles at each stage. Unfortunately, some functions don't have well-defined limits to these sums, so they have no Riemann integral. Henri **Lesbesgue** invented another method of integration to solve this problem. He first presented his ideas in *Intégrale, Longueur, Aire* (Integral, length, area) in 1902. Instead of using the areas of rectangles, a method that puts the focus on the **domain** of the function, Lebesgue turned to the **codomain** of the function for his fundamental unit of area. Lesbesgue's idea was to build first the integral for what he called simple functions— functions that take only finitely many values. Then he defined it for more complicated functions as the upper bound of all the integrals of simple functions smaller than the function in question. *Lesbesgue integration* has the beautiful property that every function with a Riemann integral also has a Lebesgue integral, and the two integrals agree. But there are many functions with a Lebesgue integral that don't have a Riemann integral. As part of the development of Lebesgue integration, Lebesgue introduced the concept of *Lebesgue measure,* which measures lengths rather than areas. Lebesgue's technique for turning a measure into an integral generalizes easily to many other situations, leading to the modern field of **measure theory**.

The Lebesgue integral was deficient in one respect. The Riemann integral had been generalized to the *improper Riemann integral* to measure functions whose domain of definition was not a **closed** interval. The Lebesgue integral integrated many of these functions (always reproducing the same answer when it did), but not all of them. The *Henstock integral* is an even more general notion of integral (based on Riemann's theory rather than Lebesgue's) that subsumes both Lebesgue integration and improper Riemann integration. However, the Henstock integral depends on specific features of the real number line and so doesn't generalize as well as the Lebesgue integral does.

interesting numbers

Clearly some numbers are of greater interest (at least to mathematicians) than are others. The number **pi**, for instance, is far more interesting than 1.283—or virtually any other number for that matter. Confining our attention to integers, can there be such a thing as an uninteresting number? It is easy to show that the answer must be "no." Suppose there were a set U of uninteresting integers. Then it must contain a least member, u. But the property of being the smallest uninteresting integer makes u interesting! As soon as u is removed from U, there is a new smallest uninteresting integer, which must then also be excluded. And so the argument could be continued until U was empty. Given that all integers are interesting, can they be ranked from least to most interesting? Again, no. To be ranked as "least interesting" is an extremely interesting property, and thus leads to another logical contradiction!

When Srinivasa **Ramanujan**, the great Indian mathematician, was ill with tuberculosis in a London hospital, his colleague G. H. **Hardy** went to visit him. Hardy opened the conversation with: "I came here in taxicab number 1729. That number seems dull to me which I hope isn't a bad omen." Ramanujan replied, without hesitation: "Nonsense, the number isn't dull at all. It's quite interesting. It's the smallest number that can be expressed as the sum of two cubes in two different ways." ($1729 = 1^3 + 12^3$ and $9^3 + 10^3$.)

International Date Line

Is it possible to assign a time to every longitude on Earth, so that each longitude has a *different* time but the times at nearby longitudes are always close? The answer is no, which is mathematically equivalent to saying that there's no way to continuously map points on a **real number** line onto a circle. This explains why an International Date Line is needed. It allows most regions on Earth to have times similar to their neighbors, though, by convention, time changes are (usually) made in chunks of one hour between adjacent time zones. Then it takes care of the inevitable discontinuity by having it happen all at once, as a jump by one whole day, on a longitude that passes mostly through open water in the Pacific. The fact that there doesn't exist any continuous **one-to-one** function from the circle onto the line follows from the **Borsuk-Ulam theorem** in one dimension.

interpolate

To estimate the value of a point that lies between two known values of a **function**. This is often done by approximating a line or a smooth curve between the values, which is the literal meaning of the word. *Inter* is the

Latin prefix for "between," *polire* translates as "to adorn or polish," so together they mean "to smooth between." *Extrapolate* was created as an extension of interpolate to suggest the smoothing of a line *outside* the known points. This operation is often done in statistics when patterns are studied over time to predict future events.

intersection

A place where two or more things meet or overlap. Two lines intersect at a point, two planes can intersect in a line, and so forth. The intersection of two or more **sets**, represented by the symbol ∩, is the set of elements that all the sets have in common; in other words, all the elements contained in every one of the sets.

invariant

(1) Something that stays the same when a particular **transformation** is carried out. (2) A value that is unchanged when a particular **function** is applied. (3) In **topology**, a number, **polynomial**, or other quantity associated with a topological object such as a **knot** or 3-**manifold**, which depends only on the underlying object and not on its specific description or presentation.

invariant theory

The study of quantities that are associated with **polynomial** equations and that are left **invariant** under transformations of the variables. For example, the discriminant $b^2 - 4ac$ is an invariant of the **quadratic** form $ax^2 + bxy + cy^2$.

inverse

(1) The inverse of a number, or a **reciprocal**, is 1 divided by the number; for example, the inverse of 8 is $\frac{1}{8}$ and the inverse of $\frac{3}{5}$ is $\frac{5}{3}$. (2) The inverse of a **function** or a **transformation** is the function or transformation that reverses the effect of the function or transformation. For example, the inverse of addition is subtraction, and of clockwise rotation is anticlockwise rotation. (3) The inverse of an element of a **set**, or a number, with respect to a particular operation, is what has to be combined with the element or number in order to obtain that operation's *identity element*.

involute

Attach a string to a point on a curve. Extend the string so that it is **tangent** to the curve at the point of attachment. Then wind the string up, always keeping it taut. The **locus** of points traced out by the end of the string is called the involute of the original curve, and the original curve is called the **evolute** of its involute. Although a curve has a unique evolute, it has infinitely many involutes corresponding to different choices of initial point.

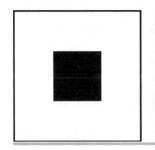 

irradiation illusion Despite appearances, the inner squares are the same size.

An involute can also be thought of as any curve **orthogonal** to all the tangents to a given curve. See also **circle involute**.

irradiation illusion

A **distortion illusion** discovered by Hermann von Helmholtz in the nineteenth century. Despite the fact that the two figures are identical in size, the white hole looks bigger than the black one.

irrational number

A **real number** that can't be written as one whole number divided by another; in other words, a real number that isn't a **rational number**. The decimal expansion of an irrational number doesn't come to an end or repeat itself (in equal length blocks), though it may have a pattern such as 0.101001000100001000001 . . . . The vast majority of real numbers are irrational, so that if you were to pick a single point on the real **number line** at random the chances are overwhelmingly high that it would be irrational. Put another way, whereas the set of all rationals is countable, the irrationals form an **uncountable set** and therefore represent a larger kind of **infinity**. Indeed, as the Harvard logician Willard Van Orman Quine pointed out: "The irrationals exist in such variety . . . that no notation whatever is capable of providing a separate name for each of them."

There are two types of irrational number: **algebraic numbers**, such as the **square root of 2**, which are the roots of algebraic equations, and the **transcendental numbers**, such as **pi** and **e**, which aren't. In some cases it isn't known if a number is irrational or not; undecided cases include 2^e, π^e, π^2, and the **Euler-Mascheroni constant**, γ. An irrational number raised to a rational power can be rational; for instance, $\sqrt{2}$ to the power 2 is 2. Also, an irrational number to an irrational power can be rational. What kind of number is $\sqrt{2}^{\sqrt{2}}$? The answer is irrational. This follows from Gelfond's theorem, which says that if *a* and *b* are roots of **polynomial**s, and *a* is not 0 or 1 and *b* is irrational, then a^b must be irrational (in fact, transcendental).

irreptile
See **rep-tile**.

Ishango bone
A bone tool handle discovered around 1960 in the African area of Ishango, near Lake Edward. It has been dated to about 9000 B.C. and was at first thought to have been a **tally** stick. At one end of the bone is a piece of quartz for writing, and the bone has a series of notches carved in groups on three rows running the length of the bone. The markings on two of these rows each add to 60. The first row is consistent with a number system based on 10, since the notches are grouped as 20 + 1, 20 − 1, 10 + 1, and 10 − 1, while the second row contains the **prime number**s between 10 and 20! A third seems to show a method for multiplying by 2 that was used in later times by the Egyptians. Additional markings suggest that the bone was also used as a lunar phase counter. The Ishango

Bone is kept at the Royal Institute for Natural Sciences of Belgium in Brussels. See also **Lebombo bone**.[45, 157]

isochrone
A set of points with the property that a given process or trajectory will take the same length of time to complete starting from any of the points. The curve formed by such a set of points is called an *isochronous curve*. See also **tautochrone problem**.

isogonal conjugate
Isogonal lines of a triangle are cevians (see **Ceva, Giovanni**) that are symmetric with respect to the angle bisector. Two points are isogonal conjugates if the corresponding lines to the vertices are isogonal (see **vertex**).

isometry
A symmetry operation, which may involve **translation**, **rotation**, and **reflection**, that preserves the distance of any two points. Each isometry of a **wallpaper group**, for example, can be represented by a 3 × 3 matrix.

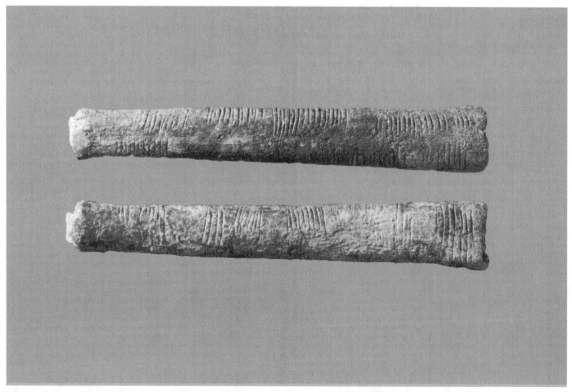

Ishango bone   The marks on this 11,000-year-old bone speak of a surprisingly sophisticated mathematical knowledge. *Museum of Natural Sciences (Brussels)*

isomorphism

In geometry, a **transformation** that doesn't alter the side lengths and the angle sizes of the figure involved. Examples of such transformations include reflections, rotations, translations, or transformations by a glide. In **set theory**, an isomorphism is a **one-to-one** correspondence between the elements of two sets such that the result of an operation on elements of one set corresponds to the result of the analogous operation on their images in the other set.

isoperimetric inequality

For any closed, three-dimensional body with volume V and surface area A, the following inequality always holds:

$$36\pi V^2 \le A^3$$

isoperimetric problem

Among all shapes of a given perimeter, which encloses the greatest area? This ancient conundrum is alluded to in the tale of Queen Dido in Virgil's *Aeneid*. Threatened by her brother, who had already murdered her father, the queen was obliged to hastily gather her valuables and flee her native city of Tyria in ancient Phoenicia. In due course, her ship landed in North Africa, where she made the following offer to a local chieftain: in return for her fortune he would give as much land as she could isolate with the skin of an ox. This was readily agreed to, whereupon the crafty queen sliced the skin into very fine threads, which she tied together and made into a giant semicircle. Combined with the natural boundary provided by the sea, this enclosed such a large area that a city—Carthage—was eventually built upon it. Two millennia later, Karl **Weierstrass** was the first to prove rigorously, using analysis and calculus, that the solution to the isoperimetric problem is a circle (something the Greeks suspected but were not able to prove geometrically). When the same question is asked in one dimension higher it becomes the **isovolume problem**.

isosceles

Having two sides of the same length, as in the case of an isosceles **triangle**. An isosceles **trapezoid** in the United States is the equivalent of a trapezium in Britain. *Isosceles* comes from the Greek *iso* ("same or equal") and *skelos* ("legs").

isotomic conjugate

Two points on the side of a triangle are *isotomic* if they are equidistant from the midpoint of that side. Two points inside a triangle are isotomic conjugates if the corresponding cevians (see **Ceva, Giovanni**) through these points meet the opposite sides in isotomic points.

isovolume problem

What surface encloses the maximum volume per unit surface? The answer is a **sphere**, for which

$$\text{volume/surface area} = (\tfrac{4}{3}\pi r^3)/(4\pi r^2) = \tfrac{1}{3}r,$$

where r is the sphere's radius. Proof of the solution to the isovolume problem came as recently as 1882 from Hermann Schwarz. See also **bubbles** and **isoperimetric problem**.

iterate

Do something repeatedly. Do something repeatedly. . . .

iteration

If at first you don't succeed, try, try, again. Then quit. There's no use being a damn fool about it.
—W. C. Fields

A **feedback** process that repeats n number of times. Iteration refers to the act of performing the calculation of a certain **function** and then picking the result, or output, as the starting value, or input, for the next calculation of the same function. The operation repeats on and on—even infinitely, despite Fields's comment!

J

Jacobi, Karl Gustav Jacob (1804–1851)

A German mathematician who did important work on **elliptic functions**, **partial differential equations**, and mechanics. Although he was preceded in many of his discoveries about elliptic functions by Carl **Gauss** (who didn't publish) and Niels **Abel**, Jacobi is nevertheless considered one of the founders of the subject. His name is probably best known from the *Jacobian*, an $n \times n$ **determinant** formed from a set of n functions in n unknowns. He wasn't the first to use it—the "Jacobian" appears in an 1815 paper of Agustin **Cauchy**—but Jacobi did write a long memoir about it in 1841, and proved that the Jacobian of n functions vanishes if and only if the functions are related (Cauchy had proved the "if" part). He also did important work on partial differential equations and their application to physics. Along with William **Hamilton**, he developed an approach to mechanics based on generalized coordinates. In this method, the total energy of a mechanical system is represented as a function of generalized coordinates and corresponding generalized momenta; for example, in a double pendulum the two generalized coordinates could be two angles. *Hamilton-Jacobi theory* is the technique of solving the system by transforming coordinates so that the transformed coordinates and momenta are constants.

Jacobi was appointed to a position at the University of Königsberg in 1826. He gained a reputation as a gifted teacher and is credited with introducing the seminar method (giving lectures on his own ongoing research) into the university. After a collapse from overwork in 1843, Jacobi was allowed to stay in Berlin with a generous allowance from the king of Prussia. Five years later, revolution swept Europe and Jacobi was persuaded to run for parliament. This proved a disaster; not only did he loose the election, but his foray into politics annoyed his royal patron, who cut off his pension. Jacobi, with a large family to support, was faced with destitution. Only his reputation as the greatest German mathematician besides Gauss saved him; faced with the prospect of losing Jacobi to the University of Vienna, the king was prevailed upon to restore the pension. Jacobi was a notoriously hard worker (indeed, he had several breakdowns due to overwork), but his death in 1851 was the result of smallpox.

Johnson solid

Any convex **polyhedron** with regular faces that is not a **Platonic solid**, an **Archimedean solid**, or a **prism** (or an antiprism). There are 92 Johnson solids, which are named after Norman W. Johnson who was the first to catalog them in 1966. They include equilateral **deltahedra** and dipyramids (two pyramids placed symmetrically base to base) and any irregular convex solid that can be made by sticking together triangles, squares, pentagons, and so on, in a way that happens to close. Some of the simpler Johnson solids are assemblages of pyramids, prisms, and antiprisms; for example, a gyrobifastigium is two triangular prisms glued together with a twist. Others are fragments of Archimedean solids; for example, a pentagonal rotunda is half an icosidodecahedron. When stretched out with a prism this makes the marvelously named elongated pentagonal orthocupolarotunda.

Johnson's theorem

If three congruent circles all intersect in a single point, then the other three points of intersection will lie on another circle of the same radius. This simple little theorem was discovered by Roger Johnson in 1916.

Jordan, (Marie-Ennemond) Camille (1838–1922)

A French mathematician who made important contributions to **group** theory. He was the first to draw attention to the work of Evariste **Galois**, which had until then been almost entirely ignored. He built on Galois's study by grasping the intimate connection between groups of permutations and the solvability of **polynomial** equations. Jordan also introduced the idea of an infinite group. He passed on his interest in group theory to two of his most outstanding pupils, Felix **Klein** and Sophus **Lie**, both of whom went on to develop the subject in new and important ways.

Jordan curve

A simple, **closed curve**.

Jordan matrix

A **matrix** whose **diagonal** elements are all equal (and nonzero) and whose elements above the principal diagonal are equal to 1, but all other elements are 0.

Johnson solid Two of the 92 Johnson solids: the bilunabirotunda (J91, right) and the snub disphenoid (J84, left). *Robert Webb, www.software3d.com; created using Webb's Stella program*

Journal of Recreational Mathematics

The only international journal devoted to the lighter side of mathematics. It was started in 1968 and is currently edited by Charles Ashbacher.

Julia set

Any of any infinite number of **fractal** sets of points on an **Argand diagram** (the complex plane) defined by a simple rule. Given two **complex numbers** z and c, and the **recursion** $z_{n+1} = z_n^2 + c$, the Julia set for any given value of c, consists of all values of z for which z, when iterated in the equation above, does not "blow up" or tend to infinity. Julia sets are closely related to the **Mandelbrot set**, which is, in a way, an index of all Julia sets. For any point on the complex plane (which represents a value of c) a corresponding Julia set can be drawn. We can imagine a movie of a point moving about the complex plane with its corresponding Julia set. When the point lies inside the Mandelbrot set the corresponding Julia set is topologically unified or connected. As the point crosses the boundary of the Mandelbrot set, the Julia set explodes into a cloud of disconnected points called *Fatou dust*. If c is on the boundary of the Mandelbrot set, but not a waist point (where two large regions of the Mandelbrot set are connected by a narrow bridge), the Julia set of c looks like the Mandelbrot set in sufficiently small neighborhoods of c.

Julia set A fractal image based on a Julia set. *Jos Leys, www.josleys.com*

Julia sets are named after the French mathematician Gaston Julia (1893–1978), whose most famous work, *Memoire sur l'iteration des fonctions rationnelles,* provides the theory for Julia sets before computers were available. In 1918, at the age of 25 Julia was severely wounded in World War I and lost his nose. He wrote his greatest trea-tise in a hospital between the painful operations necessi-tated by his wounds.

jump discontinuity

A discontinuity in a **function** where the left- and right-hand limits exist but are not equal to each other.

Kaczynski, Theodore John (1942–)

Known as the Unabomber before his true identity was discovered, a Harvard graduate with a genius IQ who abandoned a promising career as a mathematician for a life of social isolation and intermittent terrorist attacks. Kaczynski eluded the FBI for 17 years, during which time he orchestrated 16 explosions that killed 3 people and injured 23 others. He avoided execution following his arrest in 1996 through a plea bargain and was given instead 4 life sentences plus 30 years in prison. During his mathematical career in the late 1960s, Kaczynski published a doctoral dissertation and several papers in academic journals and served as an assistant math professor at the University of California, Berkeley (1967–1969).

Kakeya needle problem

A famous problem named after the Japanese mathematician Sôichi Kakeya, who first posed it in 1917. It asks: What is the smallest-area plane figure inside which a unit straight-line segment can be rotated through 180°? For some years, the answer was thought to be a **deltoid**. However, in 1928, the Russian mathematician Abram Besicovitch shocked the mathematical world by showing that the problem *had no answer* or, to be more precise, that there was no minimum area.[39] In 1917 Besicovitch had been working on a problem in **Riemann integration**, and had reduced it to the question of existence of planar **sets** of **measure** 0, which contain a line segment in each direction. He then constructed such a set, and published his construction in a Russian journal in 1920. Due to the civil war and the blockade, there was hardly any communication between Russia and the rest of the world at the time, so Besicovitch hadn't heard about the challenge posed by Kakeya. Several years later, after he left Russia and learned about the needle problem, Besicovitch modified his original construction and was able to give the startling answer that the area in question may be made arbitrarily small.

Kaluza-Klein theory

A model that seeks to unite classical gravity and electromagnetism by resorting to **higher dimensions**. In 1919 the German mathematician Theodor Kaluza (1885–1954) pointed out that if general **relativity theory** is extended to a five-dimensional **space-time**, the equations can be separated out into ordinary four-dimensional gravitation plus an extra set (which is equivalent to Maxwell's equations for the electromagnetic field) plus an extra field known as the dilaton). Thus, electromagnetism is explained as a manifestation of curvature in a **fourth dimension** of physical space, in the same way that gravitation is explained in Einstein's theory as a manifestation of curvature in the first three. In 1926 the Swedish physicist Oskar Klein (1894–1977) proposed that the reason the extra spatial dimension goes unseen is that it is compact—curled up like a ball with a fantastically small radius. In the 1980s and 1990s, Kaluza-Klein theory experienced a big revival and can now be seen as a precursor of **string theory**.

Kampyle of Eudoxus

An hourglass-shaped curve that was first studied by **Eudoxus** in an attempt to solve the classical problem of **duplicating the cube**. It is described by the Cartesian equation

$$a^2x^4 = b^4(x^2 + y^2).$$

It is also the **radial curve** of the **catenary**.

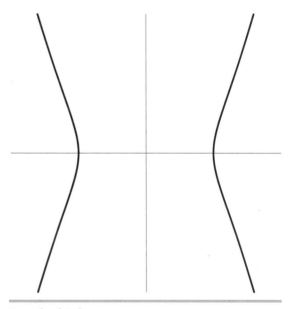

Kampyle of Eudoxus © Jan Wassenaar, www.2dcurves.com

kappa curve

Also known as *Gutschoven's curve,* named after Gérard van Gutschoven (1615–1668), who first studied it around 1662. The kappa curve, which resembles the Greek letter κ, was also investigated by Isaac **Newton** and, some years later, by Johann Bernoulli (see **Bernoulli family**). It is given by the Cartesian equation

$$y^2(x^2 + y^2) = a^2x^2.$$

Kaprekar constant

Take any four-digit number whose digits are not all identical. Rearrange the digits to make the largest and smallest 4-digit numbers possible. Subtract the smaller number from the larger. Use the resulting number and repeat the process. For example, starting with 4,731: $7,431 - 1,347 = 6,084$; $8,640 - 468 = 8,172$; $8,721 - 1,278 = 7,443$; $7,443 - 3,447 = 3,996$; $9,963 - 3,699 = 6,264$; $6,642 - 2,466 = 4,176$; $7,641 - 1,467 = 6,174$. After this, the result is always 6174. Remarkably, *every* four-digit number whose digits are not all the same will eventually reach 6,174, in at most seven steps, and then stay there. This is called the Kaprekar constant for four-digit numbers, after the Indian mathematician Dattathreya Ramachandra Kaprekar who made the discovery in 1949. The Kaprekar constant for three-digit numbers is 495, which is arrived at for any three-digit number in no more than six iterations. The same process, or algorithm, can be applied to numbers of *n*-digits, where *n* is any whole number. Depending on the value of *n*, the algorithm will result in a nonzero constant, zero (the degenerate case), or a cycle.

Kaprekar number

Take a positive whole number *n* that has *d* number of digits. Take the square of *n* and separate the result into two pieces: a right-hand piece that has *d* digits and a left-hand piece that has either *d* or $d - 1$ digits. Add these two pieces together. If the result is *n*, then *n* is a Kaprekar number. Examples are 9 ($9^2 = 81$, $8 + 1 = 9$), 45 ($45^2 = 2,025$; $20 + 25 = 45$), and 297 ($297^2 = 88,209$; $88 + 209 = 297$). The first 10 Kaprekar numbers according to this definition are 1; 9; 45; 55; 99; 297; 703; 999; 2,223; and 2,728. Kaprekar numbers can also be defined by higher powers. For example, $45^3 = 91,125$; and $9 + 11 + 25 = 45$. The first 10 numbers with this property are: 1; 8; 10; 45; 297; 2,322; 2,728; 4,445; 4,544; and 4,949. For fourth powers, the sequence begins 1; 7; 45; 55; 67; 100; 433; 4,950; 5,050; 38,212. Notice that 45 is a Kaprekar number for second, third, and fourth powers ($45^4 = 4,100,625$; and $4 + 10 + 06 + 25 = 45$)—the only number in all three Kaprekar sequences, up to at least 400,000. See also **unique number**.

Kasner, Edward (1876–1955)

An American mathematician at Columbia University best remembered for introducing the words *googol* and *googolplex* into the popular mathematical lexicon. He is also well known as the coauthor, with James **Newman**, of *Mathematics and the Imagination,* first published in 1940.[186] In a later edition (1967), he spoke about the term *mathescope,* which was coined by the science reporter Wilson Davis after listening to one of Kasner's public lectures. In Kasner's words: "It is not a physical instrument; it is a purely intellectual instrument, the ever-increasing insight which mathematics gives into the fairyland which lies beyond intuition and beyond imagination." His main field of research was **differential geometry**, which he studied in its applications to mechanics, cartography, and stereographic projections, though he also wrote papers on circle packing and on the horn angle and studied an extension of right triangles to the complex plane.

Kepler, Johannes (1571–1630)

A German astronomer and mathematician best remembered for his three laws of planetary motion. He also did important work in optics, discovered two new regular polyhedra (1619), gave the first mathematical treatment of close **packing** of equal spheres (in 1611, leading to an explanation of the shape of the cells of a honeycomb), gave the first proof of how **logarithm**s work (1624), and devised a method of finding the volumes of **solids of revolution** that (with hindsight) can be seen as contributing to the development of **calculus** (1615–1616). Although his work eventually led to our modern view of the solar system, he himself held beliefs about the arrangement of the planets that smack of Pythagorean mysticism, though ultimately they stemmed from his belief that mathematical harmony was a reflection of God's perfection. In his first cosmological model (*Mysterium cosmographicum,* 1596) he suggested that if a sphere were drawn to touch the inside of the path of Saturn, and a cube were inscribed in the sphere, then the sphere inscribed in that cube would be the sphere circumscribing the path of Jupiter. Then if a regular tetrahedron were drawn in the sphere inscribing the path of Jupiter, the insphere of the tetrahedron would be the sphere circumscribing the path of Mars, and so inward, putting the regular dodecahedron between Mars and Earth, the regular icosahedron between Earth and Venus, and the regular octahedron between Venus and Mercury. Thus the number of (then-known) planets is explained perfectly in terms of the five convex regular solids—the **Platonic solid**s.

Kepler's second work on cosmology (*Harmonices mundi* [Harmony of the world], book V, 1619) offers a more elaborate mathematical model but still with the polyhedra in place. The mathematics in this work includes the first systematic treatment of **tessellation**s, a proof that there are only 13 convex uniform polyhedra (the **Archimedean solid**s) and the first account of two non-convex regular polyhedra (see **Kepler-Poinsot solid**s).

Kepler's mathematical work was full time and continued even during the wedding to his second wife in 1613. The dedicatory letter to a book he wrote shortly after explains that at the wedding celebration he noticed that the volumes of wine barrels were estimated by means of a rod slipped in diagonally through the bunghole, which prompted him to wonder how that could work. The result was a study of the volumes of solids of revolution, *Nova Stereometria Doliorum* (New stereometry of wine barrels, 1615).

Kepler-Poinsot solids

The four regular nonconvex polyhedra that exist in addition to the five regular convex polyhedra known as the **Platonic solid**s. As with the Platonic solids, the Kepler-Poinsot solids have identical regular **polygon**s for all their faces, and the same number of faces meet at each vertex. What is new is that we allow for a notion of "going around twice," which results in faces that intersect each other. In the *great stellated dodecahedron* and the *small stellated dodecahedron*, the faces are pentagrams (five-pointed stars). The center of each pentagram is hidden inside the polyhedron. These two polyhedra were described by Johannes **Kepler** in 1619, and he deserves credit for first understanding them mathematically, though a sixteenth-century drawing by the Nuremberg goldsmith Wentzel Jamnitzer (1508–1585) is very similar to the former and a fifteenth-century mosaic attributed to the Florentine artist Paolo Uccello (1397–1475) illustrates the latter. The *great icosahedron* and *great dodecahedron* were described by Louis **Poinsot** in 1809, though Jamnitzer made a picture of the great dodecahedron in 1568. In these two, the faces (20 triangles and 12 pentagons, respectively) that meet at each vertex "go around twice" and intersect each other, in a manner that is

Kepler-Poinsot solids   From left to right: the small stellated dodecahedron, great stellated dodecahedron, great icosahedron, and great dodecahedron. *Robert Webb, www.software3d.com; created using Webb's Stella program*

the three-dimensional analog to what happens in two dimensions with a **pentagram**. Together, the Platonic solids and the Kepler-Poinsot polyhedra form the set of nine *regular* polyhedra. Augustin **Cauchy** first proved that no other polyhedra can exist with identical regular faces and identical regular vertices.

Kepler's conjecture

No **packing** of spheres of the same radius in three dimensions has a density greater than the face-centered (hexagonal) cubic packing. This claim was first published by Johannes **Kepler** in his monograph *The Six-Cornered Snowflake* (1611), a treatise inspired by his correspondence with Thomas Harriot (see **cannonball problem**). In his slender essay, Kepler asserted that face-centered cubic packing–the kind greengrocers use to stack oranges–is "the tightest possible, so that in no other arrangement could more pellets be stuffed into the same container." The question of whether Kepler's conjecture is right or not has became known, not surprisingly, as *Kepler's problem*. In the nineteenth century, Carl **Gauss** proved that face-centered cubic packing is the densest arrangement in which the centers of the spheres form a regular lattice, but he left open the question of whether an irregular stacking of spheres might be still denser. In 1953, László Tóth reduced the Kepler conjecture to an enormous calculation that involved specific cases, and later suggested that computers might be helpful for solving the problem. This was the approach taken by Thomas Hales, a mathematician at the University of Michigan at Ann Arbor, and which led him, in 1998, to claim that he had proved Kepler was right all along. Hales's proof of Kepler's conjecture remains controversial simply because of the length of the computer calculations involved and the difficulty of verifying them. The casebook on this mystery remains open. See also **monster group**.

Khayyam, Omar (1044–1123)

An outstanding Persian mathematician and astronomer, whose full name was Abu al-Fath Omar ben Ibrahim al-Khayyam, or "tent-maker"–possibly his father's profession. His work on algebra was known throughout Europe in the Middle Ages. He made several important contributions: he discovered a geometrical method to solve **cubic** equations by intersecting a **parabola** with a circle, discussed what would become known as **Pascal's triangle**, and asked if a ratio could be regarded as a number. He is best known, however, as a poet. In 1859, Edward Fitzgerald (1809–1883) translated Khayyam's *Rubaiyat*–a popular collection 600 short four-line poems. Roseships from Khayyam's tomb were germinated at Kew Gardens, London, and planted on Fitzgerald's tomb in St. Michael's Churchyard, Boulge, Suffolk, in 1893; the original plant has died, but its descendents continue to bloom.

Khintchine's constant

One of the most remarkable, yet poorly understood, constants in mathematics, which captures, in a fascinating way, the behavior of *almost all* **real numbers**. Pick a real number at random and write it down as a **continued fraction**. Almost certainly, the **geometric mean** of the terms in this continued fraction will be Khintchine's constant, which has the value 2.685452 . . . . It is important to say "almost certainly" because there are some real numbers including all **rational numbers**, roots of third-order **polynomials**, and certain other classes of number, whose continued fractions would give a different result. These exceptions, however, form a tiny minority of all real numbers.

kings problem

The problem of determining how many nonattacking kings, $K(n)$, can be placed on an $n \times n$ chessboard. For $n = 8$, the solution is 16. The general solution is $K(n) = \frac{1}{4}n^2$ if n is even and $K(n) = \frac{1}{4}(n + 1)^2$ if n is odd. The minimum number of kings needed to attack or occupy all squares on an 8×8 chessboard is nine.

kinship puzzles

Like **age puzzles and tricks**, problems to do with how family members are related go back many centuries. Some of these can be fiendishly convoluted, especially if incestual pairings are allowed (in the puzzle!). Sketching a genealogical tree is sometimes helpful. The following all involve legitimate ties.

PUZZLES

1. Brothers and sisters have I none, but that man's father is my father's son. Who is that man? (This is one of the oldest problems of the kinship variety.)

2. What is the simplest way in which two people can be an uncle to each other? (From Dudeney's *A Puzzle-Mine*.)

3. A certain family party consisted of 1 grandfather, 1 grandmother, 2 fathers, 2 mothers, 4 children, 3 grandchildren, 1 brother, 2 sisters, 2 sons, 2 daughters, 1 father-in-law, 1 mother-in-law, and 1 daughter-in-law. Twenty-three people, you will say. No; there were only seven people present. Can you show how this might be? (From *Amusements in Mathematics*.)

Solutions begin on page 369.

Kirkman, Thomas Penyngton (1806–1895)

An English rector and mathematician who did important work in **combinatorics**. His **schoolgirls problem**, which he posed in 1850, led to a more general study of certain ways of combining objects. He also explored the possibility of finding a route around the edges of a given **polyhedron** that passes through each vertex (corner) of the shape once and only once. His rather complex and ultimately unprovable ideas were later picked up and refined successfully by William **Hamilton**, but not be-fore Kirkman had managed to supply a general proof of the fact that "if a polyhedron has an odd number of vertices and each face has an even number of edges, then there is no circuit which passes through all the vertices." This result introduces the concept of a *bipartite graph*—a **graph** that can be divided into two separate sets of vertices such that every edge in one set joins a vertex.

kite

A **quadrilateral**, with two pairs of congruent adjacent sides, that is named after one of the traditional forms of toy kite. The toy itself probably draws its name from the bird commonly called a kite, or kyte, in England. The old English form of the word, cyte, may in turn derive from an early German name for an owl. A nonconvex kite is often called a *dart,* a term used by Roger **Penrose** in his proof on a nonperiodic tiling of the plane (see also **Penrose tiling**). **Proclus** referred to this shape as a "four-sided triangle" and spoke of it as a geometric paradox. John **Conway** pointed out that there is no special name for a quadrilateral that has two pairs of equal sides that, unlike in the case of a kite, are not parallel sides. He proposed the name *strombus* for such a figure, from the Greek for a spinning top.

Klein, Christian Felix (1849–1925)

A German mathematician noted for his work on **non-Euclidean geometry**, the connections between geometry and **group** theory, and the theory of functions. His Erlangen Programm (1872) for unifying the diverse forms of geometry through the study of equivalence in transformation groups was profoundly influential, especially in the United States, for over 50 years. In his *Lectures on the Icosahedron and the Solution of Equations of the Fifth Degree* (1884, tr. 1888) he showed how the rotation groups of regular solids could be applied to the solution of difficult algebraic problems. Klein was a professor of mathematics successively at the University of Erlangen, the Technical Institute in Munich, and the universities of Leipzig and Göttingen, and was a prolific writer and lecturer on the theory, history, and teaching of mathematics. See also **Klein bottle**.

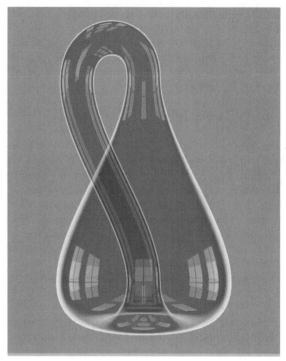

Klein bottle   The famous container, with zero volume, rendered in the improbable form of a soap bubble. © *John M. Sullivan, University of Illinois and TU Berlin*

Klein bottle

A mathematician named Klein,
Thought the Möbius band was divine,
He said, "If you glue,
The edges of two,
You'll get a weird bottle like mine."

—Leo Moser

Take a rectangle and join one pair of opposite sides to make a cylinder. Now join the other pair with a half-twist. The result is a Klein bottle. Sound easy? It is if you have access to a **fourth dimension** because that's what is needed to carry out the second step and to allow the surface to pass through itself without a hole. A true Klein bottle is a four-dimensional object. It was discovered in 1882 by Felix **Klein** when he imagined, as in the limerick, joining two **Möbius band**s together to create a single-sided bottle with no boundary.

An ordinary (three-dimensional) bottle has a crease or fold around the opening where the inside and outside of the bottle meet. A sphere doesn't have this crease or fold, but it has no opening. A Klein bottle has an opening but no crease: like a Möbius band, it is a continuous one-sided structure. Because it has no crease or fold, there's no verifiable definition of where its inside and outside begin. Therefore, the volume of a Klein bottle is considered to be zero, and the bottle has no real contents—except itself! As the joke goes: "In topological hell the beer is packed in Klein bottles." Take a coin, slide it across the surface of a Klein bottle until it returns to its starting point, and the coin, as if by magic, will be flipped over. This is because, unlike a sphere or a regular bottle, a Klein bottle is nonorientable.

Although a Klein bottle can't be "embedded" (that is, fully realized) in three dimensions, it can be "immersed" in three dimensions. Immersion is what happens when a higher dimensional object cuts through a lower dimensional one, producing a cross section. When a sphere is immersed in a plane, for example, it produces a circle. The computer-generated soap bubble structure shown here could be created by stretching the neck of a bottle through its side and joining its end to a hole in the base. Except at the side-connection (the nexus), this properly shows the shape of a four-dimensional Klein bottle. Just as the picture here is of a three-dimensional Klein bottle immersion, so an immersion in real life is like a drawing of the true four-dimensional bottle.

knights problem

To find the maximum number of knights that can be placed on an $n \times n$ chessboard in such a way that no two pieces are attacking each other. For a standard 8×8 chessboard, the answer is 32 (all on white squares or all on black). In the general, the solution is $\frac{1}{2} n^2$ if n is even and $\frac{1}{2} (n^2 + 1)$ if n is odd, giving the sequence 1, 4, 5, 8, 13, 18, ...

knight's tour

A classic **chess** puzzle: to find a sequence of moves by which a knight can visit each square of a chessboard exactly once. If the final position is a knight's move away from the first position, the tour is said to be *closed* or *reentrant*. The earliest recorded solution for a standard 8×8 chessboard was given by Abraham **de Moivre**; the earliest known reentrant solution came from the French mathematician Adrien-Marie Legendre (1752–1833). Not to be outdone Leonhard **Euler** found a reentrant tour that visits two halves of the board in turn. The problem can be generalized to an $n \times n$ board, with some surprising results; for example, a reentrant tour is not possible on a 4×4 board.

A knight's tour is called a **magic tour** if the resulting arrangement of numbers forms a **magic square**, and a *semimagic tour* if the resulting arrangement of numbers is a **semi-magic square**. It has long been known that magic knight's tours aren't possible on $n \times n$ boards if n is odd. It was also known that such tours are possible for all

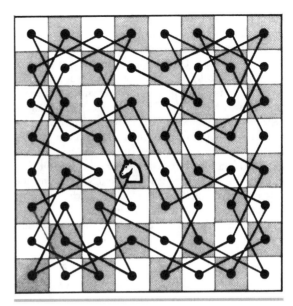

knight's tour After visiting every square of the chessboard exactly once, the knight returns to its starting point.

boards of size $4k \times 4k$ for $k > 2$. However, while a number of semi-magic knight's tours were known on the usual 8×8 chessboard, it was *not* known if any fully magic tours existed on the 8×8 board. This longstanding open problem was finally settled in the negative by an exhaustive computer enumeration of all possibilities. The software for the computation was written by J. C. Meyrignac, and a Web site was established by Guenter Stertenbrink to distribute and collect results for all possible tours. After 61.40 CPU-days, corresponding to 138.25 days of computation at 1 GHz, the project was completed on August 5, 2003. As well as netting a total of 140 distinct semi-magic knight's tours, the computation showed for the first time that no 8×8 magic knight's tour is possible.

More magical and mysterious tours can be conducted on boards on the surfaces of cubes, cylinders, toruses, and multidimensional shapes, such as hypercubes. See also **Hamilton path**.

knot

Never cut what you can untie.

—Joseph Joubert

A **closed** curve in three dimensions. The two simplest nontrivial knots are the trefoil knot, whose picture has three crossings, and the figure-eight knot, whose picture has four. To date, more than 1.7 million nonequivalent knots with pictures of 16 or fewer crossings have been identified. The mathematical theory of knots was born out of attempts to model the atom. Near the end of the nineteenth century, William Thomson (Lord Kelvin) suggested that different atoms were actually different knots tied in the ether that was believed to permeate all of space. Physicists and mathematicians set to work making a table of distinct knots, believing they were making a table of the elements. A pioneer in this effort, alongside Thomson, was Peter **Tait**. By the time the theory of the ether vanished into thin air, *knot theory* was firmly tied into mainstream mathematics. It blossomed with the development of **topology** and eventually led to important applications in DNA research and molecular biology. Today it is one of the most active areas of mathematical research. See also **tie knot, loop,** and **braid**.

Knuth's up-arrow notation

A notation for **large numbers** developed by the American mathematician Donald Knuth in 1976. A single up-arrow ($\uparrow$) is the same as exponentiation:

$$m \uparrow n = m \times m \times \ldots \times m \ (n \text{ terms}) = m^n.$$

Two up-arrows together represent a **power tower**:

$$m \uparrow\uparrow n = m^{m^{m^{\cdot^{\cdot^{\cdot^{m}}}}}} \Big\} n.$$

(a tower of height n), which is the same as the operation known as *hyper4* or *tetration*. This can very rapidly generate huge numbers. For example:

$2 \uparrow\uparrow 2 = 2 \uparrow 2 = 4$
$2 \uparrow\uparrow 3 = 2 \uparrow 2 \uparrow 2 = 2 \uparrow 4 = 16$
$2 \uparrow\uparrow 4 = 2 \uparrow 2 \uparrow 2 \uparrow 2 = 2 \uparrow 16 = 65,536$
$3 \uparrow\uparrow 2 = 3 \uparrow 3 = 27$
$3 \uparrow\uparrow 3 = 3 \uparrow 3 \uparrow 3 = 3 \uparrow 27 = 7,625,597,484,987$
$3 \uparrow\uparrow 4 = 3 \uparrow 3 \uparrow 3 \uparrow 3 = 3 \uparrow 3 \uparrow 27 = 3^{7625597484987}$

Three up-arrows together represent a still more vastly powerful operator, equivalent to *hyper5* or *pentation*, or a power tower of power towers:

$$m \uparrow\uparrow\uparrow n = m \uparrow\uparrow m \uparrow\uparrow \ldots \uparrow\uparrow m \ (n \text{ terms})$$

For example:

$2 \uparrow\uparrow\uparrow 2 = 2 \uparrow\uparrow 2 = 4$
$2 \uparrow\uparrow\uparrow 3 = 2 \uparrow\uparrow 2 \uparrow\uparrow 2 = 2 \uparrow\uparrow 4 = 65,536$
$2 \uparrow\uparrow\uparrow 4 = 2 \uparrow\uparrow 2 \uparrow\uparrow 2 \uparrow\uparrow 2 = 2 \uparrow\uparrow 65,536 = 2 \uparrow 2$
   $\uparrow \ldots \uparrow 2$ (65,536 terms)
$3 \uparrow\uparrow\uparrow 2 = 3 \uparrow\uparrow 3 = 7,625,597,484,987$
$3 \uparrow\uparrow\uparrow 3 = 3 \uparrow\uparrow 3 \uparrow\uparrow 3 = 3 \uparrow\uparrow 7,625,597,484,987 = 3$
   $\uparrow 3 \uparrow \ldots \uparrow 3$ (a power tower 7,625,597,484,987
   layers high)
$3 \uparrow\uparrow\uparrow 4 = 3 \uparrow\uparrow 3 \uparrow\uparrow 3 \uparrow\uparrow 3 = 3 \uparrow\uparrow 3 \uparrow\uparrow$
   $7,625,597,484,987 = 3 \uparrow\uparrow 3 \uparrow \ldots \uparrow 3$ (a tower
   $3 \uparrow\uparrow 7,625,597,484,987$ layers high)

Similarly,

$$m \uparrow\uparrow\uparrow\uparrow n = m \uparrow\uparrow\uparrow m \uparrow\uparrow\uparrow \ldots \uparrow\uparrow\uparrow m \ (n \text{ terms})$$

so that, for example:

$2 \uparrow\uparrow\uparrow\uparrow 2 = 2 \uparrow\uparrow\uparrow 2 = 4$
$2 \uparrow\uparrow\uparrow\uparrow 3 = 2 \uparrow\uparrow\uparrow 2 \uparrow\uparrow\uparrow 2 = 2 \uparrow\uparrow\uparrow 4 = 2 \uparrow 2 \uparrow \ldots \uparrow$
   2 (65,536 terms)
$2 \uparrow\uparrow\uparrow\uparrow 4 = 2 \uparrow\uparrow\uparrow 2 \uparrow\uparrow\uparrow 2 \uparrow\uparrow\uparrow 2 = 2 \uparrow\uparrow\uparrow 2 \uparrow 2$
   $\uparrow \ldots \uparrow 2$ (65,536 terms)
$3 \uparrow\uparrow\uparrow\uparrow 2 = 3 \uparrow\uparrow\uparrow 3 = 3 \uparrow 3 \uparrow \ldots \uparrow 3$
   (7,625,597,484,987 terms)
$3 \uparrow\uparrow\uparrow\uparrow 3 = 3 \uparrow\uparrow\uparrow 3 \uparrow\uparrow\uparrow 3 = 3 \uparrow\uparrow\uparrow 3 \uparrow 3 \uparrow \ldots \uparrow 3$
   (7,625,597,484,987 terms)
$= 3 \uparrow\uparrow 3 \uparrow 3 \uparrow \ldots \uparrow 3 \ (3 \uparrow 3 \uparrow \ldots \uparrow 3$
   (7,625,597,484,987 terms)

Even up-arrow notation becomes cumbersome, however, when one is faced with staggeringly large numbers such as **Graham's number**. For such cases, more extensive systems such as **Conway's chained arrow notation** or *Steinhaus-Moser notation* are better suited. See also the **Ackermann function**, to which up-arrow notation is closely related.

Koch snowflake

One of the most symmetric and easy to understand **fractals**; it is named after the Swedish mathematician Helge von Koch (1870–1924), who first described it in 1906. To make the snowflake, start with a straight line and split it into three equal parts. Replace the middle part with two lines, both of the same length as the first three, creating an

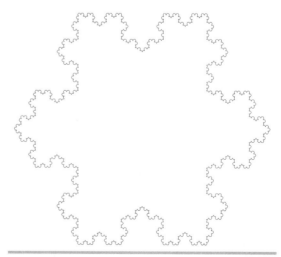

Koch snowflake   This fractal shape emerges from continually sprouting equilateral triangles. *Xah Lee, www.xahlee.org*

equilateral triangle missing the bottom line. The shape now consists of four straight lines with the same length. For each of these lines, repeat this process, then continue the transformation indefinitely. The Koch snowflake has infinite length because each iteration increases the length of a line segment one-third, and the iterations go on forever.

The same kind of process can be applied to a tetrahedron. Take a regular tetrahedron (all side lengths the same) and glue to each of its faces a smaller regular tetrahedron. (Each smaller tetrahedron is scaled down by a factor of $\frac{1}{2}$ from the larger one, and placed on each face in an inverted fashion, so that it divides the face into four equilateral triangles and covers the center one.) Then iterate this process. Intuition suggests that the end product might be a very strange-looking jagged object. But in fact, in the limit, as the number of iterations tends to infinity, the result is a perfect cube! The cube has side length $t/\sqrt{2}$, where t is the length of one of the edges of the regular tetrahedron you started with. Variations on the flat Koch snowflake include the *exterior snowflake,* the *Koch antisnowflake,* and the *flowsnake* curves.

Kolmogorov, Andrei Nikolaievich (1903–1987)

A Russian mathematician and physicist who advanced the foundations of **probability theory,** the **algorithm**ic theory of **random**ness, and made crucial contributions to the foundations of statistical mechanics, **stochastic** processes, **information theory,** fluid mechanics, and nonlinear dynamics. All of these areas, and their interrelationships, underlie **complex system**s as they are studied today. His work on reformulating probability started with a 1933 paper in which he built up probability theory in a rigorous way from fundamental axioms, similar to Euclid's treatment of geometry. Kolmogorov went on to study the motion of the planets and turbulent fluid

flows, publishing two papers in 1941 on turbulence that even today are of fundamental importance.

In 1954 he developed his work on dynamical systems in relation to planetary motion, thus demonstrating the vital role of probability theory in physics and reopening the study of apparent randomness in deterministic systems, along the lines originally conceived by Henri **Poin**-**caré.** In 1965 he introduced the algorithmic theory of randomness via a measure of complexity, now referred to as *Kolmogorov complexity.* According to Kolmogorov, the complexity of an object is the length of the shortest computer program that can reproduce the object. Random objects, in his view, were their own shortest description, whereas periodic sequences have low Kolmogorov complexity, given by the length of the smallest repeating "template" sequence they contain. Kolmogorov's notion of complexity is a measure of randomness, one that is closely related to Claude **Shannon**'s entropy rate of an information source.

Königsberg bridge problem

See **bridges of Königsberg.**

Kronecker, Leopold (1823–1891)

A German mathematician and pioneer in the field of **algebraic number**s who formulated the relationship between the theory of numbers, the theory of equations, and **elliptic function**s. He acquired a passion for **number theory** from Ernst Kummer, his instructor at the Liegnitz Gymnasium. Kronecker, who made a fortune in business before returning to his academic studies, claimed that mathematical argumentation should be based only on integers and finite procedures. He was one of Georg **Cantor**'s sternest critics and refused to accept the validity of **Weierstrass's nondifferentiable function.**

L

labyrinth
See **maze**.

Lagrange, Joseph Louis (1736–1813)
An Italian-born French mathematician who made important contributions to **number theory** and to classical and celestial mechanics. By his mid-twenties he was recognized as one of the greatest living mathematicians because of his papers on wave propagation and the maxima and minima of curves. His prodigious output included his textbook *Mécanique Analytique* (Analytical mechanics, 1788), the basis for all later work in this field. His remarkable discoveries include the *Lagrangian*, a differential operator characterizing a system's physical state, and the *Lagrangian points*, points in space where a small body in the gravitational fields of two large ones remains relatively stable. Under Napoleon, Lagrange was made both a senator and a count; he is buried in the Panthéon.

lambda calculus
A model of computation that is capable of **universal computation**. The Lisp programming language was inspired by lambda calculus.

Lamé curve
Any of a family of curves related to the **ellipse** and that was first recognized and studied in 1818 by the French physicist and mathematician Gabriel Lamé (1795–1870). The formula for the Lamé curve family is a generalization of the equation for an ellipse ($|x/a|^2 + |y/b|^2 = 1$), namely:

$$|x/a|^n + |y/b|^n = 1,$$

where n is any real number. When $n = 0$, the curve reduces to a pair of crossed lines. As n increases, the curve changes from a curved star shape to a rectangle, with diagonals a and b, when $n = 1$. The special case when $n = 2/3$ corresponds to the **astroid**. Between $n = 1$ and $n = 2$ the curve turns from a curved rectangle to an ellipse (or a circle when both a and b are 1). For values of n greater than 2, Lamé curves are known as **superellipse**s.

lamination
A decoration of a **manifold** in which some subset is partitioned into sheets of some lower dimension, and the sheets are locally parallel. It may or may not be possible to fill the gaps in a lamination to make a **foliation**.

Langley's adventitious angles
A seemingly simple problem first posed in 1922 by E. M. Langley[195] in connection with an isosceles triangle. In its original form, it is stated as follows: ABC is an isosceles triangle. B = C = 80°. CF at 30° to AC cuts AB in F. BE at 20° to AB cuts AC in E. Prove angle BEF = 30°. (No mention is made of D. Perhaps it is at the intersection of BE and CF.) A number of solutions appeared shortly after, including this one given by J. W. Mercer: Draw BG at 20° to BC, cutting CA in G. Then angle GBF = 60° and angles BGC and BCG are 80°. So BC = BG. Also angle BCF = angle BFC = 50°, so BF = BG and triangle BFG is equilateral. But angle GBE = 40° = angle BEG, so BG = GE = GF. And angle FGE = 40°, hence GEF = 70° and BEF = 30°.

Langton's ant
A type of **cellular automaton**, or a simple form of **artificial life**, named after its designer, Christopher Langton. The ant lives on an infinitely large chessboard, each square of which can be either black or white. Two pieces of information are associated with this digital insect: the

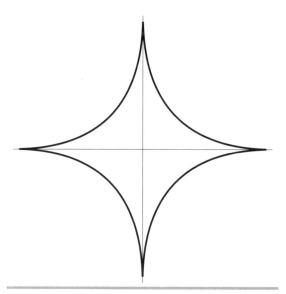

Lamé curve   One example from the family of Lamé curves. © *Jan Wassenaar, www.2dcurves.com*

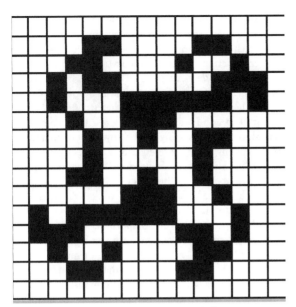

Langton's ant After 368 steps, a symmetrical pattern emerges.

direction that it's facing, and the state of the square that it currently occupies. Three simple rules govern the ant's behavior: (1) If it's on a black square, it turns left. (2) If it's on a white square, it turns right. (3) As it moves to the next square, the one it was on reverses color. Interest in Langton's ant stems from the fact that despite being a completely determined system governed by extremely simple rules, the patterns it produces are fantastically rich and complex. For the first 10,000 moves or so, the ant meanders around, building and then unbuilding structures with little pattern to them. Then, near the end of

this chaotic phase, the ant begins to construct a diagonal highway off toward one edge of the board. In fact, this pattern stems from a sequence of 104 moves that, once started, will go on forever. In the language of **chaos** theory, the pattern is a *stable attractor* for the system. Remarkably, no matter what the initial arrangement of squares— even if the white and black squares are set up randomly— the ant will end up building a highway. The ant can be allowed to wrap around the edges of a finite board, thus allowing it to intersect its own path, and it will *still* end up building the highway. Are there any initial states that don't lead to the diagonal-road-building loop? No exceptions have been found from experiments–but proving it is another matter. Most mathematicians believe there is no general analytical method of predicting the position of the ant, or of any such chaotic system, after any given number of moves. Its behavior can't be reduced to the rules that govern it. In this sense, Langton's ant is a simple demonstration of the undecidability of the **halting problem**. The British mathematician Ian **Stewart** and biologist Jack Cohen, in their book *Figments of Reality*,[321] go a step further and use Langton's ant as an analog of an essential stage in the evolution of complex systems such as life: a stage in which the existence of chaotic behavior contains the potential for the spontaneous emergence of unpredictable forms of order.

Laplace, Pierre Simon (1749–1827)

A French mathematician and astronomer who was heavily involved with the development of celestial mechanics. He made an early impact by solving a complex problem of mutual gravitation that had eluded both Leonhard **Euler** and Joseph **Lagrange**. Laplace was among the most influential scientists of his time and was called the Newton of France for his study of and contributions to the understanding of the stability of the solar system. Laplace generalized the laws of mechanics for their application to the motion and properties of the heavenly bodies. He is also famous for his great treatises *Mécanique céleste* (Celestial mechanics, 1799–1825) and *Théorie analytique des probabilités* (Analytical theory of probabilities, 1812) which were advanced in large part by the mathematical techniques that Laplace developed early in his life.

large number

> *Bigger than the biggest thing ever and then some.*
> *Much bigger than that in fact, really amazingly*
> *immense, a totally stunning size, real "wow, that's*
> *big!" time. . . . Gigantic multiplied by colossal mul-*
> *tiplied by staggeringly huge is the sort of concept*
> *we're trying to get across here.*
> —Douglas Adams, *The Restaurant at the End of the Universe*

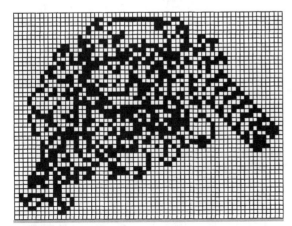

Langton's ant After 10,647 steps the "highway" is under construction.

Making, naming, and representing very large numbers is itself a big problem. A simple way to start is by adding zeros: 10; 100; 1,000; 10,000; ..., 1,000,000; .... But this quickly gets tedious and exponentiation becomes a more attractive option: 10^1, 10^2, 10^3, ..., 10^6, ... Naming the various powers of 10 follows a regular pattern of prefixes. In the United States, 10^3 is one thousand, 10^6 is one million, 10^9 is one billion, 10^{12} is one trillion, 10^{15} is one quadrillion, and so on. The "-illion" root kicks in at the sixth power of 10 prefixed by "m" for "mono"; then, every jump of three powers (factor of a thousand) comes the next prefix. Put another way, the U.S. name for 10^{3n} uses the Latinized prefix for $n - 1$. One centillion, which is the largest number with a single-word name in English, is 10^{303}. Elsewhere in the world, "billion," "trillion," and so on, can mean things other than they do in the U.S. system. A British billion, for example, is one million million, or 10^{12}, while the now largely obsolete term *milliard* was used for a thousand million. However, the American forms have become fairly standard internationally and are used without qualification in this book. It's also worth noting that, while "quadrillion," "quintillion," and so forth, are perfectly valid terms, "one thousand trillion," "one million trillion," and so forth, are generally preferred.

Exponentiating quite small numbers seems at first to be a pretty economical way of making and writing large numbers: 10^{30}, for example, is a highly effective shorthand for 1,000,000,000,000,000,000,000,000,000,000. But this method runs out of steam as the numbers get bigger and bigger. Take, for instance, the **googol** and the googolplex. One googol is the unofficial name for 10^{100}, or 1 followed by 100 zeros. This innocuous-looking number is larger than the number of atoms in the universe. What happens, therefore, if we want to represent the number that is 1 followed by *a googol number of zeros*? One way would be to write "1 followed by a googol number of zeroes"! But this is cheating because it couldn't be generalized without first giving a proper name to a fantastically large number of numbers. A better solution is to exponentiate using large numbers. Thus, 1 googolplex = $10^{\text{googol}} = 10^{10^{100}}$. This is the beginning of a **power tower**.

Are numbers as large as the googol, not to mention the googolplex, of any practical importance? Science has given us such colossi as the **Avogadro constant** (the number of molecules in a sample whose weight in grams equals its molecular weight) = 6.023×10^{23}, the **Eddington number** (astrophysicist Arthur Eddington's best estimate of the number of protons in the universe) = 1.575×10^{79}, and the Supermassive Black Hole Evaporation Time = 10^{100} years (or thereabouts)–which brings us to the level of the googol. But there's nothing known or that can be reasonably conjectured in the "real" world of physics that goes much beyond this.

Science fiction can carry us a bit further. In *The Hitchhiker's Guide to the Galaxy*[5] by Douglas Adams appears one of the largest numbers ever used in a work of fiction: 2^{260199}. These are the odds quoted against the characters Arthur Dent and Ford Prefect being rescued by a passing spaceship just after having been thrown out of an airlock. As it happens, they *are* rescued by a spaceship powered by the "infinite improbability drive." By contrast, some special numbers in mathematics make even the googolplex look tiny. **Skewes' number**, $10^{10^{10^{34}}}$, was long held up as an example of a googolplex-dwarfing number that nevertheless served a bonafide purpose in mathematics. However, even this seemingly immense integer is made to look ridiculously small by the likes of more recently described numbers, such as **Graham's number**, the *Mega*, and the *Moser*, which are so utterly vast that it takes a page or two simply to describe the various special notations used to represent them.

Just as writing out a number in full, or in place-value notation, becomes unwieldy with numbers as big as a googol, so exponentiation, in turn, endangers the world's forests if it tries to take on seriously large numbers. A more effective shorthand is *tetration—tetra* (from the Greek meaning "four") because it is the fourth dyadic operation in the series: addition, multiplication, exponentiation, tetration. Dyadic means that two numbers, or arguments, are involved in the operation. Multiplication is repeated addition (e.g., $2 \times 3 = 2 + 2 + 2$), exponentiation is repeated multiplication (e.g., $2^3 = 2 \times 2 \times 2$), and tetration is repeated exponentiation. For example, 2 tetrated to 3, represented as $^3 2$, is $2^{2^2} = 2^4 = 16$; 2 tetrated to 4, or $^4 2$, is $2^{2^{2^2}} = 2^{16} = 65,536$; and 2 tetrated to 5, or $^5 2$, is $2^{2^{2^2}} = 2^{65,536} = $ something too big to write out in full. Tetration goes by various other names including *superpower, superdegree,* and, the one used most commonly in mathematical circles and also here, *hyper4*.

Just as the exponentiation of two numbers, a and b, is represented as a^b and defined as $a \times a \times ... \times a$ (b terms), the hyper4 of a and b is represented as $a^{(4)}b$ and defined as $a^{a^{a^{...}}}$ (a power tower with b levels). Alternatively, the hyper4 operator can be represented in **Knuth's up-arrow notation** as $a \uparrow\uparrow b$. Continuing this pattern:

hyper5 of a and $b = a^{(5)}b = a^{(4)}a^{(4)} ... a^{(4)} = a \uparrow\uparrow\uparrow b$
hyper6 of a and $b = a^{(6)}b = a^{(5)}a^{(5)} ... a^{(5)} = a \uparrow\uparrow\uparrow\uparrow b$
hyper7 of a and $b = a^{(7)}b = a^{(6)}a^{(6)} ... a^{(6)} = a \uparrow\uparrow\uparrow\uparrow\uparrow b$

To get some idea of the potency of this kind of representation, consider the sequence

1

10

10,000,000,000

10,000,000,000,000,000,000,000,000,000,000,000,
000,000,000,000,000,000,000,000,000,000,000,000,
000,000,000,000,000,000,000,000,000,000 (100 zeroes)

...

Notice how big even the fourth term is? The seemingly innocuous-looking number $5 \uparrow\uparrow\uparrow\uparrow\uparrow 5$ (or $5^{(7)}5$) is so huge that it would be around the 100,000,000,000,000,000th (one hundred thousand trillionth) term of this sequence!

Since the dyadic operators discussed previously form a pattern, they can be telescoped into a single *triadic* operator that has three arguments. This can be defined as:

$$\text{hy}(a, n, b) = \begin{cases} 1 + b \text{ for } n = 0 \\ a + b \text{ for } n = 1 \\ a \times b \text{ for } n = 2 \\ a \uparrow b \text{ for } n = 3 \\ a \uparrow \text{hy}(a, 4, b - 1) \text{ for } n = 4 \\ \text{hy}(a, n - 1, \text{hy}(a, n, b - 1)) \text{ for } n > 4 \\ a \text{ for } n > 1, b = 1 \end{cases}$$

Beyond hyper are other triadic operators capable of generating large numbers even faster. The **Ackermann function** and the *Steinhaus-Moser notation* are both equivalent to a triadic operator that is somewhat more powerful than hy(a, n, b). Similarly, **Conway's chained-arrow notation** marks an evolution of Knuth's symbolism. These various techniques and notations can produce immense *finite* numbers. But beyond any of these lie the many different kinds of **infinity**.

lateral inhibition illusion
See **Hermann grid illusion**.

Latin square
An $n \times n$ square grid, or matrix, whose entries consist of n symbols such that each symbol appears exactly once in each row and each column. The following are some examples:

1 2	1 2 3	1 2 3 4	1 2 3 4	M	A	G	I	C
2 1	2 3 1	2 3 4 1	2 1 4 3	G	I	C	M	A
	3 1 2	3 4 1 2	3 4 1 2	C	M	A	G	I
		4 1 2 3	4 3 2 1	A	G	I	C	M
				I	C	M	A	G

Latin squares have a long history, stretching back at least as far as medieval Islam (c. 1200), when they were used on amulets. Abu l'Abbas al Buni wrote about them and constructed, for example, 4×4 Latin squares using letters from a name of God. In his famous etching *Melancholia*, the fifteenth-century artist Albrecht **Dürer** portrays a 4×4 **magic square**, a relative of Latin squares, in the background. Other early references to them concern the problem of placing the 16 face cards of an ordinary deck of **cards** in the form of a square so that no row, column, or diagonal contains more than one card of each suit and each rank. Leonhard **Euler** began the systematic treatment of Latin squares in 1779 and posed a problem connected with them, known as the **thirty-six officers problem**, that wasn't solved until the beginning of the twentieth century. Arthur **Cayley** continued work on Latin squares, and in the 1930s the concept arose again in the guise of multiplication tables when the theory of quasigroups and loops began to be developed as a generalization of the **group** concept. Latin squares played an important role in the foundations of finite geometries, a subject that was also in development at that time. Also in the 1930s, a large application area for Latin squares was opened by R. A. Fisher who used them and other combinatorial structures in the design of statistical experiments.

lattice
A periodic arrangement of points such as the vertices (see **vertex**) of a **tiling** of space by cubes or the positions of atoms in a crystal. More technically, a discrete Abelian subgroup (see **Abelian group**) of an n-dimensional **vector space** that is not contained in an $(n - 1)$-dimensional vector space. Lattices play a central role in the theory of **Lie group**s, in **number theory**, in error-correcting codes, and many other areas of mathematics. See also **geoboard**.

lattice path
A sequence of points in a **lattice** such that each point differs from its predecessor by a finite list of allowed steps. **Random** lattice paths are an interesting model for the random motion of a particle and lattice paths are also important in enumerative **combinatorics**.

lattice point
A point with **integer** coordinates.

latus rectum
A **chord** of an **ellipse** that passes through a **focus** and is perpendicular to the major axis of the ellipse. Its plural is latera recta.

league
An archaic unit of traveling distance. The precise value varied, but was usually around 3 miles (4.8 kilometers).

least common multiple

The smallest **integer** that is an exact multiple of every number in a set of integers.

least upper bound

The smallest number that is larger than every member of a set of numbers.

Lebesgue, Henri Leon (1875–1941)

A French mathematician who introduced the modern definition of an **integral**. Lebesgue graduated from the École Normale Supériere and, from 1921, taught at the College de France. He and Emile **Borel** founded the modern theory of **function**s of a real variable, Lebesgue's great contribution being his new general definition of an integral (1902), which became known as the *Lebesgue integral* (see **integration**). This led to important advances in calculus, curve rectification, and trigonometric series, and, in Borel's hands, marked the start of **measure theory**. Although the Lebesgue integral was an example of the power of generalization, Lebesgue himself wasn't a fan of generalization and spent the rest of his life working on very specific problems, mostly in **analysis**.

Lebombo bone

One of the oldest mathematical artifacts known, a small piece of the fibula of a baboon, found near Border Cave in the Lebombo Mountains between South Africa and Swaziland. Discovered in the 1970s during excavations of Border Cave and dated about 35,000 B.C., the Lebombo bone is marked with 29 clearly defined notches. This suggests it may have been used as a lunar phase counter, in which case African women may have been the first mathematicians, because keeping track of menstrual cycles requires a lunar calendar. Certainly, the Lebombo bone resembles calendar sticks still used by Bushmen in Namibia. See also **Ishango bone**.

left-right reversal

See **mirror reversal problem**.

Leibniz, Gottfried Wilhelm (1646–1716)

A German philosopher, mathematician, and statesman who developed differential and integral **calculus** independently of Isaac **Newton**. He also invented a calculating machine, is considered a pioneer in mathematical logic, and proposed the metaphysical theory that we live in "the best of all possible worlds." In Leibniz's philosophical view, the universe is composed of countless conscious centers of spiritual force or energy known as *monads*. Leibniz talks about the "compossible" elements of any possible world–elements that allow a logically consistent structure. Though one of the finest minds of his age, Leibniz was not immune to blunders: he thought it just as easy to throw 12 with a pair of dice as to throw 11.

Leibniz harmonic triangle

A triangle of fractions that is related to the more famous **Pascal triangle** in a very simple way. Each row of the Leibniz harmonic triangle starts with the **reciprocal** of the row number (or the row number plus one depending on whether one starts counting from 1 or 0). Every entry is the sum of the two numbers just below it. The entries can thus be computed sequentially left to right and top to bottom using subtraction instead of addition.

$$1/1$$
$$1/2 \quad 1/2$$
$$1/3 \quad 1/6 \quad 1/3$$
$$1/4 \quad 1/12 \quad 1/12 \quad 1/4$$
$$1/5 \quad 1/20 \quad 1/30 \quad 1/20 \quad 1/5$$

lemma

A short auxiliary proposition used in the proof of a larger **theorem**.

lemniscate of Bernoulli

A curve "shaped like a figure 8, or a knot, or the bow of a ribbon" in the words of Jacob Bernoulli (see **Bernoulli family**) in an article published in 1694. Bernoulli named the curve "lemniscate" after the Greek *lemniskus* for a pendant ribbon (the type fastened to a victor's garland). It has the Cartesian equation

$$(x^2 + y^2)^2 = a^2(x^2 - y^2).$$

At the time he wrote his article, Bernoulli wasn't aware that the curve he was talking about was a special case of a **Cassinian oval**, which had been described by Cassini in 1680. The general properties of the lemniscate were established by Giovanni Fagnano (1715–1797) in 1750; Leonhard **Euler**'s investigations of the length of arc of the curve (1751) led to later work on **elliptic function**s. There is a relationship between the lemniscate and the rectangular **hyperbola**. If a **tangent** is drawn to the hyperbola and the perpendicular to the tangent is drawn through the origin, the point where the perpendicular meets the tangent is on the lemniscate. See also **hippopede**.

Leurechon, Jean (c. 1591–1670)

A French Jesuit priest and mathematician who wrote *Recreations mathématiques* (1624) under the pseudonym Henrik van Etten. Much of the mathematical content centers around Claude **Bachet**'s problems and may have been copied from it or some common source. The book also gives the earliest known description of the operation of an ear trumpet.

liar paradox

"This statement is false." What do we make of this statement (call it *S*)? If *S* is true, then *S* is false. On the other hand, if *S* is false, then it is true to say *S* is false; but, because the liar sentence is saying precisely that (namely that it is false), *S* is true. So *S* is true if and only if it is false. Since *S* is one or the other, it is both! Debate about sentences like *S* has been going on among philosophers and logicians for more than 2,000 years without any clear resolution.

The roots of the liar paradox stretch back to the philosopher Epimenides in the sixth century B.C. He said: "All Cretans are liars. . . . One of their own poets has said so." Another version of this can be found in the Bible, Titus 1:12–13, "Even one of their own prophets has said, 'Cretans are always liars, evil brutes, lazy gluttons.' This testimony is true." The poet's (or prophet's) statement is sometimes wrongly considered to be paradoxical because he himself is a Cretan. But actually there's no paradox here. A "liar," in everyday language, is someone who *on occasion* knowingly gives false answers. This isn't problematic: the poet, while lying occasionally, this time spoke the truth. However, most formulations of logic define a "liar" as an entity that always produces the negation of the true answer, that is, someone who does nothing but lie. Thus, the poet's statement can't be true: if it were, then he himself would be a liar who just spoke the truth, but liars don't do that. However, no contradiction arises if the poet's statement is taken to be false: the negation of "All Cretans are liars" is "Some Cretans aren't liars," in other words: some Cretans sometimes speak the truth. This doesn't contradict the fact that our Cretan poet just lied. Therefore, the statement "All Cretans are liars," if uttered by a Cretan, is necessarily false, but not paradoxical. Even the statement "I am a liar" is not paradoxical; depending on the definition of "liar" it may be true or false. However, the statement "I am lying now," first attributed to Eubulides of Miletus in the fourth century B.C., definitely is paradoxical. It is exactly equivalent to the sentence, we started with: "This statement is false."

Various elaborations of the basic Eubulides liar paradox have appeared over the ages. In the fourteenth century, the French philosopher Jean Buridan applied it in his argument for the existence of **God**. In 1913, the English mathematician Philip Jourdain (1879–1921) offered a version that is sometimes referred to as "Jourdain's card paradox." On one side of a card is written:

The sentence on the other side of this card is true.

On the other side is written:

The sentence on the other side of this card is false.

Yet another popular version of the liar paradox, guaranteed to perplex, is given by the following three sentences written on a card:

1. This sentence contains five words.
2. This sentence contains eight words.
3. Exactly one sentence on this card is true.

Lie, Marius Sophus (1842–1899)

A Norwegian mathematician who, along with his close friend Felix **Klein**, introduced **group** theory into geometry and used it to classify geometries. Lie discovered the *contact transformation*, which maps curves into surfaces (1870), and **Lie group**s, which use continuous or infinitesimal transformations. He used these groups to classify **partial differential equation**s, making the traditional methods of solution all reduce to a single principle. Lie groups also provided a basis for the growth of modern **topology**. See also **Lie algebra**.

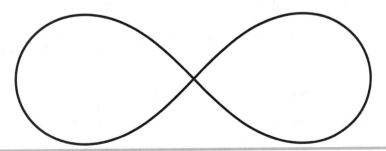

lemniscate of Bernoulli © Jan Wassenaar, www.2dcurves.com

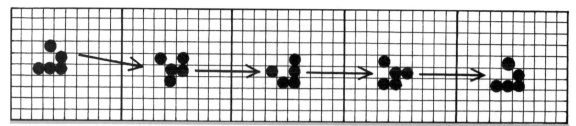

Life, Conway's game of A "glider" advances one cell down and one cell to the right.

Lie algebra

An **algebra**, named after M. Sophus **Lie**, in which multiplication satisfies properties similar to the *bracket operation* on matrices given by $[A, B] = AB - BA$, where the operation on the right-hand side involves ordinary multiplication and subtraction of matrices. The operation is not **associative**.

Lie group

A **group**, named after M. Sophus **Lie**, that is also a **manifold**. Groups of a real **matrix**, such as occur in quantum field theory, give naturally occurring examples of Lie groups. The tangent space at the identity element of a Lie group forms a **Lie algebra** in a natural way.

Life, Conway's game of

The best known example of a **cellular automaton**; it was invented by John **Conway** and first brought to public attention in Martin Gardner's *Scientific American* column in October 1970.[120] Conway's goal in creating Life was to devise a universal **Turing machine**—a sort of infinitely programmable computer. John **von Neumann** had described such a system in the 1950s, but it had very complex rules; Conway wanted to find one that was much simpler to describe and to operate. Life is played on a grid of squares on which each cell is either alive (occupied) or dead (empty). The game starts from an arbitrary initial configuration of live cells, and then progresses through generations as the life and death rules are applied. These rules are very simple: (1) A live cell survives to the next generation if it has two or three neighbors. (2) A live cell dies if it has four or more neighbors (overcrowding) or if it has only one neighbor or none (isolation). (3) A dead cell becomes a live cell in the next generation if it has exactly three neighbors (birth). The rules of Life were developed over a two-year-period during tea and coffee breaks by Conway and a group of graduate students and colleagues. Because **Go** boards and counters were used at this stage, instead of computers, it was important to have a death rule so that populations didn't tend to explode and quickly race off the board. On

the other hand, to enable sufficiently interesting behavior so the game had a chance of being a universal system, it was equally important to have a birth rule that prevented populations from dying out. The rules eventually chosen provided a balance between birth and death so that the system tended to be fairly stable yet interesting enough to study. An early sign of success was the discovery of patterns, known as "gliders," that kept their shape while drifting across the plane. This was a hopeful step toward proving universality because it showed that the system had a way to transmit information from one place to another. Conway and his group went on to build nearly all the necessary configurations for arbitrary computations: AND gates, OR gates, and so on, just like the components of an ordinary computer. What they needed next was a way of producing gliders at will–a "glider gun." At this point, Conway sent a letter describing Life and the early findings to Martin Gardner, offering a prize of $50 for a configuration whose population tended toward infinity. The resulting *Scientific American* column sparked the public's imagination and very quickly a glider gun was discovered by a group at the Massachusetts Institute of Technology led by R. W. Gosper. Within two weeks of the discovery of the glider gun, both Conway's group and the group at MIT had shown that the system was indeed universal.

limaçon of Pascal

A snail-shaped curve (*limaçon* is French for "snail") named by the French mathematician Gilles Roberval after Etienne Pascal, the father of Blaise **Pascal**. It had been discovered earlier, however; Albrecht **Dürer** gave a method for drawing it as early as 1525 in his *Underweysung der Messung*. The limaçon of Pascal is a special case of an **epitrochoid** in which the rolling circle and the rolled circle have the same radius, and is also the cata**caustic** and the **pedal curve** of the **circle**. It has the **quartic** Cartesian equation

$$(x^2 + y^2 - 2rx)^2 = k^2(x^2 + y^2),$$

where r is the radius of the rolling circle or the rolled circle and k is a constant.

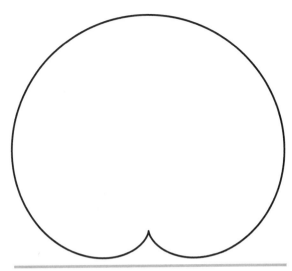

limaçon of Pascal © Jan Wassenaar, www.2dcurves.com

Sometimes the term *ordinary limaçon* is used to describe the curve when the value of k is greater than 0 and less than 1. When $k = 0$ the curve is a circle and when $k = 1$ the curve is a **cardioid**, so the ordinary limaçon is a transitional form between these two. The ordinary limaçon is also the **inverse** of the **ellipse**. For values of $k > 1$, a loop or noose appears in the curve. The inverse of a limaçon with a noose is a **hyperbola**. In fact, the constant k is the same as the eccentricity for a **conic section**. When $k = 2$, the limaçon is also called a **trisectrix**.

limerick
Why is this a limerick?:

$$((12 + 144 + 20) + 3\sqrt{4/7}) + 5 \times 11 = 9^2 + 0$$

Because, as its inventor, Jon Saxton, has pointed out:

A dozen, a gross, and a score,
Plus three times the square root of four,
Divided by seven,
Plus five times eleven,
Is nine squared and not a bit more.

limit
The target value that terms in a **sequence** of numbers get closer and closer to. This limit is not necessarily ever reached but can be approached arbitrarily close if the sequence is taken far enough.

limping triangle
A right **triangle** with two shorter sides (i.e., those other than the hypotenuse) that differ in length by one unit. An example is the 20-21-29 triangle ($20^2 + 21^2 = 29^2$).

line
The shortest distance between any two points in **Euclidean space**. A line is implicitly a *straight line;* the alternative is a **curve**. Mathematically, a line may be determined by the presence of any two points in an n-dimensional space (where n is two or more). A *line segment* is a piece of a line with definite endpoints.

linear
Having only a multiplicative factor. If $f(x)$ is a *linear function*, then $f(a + b) = f(a) + f(b)$ and $cf(x) = f(cx)$ must both be true for all values of $a, b, c,$ and x.

linear algebra
The study of **vectors** and **vector spaces**. Linear algebra has a central place in modern mathematics, is used widely in both **abstract algebra** and functional **analysis**, and finds a concrete representation in the form of **analytical geometry**. Linear algebra began as the study of vectors in two- and three-dimensions but has now been extended to a generalized n-space.

linear group
A **group** of matrices under **matrix** multiplication.

linear programming
The problem, and associated area of mathematics, of maximizing or minimizing a linear **function** on a **convex** set, especially a **polytope**. Equivalently, maximizing a linear expression in some number of variables subject to linear equalities and inequalities.

linear system
Any system whose change of values of its variables can be represented as a series of points suggesting a straight **line** on a coordinate; hence, *linear* for "line." More generally, a linear system is one in which small changes result in small effects, and large changes in large effects. In a linear system, the components are isolated and noninteractive. Real linear systems are rare in nature since living organisms and their components are not isolated and do interact. Compare with **nonlinear system**.

Liouville number
A **transcendental number** that can be approximated very closely by a **rational number**. Normally, proving that any given number is transcendental is difficult. However, the French mathematician Joseph Liouville (1809–1882) showed the existence of a large (in fact, infinitely large) class of transcendentals whose nature is easy to ascertain. An example of a Liouville number is 0.10100100000010000000000000000000000001 . . . in which the successive groups of zeros are of length 1!, 2!, 3!, 4!, and so on.

Lissajous figure

A figure or graph of the type most often seen on an oscilloscope. The simplest Lissajous figures are circles or ellipses, but they can also take the form of **lemniscates** and other, more complex shapes. They are named for the French scientist Jules Antoine Lissajous (1822–1880), who experimented with them in the 1850s, and are also known as *Bowditch curves,* because they had been written about earlier by the American astronomer and mathematician Nathanial Bowditch (1773–1838).

lituus

A **spiral** curve that was discovered by the English mathematician Roger Cotes (1682–1716) and named by the Scottish mathematician Colin Maclaurin (1698–1746) in 1722. The Latin *lituus* refers to a staff shaped like a bishop's crosier. The lituus is the locus of a point that moves in such a manner that the area of a circular sector remains constant; it has the polar equation

$$r^2 = a^2/\theta.$$

See **polar coordinates**.

Lobachevsky, Nikolai Ivanovich (1793–1856)

A Russian mathematician who was one of the pioneers of **non-Euclidean geometry**. He developed, independently of János **Bólyai**, the self-consistent system of **hyperbolic geometry** in which Euclid's **parallel postulate** is replaced by one allowing more than one parallel through the fixed point. Lobachevsky first announced his system in 1826 and subsequently wrote several expositions of it, including *Geometrical Researches on the Theory of Parallels* (originally published in 1840 in German). Lobachevsky studied and taught at the University of Kazan and eventually became rector of this institution in 1826. However, for some reason, despite serving his country and university well, he fell from favor and in 1846 was relieved by the government of his posts as professor and rector.

localized solution

A solution of a **differential equation**, or a similar mathematical object, that is confined to a small region even though it has the freedom to spread out.

loculus of Archimedes

A **dissection** game, similar to **tangrams**, which consists of 14 polygonal shapes that fit together to make a square. These pieces can be rearranged to make pictures of people, animals, and objects, or reassembled into their original form. There are many references to the game in ancient literature, including a description by the Roman poet and statesman Magnus Ausonius (A.D. 310–395).

Only two fragmentary manuscripts, one an Arabic translation and the other a Greek manuscript dating from the tenth century discovered in Constantinople in 1899, connect the puzzle to **Archimedes** by calling it *loculus Archimedius* ("Archimedes's box"). More generally, but for unclear reasons, it is known as the *ostomachion* (Greek for "stomach"), or, in Latin texts, as the *syntemachion*. In 2003, William **Cutler** used a computer program to enumerate all 536 distinct ways (barring rotations and reflections) in which the pieces can be arranged into a square.

locus

The set of all points (usually forming a curve or surface) that satisfy some condition. For example, the locus of points in the plane equidistant from a given point is a circle. The Latin word *locus* simply means "place." (The Greek equivalent is *topos,* which crops up in "topology".)

logarithm

The logarithm of a number or variable x to base b, $\log_b x$, is the **exponent** of b needed to give x. The bases most commonly used in mathematics are e and 10. A logarithm to base e, written as $\log x$ or $\ln x$, is known as a *natural logarithm*. A logarithm to base 10 is written as $\log_{10} x$ and is known as a *common logarithm*.

logarithmic spiral

A type of **spiral**, also known as an *equiangular spiral,* that is very common in the natural world. Wonderful examples are found in the shells of some mollusks, such as that of the nautilus, and in spider webs. The angle any **tangent** to the curve makes with a tangent to a circle at the same radius, known as the *pitch angle,* is constant and results in a logarithmic spiral being *self-similar:* in other words, any part of it looks like any other part (though possibly rotated). Hawks approach their prey in the form of a logarithmic spiral and their sharpest view is at an angle to their flight direction that is the same as the spiral's pitch. On an altogether different scale, the arms of spiral galaxies are roughly logarithmic spirals. Our own galaxy, the Milky Way, is believed to have four major arms, each of which is a logarithmic spiral with pitch of about 12°. Approximate logarithmic spirals with a pitch of about 17° can be generated using the **Fibonacci sequence** or the **golden ratio**.

In **polar coordinates** (r, θ) the equation of the logarithmic spiral is

$$r = a\,b^{\theta},$$

with positive real numbers a and b. Changing a rotates the spiral while b controls how tightly and in which direction it is wrapped. It can be distinguished from the

Archimedean spiral by the fact that the distance between the arms of a logarithmic spiral increase in a **geometric sequence** while in an Archimedean spiral this distance is constant. Starting at a point P and moving inward along the spiral, one has to circle the origin infinitely often before reaching it; yet, the total distance covered is finite. This was first realized by the Italian physicist Evangelista Torricelli (1608–1647) even before calculus had been invented. The total distance covered is $r/\sin\theta$ where θ is the pitch angle and r is the straight-line distance from P to the origin.

The logarithmic spiral was first described by René **Descartes** and later studied in depth by Jakob Bernoulli (see **Bernoulli family**), who called it *Spiralis mirabilis* (the wonderful spiral) and wanted one engraved on his tombstone. He did get a spiral, but unfortunately it was a rather crudely cut Archimedean type.

logic

The branch of **mathematics** concerned with how one statement can imply others, or how sets of statements can

logarithmic spiral © Jan Wassenaar, www.2dcurves.com

loculus of Archimedes The loculus of Archimedes is one of the most ancient dissection puzzles. *Kadon Enterprises, Inc., www.gamepuzzles.com*

be connected by chains of implication. Such relationships can be written down using special symbols, and different sets of rules can be studied to see how they give rise to different sorts of structures.

logical depth

A measure of the **complexity** of a system, developed in 1988 by the American computer scientist and mathematician Charles Bennett. It contrasts with another such measure, **algorithmic complexity**, but, like it, is a measure of the algorithms needed to generate the data from a system.

loop

(1) A **knot** or hitch that holds its form. (2) A degenerate edge of a **graph** that joins a vertex to itself. (3) A sequence of instructions that is repeated either a specified number of times or until a particular condition prevails. Loops lie at the heart of most computer programs. The most common type is the *iterative loop,* signified by keywords such as *for, while, do,* and *repeat,* or their equivalent, in which a given set of instructions is repeated a specified number of times. The *recursive loop* is a more powerful construct that carries out a given set of instructions, typically including **recursive** calls with modified parameters back to the instruction set itself, until a terminating condition is met. Recursive algorithms solve problems by reducing them to smaller and smaller subproblems until a solution is found, reusing the same set of instructions as often as needed.

Lorenz, Edward Norton (1917–)

A research meteorologist at the Massachusetts Institute of Technology who, in the early 1960s, using a simple system of equations to model convection in the atmosphere, ran headlong into the phenomenon of "sensitivity to initial conditions." In the process he sketched the outlines of one of the first recognized **chaotic attractors**. In Lorenz's meteorological computer modeling, he discovered the underlying mechanism of deterministic chaos: simply formulated systems with only a few variables can display highly complicated behavior that is unpredictable. Using his digital computer, culling through reams of printed numbers and simple strip chart plots of the variables, he saw that slight differences in one variable had profound effects on the outcome of the whole system. This was one of the first clear demonstrations of sensitive dependence on initial conditions. Equally important, Lorenz showed that this occurred in a simple, but physically relevant model. He also appreciated that in real weather situations, this sensitivity could mean the development of a front or pressure-system where there

never would have been one in previous models. In his famous 1963 paper, Lorenz picturesquely explained that a butterfly flapping its wings in Beijing could affect the weather thousands of miles away some days later. This sensitivity is now called the **butterfly effect**.

Lorenz system

A system of three **differential equations**, named after its discoverer Edward **Lorenz**, that was the first concrete example of **chaos** and a **chaotic attractor**.

Lovelace, Lady

See **Byron, Ada**.

loxodrome

A path on Earth's surface that is followed when a compass is kept pointing in the same direction. It is a straight line on a Mercator projection of the globe precisely because such a projection is designed to have the property that all paths along Earth's surface that preserve the same directional bearing appear as straight lines. The loxodrome isn't the shortest distance between two points on a sphere. The shortest distance is an arc of a **great circle**. But, in the past, it was hard for a ship's navigator to follow a great circle because this required constant changes of compass heading. The solution was to follow a loxodrome (from

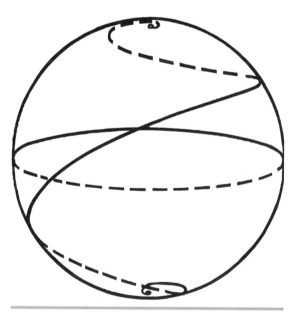

loxodrome Following a constant-compass heading between Earth's poles produces the winding path shown.

the Greek *loxos* for "slanted" and *drome* for "course"), also known as a *rhumb line*, by navigating along a constant direction. In middle latitudes, at least, this didn't lengthen the journey unduly. If a loxodrome is continued indefinitely around a sphere it will produce a spherical **spiral**, or a logarithmic spiral on a polar projection.

Loyd, Sam (1841–1911)

A great American inventor of puzzles among whose best-known creations are the Hoop-Snake Puzzle, **Get Off the Earth**, the **Pony Puzzle**, and, most famous of all, the **Fifteen Puzzle**. By age 17 he was already hailed as one of the world's leading writers of chess problems and had also just invented his deceptively simple-looking trick mules puzzle. The object of this is to cut apart three pieces that show two mules and two jockeys and then reassemble the pieces so that the jockeys are riding the mules. Loyd sold the puzzle to the showman Phineas T. Barnum (of Barnum & Bailey Circus fame) for some $10,000. This was to become his forte: devising puzzles that looked so simple to solve that people felt compelled to try them, only to find, hours later, that they were still trying to figure them out. Loyd became a full-time professional puzzlemaker only in the 1890s. He corresponded with his English counterpart Henry **Dudeney** and worked alongside his son Sam Loyd Jr. After the death of his father, Sam Loyd Jr. continued publishing puzzles that were mainly compilations of his father's work and in 1914 issued a now out-of-print and much sought-after mammoth collection of his father's puzzles called *Cyclopedia of 5,000 Puzzles, Tricks, and Conundrums.*[111, 207, 208]

lozenge

A **rhombus** with a 60° angle.

L-system

A method of constructing a **fractal** that is also a model for plant growth. L-systems use an **axiom** as a starting string and iteratively apply a set of parallel string substitution rules to yield one long string that can be used as instructions for drawing the fractal. Many fractals, including **Cantor dust**, the **Koch snowflake**, and the **Peano curve**, can be expressed as an L-system.

Lucas, (François) Edouard (Anatole) (1842–1891)

A French mathematician well known for his study of the **Fibonacci sequence** and the related **Lucas sequences** named after him. He devised methods of testing for **prime numbers**—work that was later refined by D. H. Lehmer to yield the Lucas-Lehmer test for checking **Mersenne numbers** to see if they are prime. Lucas was also interested in recreational mathematics, the **Tower of Hanoi** being his best known puzzle game. He worked at the Paris Observatory and later became a professor of mathematics in Paris.

Lucas sequences

Generalizations of the **Fibonacci sequence** first investigated by Edouard **Lucas**. One kind can be defined as follows: $L(0) = 0$, $L(1) = 1$, $L(n + 2) = PL(n + 1) + QL(n)$, where the normal Fibonacci sequence is the special case of $P = Q = 1$. Another kind of Lucas sequence begins with $L(0) = 2$, $L(1) = P$. Such sequences are used in **number theory** and in testing for **prime numbers**.

lucky number

A number in a sequence, first identified and named around 1955 by Stanislaw **Ulam**, that evades a particular type of number "sieve" (similar to the famous **sieve of Eratosthenes**), which works as follows. Start with a list of integers, including 1, and cross out every second number: 2, 4, 6, 8, ... The second surviving integer is 3. Cross out every third number not yet eliminated. This removes 5, 11, 17, 23, ... The third surviving number from the left is 7; cross out every seventh integer not yet eliminated: 19, 39, ... Repeat this process indefinitely and the numbers that survive are the "lucky" ones: 1, 3, 7, 9, 13, 15, 21, 25, 31, 33, 37, 43, 49, 51, 63, 67 . . . .

Amazingly, though produced by a sieve based solely on a number's position in an ordered list, the luckies have many properties in common with **prime numbers**. For example, there are 25 primes less than 100 and 23 luckies less than 100. In fact, primes and luckies crop up about equally often between any two given integers. Also, the gaps between successive primes and the gaps between successive luckies widen at roughly the same rate as the numbers increase, and the number of twin primes (primes that differ by 2) is close to the number of twin luckies. The luckies even have their own equivalent of the famous (still unsolved) **Goldbach conjecture**, which states that every even number greater than 2 is the sum of two primes. In the case of luckies, it's conjectured that every even number is the sum of two luckies; no exception has yet been found. Another unresolved problem is whether there are an infinite number of lucky primes. See also **Ulam spiral**.

Ludolph's number

Also known as *Ludolphine,* a name by which the number **pi** was known in Germany for many years following its evaluation to 35 digits by Ludolph von Ceulen (1540–1610).

Lyapunov fractal   A computer-generated image using Lyapunov exponents. *Radek Novak/Infojet*

lune

The portion of a **sphere** between two great semicircles (see **great circle**) having common endpoints (including the semicircles).

Lyapunov fractal

A particularly photogenic type of **fractal** that is popular with computer artists. It also represents a simple biological model of population growth in which the degree of the growth of the population can periodically alter between two values a and b.

M

Maclaurin, Colin (1698–1746)

A Scottish mathematician who developed and extended Isaac **Newton**'s work on **calculus** and gravitation, and did notable work on higher plane curves (see **Maclaurin trisectrix**). In his *Treatise of Fluxions* (1742), he gave the first systematic formulation of Newton's methods and set out a method for expanding functions about the origin in terms of series now known as *Maclaurin series*. Maclaurin also invented several devices, made astronomical observations, wrote on the structure of bees' honeycombs, and improved maps of the Scottish isles.

Maclaurin trisectrix

A curve first studied by Colin **Maclaurin** in 1742 with a view to solving one of the great geometric problems of antiquity: **trisecting an angle**. The Maclaurin trisectrix results from the Cartesian equation

$$y^2(a + x) = x^2(3a - x).$$

It is an **anallagmatic curve** that intersects itself at the origin.

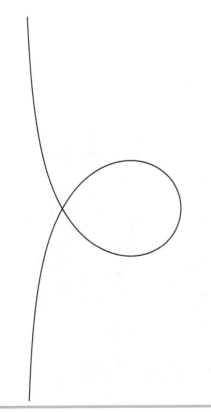

Maclaurin trisectrix *Jan Wassenaar, www.2dcurves.com*

MacMahon, Percy Alexander (1854–1929)

A British mathematician, physicist, and naval officer, born into a military family, whose leanings were evident early on when as a young child he showed a fascination with the way artillery was stacked. MacMahon later did work on missile trajectories, taking resistance into account, and on symmetric functions in the field of **combinatorics**, building on the results of James **Sylvester** and Arthur **Cayley**. His studies in symmetry led him to investigate partitions and to become a world authority on **Latin squares**. He wrote a two-volume treatise *Combinatory Analysis* (1915–1916), which became a classic, and a book on mathematical recreations called *New Mathematical Pastimes* (1921).[210] The latter shows another of the topics that intrigued MacMahon: the construction of patterns that can be repeated to fill the plane. However, much to his regret, as he wrote in the preface, "It has not been found possible to produce the book in colour." See also **MacMahon squares** and **thirty colored cubes puzzle**.

MacMahon squares

The concept of color-matching tiles based on all the permutations of colors on their edges dates from 1926, when Percy **MacMahon** invented and introduced three-color squares and four-color triangles as mathematical pastimes. MacMahon divided squares and triangles into triangles to give each edge of a piece its own color, in all possible combinations. Each set contains 24 different tiles, and MacMahon discovered that they could form a single figure with all adjacent edges matching and just one color all around the outside border. The most extensive research into these sets, over three decades, was done by the American engineer Wade Philpott (1918–1985), of Lima, Ohio, who identified all the possible symmetrical shapes that MacMahon squares and triangles could solve with both matching edge colors and uniform border color, and who calculated all the numbers of solutions for the MacMahon squares' 4 × 6 rectangle. See also **thirty colored cubes puzzle**.

Madachy, Joseph S.

An American mathematician, founder of *Recreational Mathematics Magazine*, editor of the **Journal of Recreational Mathematics** for nearly 30 years (now editor emeritus), the author of articles and books on the subject, and a well-known puzzlist.

magic cube

Similar to a **magic square** but having three dimensions instead of two. It contains the integers from 1 to n^3 and has $3n^2 + 4$ lines that sum correctly. All rows, columns, pillars, and the four triagonals (three-dimensional diagonals) must sum to the constant $\frac{1}{2}n(n^3 + 1)$. A cross section through a magic cube is a magic square.

magic square

An $n \times n$ square of the distinct whole numbers $1, 2, \ldots, n^2$, such that the sum of the numbers along any row, column, or main diagonal is the same. This sum is known as the *magic constant* and is equal to $\frac{1}{2}n(n^2 + 1)$. There is only one 3×3 magic square (not counting reflections and rotations), which was known to the Chinese as long ago as 650 B.C. as *lo-shu* and is bound up with a variety of myths. Associations between magic squares and the supernatural are also evident in early Indian and Arabian mathematics. The 3×3 square can be written as:

$$8\ 1\ 6$$
$$3\ 5\ 7$$
$$4\ 9\ 2$$

Each row, column, and main diagonal sums to 15. If the rows are read as three-digit numbers, forward and backward, and then squared, we find the interesting relation

$$816^2 + 357^2 + 492^2 = 618^2 + 753^2 + 294^2.$$

The reader may wish to see if the same rule holds for the columns and main diagonals.

magic square   A 4 × 4 magic square carved on the side of the Sagrada Familia cathedral in Barcelona.

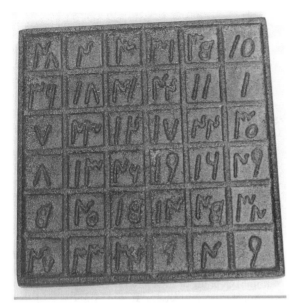

magic square A replica of a cast iron 6 × 6 magic square (vertical, horizontal, and main diagonal lines add to 111) from the Chinese Yuan dynasty (1271–1368), used as a sacred object to ward off evil spirits. *Sue & Brian Young/Mr. Puzzle Australia, www.mrpuzzle.com.au*

In the early sixteenth century Cornelius Agrippa constructed squares for n = 3, 4, 5, 6, 7, 8, and 9, which he associated with the seven "planets" then known (including the Sun and the Moon). Albrecht **Dürer**'s famous engraving of *Melancholia* (1514) includes a picture of an order-4 (4 × 4) magic square. There are 880 distinct squares of order-4 and 275,305,224 squares of order-5, but the number of larger squares is unknown. A square that fails to be magic only because one or both of the main diagonal sums don't equal the magic constant is called a *semi-magic square*. If *all* diagonals (including those obtained by wrapping around) of a magic square sum to the magic constant, the square is said to be a *pandiagonal square* (also known as a *panmagic* or *diabolical square*). Pandiagonal squares exist for all orders except 6, 10, 14, . . . , $2(2i + 1)$. There are 48 pandiagonal 4 × 4 squares. If replacing each number n_i by its square n_i^2 produces another magic square, the square is said to be a *bimagic* or *doubly magic square*. If a square is magic for n_i, n_i^2, and n_i^3, it is known as a *trebly magic square*.

A little trial and improvement is all it takes to construct the 3 × 3 magic square, but for building 4 × 4 squares and larger, a systematic method, or **algorithm**, is important. Interestingly, different algorithms are needed depending on whether the square is of an even order or an odd order. Odd order squares are the easier variety to make and there are several standard techniques, includ-

ing the Siamese (sometimes called de la Loubere's or the Staircase), the Lozenge, and de Meziriac's methods. Here is yet another approach, known as the Pyramid or extended diagonals method: (1) Draw a pyramid of same size squares as the magic square's squares, on each side of the magic square; the number of squares on the pyramid's base should be two less than the number of squares on the side of the magic square. (2) Sequentially place the numbers 1 to n^2 of the $n \times n$ magic square in the diagonals. (3) Relocate any number not in the $n \times n$ square to the opposite hole inside the square.

An *antimagic square* is an $n \times n$ array of integers from 1 to n^2 in which each row, column, and main diagonal gives a different sum such that these sums form a sequence of consecutive integers. A 4 × 4 antimagic square is a square arrangement of the numbers 1 to 16 so that the totals of the four rows, four columns, and two main diagonals form a sequence of 10 consecutive integers, for example:

$$
\begin{array}{cccc}
1 & 12 & 3 & 12 \\
15 & 9 & 4 & 10 \\
7 & 2 & 16 & 8 \\
14 & 6 & 11 & 5 \\
\end{array}
$$

The principle of magic squares can be extrapolated from two dimensions to any number of **higher dimensions**, including **magic cubes** and *magic tesseracts*, whose cross sections consist of magic cubes, and so forth.[10, 160, 162] See also **Latin squares** and **magic tour**.

magic tour

A **tour** by a **chess** piece on an $n \times n$ chessboard, whose squares are numbered from 1 to n^2 along the path of the piece, such that the resulting arrangement of numbers is a **magic square**. The tour is a *semi-magic tour* if the resulting arrangement of numbers is a **semi-magic square**. Magic **knight's tours** aren't possible on $n \times n$ boards if n is odd. They are possible for all boards of size $4k \times 4k$ for $k > 3$, but are believed to be impossible for $n = 8$.

main diagonal

In the $n \times n$ **matrix** $[a_{ij}]$, the elements a_{11}, a_{22}, . . . , a_{nn}. See also **diagonal**.

major axis

The longest **chord** of an **ellipse**.

Malfatti circles

In 1803 the Italian mathematician Giovanni Malfatti (1731–1807) posed the following problem: Given a triangle, find three nonoverlapping circles inside it such that the sum of their areas is maximal. Malfatti and many other mathematicians thought that the solution is given by the three circles, each of which is **tangent** to the other two and also to two sides of the triangle. Malfatti computed the radii of these circles, and they are now known as Malfatti's circles. Later it became clear that Malfatti's conjecture isn't true. In particular, Goldberg proved, in 1969, that the Malfatti circles *never* give a solution of the Malfatti problem! In other words, for any triangle there are three nonintersecting circles inside it, whose areas are bigger than the area of the Malfatti circles. So far as is known, the Malfatti problem hasn't been solved yet in the general case although it seems reasonable to suppose that the solution is given by what is called the *greedy algorithm:* First inscribe a circle in the given triangle; then inscribe a circle in the smallest angle of the triangle that is tangent to the first circle. The third circle is inscribed either in the same angle or in the middle angle of the triangle, depending on which of them has the bigger area.

Mancala

What is today almost universally called Mancala is, correctly speaking, the game of *Wari*. Mancala refers to a type of counting game of which Wari is a specific example, dating back to ancient times, known in Egypt and throughout Africa and Asia, and probably brought to Europe by sailors returning from those lands. Wari is played with a board and 48 markers or playing pieces, which are usually small colored stones or shells. The board consists of a piece of wood with two rows of six hollowed-out circular spaces, called *cups,* on each side, and one larger oval space at each end called a *reservoir.* The game is set up by placing four markers in each of the twelve cups. The first player picks up the markers in one of the cups on his side, then distributes the markers by placing one (and only one) in the other cups, going counterclockwise around the board. Markers are not placed in the reservoirs, which are for holding captured pieces only. Should there be enough markers that a complete circle of the board is made, the cup just emptied is skipped over. If the last marker put down goes into a cup on the opponent's side, and ends there with a total of two or three markers in that cup, then the markers in that cup are captured, and go into the reservoir at the player's right. Also, if the next cup clockwise from the captured cup has only two or three markers in it, that cup is likewise captured and the markers taken. This continues so long as the cup is clockwise to the last captured cup, has only two or three markers in it, and is an opponent's cup. However, a player can't capture all of an opponent's remaining pieces in a move, leaving him nothing to play. In capturing cups, if there's only one opponent cup left with markers in it, it can't be captured. Also, you cannot leave your opponent with all empty cups if you have a move that would put some pieces in one or more of his

cups. When one side of the board is empty, play is ended. This happens when a player has to move his last markers onto his opponent's side. At the end of play each player takes any markers left on his side of the board and adds them to his reservoir. The winner is the person with the most markers in his reservoir.

Mandelbrot, Benoît B. (1924–)

A Polish-born French mathematician, largely responsible for the present interest in **fractal** geometry. A native of Warsaw, he spent most of his early life in France. Mandelbrot was born into a family with a strong academic tradition: his mother was a doctor and his uncle Szolem Mandelbrot was a famous Parisian mathematician. His family left Poland for Paris in the 1930s to escape Hitler's regime. There, Mandelbrot was introduced to mathematics by his two uncles. Educated in France, he developed the mathematics of Gaston **Julia**, and began the (now common) graphing of equations on a computer. Mandelbrot originated what became known as fractal geometry and the object known as the **Mandelbrot set** is named after him. His work on fractals as a mathematician at IBM earned him an emeritus fellowship at the Thomas J. Watson Research Laboratories. In addition to his study of fractals in mathematics, he showed that fractals can be found in many places in nature, leading to entire new fields of exploration in **chaos theory**. He joined the faculty of Yale in 1987.

Mandelbrot set

The best known **fractal** and one of the most complex and beautiful mathematical objects known. It was discovered by Benoît **Mandelbrot** in 1980 and named after him by Adrien Douady and J. Hubbard in 1982. The set is produced by the incredibly simple **iteration** formula

$$z_{n+1} = z_n^2 + c,$$

where z and c are complex numbers and $z_0 = 0$. This can be written without complex numbers as

$x_{n+1} = x_n^2 - y_n^2 + a$, and
$y_{n+1} = 2x_n y_n + b$,

where $z = (x, y)$ and $c = (a, b)$. The Mandelbrot set consists of all the points on the **Argand diagram** for which the function $z^2 + c$ doesn't diverge under iteration. A computer is essential for carrying out the necessary calculations and for producing pictures of this remarkable structure. For the purposes of computation, the Argand diagram is broken down into pixels (picture elements), and the coordinates of each supply the constant c in $z^2 + c$. For each pixel (value of c) the function is iterated. If the function either rapidly diverges (blows up) or rapidly

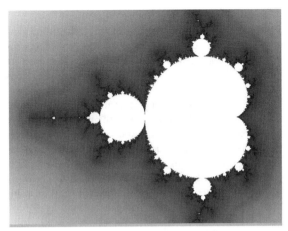

Mandelbrot set   A general view of the Mandelbrot set.

converges (collapses), the pixel is left black. If the function is more indecisive about which way it is heading, it is allowed to iterate longer. In some cases the iterations could go on for a very long time before it became clear that the function would ultimately diverge, so a limit is established, known as the *depth*, beyond which iterations are stopped. If the depth is reached without divergence, the corresponding pixel is left black as though it were in the set. At locations where divergence becomes clear prior to hitting the limit, the pixel is displayed according to a scale that represents how many iterations are needed to show divergence. The whole Mandelbrot set lies within a circle of radius 2.5 centered at the origin of the Argand diagram. Although finite in area, the set has a boundary that is infinitely long and has a **Hausdorff dimension** of 2.

The overall appearance of the Mandelbrot set is that of a series of disks. These disks have irregular borders and decrease in size heading out along the negative real axis; moreover, the ratio of the diameter of one disk to the next approaches a constant. More complex shapes branch out from the disks. One region of the Mandelbrot set containing spiral shapes is known as *Seahorse Valley* because it resembles a seahorse's tail. A computer can be used, like a microscope, to zoom in on different parts of the set. This reveals that, although the shape is infinitely complex, it also displays **self-similarity** with regions that look like the outline of the entire set. The Mandelbrot set also reveals symmetry on different levels. It is identically symmetrical about the real axis, and almost symmetrical at smaller scales. This kind of "near-but-not-quite" symmetry is one of the most unexpected properties to find in an object generated from such a simple formula and process. The Mandelbrot set was created by Mandelbrot as an index to the **Julia sets**. Each point in the Argand diagram corresponds to a different Julia set, and those points within the Mandelbrot set correspond precisely to the connected Julia sets.

manifold

A mathematical object that, in geometrical terms, is nearly "flat" on a small scale (though on a larger scale it may bend and twist into exotic and intricate forms). More precisely, a manifold is a **topological space** that looks locally like ordinary **Euclidean space**. Every manifold has a *dimension,* which is the number of coordinates needed to specify it in the local coordinate system. A circle, although curved through two dimensions, is an example of a one-dimensional manifold, or one-manifold. A close-up view reveals that any small segment of the circle is practically indistinguishable from a straight line. Similarly, a sphere's two-dimensional surface, even though it curves through three dimensions, is an example of a two-manifold. Seen locally, the surface, like that of a small portion of Earth, appears flat. A manifold that is smooth enough to have locally well-defined directions is said to be *differentiable*. If it has enough structure to enable lengths and angles to be measured, then it is called a *Riemannian manifold*. Differentiable manifolds are used in mathematics to describe geometrical objects, and are also

Mandelbrot set   A zoom view of the Mandelbrot set reveals some of the infinite variety of remarkable patterns that inhabit it.

the most natural and general settings in which to study differentiability. In physics, differentiable manifolds serve as the **phase space** in classical mechanics, while four-dimensional pseudo-Riemannian manifolds are used to model **space-time** in general relativity.

mantissa

The positive fractional part of the representation of a **logarithm**. For example, in the expression log 3,300 = 3.5185 . . . , 0.5185 is the mantissa.

many worlds hypothesis

> *Do you not think it a matter worthy of lamentation*
> *that when there is such a vast multitude of them*
> *[worlds], we have not yet conquered one?*
> —Alexander the Great

An interpretation of **quantum mechanics**, first proposed by the American physicist Hugh Everett III in 1957, according to which, whenever numerous viable possibilities exist, the world splits into many worlds, one world for each different possibility (in this context, the term *worlds* refers to what most people call "universes"). In each of these worlds, everything starts out identical, except for the one initial difference; but from this point on, they develop independently. No communication is possible between the separate universes, so the people living in them (and splitting along with them) would have no idea what was really going on. Thus, according to this view, the world branches endlessly. What is "the present" to us, lies in the pasts of an uncountably huge number of different futures. Everything that *can* happen does happen, somewhere. Until the many worlds interpretation, the generally accepted interpretation of quantum mechanics was (and perhaps still is) the *Copenhagen interpretation*. The Copenhagen interpretation makes a distinction between the observer and the observed; when no one is watching, a system evolves deterministically according to a wave equation, but when someone *is* watching, the wave function of the system "collapses" to the observed state, which is why the act of observing changes the system. The Copenhagen interpretation gives the observer special status, not accorded to any other object in quantum theory, and cannot explain the observer itself, while the many worlds hypothesis models the entire observer-observee system.

map

(1) A synonym for **function**, in which context it is also known as a *mapping*. More generally, the correspondence of elements in one **set** to elements in the same set or another set. (2) A representation, usually on a plane surface, of geographical regions. See **four-color map problem**.

Markov chain

A sequence of **random** variables in which the future variable is determined by the present variable but is independent of the way in which the present state arose from its predecessors. In other words, a Markov chain describes a chance process in which the future state can be predicted from its present state as accurately as if its entire earlier history was known. Markov chains are named after the Russian mathematician Andrei Andrevich Markov (1856–1922) who first studied them in a literary context, applying the idea to an analysis of vowels and consonants in a text by Pushkin. But his work launched the theory of **stochastic** processes and has since been applied in quantum theory, particle physics, and genetics.

Martingale system

A simple, popular, and ultimately disastrous gambling system that, on the face of it, seems like a dream come true. In the short run the player has a good chance of making a few dollars using this method. But, in the long run, two things conspire to defeat him—the table betting limits and the player's available funds. The Martingale system calls for an initial bet of, say, $2. If the player loses, he doubles his bet to $4. Another loss puts the net loss at $6 and requires a doubling of the bet to $8 to recoup the losses and show a profit. Assume the player loses five hands in a row. The sixth bet requires $64. Let's say this wins—the gambler has now won $128 but has lost $124—a net win of $4. Of course, any win, however small, if repeated over and over, could produce a fortune. The trouble is that many losing streaks run longer than 6 or 8 or 10 in a row. The Martingale quickly runs into the table limits. For example, if the player is at a $2 **blackjack** table with a $500 upper limit, he has to retire after 9 losses in a row and is down over $1,000. It would take another 500 winning hands to make up this loss! Basically, one losing streak will put the Martingale gambler in a hole he is unlikely ever to climb back out of.

Mascheroni construction

A construction done using a moveable compass alone, named after the Italian geometer Lorenzo Mascheroni (1750–1800), who, in his *Geometria del compasso* (1797), astonished the mathematical world by showing how every compass-and-straightedge construction can be done in this minimalist way. (Since straight lines can't be drawn with just a compass, it's assumed that two points, obtained by arc intersections, define a straight line.) It is now known that Georg Mohr (1640–1697) proved the same results earlier in his obscure *Euclides danicus* (1672). Mascheroni, or Mohr-Mascheroni, constructions are today primarily of interest to puzzle enthusiasts who try to improve on the older solutions by finding ones with fewer steps.

matchstick puzzle

A form of **mechanical puzzle** that involves rearranging a pattern of ordinary matches, according to a given instruction, to make a new shape or solve some other mathematical problem.

mathematical lifespan

"The mathematical life of a mathematician is short. Work rarely improves after the age of twenty-five or thirty. If little has been accomplished by then, little will ever be accomplished." Thus wrote Alfred Adler in an article titled "Mathematics and Creativity" in the *New Yorker* magazine (1972), echoing a common belief that mathematicians tend to do their best work before the age of 30, physicists before the age of 40, and biologists before the age of 50 (though there are exceptions!). The mathematical physicist Freeman Dyson put it more succinctly: "Young men should prove theorems, old men should write books." On the other hand, there are compensations for early burnout, as G. H. **Hardy** pointed out (in *A Mathematician's Apology*): "Archimedes will be remembered when Aeschylus is forgotten, because languages die and mathematical ideas do not. 'Immortality' may be a silly word, but probably a mathematician has the best chance of whatever it may mean."

mathematics

> *Pure mathematics consists entirely of assertions to the effect that if such and such a proposition is true of anything, then such and such another proposition is true of that thing. It is essential not to discuss whether the first proposition is really true, and not to mention what the anything is of which it is supposed to be true. . . . Thus mathematics may be defined as the subject in which we never know what we are talking about, nor whether what we are saying is true.*
>
> —Bertrand Russell

The science of patterns, real or imagined; *mathematics* comes from the Greek *mathema* for "knowledge" or "that which is learned." Its roots lie in the practical need to carry out commercial calculations, to measure land, and to forecast astronomical events. These activities correspond roughly to the mathematics of structure, space, and change. The investigation of structure begins with numbers—initially the **natural numbers** and **integers**. The rules governing arithmetical operations are dealt with in elementary **algebra**, while the deeper properties of whole numbers are the province of **number theory**. The study of methods to solve equations leads to **abstract algebra**, which deals with structures that generalize the properties of familiar numbers. The physically important concept of **vector**, generalized to **vector spaces** and studied in **linear algebra**, embraces both structure and space. The mathematics of space stems from **geometry**, first the **Euclidean geometry** and **trigonometry** of the everyday world, and later the various forms of **non-Euclidean geometry**. The modern fields of **differential geometry** and algebraic geometry generalize geometry in different ways. Differential geometry builds upon the concepts of coordinate system, smoothness, and direction, while algebraic geometry treats geometrical objects as sets of solutions to **polynomial** equations. **Group** theory places the concept of symmetry on an abstract footing and provides a bridge between space and structure. **Topology** links space and change through its emphasis on **continuity**. Analyzing and describing change in the physical world is a perennial theme of the natural sciences, and **calculus** was developed as a tool for doing this. The central concept used to describe a changing variable is that of a **function**. Many problems lead to relations between a quantity and its rate of change, and the methods to solve these are studied in the field of differential equations. The numbers used to represent continuous quantities are **real numbers**, and the study of their properties and the properties of real-valued functions is known as real **analysis**. For various reasons, it's convenient to generalize to **complex numbers**, which are dealt with in **complex analysis**. Functional analysis focuses attention on (typically infinite-dimensional) spaces of functions, laying the groundwork for **quantum mechanics** among many other things. In order to probe the foundations of mathematics, the fields of **set** theory, mathematical **logic**, and model theory were developed.

Since the time of the ancient Greeks, thought has been given to the ultimate nature of mathematics. What is its role and status in reality? Most crucially, is it invented or discovered? Leopold **Kronecker** was on the side of invention: "God made the integers; all the rest is the work of Man." Charles **Hermite**, by contrast, was clearly a Platonist: "There exists . . . an entire world, which is the totality of mathematical truths, to which we have access only with our mind, just as a world of physical reality exists, the one like the other independent of ourselves, both of divine creation." The German physicist Heinrich Hertz went even further: "One cannot escape the feeling that these mathematical formulas have an independent existence and an intelligence of their own, that they are wiser than we are, wiser even than their discoverers, that we get more out of them than was originally put into them." G. H. **Hardy** summed up what many mathematicians today have a tendency to believe:[151] "that mathematical reality lies outside us, that our function is to discover or observe it, and that the theorems which we prove, and which we describe grandiloquently as our 'creations,' are simply the notes of our observations." In another passage, he made this point more precisely: "317 is a prime, not because we think so, or because our minds are shaped in one way rather than another, but because it is so, because mathematical reality is built that way." We are struck by how well mathematics describes the behavior of the world in which we live. The universe, in fact, appears to have a deep mathematical infrastructure. Martin **Gardner** makes the bold claim that: "Mathematics is not only real, but it is the only reality. [The] . . . entire universe is made of matter. . . . And matter is made of particles. . . . Now what are the particles made out of? They're not made out of anything. The only thing you can say about the reality of an electron is to cite its mathematical properties. So there's a sense in which matter has completely dissolved and what is left is just a mathematical structure."

matrix

A square or rectangular array of numbers, usually written enclosed in a large pair of parentheses. Matrices, which are added and multiplied using a special set of rules, are extremely useful for representing quantities, particularly in some branches of physics. A matrix can be thought of as a **linear** operator on **vectors**. Matrix-vector multiplication can be used to carry out geometric **transformation**s such as scaling, rotation, reflection, and translation.

Matrix comes from the same Latin root that gives us *mother,* and was used to refer to the womb and to pregnant animals. It became generalized to mean any situation or substance that contributes to the origin of something. The first mathematical use of the word *matrix* was around 1850 by James **Sylvester** who saw a matrix as a way of obtaining **determinant**s, but didn't fully appreciate its potential. Within a year of his first mention of the term, he introduced the idea to Arthur **Cayley** who was the first to publish the **inverse** of a matrix and to treat matrices as purely abstract mathematical forms. The use of mathematical arrays to solve problems predates the application of the name by about 2,000 years. Around 200 B.C. in the Chinese text *Juizhang Suanshu* (Nine Chapters on the Mathematical Arts) the author solves a system of three equations in three unknowns by placing the **coefficient**s on a counting board and solving by a process that today would be called *Gaussian elimination.*

maximum

The largest of a set of values.

maze

A network of winding and interconnected passageways that a traveler must negotiate in order to reach some goal. The terms *maze* and *labyrinth* are often used interchangeably, though sometimes a distinction is made based on the layout. A labyrinth is then defined as a construction that leads from a starting point to a goal by a *single* path, with no branches or dead ends. No matter how long and twisting the route, it is predetermined by the builder: a labyrinth, according to this definition, is *unicursal.* A maze, by contrast, is *multicursal* and calls for the traveler to make a series of decisions that affect how quickly the goal is reached. In this book, *maze* and *labyrinth* mean the same thing, and refer to unicursal and multicursal mazes.

The most famous of legendary mazes was the lair (or prison) of the Minotaur at Knossos on Crete. According to Greek mythology, King Minos of Crete had his chief engineer, Daedalus, build the labyrinth in order to keep the half-human half-bovine offspring of his wife Pasiphae and a bull out of the public eye. King Aegeus of Athens was forced to pay a periodic tribute to Minos (the Athenians having earlier murdered Minos's son) in the form of seven young men and seven maidens. These unfortunates were forced to enter the maze below Minos's palace where they would get hopelessly lost and eventually be eaten by the monster. King Aegeus' son, Theseus, decided to put a stop to this and offered to take the place

maze   Possibly the world's oldest surviving labyrinth: a seven-ring labyrinth rock carving inside the Tomba del Labirinto, a Neolithic tomb at Luzzanas, Sardinia, dating somewhere between 2500 and 2000 B.C. *National University of Singapore*

maze A plan of the maze in Chartres Cathedral as drawn by Henry Dudeney.

only one way in and out is anybody's guess. But the underlying reason for the early unicursal design isn't hard to find. These mazes were intended not as intellectual puzzles but as symbolic representations of destiny as a matter of fate, beyond personal control. The world's oldest known maze-like designs, dating back about 3,500 years, have been found carved on rocks in northwest Spain and around the shores of the Mediterranean, and according to legend, were walked by fishermen before setting sail to ensure favorable winds and a good catch. In medieval times, mazes started to appear in churches, as art painted on walls or inlaid as "pavement mazes" on the floor. Some of the larger floor versions were traversed by people on their knees as a form of repentance or traversed as a substitute for an actual pilgrimage to the holy city, earning them the name *"Chemin de Jerusalem,"* or Road of Jerusalem. The oldest-known church labyrinth, at the Basilica of Reparatus at Orleansville, Algeria, dates from the fourth century A.D. and measures about 8 feet in diameter. One of the largest, built in 1288, formed part of the floor in the nave of Amiens Cathedral in France and spanned about 42 feet, but was destroyed in 1825. A splendid surviving example, however, is in Chartres Cathedral near Paris. Laid down in about 1200, this is an 11-circuit design (11 concentric windings) divided into quadrants, of the type often found in Gothic cathedrals. The only cathedral maze in Britain is at Ely, 16 miles north of Cambridge; when the Cathedral was restored around 1870, the architect, Gilbert Scott, installed a pavement maze of his own design under the west tower.

Mazes built primarily as puzzles or for recreation represent a different type of structure. For one thing, they are multicursal, offering those who enter a series of choices about which way to proceed. In his *Natural History,* the Roman historian Pliny comments that the classical type of labyrinth is quite distinct from "the mazes formed in the fields for the entertainment of children," suggesting that these diversions may have a long history. But it was in England that they really came of age. Church mazes never caught on in England, but turf mazes became tremendously popular. Ranging from 25 feet to over 80 feet across, they were constructed in or just outside villages across the countryside and were given names such as "Mizmaze," "Troy Town," "Shepherd's Race," and "Julian's Bower." A Welsh history book *Drych y Prif Oesoedd* (Mirror of the first age) published in 1740 notes the curious custom shepherds had of cutting the turf in the form of a labyrinth, which would seem to account for the origin of "Shepherd's Race." "Troy Town" probably refers to a legend that the city of Troy had seven exterior walls arranged as a maze to frustrate an attacking force.

From the turf maze it was no great leap to perhaps the most famous form of full-size maze—the topiary or hedge

of one of the sacrificial virgins. He fell in love with Ariadne, one of Minos's daughters, who gave him a *clew,* or ball of yarn, to unravel as he entered the labyrinth so that he would be able to find his way back out. (Originally, *clue* and *clew* were alternative spellings of the same word. The modern sense of *clue* as a guide to solving a problem comes from the legend of the Minotaur, while *clew* retains its ancient meaning.) In true heroic style, Theseus slew the Minotaur but then spoiled the fairytale ending by abandoning Ariadne on the voyage home. Did this labyrinth really exist? In Roman times some writers suggested that a set of winding caves at Gortyan, in southern Crete, might have formed the basis for the tale. Although this complex of natural passages sounds similar to the maze in the myth, the story definitely places the labyrinth at the ancient Cretan capital. Archaeologists have found no evidence of a labyrinth structure at Knossos, but it has been suggested that the palace was so complicated, with its many levels, stairs, and rooms, that it may itself have inspired the story.

Minoan coins from about 300 B.C. bear a round, winding design, thought to be a representation of the labyrinth. A very similar geometric pattern recurs across many different cultures and times—scratched into caves in Cornwall (possibly by visiting Phoenician seafarers), on Roman coins, and in pictures drawn by native American Indians. Almost all these designs, including the one on the Minoan coins, are unicursal. Quite why Theseus would have needed a clew to navigate a maze that had

maze. While the use of hedges in gardens dates back to Roman times, the earliest references to a topiary maze appears in thirteenth-century Belgium. By the sixteenth century the hedge maze had spread to England, as a landscape painting by Tintoretto attests. In the later part of the seventeenth century, Louis XIV had a labyrinth built as a part of the gardens at Versailles, which included 39 groups of hydraulic statuary representing the fables of Aesop. The most famous surviving historic hedge maze is that on the grounds of Hampton Court Palace in England, designed by George London and Henry Wise for William of Orange, planted between 1689 and 1694, and occupying about a third of an acre. The finest turf maze in England may be that at Bridge End Gardens in Saffron Walden, Essex, which was replanted with yew in 1838–1840, abandoned and lost by 1949, then restored in 1983. A second maze, an imitation of that at Hampton Court, is on the Common.

In August 1997, Adrian **Fisher** opened the "World's Largest Maze" at Millets Farm Centre, Frilford, Oxfordshire, England, cut from about six acres of grain. However, his 1995 effort in Shippensburg, Pennsylvania, was in a 30-acre field and the 1997 maze in Reignac sur Indre covers 37 acres.[101]

When faced with a maze, what is the best way of reaching the goal, whether this is a point in the middle of the maze or an exit that forms a second opening to the structure? A unicursal maze calls for no brainwork, only footwork, since it consists of a single winding passage with no offshoots. Multicursal mazes are a different story. The easiest solution is to place a hand on one wall at the outset and follow that wall, come what may. Each blind alley will be traversed one time in and out until the whole maze is completed, or the goal is found. This simple method isn't the most efficient and it will fail altogether if the goal lies within an island in the maze—that is, a section detached from any of the exterior walls. A classic general method of "threading a maze" is: (1) never traverse a path more than twice; (2) when arriving at a new branch point or node, select either path; (3) when arriving at an old node or at a dead end by a new path, return by the same path; and (4) when arriving at an old node by an old path, select a new path, if possible; otherwise, an old path. An

maze A handheld maze, the object of which is to get the silver ball from the starting rectangle on the left to the finishing rectangle on the right. Crafted from Tasmanian blackheart sassafrass by Kym Anderson. *Mr. Puzzle Australia, www.mrpuzzle.com.au*

explorer who follows these rules, using marks on the ground, as a record, can be assured of visiting every part of the maze.[48, 219, 240, 294]

See also **Rosamund's bower**.

mean

> *When she told me I was average, she was just being mean.*
>
> —Anonymous

A precisely calculated "typical" value of a group of numbers. There are several kinds of means. The *arithmetic mean,* which is usually what is meant when people talk about an **average**, is the sum of a set of values divided by the number of values. For example, the arithmetic mean of 3, 4, 7, and 10 is $(3 + 4 + 7 + 10)/4 = 6$. The *geometric mean* of n values is the nth root of the product of the values. For example, the geometric mean of 3, 8, and 10 is $(3 \times 8 \times 10)^{1/3}$ or the cube root of 240. The *harmonic mean* is the reciprocal of, or one over, the mean of the reciprocals of the values. For example, the harmonic mean of 3, 8, and 10 is $1/[(\frac{1}{3} + \frac{1}{8} + \frac{1}{10})/3]$. A different kind of average is *median*, which is generally the center term when a group of numbers are ordered by size. If there are an even number of values then the arithmetic mean of the center two is the median. The words *mean* and *median* both come from the Indo-European root *medhyo* meaning "middle."

measure

A way of gauging how big something is in terms of length, volume, or some other quality. One of the strangest facts in mathematics is that some objects exist that can't be measured. In the language of **sets**, the basic rules (somewhat simplified) of mathematical measures are as follows: (1) the measure of any set is a **real number**; (2) the **empty set** has measure zero; (3) if A and B are two sets with no elements in common then the measure of $A \cup B$ (the union of A and B) is equal to the measure of A plus the measure of B. The second of these rules can be very useful, for example, when integrating a **function**, since it allows us to ignore any points where the function jumps around, provided that such points are isolated. A slightly jittery function is one thing; a nonmeasurable set is a very different animal. Imagine a three-dimensional shape so fantastically intricate, so jagged and crinkled, that it is impossible to measure its volume and this gives some idea of the concept of nonmeasurability. From it flow such bizarre conclusions as the **Banach-Tarski paradox**.

measure theory

The part of **mathematics** that investigates the conditions under which **integration** can be carried out. It focuses mainly on the various ways in which the size, or **measure**, of a **set** can be estimated.

measuring and weighing puzzles

Problems that involve measuring a given quantity of liquid by pouring from one vessel into others of known capacity go back to medieval times. One of the earliest to appear in print was given by Niccoló **Tartaglia** and asks for 24 ounces of balsam to be divided into three equal portions using vessels that hold 5, 11, and 13 ounces, respectively. A similar problem was posed by a fellow traveler to the young Siméon **Poisson** while on a journey. Poisson's family had tried to steer him into careers ranging from a surgeon to a lawyer, the last on the theory that he was fit for nothing better. He seemed inept at everything he did. However, he saw the solution to the measuring problem immediately and realized his true calling. Thereafter, he threw himself into mathematics and became one of the greatest mathematicians of the nineteenth century.

The classic weighing problem was proposed by Claude-Gaspar **Bachet** and entails finding the least number of weights needed to weigh any integral number of pounds from 1 to 40 pounds inclusive, when no weights are allowed in either of the two pans. The answer is 1, 3, 9, and 27 pounds. Tartaglia had previously stated the same puzzle with the condition that the weights may only be placed in one pan, in which case the solution is 1, 2, 4, 8, 16, and 32 pounds.

PUZZLE

The following is a measuring problem from Henry **Dudeney**'s *Canterbury Puzzles,* which Dudeney claimed was the most popular of the whole collection and which the reader may like to try:

> Here be a cask of fine London ale, and in my hands do I hold two measures—one of five pints, and the other of three pints. Pray show how it is possible for me to put a true pint into each of the measures.

Solutions begin on page 369.

---

mechanical puzzle

A puzzle, involving several objects or a single object composed of one or more movable parts, whose solution requires moving from an initial state to a predefined final state. Mechanical puzzles were first classified by Louis **Hoffmann** in *Puzzles Old and New*[171] (1893). A modified form of his scheme is shown in the table "Mechanical Puzzles" on the following page.

medial triangle

The **triangle** whose vertices are the midpoints of the sides of a given triangle (see **vertex**).

Mechanical Puzzles

Type	Subtype	Examples
Assembly	2-dimensional assembly	**Tangrams, T-puzzle**, jigsaw
	3-dimensional assembly (noninterlocking)	**Soma cube**
	Matchstick puzzles	
	Miscellaneous	**Puzzle rings**
Disassembly	Trick or secret opening	**Puzzle jug**
	Miscellaneous	Trick locks, keys, etc.
Interlocking solid	**Burr puzzle**	
	3-dimensional jigsaws	
	Miscellaneous	Cubes, other objects
Disentanglement and entanglement	Wire puzzles	**Chinese rings**
	String puzzles	Cat's cradle
	Miscellaneous	
Sequential movement	**Peg solitaire** (peg removal)	
	Other counter (peg rearrangement)	
	Sliding-piece puzzles	**Fifteen Puzzle**
	Miscellaneous	**Tower of Hanoi**
Puzzle vessels	**Puzzle jug**	
	Miscellaneous	Bottom-fill teapots, pitchers
Vanishment puzzle		**Get Off the Earth**
Folding	**Origami**	
	Flexagon	
Impossible figure		**Penrose stairway**
		Penrose triangle
		Impossible trident

median
1. The line from a **vertex** of a **triangle** to the midpoint of the opposite side. 2. See **mean**.

Menaechmus (c. 380–c. 320 B.C.)
A Greek mathematician, thought to have been a pupil of **Eudoxus**, who is famed for his discovery of the **conic sections** and for being the first to show that **ellipses**, **parabolas**, and **hyperbolas** are formed by cutting a cone in a plane that is not parallel to the base. Menaechmus made his discoveries on conic sections while attempting, unsuccessfully, to solve the problem of **duplicating the cube**. It has also been suggested that he served as a tutor to Alexander the Great.

Menger sponge
A famous **fractal** solid that is the three-dimensional equivalent of the **Sierpinski carpet** (which, in turn, is the

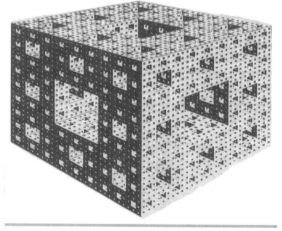

Menger sponge

two-dimensional equivalent of **Cantor dust**). To make a Menger sponge, take a cube, divide it into 27 (= $3 \times 3 \times 3$) smaller cubes of the same size and remove the cube in the center and the six cubes that share faces with it. What's left are the eight small corner cubes and twelve small edge cubes holding them together. Now, imagine repeating this process on each of the remaining 20 cubes. Repeat it again. And again . . . ad infinitum. The Menger sponge was invented in 1926 by the Austrian mathematician Karl Menger (1902–1985).

Mersenne, Marin (1588–1648)

A French monk, philosopher, and mathematician best remembered for his work to find a formula to generate **prime number**s based on what are now known as **Mersenne number**s. However, in addition to being a mathematician, he wrote about music theory and other subjects; edited works of Euclid, Archimedes, and other Greek mathematicians; but most importantly, corresponded extensively with mathematicians and other scientists in many countries. At a time before scientific journals existed, Mersenne was at the heart of a network for information exchange.

Mersenne number

A number of the form $2^n - 1$ (one less than a power of 2), where n is a positive integer. Mersenne numbers are named after Marin **Mersenne** who wrote about them in his *Cogita Physico-Mathematica* (Physical Mathematics Knowledge 1644) and wrongly conjectured that they were prime for $n = 2, 3, 5, 7, 13, 17, 19, 31, 67$, and 257, and composite for $n < 257$. See also **Mersenne prime**.

Mersenne prime

A **prime number** of the form $2^p - 1$, where p is prime. A prime exponent is necessary for a **Mersenne number** to be prime but is not sufficient; for example, $2^{11} - 1 = 2,047 = 23 \times 89$. In fact, after an early clustering of Mersenne primes for fairly small values of p, further occurrences become increasingly rare. At the time of writing there are 40 known Mersenne primes, corresponding to values for p of 2; 3; 5; 7; 13; 17; 19; 31; 61; 89; 107; 127; 521; 607; 1,279; 2,203; 2,281; 3,217; 4,253; 4,423; 9,689; 9,941; 11,213; 19,937; 21,701; 23,209; 44,497; 86,243; 110,503; 132,049; 216,091; 756,839; 859,433; 1,257,787; 1,398,269; 2,976,221; 3,021,377; 6,972,593; 13,466,917; and 20,996,011. However, it isn't known if the current largest Mersenne prime is the fortieth in order of size because not all lower exponents have been checked. Mersenne primes rank among the largest of all known

primes because they have a particularly simple test for primality, called the *Lucas-Lehmer test*.

The search for Mersenne primes has been going on for centuries. They are named after Marin **Mersenne** who, in 1644, helped the search gain wide recognition by writing to many mathematicians of his conjecture about which small exponents yield primes. Around the time that Mersenne's conjecture was finally settled, in 1947, digital computers gave a new impetus to the hunt for Mersenne primes. As time went on, larger and larger computers found many more Mersennes and, for a while the search belonged exclusively to those with the fastest computers. This changed in 1995 when the American computer scientist George Woltman began the *Great Internet Mersenne Prime Search* (*GIMPS*) by providing a database of what exponents had been checked, an efficient program based on the *Lucas-Lehmer test* that could check these numbers, and a way of reserving exponents to minimize the duplication of effort. Today GIMPS pools the combined efforts of dozens of experts and thousands of amateurs. This coordination has yielded several important results, including the discovery of the Mersennes $M_{3021377}$, $M_{2976221}$, and $M_{20996011}$ and the proof that M_{756839}, M_{859433}, and $M_{3021377}$ are the thirty-second, thirty-third, and thirty-fourth Mersennes.

meter

The basic unit of length adopted under the System International d'Unites (SI units). Over the years the definition of the meter has changed several times. Throughout all these definition changes the length of the meter hasn't changed, but the precision by which it is measured was improved. In 1793, the meter was defined to be 1/10,000,000 of the distance from the pole to the equator. During the nineteenth century, the definition was in terms of the length of standard bars of platinum kept under controlled conditions. In 1983, the current definition was adopted of the distance traveled by light in a vacuum in 1/299,792,458 of a second.

method of exhaustion

Finding an **area** by approximating it by the areas of a sequence of **polygon**s; for example, filling up the interior of a circle by inscribing polygons with more and more sides.

metric

Any **function** $d(x, y)$ that describes the *distance* between two points. Distance is formally defined as a single number with the following properties: (1) $d(x, y) = 0$ if and only if $x = y$; (2) $d(x, y) = d(y, x)$; (3) $d(x, y) + d(y, z) \geq d(x, z)$ (the triangle inequality).

metric space

A **set** that has a **metric**; in other words, a kind of space in which the concept of distance has meaning. Compare with **topological space**.

metrizable

For a **topological space**, the property that there exists a **metric** compatible with the topology. To say that a topological space is metrizable is to treat it as a **metric space**, but without distinguishing any specific or preferred distance function.

Michell, John (1724–1793)

An English natural philosopher and clergyman, educated at Queen's College, Cambridge, who discovered that the force between magnetic poles varies as $1/r^2$. In 1767, Michell became rector of Thornhill, now a suburb of Dewsbury, Yorkshire. Sometime in the early 1770s, he played music with the great astronomer William Herschel and gave Herschel his first telescope. In 1784, Michell deduced the existence of what are now called black holes from Newton's corpuscular theory of light, and suggested that some stars might have dark companions. He also devised and built the torsion balance for determining the universal gravitational constant G, but died before having the chance to use it. The balance was passed to the Cambridge physicist F. J. H. Wollaston, and from him to Henry Cavendish, who used it at his house in London and is often, mistakenly, identified as its inventor.

midpoint

The point M is the midpoint of line segment AB if $AM = MB$. That is, M is halfway between A and B.

mile

A measure of distance, the name of which is an abbreviation of the Latin *mille passes* or "one thousand paces." Since the paces (one step with each foot) of the Roman Army were supposed to be two steps, each 2.5 feet long, 1,000 paces is very close to the length now called a statute mile (5,280 feet). A *nautical mile* was developed to be a distance equal to 1 minute of arc ($\frac{1}{60}$ of a degree) distance along a great circle and is equal to 6,076 feet.

million

A thousand thousand, 1,000,000, or 10^6. The word comes from the Latin *mille* for "thousand" (which is also the root for *mile* and *millennium*) and the suffix *ion* that implies "large" or "great;" thus a million is literally a "great thousand." Although *million* seems to have come into use as early as the middle of the fourteenth century, most mathematicians would use the phrase "thousand thousands" to avoid confusion and it was not until the 1700s that "million" caught on. It appears in the King James Version of the Bible (Genesis 24:60) and in Shakespeare (*Hamlet* act II scene II)– "for the play, I remember, pleased not the million."

minimal prime

A **prime number** that is a substring of another prime when written in **base** 10. A string a is a substring of another string b, if a can be obtained from b by deleting zero or more of the characters in b. For example, 392 is a substring of 639,802. The minimal primes are:

2; 3; 5; 7; 11; 19; 41; 61; 89; 409; 449; 499; 881; 991; 6,469; 6,949; 9,001; 9,049; 9,649; 9,949; 60,649; 666,649; 946,669; 60,000,049; 66,000,049; 66,600,049

minimal surface

A surface that, bounded by a given closed curve or curves, has the smallest possible area. A minimal surface has a mean **curvature** of zero. Finding and classifying minimal surfaces and proving that certain surfaces are minimal have been major mathematical problems for over 200 years. If the closed curve is planar then the solution is trivial; for example, the minimal surface bounded by a circle is just a disk. But the problem becomes much more difficult if the bounding curve is nonplanar–in other words, is allowed to move up and down in the third dimension. The first nontrivial examples of minimal surfaces, the **catenoid** and the **helicoid**, were discovered by the French geometer and engineer Jean Meusnier (1754–1793) in 1776, but there was then a gap of almost 60 years before the German Heinrich Scherk found some more. In 1873 the Belgian physicist Joseph **Plateau** carried out experiments that led him to conjecture that soap **bubbles** and soap films always form minimal surfaces. Proving mathematically this was true became known as the **Plateau problem**. Most minimal surfaces are extremely hard to construct and visualize, in part because the majority of them are self-intersecting. However, the development of high-performance computer graphics has provided mathematicians with a powerful tool and the last couple of decades have seen a huge increase in the number of such surfaces that have been defined and investigated.

minimax theorem

A theorem that says there is always a rational solution to a precisely defined conflict between two people whose interests are completely opposite. It is rational in that both parties can convince themselves that they can't expect to do any better, given the nature of the conflict.

minimum

The smallest of a set of values.

minimal surface A Scherk Surface—a type of minimal surface—portrayed as a membrane. *Anders Sandberg*

Minkowski, Hermann (1864–1909)

A German mathematician, born in Lithuania, who played an important part in the early development of **relativity theory**. Minkowski was the first to realize that the work of Hendrik Lorentz and Albert **Einstein** could be best understood if space and time, formerly thought to be separate entities, were treated as part of a four-dimensional **space-time** with a **non-Euclidean geometry**. The concept of the space-time continuum, which provided a framework for all later mathematical work in relativity, appeared in Minkowski's book *Raum und Zeit* (Space and Time 1907). From 1896 to 1902, Minkowski taught at the Zurich Federal Institute of Technology when Einstein was a student. In fact, Einstein attended several of the courses he gave but didn't create a good impression at the time. Minkowski described him as a "lazy dog" who "never bothered about mathematics at all." In 1902, Minkowski accepted a chair at the University of Göttingen, where he stayed for the rest of his life. His main interest was in pure mathematics, including number theory and geometry, and it was through his understanding of the more abstract side of mathematics and geometry in more than three dimensions that he developed the idea of four-dimensional space-time.

Minkowski space

A finite-dimensional **vector space**, especially a four-dimensional one, together with an indefinite inner product with one positive or timelike direction and many negative or spacelike directions. In particular, Minkowski space is ordinary **space-time** in the special **relativity theory**.

minor axis

The smallest **chord** of an **ellipse**.

minute

(1) One sixtieth of an hour or of a **degree** of arc or angle. (2) One sixtieth of an hour. As a measure of both time and angle, it equals 60 **seconds**.

mirror reversal problem

Why does a mirror reverse right and left, but not up and down? This question crops up perennially in the letter and query columns of magazines and newspapers. It was the inspiration for Lewis **Carroll**'s *Alice through the Looking Glass.* Alice Raikes (not to be confused with Alice Liddell, after whom the fictional Alice was modeled) was another of Carroll's young friends. On one occasion, in 1868, Carroll put an orange in her right hand and then asked her to stand in front of a mirror and say which hand showed the reflection of the orange. She said the left hand and Carroll asked her to explain. She finally replied "If I was on the other side of the glass, wouldn't the orange still be in my right hand?" Carroll said this was the best answer he'd had and later said it gave him the idea for his book.

Others have struggled harder but not always convincingly to explain the phenomenon, appealing variously to gravity, the psychology of perception, and philosophy. Why *does* a mirror reverse right and left, but not up and down? A frequently given answer is that a mirror doesn't reverse right and left. It reverses *front and back*. This is certainly true: the looking glass you is facing in the opposite

mirror reversal problem A looking-glass world beckons the curious Alice in Lewis Carroll's book *Alice through the Looking Glass.*

direction to the "real" you. But this short, crisp explanation doesn't completely dispel the mystery. If you imagine that the mirror is not there and that instead you are looking at a flesh-and-blood twin of yourself, that twin is differently handed. If you have a watch on your left wrist, the person you are facing has his/her watch on the right wrist. The mirror *has* done a left-right swap, surely! At any rate, something has happened to left and right that hasn't happened to up and down. To be more convinced of this, hold this book up to the mirror and try to read it. If no left-right swap has happened, why is the reflected writing so hard to read? First, remember that you are only looking at an image! The mirror hasn't (Carrollian fantasies aside) created some*thing* of opposite handedness. Secondly, appreciate how the writing appears in the mirror's frame of reference. This is easy to do by looking at the writing from the other side of the page (i.e., back to front, thus undoing the back to front reversal caused by the reflection). From the mirror's point of view the writing looks perfectly normal.

missing dollar problem

A version of this problem first appeared in R. M. Abraham's *Diversions and Pastimes* in 1933.[2] See also **nine nooms paradox.**

Mittag-Leffler, (Magnus) Gösta (1846–1927)

A Swedish mathematician who, in 1882, founded the international journal *Acta Mathematica,* and was its chief editor for 45 years. He studied in Paris under Charles **Hermite** and in Berlin under Karl **Weierstrass**, and made significant contributions to **analysis**. His best known work concerned the analytic representation of a one-valued **function**s and culminated in the *Mittag-Leffler theorem.* Since he took a special interest in Georg **Cantor**'s discoveries, much of Cantor's work was published in *Acta Mathematica.* Inscribed on the mantlepiece of Mittag-Leffler's home—now a research institute—in Djursholm is the epitaph: "Number is the beginning and end of thought. Thought gave birth to number but reaches not beyond. ML 1903"

mixed strategy

In **game theory**, a **strategy** that uses **random**ness by employing different actions in identical circumstances with different probabilities.

Möbius band

A simple and wonderfully entertaining two-dimensional object, also known as the Möbius strip, that has only one surface and one edge. It is named after the German mathematician and theoretical astronomer August Ferdinand Möbius (1790–1868), who discovered it in September 1858, although his compatriot and fellow mathematician Johann Benedict Listing (1808–1882) independently devised the same object in July 1858. Making a Möbius band is simple: take an ordinary sheet of typing paper, cut an 11″ × 1″ rectangle, bring the two long ends together, twist one of the ends 180°, and tape the two ends together. To prove that the band is single-sided, take a pen and start drawing a line around the band's circumference. When drawing the line, never take the pen off the paper; just

Möbius band   A sculpture of a Möbius band outside the Fermi National Accelerator Laboratory. *FNAL*

keep drawing the line until the starting point is reached. Once you are finished, look at both sides: there should be a line on both sides, thus proving that it is all the same side because you never took the pen off the paper.

The Möbius band has a lot of curious properties. If you cut down the middle of the band, instead of getting two separate strips, it becomes one long strip with two half-twists in it. If you cut this one down the middle, you get two strips wound around each other. Alternatively, if you cut along the band, about a third of the way in from the edge, you will get two strips; one is a thinner Möbius band, the other is a long strip with two half-twists in it. Other interesting combinations of strips can be obtained by making Möbius bands with two or more flips in them instead of one. Cutting a Möbius band, giving it extra twists, and reconnecting the ends produces unexpected figures called *paradromic rings*.

The Möbius band has provided inspiration both for sculptures and for graphical art. M. C. **Escher** was especially fond of it and based many of his lithographs on it. It is also a recurrent feature in science fiction stories, such as Arthur C. Clarke's *The Wall of Darkness*. A common fictional theme is that our universe might be some kind of generalized Möbius band. There have been technical applications; giant Möbius bands have been used as conveyor belts (to make them last longer, since "each side" gets the same amount of wear) and as continuous-loop recording tapes (to double the playing time).

A closely related strange geometrical object is the **Klein bottle**, which can be produced by gluing two Möbius bands together along their edges; however, this can't be done in ordinary three-dimensional Euclidean space without creating self-intersections.[116]

mode
The most frequently occurring value in a sequence of numbers.

model of computation
An idealized version of a computing device that usually has some simplifications such as infinite memory. A **Turing machine** and the **lambda calculus** are models of computation.

model theory
The study of mathematical structures that satisfy a particular set of **axiom**s, especially in the field of **logic**.

modulo
The integers a and b are said to be congruent modulo m if $a - b$ is divisible by m.

Moiré pattern
A radiating curved pattern created when two repetitive patterns overlap and interfere with one another. A Moiré pattern is seen, for example, when someone on TV wears a herringbone jacket. *Moiré* is the French word for "silk", and *silk moiré,* introduced from China to France in 1754, is the fabric that shows the familiar shifting patterns.

Moivre, Abraham de
See **de Moivre, Abraham**.

monad
A central concept in the Pythagorean worldview (see **Pythagoras of Samos**), in which it is regarded as the first thing that came into existence. Following the monad came, in order: the *dyad,* numbers, lines, two-dimensional entities, three-dimensional entities, bodies, the four elements (earth, air, fire, and water), and the rest of the world. The monad plays a similarly fundamental role in the metaphysics of Leibnitz as an indivisible, impenetrable unit of mental experience. It also has several different technical meanings in modern mathematics. For example, in **nonstandard analysis**, a monad consists of all those numbers infinitesimally closer to a given number. The word comes from the Latin *monas* (single) and Greek *monos* (unit).

Monge, Gaspard (1746–1818)
A French mathematician and physicist who put descriptive geometry, introduced by Albrecht **Dürer**, on a firm mathematical footing. He became professor of mathematics at Mézières (1768) and of hydraulics at the Lycée in Paris (1780), and published his groundbreaking treatise on the application of geometry to the arts of construction in 1795.

monkeys and typewriters
Six monkeys pounding away on typewriters would, by pure chance, if given enough time, be bound to write out all the works held in the British Library (or in all other libraries, for that matter). This idea was first suggested by the biologist Julian Huxley (1887–1975); it was discussed by the physicist James Jeans (1877–1946) in his *Mysterious Universe* (1930); and it has been restated in various forms over the years, in terms of chimpanzees, Shakespeare's sonnets, and the like. It was also the subject of Russell Maloney's short story "Inflexible Logic," first published in the *New Yorker* magazine (1940) and reprinted in Clifton **Fadiman**'s *Fantasia Mathematica*,[96] which tells the tragic tale of what happens when the fantastically improbable comes true. See also **Universal Library**.

monochromatic triangle

A triangle whose vertices are all colored the same (see **vertex**).

monomial

An algebraic expression that consists of just one term.

monotonic

The property of a **function** that is always strictly increasing or strictly decreasing, but never both.

Monster curve

See **Peano curve**.

Monster group

The largest, most fascinating, and most mysterious of the so-called sporadic **groups**; it was constructed by Robert Griess at Princeton in 1982, having been predicted to exist by him and Bernd Fischer in 1973, and was named the Monster by John **Conway**. Think of the Monster group as a preposterous snowflake with more than 1,050 symmetries that exists in a space of 196,883 dimensions. It contains the following number of elements:

$$2^{46} \times 3^{20} \times 5^9 \times 7^6 \times 11^2 \times 13^3 \times 17 \times 19 \times 23 \times 29 \times 31$$
$$\times 41 \times 47 \times 59 \times 71$$
$$= 808{,}017{,}424{,}794{,}512{,}875{,}886{,}459{,}904{,}961{,}710{,}757{,}$$
$$005{,}754{,}368{,}000{,}000{,}000$$
$$\approx 8 \times 10^{53} \text{ (more than the number of quarks in the Sun)}.$$

Despite these impressive credentials, however, it is still classified as a *simple group*, meaning that it doesn't have any normal subgroups other than the identity element and itself. All 26 simple groups have now been classified and the Monster is far and away the biggest. At first, it seemed that the Monster was just a curiosity–a Guinness Book record of pure math. Its only "useful" application seemed to be to give the best way for packing spheres in 24 dimensions! In ordinary three-dimensional space (and also four and five dimensions), the grocer's way of stacking oranges in a hexagonal lattice is thought to be the tightest possible (see **Kepler's conjecture**). But as the number of dimensions increases, the optimal packing method changes. A 24-dimensional grocer would get the most efficient arrangement of his 24-dimensional oranges by using the same symmetry as that of the Monster. This is unlikely to be immediately useful. Much more interesting, however, is the connection that has been found between the symmetry of the Monster and one of the most promising unifying theories in physics–**string theory**–which has been revealed by the **Monstrous Moonshine conjecture**.[70]

Monstrous Moonshine conjecture

An outrageous idea that stemmed from an observation made by John McKay of Concordia University in 1978. McKay was leafing through a table of abstruse mathematical data, giving possible values for coefficients of the j-function of certain elliptic curves, when he noticed the number 196,884 in the expression $j(q) = q - 1 + 196884q + 21493760q^2 + \ldots$. In a moment of inspiration, he recognized this number as being one more than the number of dimensions in which the **Monster group** can be most simply represented. Looking into this "coincidence" more closely, he found that it was no coincidence at all. In fact, *all* the coefficients of the j-function were simple combinations of the degrees of possible representations of the Monster. This pointed to some deep connection between two seemingly unrelated areas of mathematics. On the one hand were the coefficients of what is called an elliptic modular function–exactly the kind of function that would play a key role in the proof of **Fermat's last theorem**. On the other was the number of dimensions, and combinations of degrees, of a crystal lattice whose symmetry rotations and reflections formed the Monster. Subsequently, McKay and a few other mathematicians, including John **Conway** and Simon Norton, drew out the link between elliptic modular functions and the Monster in a proposition christened, because of its fantastic nature, the Monstrous Moonshine. In 1998, this conjecture was proved by Richard Borcherds (a former student of Conway's) at the University of California at Berkeley. Astonishingly, Borcherds's proof reveals a deep relationship between elliptic curves, the Monster Group, and **string theory**–the most promising theory on offer to unify our understanding of nature at the subatomic level. Borcherds showed that the Monster is the group of symmetries of 26-dimensional strings expressed in a form known as *vertex algebra*. Some people believe the connection may run even deeper and that Monstrous Moonshine may hold clues to the very existence of the reality in which we live.[69]

Monte Carlo method

A method of estimating the true value of a quantity by carrying out a lot of **random** samples. For example, suppose we want to know the probability of getting a double six when we roll two dice. We could roll a pair of dice a thousand times, and count how many times, n, a double six came up; the estimated probability would then be $n/1{,}000$. A famous example of using the Monte Carlo method is to calculate **pi**. Get a computer to generate two random numbers x and y, each in the range -1 to 1, so that the point (x, y) lies somewhere randomly inside a square of side 2 units. Do this thousands of times, and count up what proportion of the points also lie inside the circle that inscribes

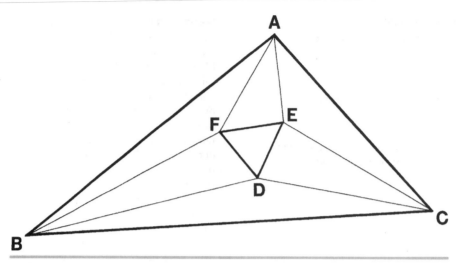

Morley's miracle The trisectors of the angles of triangle *ABC* meet the corners of a inner equilateral triangle *FED* known as Morley's triangle.

the square (you can tell whether a point does or not by working out whether $x^2 + y^2 < 1$). The proportion inside the circle is an approximation to $\pi/4$ (because the circle has area π but the square has area 4); millions of points are needed to obtain a good estimate.

This method was developed by researchers working on the Manhattan Project during World War II. To answer some of their scientific questions, they would repeatedly sample from their best estimates of the partial results, then apply the math they knew to the interactions and study the range of results. This process, which they named after the famous Monaco casino town of Monte Carlo, was created by John **von Neumann** and Stanislaw **Ulam**. The term and a description of the method seems not to have been published until some time after the war.

Montucla, Jean Etienne (1725–1799)

A French writer, mathematician, and scientist, who wrote several important early works on the history of mathematics. His *Histoire des mathématiques* (1758) was published in two volumes, the first of which covers the subject from ancient times to 1700, while the second is entirely devoted to seventeenth-century mathematics. It is considered the first attempt at a history of mathematical ideas and problems, in contrast to earlier works that were mostly lists of names, titles, and dates. Montucla had intended to produce a third volume covering the first half of the eighteenth century but the amount of new developments that had appeared during this time,

and the difficulties of putting recent work into its historical context, led him to abandon this aim. A few years later he published another text for which he is famed—a new, greatly expanded and improved edition of Jacques **Ozanam**'s *Récréations mathématiques et physiques* (1778). Montucla's edition was particularly influential in popularizing geometric **dissection** problems. Charles Hutton translated it into English in 1803 and Riddle's edition was published in 1844, called *Recreations in science and natural philosophy*.

Monty Hall problem

A puzzle in probability that was inspired by the American game show *Let's Make a Deal*, hosted by Monty Hall. In its original form it goes like this: at the end of the show, you, the player, are shown three doors. Behind one of them is a new car, behind the other two are goats. Monty knows where the car is, but you don't. You choose a door. Before that door is opened however, Monty opens one of the two other doors with a goat behind it. He then gives you the option of switching to the other closed door. Should you switch or stick? At first glance, it seems as if it shouldn't make any difference. But the answer is surprising.

Suppose you stick. Your original choice made when all three doors were equally likely gives you a probability of winning the car of $\frac{1}{3}$. Now suppose you switch. In other words, you choose a door, wait for Monty to expose a goat, then switch to the other remaining door. This means that you win if the door you chose to begin with had a goat behind it. The odds that your initial choice had a goat is two thirds, so you are twice as likely to win the car if you switch! This can be hard to grasp. To make it easier, suppose there are 100 doors to choose from, but still only one car. You pick a door, Monty opens 98 that have goats behind them, then he gives you the option of switching to the other remaining closed door. Should you? Of course—it's almost certain that the car is behind the other door, and very unlikely that it's behind your original choice.

In a generalization of the original problem there are n doors. In the first step, you choose a door. Monty then opens some other door that's a loser. If you want, you may then switch your choice to another door. Monty will then open an as yet unopened losing door, different from your current preference. Then you may switch again, and so on. This carries on until there are only two unopened doors left: your current choice and another one. How many times should you switch, and when, if at all? The answer is: stick all the way through with your first choice but then switch at the very end.

In another variation of the problem, consider that in the actual game show there were two contestants. Both of them were allowed to pick a door but not the same one. Monty then eliminated a player with a goat behind his door (if both players had a goat, one was eliminated randomly, without letting either player know about it), opened the loser's door, and then offered the remaining player a chance to switch. Should the remaining player switch? The answer is *no*. The reason: a switcher in this game will lose if and only if either of two initial choices of the two contestants was correct. How likely is that? Two-thirds. A sticker will win in those $2/3$ of the cases. So stickers will win twice as often as switchers.

Morley's miracle

A remarkable theorem, discovered in 1899, by Frank Morley, then professor of Mathematics at Haverford College. Take any triangle. Mark the three points that are the intersections of adjacent angle trisectors. Then, no matter what triangle you start with, these three points will form an equilateral triangle. That such a simple and elegant result was not known to the ancient Greeks may be because it is quite hard to prove.

One of the interesting auxiliary results of some of the proofs is that the side of the equilateral triangle is equal to $8r\sin(A/3)\sin(B/3)\sin(C/3)$, where A, B, and C are the angles of the larger triangle, and r is the radius of the circumcircle. A surprise awaits anyone who takes the intersections of the exterior, as well as the interior, angle trisectors. In addition to the interior equilateral triangle, four exterior equilateral triangles appear, three of which have sides that are extensions of a central triangle.

Moscow papyrus
See **Rhind papyrus**.

mousetrap
See **Cayley's mousetrap**.

moving sofa problem
In Douglas Adams's book *Dirk Gently's Holistic Detective Agency*, the character Richard MacDuff says at one point, "It would be really useful to know before you buy a piece of furniture whether it's actually going to fit up the stairs or around the corner." Mathematicians call this the moving sofa problem and it has been tackled in various forms over the past few decades. One version of it, formulated by Leo Moser in 1966, asks: What is the largest sofa (in terms of area) that can be moved around a right-angled corner in a hallway of unit width? The sofa can be any shape and doesn't even have to resemble a piece of furniture! The question simply asks for the biggest, unbendable area that can be maneuvered around the corner. Several different approaches suggest that the answer is about 2.21 square units. Variations on the problem involve negotiating pianos and other items around different types of bends and passageways.[133]

Mrs. Perkins's quilt
A square **dissection** problem first posed by Henry **Dudeney** in his *Amusements in Mathematics* (1917):[88]

PUZZLE
It will be seen that in this case the square patchwork quilt is built up of 169 pieces. The puzzle is to find the smallest possible number of square portions of which the quilt could be composed and show how they might be joined together. Or, to put it the reverse way, divide the quilt into as few square portions as possible by merely cutting the stitches.
Solutions begin on page 369.

Dudeney's problem can be generalized to the dissection of a square of side n into a number S_n of smaller squares. Unlike a perfect **squaring the square** problem, the smaller squares needn't be all different sizes. In addition, only prime dissections are considered so that patterns that can be dissected on lower order squares aren't allowed. The smallest number of relatively prime dissections of an $n \times n$ quilt for $n = 1, 2, \ldots$, are 1, 4, 6, 7, 8, 9, 9, 10, 10, 11, 11, 11, 11, 12, ...[66]

Müller-Lyer illusion
A **distortion illusion** in which the orientation of arrowheads makes one line segment look longer than another.

Mrs. Perkins's quilt Dudeney's drawing of one of his most famous puzzles.

multigrade

A set of equations in which the sums of powers of two different sets of numbers are the same for several different exponents. The simplest example is:

$$1 + 6 + 8 = 2 + 4 + 9$$
$$1^2 + 6^2 + 8^2 = 2^2 + 4^2 + 9^2$$

Another multigrade is:

$$1 + 8 + 10 + 17 = 36 = 2 + 5 + 13 + 16$$
$$1^2 + 8^2 + 10^2 + 17^2 = 454 = 2^2 + 5^2 + 13^2 + 16^2$$
$$1^3 + 8^3 + 10^3 + 17^3 = 6426 = 2^3 + 5^3 + 13^3 + 16^3$$

Remarkably, if any integer is added to all the terms of a multigrade it will still hold. Adding 1 to the example above, gives the multigrade (2, 9, 11, 18); (3, 6, 14, 17) ($n = 1, 2, 3$). Some high-order multigrades include: (1, 50, 57, 15, 22, 71); (2, 45, 61, 11, 27, 70); (5, 37, 66, 6, 35, 67) ($n = 1, 2, 3, 4, 5$), and (1, 9, 25, 51, 75, 79, 107, 129, 131, 157, 159, 173); (3, 15, 19, 43, 89, 93, 97, 137, 139, 141, 167, 171) ($n = 1, 3, 5, 7, 9, 11, 13$).

Müller–Lyer illusion   Which of the horizontal lines is longer?

multiple

The integer b is a multiple of the integer a if there is an integer d such that $b = da$.

multiplication

A **binary operation** that is the equivalent of repeated addition and the **inverse** of **division**.

Mydorge, Claude (1585–1647)

A French mathematician who trained as a lawyer but was wealthy enough that he didn't have to work for a living. He was interested in mathematical puzzles and his book *Examen du livre des récréations mathématiques* (Study of the book of recreational mathematics, 1630) formed the basis for later works such as that by Denis Henrion (1659). Mydorge edited *Récréations mathématique* and left an unpublished manuscript of over 1,000 geometric problems and their solutions. He was also interested in optics and made a large number of instruments for his close friend René **Descartes**; the two shared a strong interest in explaining vision and the instruments and lenses were designed to help test their theories.

myriad

A term which today is normally synonymous with "very large number." Its origins go back to the Greek word *murious,* meaning "uncountable". The plural of this, *murioi,* evolved into the Latin *myriad,* which the Romans used to represent ten thousand. *Myriapod* is a general name for any many-legged anthropod, such as a millipede or a centipede

N

Nagel point

A point in a triangle where the lines from the vertices (see **vertex**) to the points of contact of the opposite sides with the excircles to those sides meet. (An excircle touches one side of a triangle and also touches the lines extended from the other two sides.)

nano-

Prefix for billionth (10^{-9}), from the Greek *nanos,* meaning "dwarf."

Napier, John (1550–1617)

A Scottish mathematician and theological writer who invented **logarithms** and wrote *Mirifici logarithmorum canonis descriptio* (The description of the wonderful canon of logarithms, 1614), which contains the first logarithmic table and the first use of the word *logarithm*. He also introduced the decimal point in writing numbers. His *Rabdologiae* (1617) describes various shortcuts for carrying out arithmetical calculations. One method of multiplication uses a system of numbered rods called *Napier's rods*, or *Napier's bones*—a major improvement on the ancient system of counters then in use. In 1619, after Napier's death, his *Mirifici logarithmorum canonis constructio,* which gave the method of construction of his logarithms, was published by his son Robert and edited by Henry **Briggs**.

Napoleon Bonaparte (1769–1821)

> *The advancement and perfection of mathematics are intimately connected with the prosperity of the state.*

Emperor of France and a very good amateur mathematician, having excelled in this subject as a student at school and at military college. Even after becoming first consul he was proud of his membership in the Institute de France (the nation's leading scientific society), and was close friends with several mathematicians and scientists, including Joseph **Fourier**, Gaspard **Monge**, Pierre Simon **Laplace**, Chaptal, and Berthollet. Indeed, in his grand expedition to Egypt in 1798 Napoleon brought along (in addition to 35,000 troops) over 150 experts in various fields, among them Monge, Fourier, and Berthollet, not to mention a complete *encyclopedie vivante* with libraries and instruments. One result of the expedition was that Fourier served for a time as the gov-

ernor of lower Egypt. Likewise Laplace (who interviewed the young Napoleon for admission to the artillery) received titles and high office as a result of his friendship with Bonaparte. However, Laplace was relieved of his duties as the minister of the interior after only six weeks, and Napoleon later commented that Laplace had "sought subtleties everywhere, had only doubtful ideas, and carried the spirit of the infinitely small into administration." The most famous exchange between these two men occurred after Laplace had given Napoleon a copy of his great work, *Mecanique Celeste.* Napoleon looked it over, and remarked that in this massive volume about the universe there was not a single mention of **God**. Laplace replied "Sire, I had no need of that hypothesis." Regarding the idea that Napoleon might have discovered what is now called *Napoleon's theorem* (if equilateral triangles are constructed on the sides of any triangle (all outward or all inward), the centers of these equilateral triangles themselves form an equilateral triangle), Harold **Coxeter** and Samuel Greitzer have said that "The possibility of [Napoleon] knowing enough geometry for this feat is as questionable as the possibility of his knowing enough English to compose the famous palindrome, ABLE WAS I ERE I SAW ELBA."

nappe

Either of the two parts into which a **cone** is divided by the **vertex**. "Nappe" is the French for "tablecloth," which in turn comes from the Latin *mappa* (napkin).

narcissistic number

Also known as an *Armstrong number* or a *plus perfect number,* an n-digit number equal to the sum of its digits raised to the nth power. For instance, 371 is narcissistic because $3^3 + 7^3 + 1^3 = 371$, and 9474 is narcissistic because $9^4 + 4^4 + 7^4 + 4^4 = 9474$. The smallest narcissistic number of more than one digit is $153 = 1^3 + 5^3 + 3^3$. The largest narcissistic number (in base 10) is 115,132,219,018,763,992,565,095,597,973,971,522,401, which is the sum of the thirty-ninth powers of its digits. The reason there are no larger numbers is related to the fact that, as the number of digits increases, more and more nines are required to get a sum that has n digits. For example, $10^{70} - 1$ is a number consisting of 70 nines in a row, and the sum of the seventieth powers of its digits is $70 \times 9^{70} \cong 4.386051 \times 10^{68}$, which is only 69 digits long. So there is no way any 70-digit number can be equal to the sum of the seventieth powers of its digits. The reason we see the last number occur at 39 digits is because, as the limit is approached, the number of big digits like eights and nines has to increase to make sure the sum will be big enough, but this means that there are a lot fewer combinations of digits to choose from.

Nash, John Forbes Jr. (1928–)

An American mathematician not very accurately portrayed in the Oscar-winning film *A Beautiful Mind* (2001), loosely based on the biography of the same name by Sylvia Nasar (1998).[229] Nash, who worked in **game theory** and **differential geometry**, shared the 1994 Nobel prize for economics with two other game theorists, Reinhard Selten and John Harsanyi. After a promising start to his mathematical career, Nash began to suffer from schizophrenia around the age of 30 and battled with the illness for the next quarter of a century. His Ph.D. dissertation, entitled "Non-cooperative Games," contained the definition and properties of what would later be called **Nash equilibrium** and the basis of the work that, 44 years later, would make him a Nobelist. Between 1966 and 1996, Nash published nothing. However, as his mental health slowly began to improve in the mid-1990s, his ability to tackle mathematical problems returned, and he also became interested in computer programming.

Nash equilibrium

In **game theory**, a pair of strategies (see **strategy**) for a game such that neither player can improve his outcome by changing his strategy. A Nash equilibrium sometimes takes the form of a **saddle** structure. In other cases, when a strategy is at a Nash equilibrium with itself, the strategy resembles an evolutionary stable strategy.

natural logarithm

Also called a *Naperian logarith,* a **logarithm** to base **e**. For example, $\log_e 10$ (also written as $1n\ 10$) is approximately 2.30258.

natural number

A number used for counting: 1, 2, 3, ... The debate about whether zero should also be included as a natural number has been going on for hundreds of years, and there's no general agreement even today. To avoid confusion, 0, 1, 2, 3, . . . , are often referred to as *nonnegative integers* or *whole numbers,* while 1, 2, 3, . . . , are called *positive integers.* Adding or multiplying natural numbers always produces other natural numbers. However, sub-

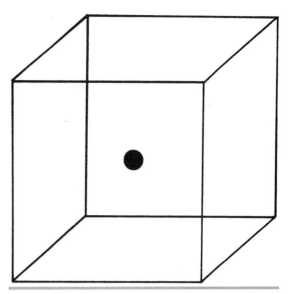

Necker cube   Is the circle on the front face of the cube or the back?

tracting them can produce zero or negative integers, while dividing them produces **rational numbers**. An important property of the natural numbers is that they are *well-ordered,* in other words, every set of natural numbers has a smallest element. The deeper properties of the natural numbers, such as the distribution of **prime numbers**, are studied in **number theory**. Natural numbers can be used for two purposes: to describe the position of an element in an ordered sequence, which is generalized by the concept of **ordinal number**, and to specify the size of a finite set, which is generalized by the concept of **cardinal number**. In the finite world, these two concepts coincide; however, they differ when it comes to infinite sets (see **infinity**).

Necker cube

A classic example of an **ambiguous figure**. In 1832, the Swiss crystallographer Louis Necker noticed, while examining crystals, that three-dimensional objects can fluctuate in appearance. He published pictures of an unusual **cube** that appeared to assume different orientations as one looked at it. The effect works because the drawing of the cube (an orthographic projection) carefully eliminates all depth cues. In attempting to fit the expected model of a cube to the picture, our brain must resolve the ambiguity as to which corner of the cube is closer.

negative base

The use of a negative **base** to represent numbers gives rise to some intriguing possibilities. Consider "negadecimal," for example, in which the base is *minus* 10 instead of the familiar positive 10. In this system, the number 365 is equivalent to the decimal number 5 + (6 × −10) + (3 × −10 × −10), = 245, while 35 in negadecimal is equivalent to 5 + (3 × −10), = −25, in ordinary decimal. This points to an interesting fact: the negadecimal equivalent of any positive or negative decimal number is always positive and therefore doesn't need to be accompanied by a sign. The Polish UMC-1, of which a few dozen were built in the late 1950s and earlier 1960s, is the only computer ever to use "nega**binary**" (base 2 arithmetic).

negative number

Long denied legitimacy in mathematics, negative numbers are nowhere to be found in the writings of the Babylonians, Greeks, or other ancient cultures. On the contrary, because Greek mathematics was grounded in geometry, and the concept of a negative distance is meaningless, negative numbers seemed to make no sense. They surface for the first time in bookkeeping records in seventh-century India and in a chapter of a work by the Hindu astronomer **Brahmagupta**. Their earliest documented use in Europe is in 1545 in the *Ars magna* of Girolamo **Cardano**. By the early seventeenth century, Renaissance mathematicians were explicitly using negative numbers but also meeting with heavy opposition. René **Descartes** called negative roots "false roots," and Blaise **Pascal** was convinced that numbers "less than zero" couldn't exist. Gottfried **Leibniz** admitted that they could lead to some absurd conclusions, but defended them as useful aids in calculation. By the eighteenth century, negative numbers had become an indispensable part of algebra.

Neile's parabola

Also known as the semi-cubical **parabola**, a curve discovered by the English mathematician William Neile (1637–1670) in 1657; it is the first **algebraic curve** to have its arc length calculated. (Before this, only the arc lengths of transcendental curves such as the **cycloid** and the **logarithmic spiral** had been calculated.) Neile's parabola is described in **Cartesian coordinate**s by the formula

$$y^3 = ax^2.$$

Christiaan **Huygens** showed that this curve satisfies the requirement requested by Gottfried **Leibniz** in 1687, namely, the curve along which a particle may descend under gravity so that it moves equal vertical distances in equal times. Neile's parabola is the **evolute** of a parabola.

nephroid

A type of curve often seen on the surface of a cup of coffee in the sunshine–a crescent of light formed by sun-

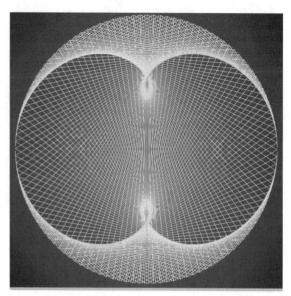

nephroid   A nephroid curve spun by thread on a computer loom. *Jos Leys, www.josleys.com*

light reflecting off the inside of the cup onto the surface of the drink. More generally, it is the shape made by parallel rays of light reflecting from the inside of any semicircle. In mathematical terms, this means that the nephroid is the cata**caustic** of a **circle** when the light source is at infinity, a fact first demonstrated by Christiaan **Huygens** in 1678 and published by him in his *Traité de la Lumière* (Treatise on light, 1690). A physical explanation wasn't forthcoming, however, until 1838 when George Airy gave a proof in terms of the wave theory of light. The name *nephroid* (from the Latin for "kidney-shaped") was introduced in 1878 by the English mathematician Richard Proctor in his book *The Geometry of Cycloids*. Prior to that it was known as a two-cusped **epicycloid**. Specifically, the nephroid is the epicycloid formed by a circle of radius *a* rolling around the outside on a fixed circle of radius 2*a*. It has a length of 24*a*, an area of $12\pi^2$, and is given by the parametric equations:

$$x = a(3\cos(t) - 3\cos(3t))$$
$$y = a(3\sin(t) - \sin(3t)).$$

The nephroid is the **involute** of **Cayley's sextic** and is also the **envelope** of circles with their centers on a given circle, touching a given diameter of that circle. The nephroid has been described as the perfect shape for a multiseat dining table. *Freeth's nephroid*, not to be mistaken for the ordinary nephroid just described, is named after the English mathematician T. J. Freeth (1819–1904) who first wrote about it in a paper published by the London Mathematical Society in 1879. Freeth's nephroid is the **strophoid** of a circle and has the polar equation $r = a(1 + 2\sin(\theta/2))$. Freeth's nephroid is also the name of a group of mathematicians, mostly from Royal Holloway College, London, who gather weekly in a pub called the Beehive and compete in games of trivial pursuit.

net

A drawing of a **polyhedron** unfolded along its **edges**, so as to lie flat in a plane. The earliest known examples of nets to represent polyhedra are by Albrecht **Dürer**.

neural network

An electronic automaton, similar in some ways to a **cellular automaton**, that offers a highly simplified model of a brain. As such, a neural network is a device for machine learning that is based on associative theories of human cognition. Using various algorithms and weightings of different connections between "neurons," neural networks are set up to learn how to recognize a pattern in applications such as voice recognition, visual pattern recognition, robotic control, symbol manipulation, and decision making. Generally, they consist of three layers: input neurons, output neurons, and a layer in between where information from input to output is processed. Initially the network is loaded with a random program, then the output is measured against a desired output which prompts an adjustment in the weights assigned to the connections in response to the discrepancy between the actual and desired output. This is repeated many times so that the network effectively learns as a child does: in a sense, the net discovers its own rules. Changing the rules of interaction between the "neurons" in the net can lead to interesting emergent behavior, so that neural networks have become another tool for investigating **emergence** and **self-organization**.

Neusis construction

A geometric construction that breaks the strict rules of classical Greek straightedge-and-compass construction (see **constructible**) by allowing a marked ruler to be slid into different positions. Neusis construction makes possible **duplicating the cube** and **trisecting an angle**. John **Conway** and Richard **Guy** have also shown how Neusis constructions, based on angle trisection, can be used to draw regular polygons with 7, 9, and 13 sides.

Newcomb's paradox

One of the most simply stated but astonishing of the so-called prediction paradoxes that bear on the problem of free will. It was devised in 1960 by William Newcomb, a theoretical physicist at the Lawrence Livermore Laboratory, while contemplating the **prisoner's dilemma**. A superior being, with super-predictive powers that have never been known to fail, has put \$1,000 in box A and either nothing or \$1 million in box B. The being presents you with a choice: (1) open box B only, or (2) open both box A and B. The being has put money in box B only if it predicted you will choose option (1). The being put nothing in box B if it predicted you will do anything other than choose option (1) (including choosing option (2), flipping a coin, etc.). The question is, what should you do to maximize your winnings? You might argue that since your choice now can't alter the contents of the boxes you may as well open them both and take whatever's there. This seems reasonable until you bear in mind that the being has never been known to have made an incorrect prediction. In other words, in some peculiar way, your mental state is highly correlated with the contents of the box: your choice is linked to the probability that there's money in box B. These arguments and many others have been put forward in favor of either choice. The fact is there is no known "right" answer, despite the concerted attentions of many philosophers and mathematicians over several decades.[123]

Newton, Isaac (1642–1727)

One of the great intellectual giants in human history; an English mathematician, physicist, and sometime head of the Royal Mint, Newton was one of the chief architects of **calculus** (though not in its modern form), discovered the **binomial theorem**, established the principles of universal gravitation, and saw through numerous other developments, any one of which would have brought a lesser person fame. Interestingly, his most productive period was 1665–1666, his so-called "miraculous year," when Cambridge University was closed because of the plague and Newton had to work at home. He was born in the same year that Galileo died. After publishing his *Principia* (1687), the most important and influential scientific book ever written, his interests drifted toward theology, politics, and, for a while, alchemy. His last two decades were largely spent in acrimonious debate with Gottfried **Leibniz**, over priority in the discovery of calculus, and with the astronomer John Flamsteed.

Newton's method

An iterative method for finding the zeros of a **function**.

n-gon

A **polygon** with n sides.

Nim

A **game**, of which there are many different versions, that involves two players alternately removing at least one item from one of two or more piles or rows. The person who picks up the last item wins. In one form of the game, five rows of matches are laid out in such a way that there is one match in the first row, two matches in the second, and so on, down to five matches in the bottom row. Players take turns to remove any nonzero number of matches from any one row. The game may have originated in China. The name "Nim" was coined by Charles Bouton, an associate professor of mathematics at Harvard at the turn of the twentieth century, who took it from an archaic English word meaning to steal or to take away. In 1901 he published a full analysis of Nim and proof of a winning strategy.[50] The first Nim-playing computer, the Nimatron, a 1-ton behemoth, was built in 1940 by the Westinghouse Electrical Corporation and was exhibited at the New York World's Fair. It played 100,000 games against spectators and attendants, and won an impressive 90% of the time; most of its losses came at the hands of attendants who were instructed to reassure incredulous onlookers that the machine could be beaten! In 1951 a Nim-playing robot, the Nimrod, was shown at the Festival of Britain, and later at the Berlin trade fair. It was so popular that spectators entirely ignored a bar at the other end of the room where free drinks were being offered. Eventually the local police had to be called in to control the crowds.[110]

nine

A number long considered to have strange, mystic properties. A phrase in a book written during the Dark Ages gave rise to the superstition that cats have nine lives. English author and satirist William Baldwin wrote in his *Beware the Cat,* "It is permitted for a witch to take her cat's body nine times." There were nine Muses, nine rivers of Hades, and nine heads on the Hydra. It took nine days for Vulcan to fall from the heavens. The phrase "nine days' wonder" comes from the proverb "a wonder lasts nine days and then the puppy's eyes are open." A cat-o'-nine-tails is a whip, usually made of nine knotted lines or cords fastened to a handle that produces scars like the scratches of a cat. Being on "cloud nine" may have its origin in Dante's ninth heaven of Paradise, whose inhabitants are blissful because they are closest to God.

The term "the whole 9 yards" came from World War II fighter pilots in the Pacific. When arming their planes on the ground, the .50-caliber machine gun ammo belts measured exactly 27 feet, before being loaded into the fuselage. If the pilots fired all their ammo at a target, it got "the whole 9 yards." Less certain—though there is no shortage of theories—is the source of the expression "dressed to the nines."

Nine is the largest single-digit number and the one that occurs least frequently in most situations; an exception is the tendency of businesses to set prices that end with one or more nines. Because nine is one less than the base of

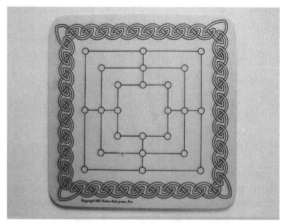

nine men's morris A modern version of the playing board.
Kadon Enterprises, Inc., www.puzzlegames.com

our number system, it is easy to see if a number is divisible by 9 by adding the digits (and repeating on the result if necessary: the result should be nine). This process is sometimes called **casting out nines**. Similar processes can be developed for divisibility by 99, 999, etc. or any number that divides one of these numbers. Nine has many other interesting properties. For example, write down a number containing as many digits as you like, add these digits together, and deduct the sum from the first number. The sum of the digits of this new number will always be a multiple of nine.

nine holes
See **three men's morris**.

nine men's morris

> *The nine-men's morris is filled up with mud;*
> *And the quaint mazes in the wanton green,*
> *For lack of tread, are indistinguishable.*
> —*Midsummer Night's Dream* (Act 2, scene 1),
> Shakespeare

One of the oldest of board games, known by different names and played with variations of rules in different places and periods. In France it is Marelle, in Austria it is Muhle, and in England it was known as Peg Meryll, Meg Marrylegs, and other names, all referring to a "mill" because that is the name of a run of three counters in the game. Versions of it have been found etched into the roof of the Temple of Kurna in Egypt (dated to about 1400 B.C.), cut into the oak planks that form the deck of the great Viking ship discovered at Gokstad in 1880, and carved in the choir stalls of several English cathedrals. A typical board layout is shown in the accompanying figure; diagonal lines may or may not be included. Joseph Strutt in *The Sports and Pastimes of the People of England* (1801) described the rules in this way:

> Two persons, having each of them nine pieces, or men, lay them down alternately, one by one upon the spots; and the business of either party is to prevent his antagonist from placing three of his pieces so as to form a row of three, without the intervention of an opponent piece. If a row be formed, he that made it is at liberty to take up one of his competitor's pieces from any part he thinks most to his advantage; excepting he has made a row, which must not be touched if he have another piece upon the board that is not a component part of that row. When all the pieces are laid down, they are played backwards and forwards, in any direction that the lines run, but only can move from one spot to another (next to it) at one time. He that takes off all his antagonist's pieces is the conqueror.

In 1996, the German mathematician Ralph Gasser used a computer to prove that nine men's morris is a guaranteed draw if both players make optimal moves from the outset. He programmed the computer to figure out and tabulate 10 billion positions that were known to be a win for one side or the other, then worked forward 18 moves from the beginning of the game until his opening analysis met his endgame analysis. As a result he showed that every potentially winning position could be countered by the opponent in the early stages of the game. See also **three men's morris**.[132]

nine rooms paradox

A puzzle that was first published in *Current Literature* vol. 2, April 1889. It takes the form of a poem and is similar to that of the **missing dollar**.

PUZZLE

> Ten weary, footsore travellers,
> All in a woeful plight,
> Sought shelter at a wayside inn
> One dark and stormy night.
>
> 'Nine rooms, no more,' the landlord said
> 'Have I to offer you.
> To each of eight a single bed,
> But the ninth must serve for two.'
>
> A din arose. The troubled host
> Could only scratch his head,
> For of those tired men not two
> Would occupy one bed.

> The puzzled host was soon at ease—
> He was a clever man—
> And so to please his guests devised
> This most ingenious plan.
>
> In a room marked A two men were placed,
> The third was lodged in B,
> The fourth to C was then assigned,
> The fifth retired to D.
>
> In E the sixth he tucked away,
> In F the seventh man.
> The eighth and ninth in G and H,
> And then to A he ran,
>
> Wherein the host, as I have said,
> Had laid two travellers by;
> Then taking one—the tenth and last—
> He logged him safe in I.
>
> Nine single rooms—a room for each—
> Were made to serve for ten;
> And this it is that puzzles me
> And many wiser men.

How has the host managed to bamboozle his patrons?

Solutions begin on page 369.

---

nine-point circle

Draw a triangle, any triangle (although it may be best to start with an acute triangle). Mark the midpoints of each side. Drop an altitude from each vertex to the opposite side, and mark the points where the altitudes intersect the opposite side. (If the triangle is obtuse, an altitude will be outside the triangle, so extend the opposite side until it intersects.) Notice that the altitudes intersect at a common point. Mark the midpoint between each vertex and this common point. No matter what triangle you start

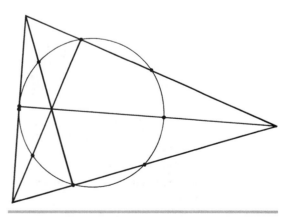

nine-point circle

with, these nine points all lie on a perfect circle! This result was known to Leonhard **Euler** in 1765, but was rediscovered by the German mathematician Karl Feuerbach (1800–1834) in 1822.

node
See **crunode**.

Noether, Emmy (Amalie) (1882–1935)
A German mathematician, one of the most talented of the early twentieth century. The crucial result now known as *Noether's theorem,* which is important in other symmetries in natural systems, is of great importance in physics. She received her doctorate in 1907 and rapidly built an international reputation, but the University of Göttingen refused to let her teach, and her colleague David **Hilbert** had to advertise her courses in the university's catalog under his own name. A long controversy ensued, with her opponents asking what the country's soldiers would think when they returned home and were expected to learn at the feet of a woman. Allowing her on the faculty would also mean letting her vote in the academic senate. Said Hilbert, "I do not see that the sex of the candidate is against her admission as a privatdozent. After all, the university senate is not a bathhouse." She was finally admitted to the faculty in 1919. A Jew, Noether was forced to flee Nazi Germany in 1933 and joined the faculty at Bryn Mawr in the United States.

non-Abelian
Non**commutative** or order-dependent. For example, the group of manipulations of **Rubik's cube** is non-Abelian because the state of the cube depends greatly on the order of the moves performed on it. See also **Abelian group**.

nonagon
A **polygon** with nine sides. A *nonagonal number* is a number of the form $n(7n - 5)/2$.

nonconvex uniform polyhedron
A **uniform polyhedron** of a type obtained by relaxing the conditions used to produce the **Archimedean solids** (which have regular convex faces and identical convex vertices) to allow both nonconvex faces and vertex types, as in the case of the **Kepler-Poinsot solids**. The condition that every vertex must be identical, but the faces need not be, gives rise to 53 nonconvex uniform polyhedra. An example is the *great truncated dodecahedron,* obtained by truncating the corners of the great dodecahedron at a depth which gives regular decagons.

non-Euclidean geometry
Any **geometry** in which Euclid's **parallel postulate** doesn't hold. (One way to state the parallel postulate is: given a straight line and a point A not on that line, there is only one exactly straight line through A that never intersects the original line.) The two most important types of non-Euclidean geometry are **hyperbolic geometry** and **elliptical geometry**. The different models of non-Euclidean geometry can have positive or negative **curvature**. The sign of curvature of a surface is indicated by drawing a straight line on the surface and then drawing another straight line perpendicular to it: both these lines are **geodesics**. If the two lines curve in the same direction, the surface has a positive curvature; if they curve in opposite directions, the surface has negative curvature. Elliptical (and spherical) geometry has positive curvature whereas hyperbolic geometry has negative curvature.

The discovery of non-Euclidean geometry had immense consequences. For more than 2,000 years, people had thought that **Euclidean geometry** was the only geometric system possible. Non-Euclidean geometry showed that there are other conceivable descriptions of space—a realization that transforms mathematics into an altogether more abstract science. Thereafter, it was clear that in mathematics, one could start out with *any* set of self-consistent

Types of Geometry

	Euclidean	Elliptical	Hyperbolic
Curvature	Zero	Positive	Negative
Given a line m and a point P not on m, the number of lines passing through P and parallel to m	1	0	Many
Sum of interior angles of a triangle	180°	> 180°	< 180°
Square of hypotenuse of a right triangle with sides a and b	$a^2 + b^2$	$< a^2 + b^2$	$> a^2 + b^2$
Circumference of a circle with diameter 1	π	$< \pi$	$> \pi$

postulates and follow through their ramifications. The discovery of non-Euclidean geometry has been compared with Copernicus's theory and Einstein's **relativity theory**, the analogy being that each freed people from long-held models of thought. In fact, Einstein said about non-Euclidean geometry: "To this interpretation of geometry, I attach great importance, for should I have not been acquainted with it, I never would have been able to develop the theory of relativity."

It's important to realize that both Euclidean and non-Euclidean geometry are consistent in that the assumptions on which they rest don't involve any contradictions. In response to a question as to which geometry is true, Henri **Poincaré** said: "One geometry cannot be more true than the other; it can only be more convenient." Which geometry is valid in the physical space in which we live? On a small scale, and for all practical purposes on Earth, Euclidean geometry works just fine. But on larger scales this is no longer true. Einstein's general theory of relativity uses non-Euclidean geometry as a description of **space-time**. According to this idea, space-time has a positive curvature near gravitating matter and the geometry is non-Euclidean. When a body revolves around another body, it appears to move in a curved path due to some force exerted by the central body, but it is actually moving along a **geodesic**, without any force acting on it. Whether all of space-time contains enough matter to give itself an overall positive curvature is one of the many unanswered question in physics today, but it is generally accepted that the geometry of space-time is more non-Euclidean than Euclidean. It is proposed that if space-time does happen to have an overall positive curvature, then the universe will stop expanding after a fixed amount of time and start to shrink resulting in a "big crunch," as opposed to the "big bang" that resulted in its creation. At the moment, astronomical observations seem to favor a universe that is "open" and has a hyperbolic geometry. Another consequence of non-Euclidean geometry is the possibility of the existence of a **fourth dimension**. Just as the surface of the sphere curves in the direction of the third dimension, i.e., perpendicular to its surface, it is believed that space-time curves in the direction of the fourth dimension. Non-Euclidean geometry has applications in other areas of mathematics, including the theory of elliptic curves, which was important in the proof of **Fermat's last theorem**.[72, 140] (See table, "Types of Geometry.")

nonlinear system

Any system in which the data points coming from the measurement of the values of its variables can be represented as a curvilinear pattern on a coordinate plane; hence, "nonlinear" for "not-a-line." More generally, a system in which small changes can result in large effects, and large changes in small effects. Thus, sensitive dependence on initial conditions (see **butterfly effect**) in chaotic systems illustrates the extreme nonlinearity of these systems. In a nonlinear system the components are interactive, interdependent, and exhibit feedback effects.

nonstandard analysis

In a broad sense, the study of the infinitely small; more specifically, the study of **hyperreal numbers**, their functions and properties. Nonstandard analysis, which was pioneered by Abraham **Robinson** in the 1960s, puts the concept of **infinitesimal**s on a firm mathematical footing and is, for many mathematicians, more intuitive than real **analysis**.

normal

A line that is perpendicular to a given line or plane.

normal number

A number in which digit sequences of the same length occur with the same frequency. A constant is considered normal to base 10 if any single digit in its decimal expansion appears one-tenth of the time, any two-digit combination one-hundredth of the time, any three-digit combination one-thousandth of the time, and so on. In the case of **pi**, the digit 7 is expected to appear 1 million times among the first 10 million digits of its decimal expansion. It actually occurs 1,000,207 times—very close to the expected value. Each of the other digits also turns up with approximately the same frequency, showing no significant departure from predictions. A number is said to be *absolutely normal* if its digits are normal not only to base 10 but also to every integer base greater than or equal to 2. In base 2, for example, the digits 1 and 0 would appear equally often. Émile **Borel** introduced the concept of normal numbers in 1909 as a way to characterize the resemblance between the digits of a mathematical constant such as π and a sequence of **random numbers**. He quickly established that there are lots of normal numbers, though finding a specific example of one proved to be a major challenge. The first to be found was **Champernowne's number**, which is normal to base 10. Analogous normal numbers can be created for other bases. To date, no specific "naturally occurring" **real number** has been proved to be absolutely normal, even though it is known that almost all real numbers are absolutely normal! However, in 2001, Greg Martin of the University of Toronto found some examples of the opposite extreme—real numbers that are normal to no base whatsoever. To start with, he noted that every

rational number is absolutely abnormal. For example, the fraction $^1/_7$ can be written in decimal form as 0.1428571428571.... The digits 142857 just repeat themselves. Indeed, an expansion of a rational number to any base b or b^k eventually repeats. Martin then focused on constructing a specific *irrational* absolutely abnormal number. He nominated the following candidate, expressed in decimal form, for the honor:

$$\alpha = 0.6562499999956991\underline{999999\ldots999999}8528404201690728\ldots$$

The middle portion (underlined) of the given fragment of α consists of 23,747,291,559 nines. Martin's formulation of this number and proof of its absolute abnormality involved so-called **Liouville numbers**.[216]

nothing

The absence of anything; nonexistence. Nothing is not the same as the **empty set**, which exists as the set that mathematically denotes nothing, nor is it the same as **zero**, which exists as the number that denotes how many members the empty set contains. In physics, nothing is not a **vacuum**, because a vacuum not only contains energy but exists in space and time; nor is it a **singularity**, which contains a great deal of concentrated matter and energy. So, can there be nothing? No. "To be" implies existence of some sort: the one thing we can be absolutely sure has never existed, or will exist, is nothing.

noughts and crosses
See **tic-tac-toe**.

NP-hard problem

A mathematical problem for which, even in theory, no shortcut or smart **algorithm** is possible that would lead to a simple or rapid solution. Instead, the only way to find an optimal solution is a computationally intensive, exhaustive analysis in which all possible outcomes are tested. Examples of NP-hard problems include the **traveling salesman problem** and the popular game **Tetris**. NP stands for "non-deterministic polynomial-time."

nucleation

A process in a physical system, or a mathematical model such as a **cellular automaton** or a statistical model, whereby a bubble or other structure appears spontaneously at a random or unpredictable spot.

null hypothesis

The **hypothesis** that is being tested in a hypothesis-testing situation.

null set
See **empty set**.

number

An abstract measure of quantity. The most familiar numbers are the **natural numbers**, 0, 1, 2, ..., used for counting. If negative numbers are included, the result is the **integers**. Ratios of integers are called **rational numbers**, which can be expressed as terminating or repeating decimals. If all infinite and nonrepeating decimal expansions are thrown in as well, the scope of numbers extends to all **real numbers**, which can be extended to the **complex numbers** in order to include all possible solutions to algebraic equations. More recent developments are the **hyperreal numbers** and the **surreal numbers**, which extend the real numbers by adding infinitesimal and infinitely large numbers. For measuring the size of infinite **sets**, the natural numbers have been generalized to the **ordinal numbers** and to the **cardinal numbers**. See also **numeral** and **number system**.

number line

A way of representing numbers by thinking of them as the positions of points on a line.

number system

A way of counting using a particular **base**. The familiar **decimal** number system is a base 10 system, because it uses 10 different digits, 0, 1, 2, 3, 4, 5, 6, 7, 8, 9, and combinations of these. Other number systems, used for special purposes or by certain cultures, are the **binary** (base 2), trinary (base 3), hexadecimal (base 16), **vigesimal** (base 20), and **sexagesimal** (base 60). In a *positional number system* of base b, b basic symbols (or digits) corresponding to the first b **natural numbers**, including zero, are used. To generate the rest of the numbers, the position of the symbol in the figure is used. The symbol in the last position has its own value, and as it moves to the left its value is multiplied by b. In this way, with only finitely many different symbols, every number can be expressed. This is unlike systems, such as that of **Roman numerals**, which use different symbols for different orders of magnitude.

number theory

The study of the whole numbers and their properties and relationships. Often, a problem in number theory can be restated in terms of finding the solution, or showing that there is no solution, to a **Diophantine equation**. A Diophantine equation is one where the coefficients are whole numbers and where the solution is also constrained to be a whole number. Sometimes, what appears to be a simple Diophantine equation can lead to an **elliptic curve**.[30]

numeral

A symbol, or combination of symbols, that describes a number. See **Arabic numeral** and **Roman numeral**.

numerator

The number above the fraction bar that indicates the number of parts of the whole there are in a **rational number**.

numerical analysis

The study of methods for approximating the solutions of various classes of mathematical problems including *error analysis*.

obelus

The name of the symbol "÷", used as a sign for division. It comes from the Greek *obelos* meaning a pointed stick (a spit) used for cooking. This root word also gave rise to *obelisk* for a pointed stone pillar. The "÷" symbol was originally used as an editing mark in early manuscripts, sometimes only as a line without the two dots, to point out material that the editor thought needed cutting. It was also used occasionally as a symbol for subtraction. As a division symbol it was first employed by the Swiss mathematician Johann Rahn (1622–1676) in his *Teutsche Algebra* in 1659. By a misunderstanding of a credit to John Pell about other material in the book, many English writers started using the symbol and calling it "Pell's notation." Although it appears regularly in literature produced in Britain and the United States, it is virtually unknown in the rest of the world.

oblate spheroid

An **ellipsoid** produced by rotating an **ellipse** through 360° about its **minor axis**.

oblique

Slanted or not perpendicular. *Oblique coordinates* are measured on a plane coordinate system whose axes are not perpendicular. An *oblique angle* is any angle except a **right angle** and an *oblique triangle* is any triangle that doesn't contain a right angle. The first reference to "oblique angled triangles" may have been by Thomas Blundevil in his *Exercises* in 1594.

oblong

An alternate name for any **rectangle** that is not a square. The word comes from the Latin *ob* ("excessive") and *longus* ("long").

oblong number

Any positive integer that is not a **perfect square**.

obtuse

An *obtuse angle* is an **angle** greater than 90° but less than 180°. An *obtuse triangle* is a **triangle** that contains an obtuse angle. The word comes from the Latin *ob* ("against") and *tundere* ("to beat") and thus refers to things that are blunt, dull, or rounded.

octa-

The Greek prefix meaning "eight."

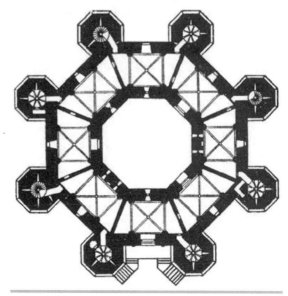

octa- An aerial view of Castel del Monte in Italy's Apulien region.

octagon

An eight-sided **polygon**. One of the most remarkable examples of octagonal design in architecture is the Castel del Monte, in southern Italy, which consists of a central octagonal core containing an inner octagonal courtyard that is surrounded by eight tall, perfectly octagonal towers. Another famous octagon is Oxford University's Radcliffe Observatory tower, built from 1772 to 1794, vaguely based on the Tower of the Winds in Athens, and considered architecturally the finest observatory in Europe.[139]

octahedron

A **polyhedron** with eight faces, each of which is an equilateral triangle. A regular octahedron, whose sides are all equilateral triangles, is one of the **Platonic solids**; it looks like two square pyramids with their square bases stuck together. See also **octa-**.

octant

Any one of the eight portions of space determined by the three coordinate planes.

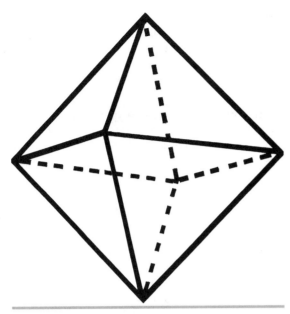

octahedron   One of the Platonic solids.

octonion
Also known as a *Cayley number.* Octonions are a non**associative** generalization of the **quaternions** and the **complex numbers** involving numbers with one real **coefficient** and seven imaginary coefficients.

odd
Not divisible by two. An *odd function* is a **function** $f(x)$ with the property that $f(-x) = -f(x)$ for any value of x; an example is $\sin(x)$.

officer problem
See **thirty-six officers problem.**

omega
See **Chaitin's constant.**

one
The first positive **integer** and the first odd number; it is also known as *unity.* From the time of Euclid to the late 1500s, one wasn't generally considered to be a number but instead was thought of as the unit of which bonafide numbers were composed. The Old English (c. 550– c. 1100) *ane* served both for counting and as the indefinite article. Toward the end of the Old English period and the beginning of Middle English (c. 1100–c. 1500), *ane* developed two pronunciations, the first being used for 1 and the other for the indefinite article *an, a.* The

existence of different words for the number one and the indefinite article seems to be unique to the English language. "One" can be traced back to the Latin *unus* and the Greek *oine* but probably came into English from the German *eine.*

153
A number with some very curious properties. It is the smallest number that can be expressed as the sum of the cubes of its digits: $153 = 1^3 + 5^3 + 3^3$. It is equal to the sum of the factorials of 1 to 5: $153 = 1! + 2! + 3! + 4! + 5!$ The sum of the digits of 153 is a perfect square: $1 + 5 + 3 = 9 = 3^2$. The sum of the **aliquot parts** of 153 is also a perfect square: $1 + 3 + 9 + 17 + 51 = 81 = 9^2$. On adding the number 153 to its reverse (351), the result is 504 whose square is the smallest square that can be expressed as the product of two different numbers that are the reverse of one another: $153 + 351 = 504$; $504^2 = 288 \times 882$. It can be expressed as the sum of all integers from 1 to 17. In other words, 153 is the seventeenth triangular number; its reverse is also a triangular number. In addition, 153 and its reverse are **Harshad numbers.** It can be expressed as the product of two numbers formed from its own digits: $153 = 3 \times 51$. A reference to 153 occurs in the New Testament: the net that Simon Peter drew from the Sea of Tiberias held 153 fishes.

one-to-one
A **function** or map that for every possible output has only one input that yields that particular output; if $f(a) = f(b)$, then $a = b$.

open
An *open interval* is a piece of a straight line that doesn't contain its endpoints; a *half-open interval* contains one endpoint. An *open set* is one in which every point in the set has a neighborhood lying in the set.

operator
Something that acts on a **function** to give another function.

optical illusion
A picture or figure that deceives or confuses the eye and/or brain. Categorization of optical illusions is difficult because several underlying mechanisms may contribute to an effect, or the cause of the illusions may not be completely understood. For the purposes of this book illusions are grouped as shown in the table "Some Types and Examples of Optical Illusions." See individual entries for details. See also **anamorphosis.**

Some Types and Examples of Optical Illusions

Type	Examples
Distortion illusion	Fraser spiral
	Müller-Lyer illusion
	Orbison's illusion
	Poggendorff illusion
	Titchener illusion
	Zöllner illusion
Impossible figures and objects	Freemish crate
	Penrose triangle
	Penrose stairway
	Tribar illusion
Ambiguous figure	Ames room
	Necker cube
	Schröder's reversible staircase
	Thiery figure
Lateral inhibition illusion	Hermann grid illusion
Antigravity houses and hills	

Orbison's illusion

A **distortion illusion** in which a background of radiating lines appears to distort the shape of a superimposed figure, such as a square or circle.

orbit

(1) The path, in the form of a **conic section**, that an object takes when under the gravitational influence of another body. (2) More generally, the trajectory of a **differential equation**.

order

A word with many different meanings in mathematics. Among these are: (1) the sequence in which a set of objects or numbers is placed; (2) the number of elements in a set; (3) the number of times a shape can be fitted back onto its own outline during a complete turn (*order of symmetry*); (4) the highest order **power** in a one-variable **polynomial** (*order of a polynomial,* also known as its *degree*); (5) the type of curve described by such a polynomial (*order of a curve*).

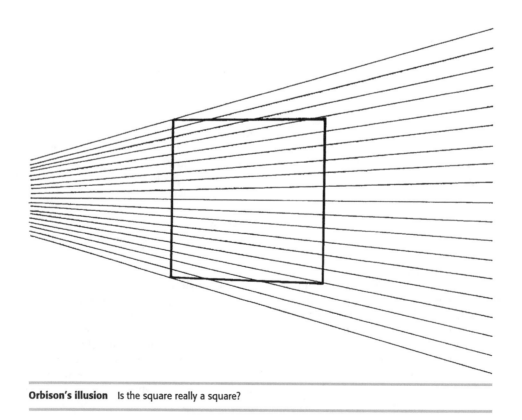

Orbison's illusion   Is the square really a square?

ordered pair

A collection of two objects such that one can be distinguished as the *first element* and the other as the *second element*. An ordered pair with first element a and second element b is usually written as (a, b). Two such ordered pairs (a_1, b_1) and (a_2, b_2) are equal if and only if $a_1 = a_2$ and $b_1 = b_2$. *Ordered triples* and *ordered n-tuples* (ordered lists of n terms) are defined in the same way. An ordered triple (a, b, c) can be defined as $(a, (b, c))$, that is, as two nested pairs.

ordinal number

A number used to give the position in an ordered sequence: first, second, third, fourth, ...Ordinal numbers are distinct from **cardinal numbers** (one, two, three, four, . . .), which describe the *size* of a collection. The mathematician Georg **Cantor** showed in 1897 how to extend the concept of ordinals beyond the **natural numbers** to the infinite and how to do arithmetic with the resulting *transfinite ordinals* (see **infinity**).

ordinary differential equation

Any equation relating a **function** of one variable to its **derivatives**. Compare with **partial differential equation**.

ordinate

The y-coordinate, or vertical distance from the x-axis, in a system of **Cartesian coordinates**. Compare with **abscissa**.

origami

The Japanese art of the paper-folding, without cutting and joining, to form definite objects.

origin

The point $(0, 0)$ on the coordinate axes.

orthic triangle

The triangle whose vertices (see **vertex**) are the feet of the **altitudes** of a given triangle.

orthocenter

The point of intersection of the **altitudes** of a **triangle**.

orthogonal

At right angles to; independent of. *Orthogonal curves* are two families of curves with the property that each member of one family meets members of the other family at right angles.

osculating

To share the same **tangent** and **curvature** and at a given point, as in the case of two *osculating curves*. For example the curve $y = x^3$ and the x-axis osculate at the origin.

Oughtred, William (1574–1660)

An English clergyman and mathematician who invented the **slide rule** and who, in the first English edition of his *Clavis Mathematicae* in 1647, first used the name "**pi**" for the number 3.141 . . . . He wrote it $\pi.\delta$, where π stood for the English word *periphery* (what we would call circumference), the dot was his symbol for division, and δ stood for the English word *diameter*. The use of π to represent words starting with the letter p, like *periphery*, was not uncommon. Before π = 3.14 . . . caught on, π was variously used to indicate a point, a polygon, a positive number, a power, a proportion, the number of primes in a series, and a factorial (which is a product). Oughtred used the same notation in all later English and Latin editions of his book, but not in the earlier first Latin edition. He was a prodigious inventor of mathematical symbols, though most of them have not survived. He did, however, introduce a couple of other symbols we still use: $\times$ for multiplication (as distinct from the letter x, which he also used for this purpose) and $\pm$ for "plus or minus." Oughtred served as rector of Albury from 1608 or 1610 until his death. There he tutored many young mathematicians of the time, including John **Wallis**, Seth Ward, Charles Scarburgh, and Christopher Wren. Moreover, all English mathematicians for the next century, including Isaac **Newton**, learned algebra from *Clavis Mathematicae* (first published in 1631).

oval

A curve that looks like a squashed **circle** but, in contrast with the **ellipse**, doesn't have a precise mathematical definition. The word *oval* comes from the Latin *ovus* for "egg." Unlike ellipses, ovals sometimes have only a single axis of reflection symmetry (instead of two).

Ovid's game

A board game, described by the Roman poet Ovid in Book III of his *Art of Love,* that was popular in ancient Greece, Rome, and China. Each player has three counters, which they place alternately on to nine points laid out in a 3×3 grid, with the object of getting three in a line and so winning. This may sound like **tic-tac-toe**, and there is no doubt that both Ovid's game and **nine men's morris** are antecedents of the familiar noughts and crosses. But in Ovid's game play continues after the six counters are down, if no one has yet won, by moving a single counter on each turn, though not diagonally, to any adjacent square. As in the case of tic-tac-toe, two experts (i.e., "rational" players) will always draw. Because the first player is ensured a win by covering the center square, this move is usually not allowed. Ovid advised women to master the game in order to gain the attention of men!

Ozanam, Jacques (1640–1717)

A French mathematician, scientist, and writer best remembered for his book on mathematical and scientific puzzles *Récréations Mathématiques et Physiques* (four volumes, 1694), which later went through 10 editions. Based on earlier works by Claude **Bachet**, Claude **Mydorge**, Jean **Leurechon**, and Daniel Schwenter (1585–1636), it was later revised and enlarged by Jean **Montucla**, then translated into English by Charles Hutton (1803, 1814). Edward Riddle edited a new edition, which was published in 1844, removing some old material and adding new material so that "[T]he work might continue to be to the present generation a useful manual of scientific recreation, as its predecessors have been to the generation which has passed." Ozanam's original edition contained an early example of a problem about orthogonal **Latin square**s: "Arrange the 16 court cards so that each row and each column contains one of each suit and one of each value."

P

Pacioli, Luca (1445–1517)

An Italian mathematician and Franciscan monk who wrote several influential books. His encyclopedic *Summa de Arithmetica, Geometria, Proportioni et Proportionalita* (Summary of arithmetic, geometry, proportion and proportionality, 1494) summarized what was known about contemporary arithmetic, algebra, geometry, and trigonometry and gave a basis for the major progress in mathematics which subsequently took place in Europe. *Divina proportione* (1509), with drawings by none other than Leonardo da Vinci (surely no mathematical text was more impressively illustrated!), deals with the **golden ratio**, a subject that Pacioli treats from an architectural standpoint in a second volume. At his death, he left a major book unpublished, *De Viribus Amanuensis,* on recreational problems, geometrical problems, and proverbs. It makes frequent reference to Leonardo, who assisted him with the project: many of the problems in this treatise are also in Leonardo's notebooks. Again it is a work for which Pacioli claimed no originality, describing it as a compendium.

packing

A way to place objects of the same kind so that they touch in some specified way, often inside a container with specified properties. The objects to be packed may be polyhedra, polygons, spheres, ellipsoids, hyperspheres, or any other type of shape, and the number of dimensions involved may range upward from two. The fraction of a space filled by a given collection of objects is called the *packing density*. The densest packing of circles in the plane is the hexagonal lattice of the bee's honeycomb, which has a packing density of 0.9069 . . . . In 1611, Johannes **Kepler** proposed that hexagonal, or face-centered cubic, packing is also the densest possible way to arrange spheres in three dimensions—an assertion known as **Kepler's conjecture**. Currently, the worst known **convex** packer in two-dimensions is the smoothed octagon, with a packing density of about 0.902. Stanislaw **Ulam** conjectured that the sphere was the worst packing object in three-dimensional space.

Padovan sequence

See **plastic number**.

palindrome

A word, verse, sentence, or passage that reads the same forward or backward; the term comes from the Greek *palindromos* for "running back again." Well-known examples include: "Madam, I'm Adam;" "A man, a plan, a canal–Panama!;" and "Able was I ere I saw Elba." A slightly longer one, devised by Peter Hilton, a codebreaker on the British team that cracked the German Enigma, is "Doc, note. I dissent. A fast never prevents a fatness. I diet on cod." Credit for inventing the palindrome is often given to Sotades the Obscene of Maronea (third century B.C.). Though only eleven lines of his work have survived, he is thought to have recast the entire Iliad in palindromic verse. Sotades also wrote lines (now sometimes called Sotadic verses) that, when read backward, had the opposite meaning. His acid tongue eventually landed him in jail by order of Ptolemy II, though worse was to follow. Sotades escaped but was captured by Ptolemy's admiral Patroclus, who sealed him in a leaden chest and tossed him into the sea. A musical palindrome is formed by Haydn's Symphony No. 47 in G, sometimes referred to as *The Palindrome,* because in both the minuet and the trio the orchestra plays the music twice forward and then twice backward to arrive at the beginning. See also **palindromic number**.

palindromic number

A number such as 1,234,321 that reads the same forward and backward; more generally, a symmetrical number written in some **base** a as $a_1\,a_2\,a_3\ldots|\ldots a_3\,a_2\,a_1$. In the familiar base 10 system, there are nine two-digit palindromic numbers: 11, 22, 33, 44, 55, 66, 77, 88, 99; there are 90 palindromics with three digits: 101, 111, 121, 131, 141, . . . , 959, 969, 979, 989, 999; and there are 90 palindromics with four digits: 1,001; 1,111; 1,221; 1,331; 1,441; . . . , 9,559; 9,669; 9,779; 9,889; 9,999, giving a total of 199 palindromic numbers below 10^4. Below 10^5 there are 1,099 palindromics and for other exponents of 10^n there are 1,999; 10,999; 19,999; 109,999; 199,999; 1,099,999; . . . . It is conjectured, but has not been proven, that there are an infinite number of palindromic **prime numbers**. With the exception of 11, palindromic primes must have an odd number of digits. A normally quick way to produce a palindromic number is to pick a positive integer of two or more digits, reverse the digits, and add to the original, then repeat this process with the new number, and so on. For example, 3,462 gives the sequence 3,462; 6,105; 11,121; 23,232. Does the series formed by adding a number to its reverse always end in a palindrome? It used to be thought so. However, this conjecture has been proven false for bases 2, 4, 8, and other powers of 2, and seems to be false for base 10 as well. Among the first 100,000 numbers, 5,996 numbers are known that have not produced palindromic numbers by the add-and-reverse method in calculations carried out to date. The first few of these are 196; 887; 1,675; 7,436; 13,783; 52,514; . . . . A proof that these numbers never produce palindromes, however, has yet to be found. The largest known palindromic prime, containing 30,913 digits, was found by David Broadhurst in 2003.

pandiagonal magic square
A **magic square** in which all the broken diagonals as well as the main diagonals add up to the magic constant.

pandigital number
An integer that contains each of the digits from zero to nine exactly once and whose leading digit is nonzero. The smallest pandigital numbers are 1,023,456,789; 1,023,456,798; 1,023,456,879; 1,023,456,897; and 1,023,456,978. Like all pandigitals these are divisible by nine. The first few "zeroless" pandigitals are 123,456,789; 123,456,798; 123,456,879; 123,456,897; 123,456,978; and 123,456,987; and the first few zeroless pandigital primes are 1,123,465,789; 1,123,465,879; 1,123,468,597; 1,123,469,587; and 1,123,478,659. The sum of the first 32,423 (a **palindromic number**) consecutive primes is 5,897,230,146, which is pandigital. No other palindromic number shares this property. Examples of palindromic numbers that are the product of pandigital numbers are 2,970,408,257,528,040,792 (= 1,023,687,954 × 2,901,673,548) and 5,550,518,471,748,150,555 (= 1,023,746,895 × 5,421,768,309), both found in 2001. See also **pandigital product**.

pandigital product
A product in which the digits of the multiplicand, multiplier, and product, taken together, form a **pandigital number**.

pangram
A phrase or sentence that contains every letter of the alphabet at least once. The best known example in English and a familiar typing test is "The quick brown fox jumps over the lazy dog." One of the shortest known that still makes some kind of sense is "The five boxing wizards jump quickly."

Pappus of Alexandria (c. A.D. 300)
The last of the great Greek geometers whose eight-volume *Mathematical Collection* summarized the bulk of mathematics known at that time. In this compendium, Pappus added a considerable number of his own explanations and amplifications of the earlier work of **Euclid**, **Archimedes**, **Apollonius**, and others.

parabola
One of the **conic sections** and one of the most studied curves in the history of mathematics. A parabola is the outline of the figure obtained if a right circular **cone** is cut by a plane that is exactly parallel to the cone's side. Just as the circle is a limiting case of the **ellipse** when the two foci coincide, the parabola is a limiting case of the ellipse when one of the foci is moved to infinity. As the French mathematician Henri Fabre eloquently put it, the parabola is an ellipse that "seeks in vain for its second, lost center." Johannes **Kepler** drew this further connection between the parabola and other conic sections: "Because of its intermediate nature the parabola occupies a middle position [between the ellipse and the **hyperbola**]. As it is produced it does not spread out its arms like the hyperbola but contracts them and brings them nearer to parallel, always encompassing more, yet always striving for less—whereas the hyperbola, the more it encompasses, the more it tries to obtain."

A parabola is the **locus** of all points in a plane that are equidistant from a given line, known as the *directric*, and

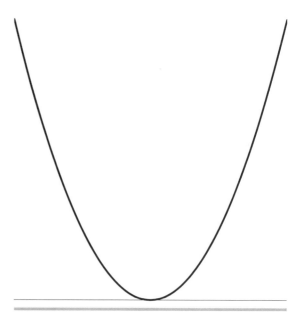

parabola *Jan Wassenaar, www.2dcurves.com*

a given point not on the line, known as the focus. The Cartesian equation of a parabola that opens upward and has its vertex (turning point) at the origin is $y = 4ax^2$, where a is the distance from the vertex to the focus, and the quantity $4a$ is known as the *latus rectum*. More generally, any **quadratic** equation of the form $y = ax^2 + bx + c$, where a is not zero, graphs a parabola. The simplest form of this, when $a = 1$, and both b and c are zero, is $y = x^2$. Like the circle, but unlike the ellipse and the hyperbola, the parabola has only one distinct shape. In other words, any parabola can be superimposed exactly on any other parabola simply by rotating, translating (sliding), and/or enlarging or shrinking it.

Euclid dealt with the parabola in his *Conic Sections,* and, although this treatise was lost, it provided a foundation for the first four books by **Apollonius** of the same name. Galileo (1564–1642) discovered that a cannonball, or any other projectile launched at an angle to ground, follows a parabolic path: a result that immediately grabbed the attention of not only scientists but of monarchs and military leaders. René **Descartes**, in writing *La Géométrie* (1637), chose the parabola to illustrate his innovative **analytical geometry**. In 1992, Rudolph Marcus of the California Institute of Technology won the Nobel Prize in Chemistry for his work showing that parabolic reaction surfaces can be used to calculate how fast electrons travel in molecules. His most famous theoretical result, an inverted rate-energy parabola, predicts electron transfer will slow down at very high-reaction free energies.

paraboloid

The surface of revolution of the **parabola**. It is a **quadratic** surface described by the equation $z = a(x^2 + y^2)$.

paradox

Please accept my resignation. I don't want to belong to any club that will accept me as a member.
    —Groucho Marx (1895–1977)

A statement that seems to lead to a logical self-contradiction, or to a situation that contradicts common intuition. The word *paradox* comes from the Greek *para* ("beyond") and *doxa* ("opinion" or "belief"). The identification of a paradox based on seemingly simple and reasonable concepts has often led to significant advances in science, philosophy, and mathematics. See **Allais paradox, Arrow paradox, Banach-Tarski paradox, Berry's Paradox, birthday paradox, Burali-Forti paradox, coin paradox, grandfather paradox, Grelling's paradox, liar paradox, Newcomb's paradox, nine rooms paradox, Parrondo's paradox, raven paradox, Russell's paradox, St. Petersburg paradox, Siegel's paradox, unexpected hanging,** and **Zeno's paradoxes.**

parallel

Said of two or more things, such as lines or planes, that are equally distant from one another at all points.

parallel postulate

The fifth and most controversial of **Euclid's postulates** set forth in the Greek geometer's great work, *Elements.* To later mathematicians, the parallel postulate seemed less obvious than the other four and many attempts were made to derive it from them, but without success. In 1823, János **Bólyai** and Nikolai **Lobachevsky** independently realized that entirely self-consistent types of **non-Euclidean geometry** could be created in which the parallel postulate doesn't hold. Carl **Gauss** had made the same discovery earlier but kept the fact secret.

parallelepiped

A **polyhedron** with six faces bounded by three pairs of parallel planes, so that all its faces are **parallelogram**s. It is also a **prism** with a parallelogram for a base. A *rectangular parallelepiped* has the shape of a shoebox. The word comes from the Greek *parallelepipedon* for the same shape, which in turn comes from the roots *para* ("beside"), *allel* ("other"), *epi* ("on"), and *pedon* ("ground"). *Parallelepipedon* may have been first used in English by Billingsley in his 1570 translation of Euclid. It seems to have then given way to *parallelepiped* in the last quarter of the nineteenth century.

parallelogram

A **quadrilateral** (four-sided figure) whose opposite sides are parallel, and whose opposite angles, therefore, are equal. The diagonals of a parallelogram bisect each other. A parallelogram of base b and height h has an area

$$bh = ab \sin A = ab \sin B.$$

The height of a parallelogram is $h = a \sin A = a \sin B$. The sides a, b, c, and d and diagonals p and q of a parallelogram satisfy the equality

$$p^2 + q^2 = a^2 + b^2 + c^2 + d^2.$$

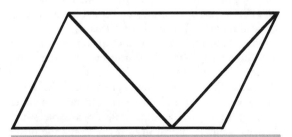

parallelogram The two lines drawn inside this parallelogram have the same length.

Special cases of a parallelogram are a **rhombus**, which has sides of equal length; a **rectangle**, which has two sets of parallel sides that are perpendicular to each other; and a **square**, which meets the conditions of both a rectangle and a rhombus.

In the *parallelogram illusion* shown, despite appearances to the contrary, the interior diagonal lines are of equal length.

parameter

An independent variable; one of the inputs for a **function**. *Parametric equations* are equations in which the variables of interest are given in terms of another variable.

parity

(1) The sum of the digits in a **binary** representation of a number. (2) An indication of whether two numbers are both even or both odd (same parity), or are such that one is even and one is odd (opposite parity). This concept, of same or opposite parity, can be applied to any things that can have two different states–the sense in which a knot is tied, the color of a chessboard square, and so on.

Parrondo's paradox

Two losing gambling games can be set up so that when they are played one after the other, they become winning. This paradox is named after the Spanish physicist Juan Parrondo who, in the late 1990s, discovered how to construct such a scenario. The simplest way is to use three biased coins. Imagine you are standing on stair zero, in the middle of a long staircase with 1001 stairs numbered from −500 to 500. You win if you can get to the top of the staircase, and the way you move depends on the outcome of flipping one of two coins. Heads you move up a stair, tails you move down a stair. In game 1, you use coin *A*, which is slightly biased and comes up heads 49.5% of the time and tails 50.5%. Obviously, these are losing odds. In game 2, you use two coins, *B* and *C*. Coin *B* comes up heads only 9.5% of the time, tails 90.5%. Coin *C* comes up heads 74.5% of the time, tails 25.5%. In game 2 if the number of the stair you are on at the time is a multiple of 3 (that is, . . . , −9, −6, −3, 0, 3, 6, 9, 12, . . .), then you flip coin *B;* otherwise you flip coin *C*. Game 2, it turns out, is also a losing game and would eventually take you to the bottom of the stairs. What Parrondo found, however, is that if you play these two games in succession in random order, keeping your place on the staircase as you switch between games, you will steadily rise to the top of the staircase! This kind of process has been called a *Brownian motion*.[153]

partial differential equation

An equation that involves **derivatives** with respect to more than one variable. Many of the equations used to model the physics of the real world are partial differential equations.

partition number

A number that gives the number of ways of placing n indistinguishable balls into n indistinguishable urns. For example:

1: (*)
2: (**) (*)(*)
3: (***) (**)(*) (*)(*)(*)
5: (****) (***)(*) (**)(**) (**)(*)(*) (*)(*)(*)(*)
7: (*****) (****)(*) (***)(**) (***)(*)(*) (**)(**)(*)
   (**)(*)(*)(*)(*)(*)(*)(*)(*)
11: (******) (*****)(*) (****)(**) (****)(*)(*)
   (***)(***) (***)(**)(*)(***)(*)(*)(*)
   (**)(**)(**) (**)(**)(*)(*) (**)(*)(*)(*)(*)
   (*)(*)(*)(*)(*)(*)

The sequence runs: 1, 2, 3, 5, 7, 11, 15, 22, 30, 42, 56, 77, 101, 135, 176, 231, 297, 385, ... If the urns are distinguishable, the number of ways is 2^n. If the balls are distinguishable, the number of ways is given by the nth **Bell number**.

Pascal, Blaise (1623–1662)

A French mathematician, philosopher, and pioneer of **probability theory** whose short life was rich with mathematical invention but whose name, ironically, is most familiar from its association with an array of numbers known as **Pascal's triangle** (which he didn't discover, though he did important work on it). Educated by his father (after whom the **limaçon of Pascal** is named), Pascal showed his intellectual prowess early on by proving one of the most important theorems in **projective geometry** at the age of 16. Three years later he devised the world's second mechanical calculating machine (the first was made by Wilhelm Schickard in 1623) to help with his father's business; he sold about 50 of these "Pascalines," several of which survive. In 1654, he and Pierre de **Fermat**, in an exchange of correspondence, laid the foundation for **probability theory**. They considered the dice problem, already studied by Girolamo **Cardano**, and the problem of points also considered by Cardano and, around the same time, by Luca **Pacioli** and Niccoló **Tartaglia**. The dice problem asks how many times one can expect to throw a pair of dice before getting a double six; the problem of points asks how to divide the stakes if a game of dice is incomplete. Pascal and Fermat solved the problem of points for a two-player game but didn't develop powerful enough methods to solve it for three or more players. In the same year, Pascal was almost killed in an incident in which the horses pulling his carriage bolted and the carriage was left hanging over a bridge above the river Seine. Though rescued unharmed, he shortly after converted to the rigorous Jansenist sect of the Catholic Church. His philosophical work *Pensées*, written between 1656 and 1658 contains his famous argument, often called Pascal's wager, for belief in **God**. Having suffered poor health for most of his adult life, he died in great pain of cancer at the age of 39.

Pascal's mystic hexagon

If a **hexagon** *ADBFCE* (not necessarily convex) is inscribed in a **conic section** (in particular a circle), then the points of intersections of opposite sides (*AD* with *FC, DB* with *CE* and *BF* with *EA*) are collinear. This line is called the *Pascal line* of the hexagon. A special case is when the conic degenerates into two lines; the theorem still holds but is then usually called *Pappus's theorem*.

Pascal's triangle

A triangular pattern of numbers in which each number is equal to the sum of the two numbers immediately above it:

$$
\begin{array}{c}
1 \\
1 \quad 1 \\
1 \quad 2 \quad 1 \\
1 \quad 3 \quad 3 \quad 1 \\
1 \quad 4 \quad 6 \quad 4 \quad 1 \\
1 \quad 5 \quad 10 \quad 10 \quad 5 \quad 1 \\
1 \quad 6 \quad 15 \quad 20 \quad 15 \quad 6 \quad 1 \\
1 \quad 7 \quad 21 \quad 35 \quad 35 \quad 21 \quad 7 \quad 1 \\
1 \quad 8 \quad 28 \quad 56 \quad 70 \quad 56 \quad 28 \quad 8 \quad 1 \\
\vdots
\end{array}
$$

Although named after Blaise **Pascal**, who studied it, this arithmetic triangle has been recognized since the twelfth century and has a variety of other names. In Italy it is called *Tartaglia's triangle* and in many parts of Asia it is referred to as *Yang Hui's triangle*. Yang Hui was a minor Chinese official who wrote two books, dated 1261 and 1275, that use decimal fractions (long before they appeared in the West) and contain one of the earliest accounts of the triangle. At about the same time, Omar **Khayyam** also wrote about it. The Chinese triangle appears again in 1303 on the front of Chu Shi-Chieh's *Ssu Yuan Yü Chien* (Precious mirror of the four elements), a book in which Chu says the triangle was known in China more than two centuries before his time.

The numbers in Pascal's triangle give the number of ways of picking *r* unordered outcomes from *n* possibilities. This is equivalent to saying that the numbers in each row are the **binomial coefficient**s in the expansion of $(x+y)^n$:

$$
\begin{aligned}
(x+y)^0 &= 1 \\
(x+y)^1 &= 1x + 1y \\
(x+y)^2 &= 1x^2 + 2xy + 1y^2 \\
(x+y)^3 &= 1x^3 + 3x^2y + 3xy^2 + 1y^3 \\
(x+y)^4 &= 1x^4 + 4x^3y + 6x^2y^2 + 4xy^3 + 1y^4
\end{aligned}
$$

and so on. In addition, the shallow diagonals of the triangle sum to give the **Fibonacci sequence**.

path

See **trajectory**.

pathological curve

A curve often specifically devised to show the falseness of certain intuitive concepts. In particular, the image of **continuity** as a smooth curve in our mind's eye severely misrepresents the situation and is the result of a bias stemming from an overexposure to the much smaller class of differentiable functions (see **differentiation**). A chief lesson of pathological curves is that continuity is a weaker notion than differentiability. Many pathological curves are **fractals**, such as **Cantor dust**, including **space-filling curves**, such as the **Peano curve**. The earliest known example is the graph of **Weierstrass' nondifferentiable function**.

payoff

In **game theory**, the amount that a player wins, given the player's and his opponent's actions.

Peano, Guiseppe (1858–1932)

An Italian pioneer of mathematical **logic** and the axiomatization of mathematics. In 1889, he published his first version of a system of mathematical logic in *Arithmetics Principia*, which included his famous **axiom**s of natural numbers, now known as *Peano's axioms*. Two years later, he established a journal, the *Rivista*, in which he proposed the symbolizing of all mathematical propositions into his system. The project, which became known as the *Formulario*, was his chief focus for the next fifteen years.

Peano curve

The first known example of a **space-filling curve**. Discovered by Guiseppe **Peano** in 1890, its effect was like that of an earthquake on the traditional structure of mathematics. Commenting in 1965 on the impact of the curve in Peano's day, N. Ya Vilenkin said: "Everything has come unstrung! It's difficult to put into words the effect that

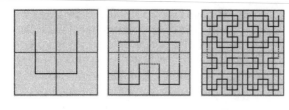

Peano curve A simple iterative process produces a remarkable space-filling shape.

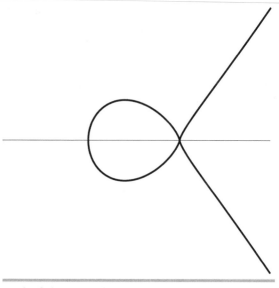

pearls of Sluze *Jan Wassenaar, www.2dcurves.com*

Peano's result had on the mathematical world. It seemed that everything was in ruins, that all the basic mathematical concepts had lost their meaning." Today, the Peano curve is recognized as just one of an infinite class of familiar objects known as **fractals**. But at the end of the nineteenth century it was an extravagant, completely counterintuitive thing; indeed, it was something that had been believed impossible. Writing of Peano's result in *Grundzüge der Mengenlehre* (Basic features of set theory) in 1914, Felix **Hausdorff** said: "This is one of the most remarkable facts of set theory." Originally, the Peano curve was derived purely analytically, without any kind of drawing or attempt at visualization. But the first few steps in drawing it, as shown in the diagrams, are easy enough, even though the finished product is unattainable in this way—and totally unimaginable. To fill the unit square, as the Peano curve does, without leaving any holes, the curve has to be both continuous and self-intersecting.

pearls of Sluze

Curves that are generated by the Cartesian equation

$$y^n = k(a - x)^p x^m,$$

where *n*, *m*, and *p* are integers. They were first studied by the French mathematician René de Sluze (1622–1685) and named the pearls of Sluze by Blaise **Pascal**.

pedal curve

The locus of the feet of the perpendiculars from a given point to the **tangents** to the given figure. The "given point" is known as the *pedal point*.

pedal triangle

The pedal triangle of a point *P* with respect to a triangle *ABC* is the triangle whose vertices (see **vertex**) are the feet of the perpendiculars dropped from *P* to the sides of triangle *ABC*.

peg solitaire

Also known as Hi-Q, a game that, in its commonest form, is played with 32 pegs or marbles on a rectangular grid, the middle position of which starts out empty. A peg may jump horizontally or vertically, but not diagonally, over a peg in an adjacent hole into a vacant hole immediately beyond. The peg that was jumped over is then removed. The object is to be left with a single peg in the center position.

The quickest solution of peg solitaire was found by Ernest Bergholt in 1912 and was proved to be minimal by John Beasley in 1964. Suppose the rows and columns of the board are each labeled 1 to 7 so that, for example, a peg on the fourth row and the third column is in position 43. The quickest solution is: (1) 46→44. (2) 65→45.

peg solitaire A version of peg solitaire made of tin with marbles by the British firm MAR Toys in the 1950s. *Sue & Brian Young/Mr. Puzzle Australia, www.mrpuzzle.com.au*

(3) 57→55. (4) 54→56. (5) 52→54. (6) 73→53. (7) 43→63. (8) 75→73→53. (9) 35→55. (10) 15→35. (11) 23→43→63→65→45→25. (12) 37→57→55→53. (13) 31→33. (14) 34→32. (15) 51→31→33. (16) 13→15→35. (17) 36→34→32→52→54→34. (18) 24→44.[28]

Pell equation

An equation of the form $y^2 = ax^2 + 1$, where a is any positive whole number except a square number. The name comes from the English mathematician John Pell (1611–1685); however, he was wrongly credited. In writing about some of the work done on this type of equation, Leonhard **Euler** gave priority to Pell whereas, in fact, Pell had done no more than copy it in his papers from some of Pierre de **Fermat**'s letters. Fermat had been the first to state that an equation of this form always has an unlimited number of integer solutions. For example, the equation $y^2 = 92x^2 + 1$, has the solutions $x = 0, y = 1$; $x = 120, y = 1,151$; $x = 276,240, y = 2,649,601$; and so on. Each successive solution is about 2,300 times the previous solution. In fact, the solutions are every eighth partial fraction (where x is the numerator and y is the denominator) of the **continued fraction** for $\sqrt{92}$. A Pell equation was used in finding the solution to **Archimedes's cattle problem**.

Pell numbers

Numbers, named after John Pell (1611–1685), that are similar to the **Fibonacci sequence** and are generated by the formula $A_n = 2A_{n-1} + A_{n-2}$. The sequence runs: 1; 2; 5; 12; 29; 70; 169; 408; 985; 2,378; 5,741; 13,860; 33,461; 80,782; 195,025; . . . . The ratio of successive terms approaches 1 plus the **square root of 2**.

Penrose, Roger (1931–)

An English mathematical physicist famous for his important contributions to cosmology and the physics of black holes, for his controversial views on the nature of human consciousness and its relationship to **quantum mechanics**, and for his work in the field of recreational mathematics. The **Penrose tiling** and the **Penrose triangle** are named after him (but not the **Penrose stairway**, which is named after his father). In his book *The Emperor's New Mind* he argues that there must be errors in the known laws of physics, notably in quantum mechanics, and that true **artificial intelligence** is impossible.[242] The latter claim is based on his assertion that humans can do things outside the power of formal logic systems, such as knowing the truth of unprovable statements, or solving the **halting problem** (claims that were originally made by the philosopher John Lucas of Merton College, Oxford). These are controversial views, with which most of the mathematical and computer science communities disagree.

Penrose stairway

An **impossible figure** named after the British geneticist Lionel Penrose (1898–1972), father of Roger **Penrose**. It served as an inspiration for the staircase in M. C. **Escher**'s famous print "Ascending and Descending." Although the Penrose stairway cannot be realized in

Penrose stairway Steps that ascend and descend simultaneously. *Jos Leys, www.josleys.com*

three dimensions, this impossibility is not immediately perceived and, in fact, the paradox is not even apparent to many people at a quick glance. Although Escher and the Penroses made the Stairway famous, it was, unbeknownst to them, independently discovered and refined years before by the Swedish artist Oscar **Reutersvärd**. In

the 1960s the Stanford psychologist Roger Shepard created an auditory analogue of the Stairway. See also **Penrose triangle**.

Penrose tiling

A kind of **aperiodic tiling** discovered by Roger **Penrose**. In 1973 he announced a tiling made from a set of six tiles and then, by slicing and re-gluing, was able to reduce the number of tiles to just two. The most elegant of Penrose tilings use two rhombi (see **rhombus**), a thick one called a "kite" and a thin one called a "dart," which are fitted together so that no two tiles are aligned to form a single parallelogram (otherwise, a single rhombus could be used to make a periodic tiling). All angles are multiples of $\pi/5$ radians (36°, or one-tenth of a circle). Each tile has four sides with a length of one unit. One tile has four corners with the angles 72°, 72°, 108°, and 108° (2, 2, 3, and 3 multiples of 36°); the other has angles of 36°, 36°, 144°, and 144° (1, 1, 4, and 4 multiples of 36°). On each tile one of the vertices (corners) is colored black and two of the sides are marked with arrows. The only rules for assembling the tiles to ensure an aperiodic tiling are that two adjacent vertices must be of the same color, and two adjacent edges must have arrows pointing in the same direction or have no arrows at all. These rules ensure that, taken over a large enough area of the plane, the pattern of tiles doesn't repeat. In a correct Penrose tiling, the ratio of kites to darts is always the same and is equal to the **golden ratio**. Although Penrose tilings started out as nothing more than an interesting mathematical diver-

Penrose tiling Two of an infinite variety of nonperiodic patterns possible using Penrose tiles. *Jos Leys, www.josleys.com*

Penrose triangle

sion, they have turned out to mimic the arrangements of atoms in some newly discovered materials, known as **quasicrystals**.[127]

Penrose triangle
The most famous and one of the simplest **impossible figures**. Its roots go back to 1934 when Oscar **Reutersvärd** made the first recognizable impossible triangle out of a strange two-dimensional representation of cubes; this artwork appeared on a Swedish postage stamp issued in 1982. In 1954, Roger **Penrose**, after attending a lecture by the artist M. C. **Escher**, rediscovered the impossible triangle and drew it in its most familiar form, which he published in a 1958 article in the *British Journal of Psychology*, coauthored with his father Lionel.[241] Penrose was unfamiliar with the work of Reutersvärd, Giovanni Piranesi, and others who had previously created impossible figures. Penrose's impossible triangle, unlike Reutersvärd's earlier version, was drawn in perspective, which added an additional size paradox to the object. In 1961, Escher, inspired by Penrose's version of the impossible triangle (he was sent a copy of the article by the Penroses), incorporated it into his famous lithograph *Waterfall*.

penta-
The Greek prefix meaning "five." A *pentagon* is a five-sided **polygon** and a *pentahedron,* such as the Great Pyramid of Gisa, is a five-sided **polyhedron**. A *pentagram,* or *pentacle*, is a five-pointed star formed by the pairwise extension of the sides of a regular polygon so that each side meets not with the next side but with the one after that. The Pythagoreans (see **Pythagoras of Samos**) used the pentagram as a secret identification emblem; later it became a trademark of alchemists and, perhaps because

of its repeating properties, a sign of the occult. According to legend, a pentagram was used by Doctor Faustus to exorcise Mephistopheles. *Pentagonal numbers* are **figurate numbers** (numbers that can be represented by a regular geometric arrangement of equally spaced points) of the form $n(3n - 1)/2$; the first few are 1, 5, 12, 22, ...

pentomino
A five-square **polyomino**.

Perelman, Yakov Isidorovitch (1882–1942)
A Russian scientist and exponent of recreational mathematics whose stature in his own country was, and remains, similar to that of Martin **Gardner** in the United States. Between 1913 and his death, Perelman authored a dozen books and scores of articles covering many different aspects of popular mathematics, physics, and astronomy. His books have been republished so often in Russia that Perelman is universally known among both amateur and professional mathematicians across several generations. He was also a leading proponent of the ideas of the spaceflight visionary Konstantin Tsiolkovsky.

perfect cube
An integer of the form n^3 where n is an integer.

perfect cuboid problem
See **cuboid**.

perfect number
A whole number that is equal to the sum of all its factors except itself. For example, 6 is a perfect number because its factors, 1, 2, and 3 add to give 6. The next smallest is 28 (the sum of $1 + 2 + 4 + 7 + 14$). Augustine (354–430) argued that God could have created the world in an instant but chose to do it in a perfect number of days, 6. Early Jewish commentators saw the perfection of the universe in the Moon's period of 28 days. The next in line are 496, 8,128, and 33,550,336. As René **Descartes** pointed out: "Perfect numbers like perfect men are very rare." All end in six or eight, though what seems to be an alternating pattern of sixes and eights for the first few perfect numbers doesn't continue. All are of the form $2^{n-1}(2^n - 1)$, where $2^n - 1$ is a **Mersenne prime**, so that the search for perfect numbers is the search for Mersenne primes. The largest one found is $2^{3021376}(2^{3021377} - 1)$. It isn't known if there are infinitely many perfect numbers or if there are any odd perfect numbers. A *pseudoperfect number* or *semi-perfect number* is a number equal to the sum of some of its divisors, for example, $12 = 2 + 4 + 6$, $20 = 1 + 4 + 5 + 10$. An *irreducible semi-perfect number* is a semi-perfect number, none of whose factors is semi-perfect, for example, 104. A *quasi-perfect number*

would be a number n whose divisors (excluding itself) sum to $n + 1$, but it isn't known if such a number exists. A *multiply perfect number* is a number n whose divisors sum to a multiple of n. An example is 120, whose divisors (including itself) sum to $360 = 3 \times 120$. If the divisors sum to $3n$, n is called multiply perfect of order 3, or *tri-perfect*. Ordinary perfect numbers are multiply perfect of order 2. Multiply perfect numbers are known of order up to 8. See also **abundant number**.

perfect power
An integer of the form m^n where m and n are integers and $n > 1$.

perfect square
A number that is the product of two equal whole numbers; for example, $1 = 1 \times 1$, $4 = 2 \times 2$, $9 = 3 \times 3$, $16 = 4 \times 4$.

Perigal, Henry (1801–1898)
An English amateur mathematician best known for his elegant **dissection** proof of **Pythagoras's theorem**, a diagram of which is carved on his gravestone. Perigal also discovered a number of other interesting geometrical dissections and, though employed modestly for much of his life as a stockbroker's clerk, was well known in British scientific society.

perimeter
The distance around a two-dimensional shape.

period, of a decimal expansion
The length of the smallest block of repeating digits in the decimal expansion of a **rational number** that does not terminate. For example:

$\frac{1}{3} = 0.333333333333\ldots$
  repeating block = 3, period = 1
$\frac{5}{7} = 0.71428571428571\ldots$
  repeating block = 714285, period = 6
$\frac{89}{26} = 3.4230769230769\ldots$
  repeating block = 230769, period = 6

periodic
Refers to motion or to an entity that goes through a finite number of regions, returns to a previous state, and repeats the same fixed pattern forever.

periodic attractor
Also called a *limit cycle attractor*, an **attractor** that consists of a periodic movement back and forth between two or more values. The periodic attractor represents more possibilities for system behavior than does the **fixed-point**

attractor. An example of a period two attractor is the oscillating movement of a metronome or, in psychiatry, a bipolar disorder that causes a person's mood to shift back and forth from elation to depression.

periodic tiling
A **tiling** in which a region can be outlined that tiles the plane by translation, that is, by shifting the position of the region without rotating or reflecting it. M. C. **Escher** is famous for his many pictures of periodic tilings with shapes that resemble living things. An infinity of shapes—for instance, the regular hexagon—tile only periodically, though all these fall into 17 distinct **wallpaper group**s. An infinity of other shapes tile both periodically and aperiodically. But it was only quite recently that the first **aperiodic tilings** were discovered.

periphery
A curved **perimeter**.

permutable prime
Also known as an *absolute prime*, a **prime number** with at least two distinct digits, which remains prime on every rearrangement (permutation) of the digits. For example, 337 is a permutable prime because 337, 373, and 733 are all prime. Most likely, in base 10, the only permutable primes are 13, 17, 37, 79, 113, 199, 337, and their permutations. Permutable primes can't have any of the digits 2, 4, 6, 8 or 5, nor can they have all four of the digits 1, 3, 7, and 9 simultaneously.

permutation
A particular ordering of a collection of objects. For example, if an athlete has won three medals, a bronze one (B), a silver one (S), and a gold one (G), there are six ways they can be permuted or lined up: BSG, BGS, SBG, SGB, GBS, and GSB. If six people want to sit on the same park bench, there are 720 ways in which they can organize themselves. In general, n things can be permuted in $n \times (n - 1) \times (n - 2) \times \ldots \times 2 \times 1 = n!$ ways (where "!" is the symbol for **factorial**). How about if there are n distinct objects but we want to permute them in groups of k (where $k \leq n$): how many ways can that be done? The first member of the group can be picked in n ways because there are n objects to pick from. The second member can be filled in $(n - 1)$ ways since one of the n elements has already been taken. The third member can be filled in $(n - 2)$ ways since 2 elements have already been used, and so on. This pattern continues until k things have been chosen. This means that the last member can be filled in $(n - k + 1)$ ways. Therefore a total of $n(n - 1)(n - 2) \ldots (n - k + 1)$ different permutations of k

objects, taken from a pool of n objects, exist. If we denote this number by $P(n, k)$, we can write $P(n, k) = n! / (n - k)!$

perpendicular
At right angles. Two lines, planes, etc., are said to be perpendicular if they are 90° apart.

Perrin sequence
The sequence of integers defined by the recurrence relation $P(n) = P(n - 2) + P(n - 3)$, with the initial conditions $P(0) = 3$, $P(1) = 0$, $P(2) = 2$. Although the sequence is named after R. Perrin who studied it in 1899, it had been explored earlier, in 1876, by Édouard **Lucas**. As in the case of the similar *Padovan sequence,* the ratio of consecutive Perrin numbers tends toward the **plastic number**. More importantly, it appears that n divides $P(n)$ exactly if and only if n is a **prime number**. For example, 19 is prime, $P(19) = 209$ and $^{209}/_{19} = 11$; on the other hand, 18 is composite, $P(18) = 158$ and $^{158}/_{18} = 8.777$, which is not a whole number. Lucas conjectured that this is true for all values of n so that the Perrin sequence can be used as a test for non-primality: any number n that does not divide $P(n)$ is composite. Whether this conjecture is correct remains an open question. No one has ever found a composite n that divides $A(n)$, nor has anyone been able to prove that such numbers, known as *Perrin pseudoprimes,* don't exist. In 1991 Steven Arno of the Supercomputing Research Center in Bowie, Maryland, proved that Perrin pseudoprimes must have at least 15 digits.[15] The conjecture that no Perrin pseudoprimes exist is important, because the remainder on dividing $P(n)$ by n can be calculated very rapidly.

perturbation
A slight nudge that, temporarily or permanently, displaces an object or system out of **equilibrium**.

peta-
Prefix for 10^{15}, from the Greek *pentakis,* meaning "five times."

phase space
The mathematical space of all possibilities in a given situation. A motion is then described by a *path, trajectory,* or *orbit* in this space. This is not the usual kind of path laid out on the ground, but a series of locations in phase space, describing motion or change over a period of time. The terms do, however, recall the origins of qualitative dynamics in Henri **Poincaré**'s study of planetary motion. The *dimension* of the phase space is the number of initial conditions needed to uniquely specify a path and is equal to the number of variables in the dynamical system. The temporal behavior of the system is viewed as the succession of states in the system's state space. In the case of a simple pendulum, for example, the instantaneous configuration is given by just two numbers—the position of the pendulum bob and its velocity—which completely describe the system's state. For more complex systems, such as a chain of n pendulums coupled together, the state of the system is much larger. It requires, in this case, $2n$ numbers to specify the state of the entire system. This collection of all possible configurations is the phase space.

phase transition
In physics, a change from one state of matter to another. In **dynamical system**s theory, a change from one mode of behavior to another.

phi
See **golden ratio**.

Phillips, Hubert (1891–1964)
An English compiler of crosswords and word puzzles who wrote under the pseudonyms "Dogberry" in the *News Chronicle* magazine and "Caliban" in the *New Statesman.* He was also a prolific writer of epigrams, parodies, and satirical verse, and appeared on many radio quizzes. Phillips earned a first class degree in history at Oxford, served in the army throughout World War I, taught economics at Bristol University, and was active in the British Liberal Party. He was an accomplished player of contract bridge, captaining England in 1937 and 1938. Among his many publications are *Caliban's Problem Book, The Complete Book of Card Games, Brush Up Your Wits, My Best Puzzles in Mathematics,* over a hundred crime-problem stories, and a novel, *Charteris Royal.*[249, 250, 251]

Phutball
Also known as *philosopher's football,* a two-player board **game** first described by Elwyn Berlekamp, John **Conway**, and Richard **Guy**, in their book *Winning Ways for your Mathematical Plays.* Phutball is played on the intersections of a 19×15 grid using one white piece and as many black pieces as necessary. The objective is to score goals by using the men (black pieces) to move the football (white piece) onto or over the opponent's goal line, that is, either rows 1 or 0 or rows 19 and 20 (rows 0 and 20 being off the board). At the start of the game the football is placed on the central point, although a handicapping scheme exists where the ball can start nearer to the stronger player's goal line. Players alternate making moves, which consist either of adding a man to any vacant point on the board or of moving the ball. There is no difference between men played by the two opponents. The football is moved by a series of jumps of adjacent men. One jump is from the football's c

point to the first vacant point in a straight line orthogonally or diagonally over one or more men. The jumped men are then removed from the board (before any subsequent jump occurs). Jumping is optional and can be repeated for as long as there are men available to be jumped and the player desires to do so. In contrast to checkers, multiple men in a row are jumped and removed as a group. If the football ends the move on or over the opponent's goal line then a goal has been scored. If the football passes through a goal line, but ends up elsewhere due to further jumps, the game continues. The game is sufficiently complex that, although theoretically one of the players has a winning strategy, this strategy is not known.

pi (π)

The ratio of the circumference to the diameter of a **circle**. Pi, sometimes known as *Archimedes' constant,* is one of the most important and ubiquitous numbers in mathematics, popping up in all kinds of seemingly unrelated areas. Its approximate value is 3.1419 but, being an **irrational number**, it can't be written as a terminating or recurring decimal. The fact that π can't be expressed as the ratio of two integers was proved by Johann Lambert in 1761. Then, in 1881, Ferdinand Lindemann showed that it is a **transcendental number**, meaning that there is no polynomial with integer or rational coefficients of which π is a root. Because of this, it's impossible to write π in terms of *any* finite number of integers or fractions or their roots. From the outset, this dooms any attempt at **squaring the circle**, that is, constructing a square whose area is equal to the area of a given circle, using straightedge and compass alone.

The association of the Greek letter π with the number 3.141 . . . started with William **Oughtred** in 1647 who used π.δ as a stand-in for "periphery-diameter" in his *Clavis mathematicae.* The Welsh mathematician William Jones was the first to use π as a symbol on its own in his 1706 *Synopsis palmariorum matheseos.* But it fell to the great Leonhard **Euler** to popularize the notation 30 years later.

Calculating π with greater and greater precision became a popular pursuit. Around 1600, Ludolph van Ceulen computed the first 35 decimals and was so proud of his accomplishment that he had the number inscribed ⌐n his tombstone. In 1789, the Slovene mathematician ⌐nded known π to 140 places (of which only while, in 1873, William Shanks pushed to 707 places—a record that stood until first electronic computers appeared on ptember 2002, Yasumasa Kanada and his ⌐kyo University used 400 hours of super- / to calculate π to 1.24 trillion places, beat-

ing their previous best, set in 1999, of 206 billion places. Although such an exercise may seem pointless, it serves as a benchmark for new high-speed computers and algorithms, and also to test the long-standing, but still unproven assertion that the distribution of digits in π is completely random.

For the record, the first 100 digits of π are: 3.1415926535 8979323846 2643383279 5028841971 6939937510 5820974944 5923078164 0628620899 8628034825 3421170679. The wonderfully pointless pastime of "piphilology" is concerned with devising mnemonics for remembering some of these: the number of letters in each word of the mnemonic gives the corresponding digit. A famous example is Isaac Asimov's "How I want a drink, alcoholic of course, after the heavy lectures involving quantum mechanics!" For the more ambitious, there is Michael Keith's 1995 poem that supplies the first 42 digits:

> Poe, E. Near A Raven
> Midnights so dreary, tired and weary,
> Silently pondering volumes extolling all by-now
>   obsolete lore.
> During my rather long nap, the weirdest tap!
> An ominous vibrating sound disturbing my cham-
>   ber's antedoor.
> "This," I whispered quietly, "I ignore."

There are some reasonably good approximations to π in the form of ordinary (rational) fractions. The best known is $^{22}/_7$, but this is only accurate to two decimal places. A fraction with a larger denominator offers a better chance of getting a more refined estimate. There is also $^{333}/_{106}$, which is good to 5 places. But an outstanding approximation is $^{355}/_{113}$ which is accurate to 6 places; in fact, there is no better approximation among all fractions with denominators less than 30,000. See also **continued fraction**.

In 1897 the General Assembly of the state of Indiana tried to pass legislation to the effect that the exact value of π is 3.2. The bill was referred, for some bizarre reason, to the House Committee on Canals. But then the true motive for the attempted change to the law became clear. By chance a professor of mathematics happened to be present during a debate and heard an ex-teacher saying "The case is perfectly simple. If we pass this bill, which establishes a new and correct value for π, the author offers to our state without cost the use of his discovery and its free publication in our school textbooks, while everyone else must pay him a royalty." Fortunately, the professor was able to teach the senators a little math and the bill was stopped in its tracks.

Any formulas to do with circles or spheres, not surprisingly, have π in them. For example, the circumference of a circle of radius r is $2\pi r$, the area of a circle is πr^2, the volume of a sphere is $(\frac{4}{3})\pi r^3$, and the surface area of a sphere is $4\pi r^2$. But π also has a strange habit of appearing in the most unexpected places. In the eighteenth century, the French naturalist Comte de Buffon showed that you could estimate π by experiment using the unlikely apparatus of a repeatedly dropped needle and a shove-halfpenny board (see **Buffon's needle**). Two of the most important, but (as far as is known) unrelated, equations in modern physics, Heisenberg's uncertainty principle ($\Delta x\ \Delta p = h/4\pi$) in **quantum mechanics** and Einstein's field equation ($R_{ik} - \frac{1}{2}\ g_{ik}\ R + \Lambda\ g_{ik} = 8\pi G/c^4\ T_{ik}$) in general **relativity theory** include π. The ever-present constant also emerges as a result of various remarkable **infinite series**. These include:

$$\frac{1}{1^2} + \frac{1}{2^2} + \frac{1}{3^2} + \frac{1}{4^2} + \ldots = \pi^2/6$$
(found by Leonhard **Euler**)

$$\frac{1}{1} - \frac{1}{3} + \frac{1}{5} - \frac{1}{7} + \frac{1}{9} - \ldots = \pi/4$$
(found by Gottfried **Leibniz**)

$$\frac{2}{1} \times \frac{2}{3} \times \frac{4}{3} \times \frac{4}{5} \times \frac{6}{5} \times \frac{6}{7} \times \frac{8}{7} \times \frac{8}{9} \times \ldots = \pi/2$$
(found by John **Wallis**)

Among many other formulas that give rise to or contain π are

the integral
$$\int_{-\infty}^{\infty} e^{-x^2}\ dx = \sqrt{\pi},$$

Stirling's formula, $n! \sim \sqrt{2\ \pi\ n}\ (n/e)^n$ and, the most intriguing equation in mathematics, $e^{i\pi} + 1 = 0$.

In **number theory**, the probability that two randomly chosen integers have no common divisors (in other words, that they're *relatively prime*) is $6/\pi^2$ or 1 in $1.644934\ldots$, and the average number of ways to write a positive integer as the sum of two **perfect squares** is $\pi/4$. Both of these facts are astonishing because it is hard to see how a basically geometric constant to do with circles has any bearing at all on how various types of numbers are distributed. Deep truths are buried here!

In 1995, David Bailey, Peter Borwein, and Simon Plouffe of the University of Quebec at Montreal discovered a new formula for π as an infinite series:[21]

$$\pi = \sum_{k=0}^{8} \frac{1}{16^k}\left[\frac{4}{8k+1} - \frac{2}{8k+4} - \frac{1}{8k+5} - \frac{1}{8k+6}\right]$$

What is so remarkable about this is that it enables the calculation of *isolated digits* of π–say, the trillionth digit–without computing and keeping track of all the preceding digits. How such a formula could possibly arise constitutes a mystery in itself. The only catch is that the formula works for base 2 (**binary**) and 16 (**hexadecimal**) but not base 10. So, it's possible to use the formula to determine, for example, that the five-trillionth binary digit of π is 0, but there's no way to convert the result into its decimal equivalent without knowing all the binary digits that come before the one of interest. The new formula allows the calculation of the nth base 2 or base 16 digit of π in a time that is essentially linear in n, with memory requirements that grow logarithmically (very slowly) in n. One possible use of the Bailey-Borwein-Plouffe formula is to help shed light on whether the distribution of π's digits are truly random, as most mathematicians suppose.[29, 42, 68, 114] See also **pyramid**.

Pick's theorem

First published in 1899, a theorem that was brought to broad attention as recently as 1969 through Hugo **Steinhaus**'s popular book *Mathematical Snapshots*.[316] Pick's theorem gives an elegant formula for the area of *lattice polygons*–**polygon**s that have vertices located at the integral nodes of a square grid or lattice that are spaced a unit distance from their immediate neighbors. Pick's theorem says that the area of such a polygon can be found simply by counting the lattice points on the interior and boundary of the polygon. The area is given by

$$i + (b/2) - 1,$$

where i is the number of interior lattice points and b is the number of boundary lattice points. The Austrian mathematician Georg Pick (1859–1942) after whom the result is named, was born in Vienna and perished during World War II in the Theresienstadt concentration camp. Over the past few decades, beginning with a paper by J. E. Reeve in 1957, various generalizations of Pick's theorem have been made to more general polygons, to higher-dimensional polyhedra, and to lattices other than square lattices. Most recently, mathematicians have become interested in the theorem because it provides a link between traditional Euclidean geometry and the modern subject of digital (discrete) geometry.

pico-

Prefix meaning a trillionth (10^{-12}), from the Italian *piccolo*, meaning "small."

pitch drop experiment

The world's longest running experiment. It started in 1927 when Thomas Parnell, the first professor of physics at the University of Queensland in Brisbane, Australia, heated a sample of pitch (a derivative of tar) and poured it into a glass funnel with a sealed stem. Three years were allowed for the pitch to settle, then, in 1930, the sealed stem was cut. From that date on the pitch has slowly dripped out of the funnel–so slowly that, up to the present time, only eight drops have fallen! The experiment

pitch drop experiment   Professor John Mainstone alongside the world's longest-running experiment. *John Mainstone, University of Queensland*

stands in a display cabinet in the foyer of the Department of Physics at the University of Queensland demonstrating for all to see the fact that pitch, though it feels like a solid and is brittle enough to smash with a hammer, is really a fluid of very high viscosity (about 100 billion times as viscous as water).[91]

pizza
Order a pizza, pick an arbitrary point in it, and cut the pizza into eight slices by slicing at 45° angles through the chosen point. Color the alternate pieces red, with ketchup, and yellow, with mustard. Measure the area of the red slices and the yellow slices. Surprisingly, it will be found that these areas are the same! This fact holds true for any multiple of four slices cut by using equal angles through a fixed arbitrary point in the pizza. See also **tautology**.

place-value system
A **number system** in which the value of a number symbol depends not only on the symbol itself but also on the position where it occurs.

Planck time
In **quantum mechanics**, the shortest meaningful period of **time**; any two events that are separated by less than this amount of time can be considered simultaneous. It

has the value 5.390×10^{-44} second. Related to this is the *Planck length* of 6.160×10^{-35} meters, which is the distance that light can travel in the Planck time.

plane
A flat surface such that a straight **line** joining any two points on it will also lie entirely on the surface.

plane partition
A stack of unit cubes in a rectangular box or in the positive **octant** in space such that to the left, behind, and below every cube lies either another cube or a wall. A plane partition in a box is equivalent to a **lozenge** tiling of a hexagon in the plane.

plastic number
A little-known number that has much in common with the **golden ratio** in that it is closely linked to architecture and to aesthetics. The concept of the plastic number was first described by the Dutchman Hans van der Laan (1904–1991) in 1928, shortly after he had abandoned his

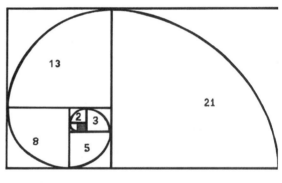

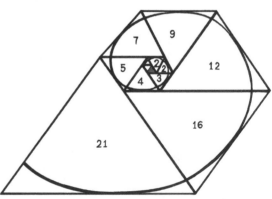

plastic number   Spiraling systems illustrate the Fibonacci numbers (top) and the Padovan sequence.

architectural studies and become a novice monk, and has subsequently been explored by the English architect Richard Padovan (1935–). It is derived from a **cubic** equation, rather than a **quadratic** in the case of the golden ratio, and is intimately linked to two ratios, approximately 3:4 and 1:7, which van der Laan considered fundamental in the relationship between human perception and shape and form. These ratios, he believed, express the lower and upper limits of our normal ability to perceive differences of size among three-dimensional objects. The lower limit is that at which things differ just enough to be of distinct types of size. The upper limit is that beyond which they differ too much to relate to each other; they then belong to different orders of size. According to van der Laan, these limits are precisely definable. The mutual proportion of three-dimensional things first becomes perceptible when the largest dimension of one thing equals the sum of the two smaller dimensions of the other. This initial proportion determines, in turn, the limit beyond which things cease to have any perceptible mutual relation.

In mathematical terms, the plastic number is the unique **real number** solution to the equation $x^3 - x - 1 = 0$ and has the approximate value 1.324718 . . . . Just as the golden ratio is approximated better and better by successive terms of the **Fibonacci sequence**, $F(n + 1) = F(n) + F(n - 1)$, where $F(0) = F(1) = 1$, so the plastic number arises as the limit of the ratio of successive numbers in the sequence $P(n + 1) = P(n - 1) + P(n - 2)$ where $P(0) = P(1) = P(2) = 1$. This sequence has been called the *Padovan sequence,* and its members the *Padovan numbers.* The Padovan sequence increases much more slowly than does the Fibonacci sequence. Some numbers, such as 3, 5, and 21, are common to both sequences; however, it is not known if there are finitely many or infinitely many such pairs. Some Padovan numbers, such as 9, 16, and 49, are perfect squares; the square roots of these (3, 4, and 7) are also Padovan numbers, but it is not known if this is a coincidence or a general rule. Another way to generate the Padovan numbers is to mimic the use of squares for Fibonacci numbers, but with cuboid structures—boxes with rectangular faces. A kind of three-dimensional spiral of boxes emerges. Start with a cube of side 1 and place another adjacent to it: the result is a $1 \times 1 \times 2$ cuboid. On the 1×2 face, add another $1 \times 1 \times 2$ box to produce a $1 \times 2 \times 2$ cuboid. Then on a 2×2 face, add a $2 \times 2 \times 2$ cube to form a $2 \times 2 \times 3$ cuboid overall. To a 2×3 face, add a $2 \times 2 \times 3$ to get a $2 \times 3 \times 4$ box overall, and so on. Continue this process, always adding cuboids in the sequence east, south, down, west, north, and up. At each stage the new cuboid formed will have three consecutive Padovan numbers as its sides. Moreover, if successive square faces of the added cuboids are connected

by straight lines, the result is a spiral that lies in a plane. The Padovan sequence is very similar to the **Perrin sequence**.[319]

Plateau, Joseph Antoine Ferdinand (1801–1883)

A Belgian physicist who did pioneering experimental work on soap bubbles and soap films, which stimulated the mathematical study of bubbles as **minimal surface**s. In 1829 Plateau carried out an optical experiment that involved looking at the Sun for 25 seconds: this damaged his eyes, and he eventually became blind.

Plateau curves

A set of curves, named after Joseph **Plateau**, that are described by the parametric equations:

$$x = a \sin(m + n)t / \sin(m - n)t$$
$$y = 2a \sin(mt) \sin(nt) / \sin(m - n)t$$

If $m = 2n$ the Plateau curves become a circle with a center at $(1, 0)$ and a radius of 2.

Plateau problem

The general problem of determining the shape of the **minimal surface** constrained by a given boundary. It is named after Joseph **Plateau** who noticed that a handful of simple patterns seemed to completely describe the geometry of how soap **bubble**s fit together. Plateau claimed that soap bubble surfaces always make contact in one of two ways: either three surfaces meet at 120° angles along a curve; or six surfaces meet at a vertex, forming angles of about 109°. For instance, in a cluster of bubbles, two intersecting bubbles (of possibly different sizes) will have a common dividing wall (the third surface) that meets the outer surfaces of the bubbles in 120° angles. On the other hand, the edges of the six soap-film faces that emerge within a tetrahedral wire frame, when dipped in a soapy solution, form angles of roughly 109° at a central vertex. Until the American mathematician Jean Taylor came along in the mid-1970s, Plateau's patterns were just a set of empirical rules. However, as a follow-up to her doctoral thesis, Taylor was able to prove that Plateau's rules were a necessary consequence of the energy-minimizing principle—no other yet unobserved configurations were possible—thus settling a question that had been open for more than a century. The forces acting along the surface of a soap bubble all have the same magnitude in all directions. In crystals this is not the case (magnitudes of surface forces differ in different directions, though they may exhibit a grain, analogous to that in a piece of wood), but they still require the least energy to enclose a given volume. Minimal surfaces that model these conditions, like a cube with its corners chopped off or the bottom half of a cone

Platonic solid Clockwise from the extreme right: the cube, tetrahedron, octahedron, icosahedron, and dodecahedron. *Robert Webb, www.software3d.com; created using Webb's Stella program*

mounted on a cylinder, are known as *Wulff Shapes*, and provide fertile ground for mathematical study today.[328]

Platonic solid

Any of the five regular **polyhedron**s–solids with regular **polygon** faces and the same number of faces meeting at each corner–that are possible in three dimensions. They are the **tetrahedron** (a pyramid with triangular faces), the **octahedron** (an eight-sided figure with triangular faces),

the **dodecahedron** (a 12-sided figure with pentagonal faces), the **icosahedron** (a 20-sided figure with triangular faces), and the **cube** (see table, "The Platonic Solids"). They are named after **Plato** who described them in one of his books, though it was **Euclid** who proved that there are only five regular polyhedra. A regular solid with hexagonal faces cannot exist because if it did, the sum of the angles of any three hexagonal corners that meet would already equal 360°, so such an object would be

The Platonic Solids

Name	Number of			Schläfi Symbol
	Faces	Edges	Vertices	
Tetrahedron	4	6	4	{3,3}
Cube	6	12	8	{4,3}
Octagon	8	12	6	{3,4}
Dodecahedron	12	30	20	{5,3}
Icosahedron	20	30	12	{3,5}

planar. See also **Archimedean solid**, **Catalan solid**, and **Johnson solid**.

Platonism
The belief that mathematical objects exist independent of physical models. It is, at the very least, a useful pretense in mathematics, especially in **geometry**.

pleated surface
A surface in **Euclidean space** or hyperbolic space (see **hyperbolic geometry**) that resembles a **polyhedron** in the sense that it has flat faces that meet along edges. Unlike a polyhedron, a pleated surface has no corners, but it may have infinitely many edges that form a **lamination**.

Poggendorff illusion
A **distortion illusion** in which the two ends of a straight line passing behind a rectangle appear offset when, in fact, they are aligned. It was discovered in 1860 by the physicist J. C. Poggendorff, editor of *Annalen der Physik und Chemie* (Annals of physics and chemistry), after receiving a letter from the astronomer Friedrich Zöllner. In this letter, Zöllner described an illusion (see **Zöllner illusion**) he had noticed on a fabric design in which parallel lines intersected by a pattern of short diagonal lines appear to diverge.[134]

Poincaré, (Jules) Henri (1854–1912)
A French theoretical physicist and mathematician who almost discovered **relativity theory** before Einstein and who is most famous in mathematics because of his hypothesis known as the **Poincaré conjecture**. Poincaré, who taught for most of his life at the University of Paris, was sometimes referred to as "the last universalist," because of his huge published output on a wide variety of mathematics and mathematical physics. Strangely, he was also clumsy, absent-minded, and inept at simple

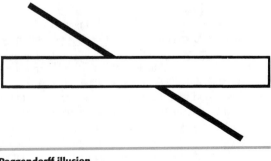

Poggendorff illusion

arithmetic. In order to translate problems in **topology** into questions in **algebra**, he devised (or discovered) *homotopy groups*—quantities that capture the essence of multidimensional spaces in algebraic terms, and have the power to reveal similarities between them.

Poincaré conjecture
A proposition in **topology** put forward by Henri **Poincaré** in 1904. Poincaré was led to make his conjecture during his pioneering work in topology, the mathematical study of the properties of objects that stay unchanged when the objects are stretched or bent. In loose terms, the conjecture is that every three-dimensional object that has a set of spherelike properties (i.e., is topologically equivalent to a sphere) can be stretched or squeezed until it is a three-dimensional sphere (a 3-sphere) without tearing (i.e., making a hole in) it. Strictly speaking, the conjecture says that every **closed**, **simply connected** three-**manifold** is **homeomorphic** to the 3-sphere.

Poincaré proved the two-dimensional case and he guessed that the principle would hold in three dimensions. Determining if the Poincaré conjecture is correct has been widely judged the most important outstanding problem in topology—so important that, in 2000, the Clay Mathematics Institute in Boston named it as one of seven Millennium Prize Problems and offered a $1 million prize for its solution. Since the 1960s, mathematicians have shown by various means that the generalized conjecture is true for all dimensions higher than three—the four-dimensional case finally falling in 1982. But none of these strategies work in three dimensions. On April 7, 2002, came reports that the Poincaré conjecture might have been proved by Martin Dunwoody of Southampton University, but within a few days a fatal flaw was found in his proof. Then, in April 2003, what appears to be a genuine breakthrough emerged during a series of lectures delivered at the Massachusetts Institute of Technology by the Russian mathematician Grigori Perelman of the Steklov Institute of Mathematics (part of the Russian Academy of Sciences in St. Petersburg). His lectures, entitled "Ricci Flow and Geometrization of Three-Manifolds," constituted Perelman's first public discussion of important results contained in two earlier preprints. Mathematicians will now scrutinize the validity of Perelman's work (which does not actually mention the Poincaré conjecture by name). In any event, the Clay Institute requires a two-year cooling-off period before the prize can be awarded.

Poincaré disk
The region inside (but not including) a bounding circle, where straight lines are defined to be either diameters of the bounding circle, or arcs of circles that are perpendicular to the bounding circle. The Poincaré disk is a model

Poincaré disk   A pattern on a hyperbolic surface. *Jos Leys, www.josleys.com*

of **hyperbolic geometry**. The great flexibility it allows in specifying the angles of a triangle leads to an infinite variety of ways in which the disk can be tiled.

Poinsot, Louis (1777–1859)

A French mathematician who invented *geometrical mechanics* while investigating how a system of forces acting on a rigid body can be resolved into a single force and a couple (a pair of equal and oppositely directed forces, as when you try to unscrew a bottle cap). Together with Gaspard **Monge**, he helped geometry regain its leading role in mathematical research in France in the eighteenth century. He wrote an important work on polyhedra in 1809, discovering four new regular polyhedra, which are now known as the **Kepler-Poinsot solids**. Two of these had been found by Johannes **Kepler** in 1619 but Poinsot was unaware of this; the two additional ones that Poinsot discovered were the *great dodecahedron* and the *great icosahedron*. In 1810 Augustin **Cauchy** proved that, with this definition of regular, the enumeration of regular polyhedra is complete (although a mistake was discovered in Poinsot's, and hence Cauchy's, definition in 1990 when an internal inconsistency became apparent). Poinsot also worked in number theory, studying **Diophantine equations** with a view to expressing numbers as the difference of two squares and primitive roots.

point

A dimensionless geometric object having no properties other than location or place. More generally, an element in a geometrically described **set**.

point-set topology

Also known as *general topology*, a branch of **topology** concerned with how to put a structure on a **set** in such a way as to generalize the idea of **continuity** for maps from the real numbers to itself. A topology on a set X is a certain set of *open subsets* of the set X which satisfy various axioms. The set X together with this topology is called a **topological space**.

Poisson, Siméon Denis (1781–1840)

A French mathematician whose main interest lay in the application of mathematics to physics, especially in electrostatics and magnetism. He developed a two-fluid theory of electricity and provided theoretical support for the experimental results of others, notably Charles de Coulomb. Poisson also made important contributions to mechanics, especially the theory of elasticity; optics; calculus, especially definite integrals; differential geometry; and probability theory. In all, he wrote more than 300 papers on mathematics, physics, and astronomy, and his *Traité de Mécanique* (1811) was long a standard work.

polar coordinates

A **coordinate** system in which distances are measured from a fixed reference point (the pole) and angles from a fixed reference line. A *polar equation* is one that uses polar coordinates.

pole

(1) The **origin** of a system of **polar coordinates**. (2) In **complex analysis**, a function of a certain simple type of **singularity**. (3) One of the two points, where the axis of rotation of a rotating body, such as Earth, passes through the surface. (4) An old unit of length, also called a *rod*, equal to 5.5 yards.

Pólya's conjecture

A hypothesis put forward by the Hungarian mathematician George Pólya (1887–1985) in 1919. A positive integer is said to be of *even type* if it factorizes into an even number of **prime numbers**; otherwise it is said to be of *odd type*. For example, 4, $= 2 \times 2$, is of even type, whereas 18, $= 2 \times 3 \times 3$, is of odd type. Let $O(n)$ be the number of odd type and $E(n)$ be the number of even type integers in the first n integers. Pólya's conjecture says that $O(n) \geq E(n)$ for all $n \geq 2$. After the conjecture had been checked for all values of n up to 1 million, many mathematicians assumed it was probably true. However, in 1942 A. E. Ingham came up with an ingenious method to show how a counterexample could be constructed, even though there wasn't enough computing power around at the time to do the necessary calculations.[177] Twenty years later, R. S. Lehman ran Ingham's method on a computer to find a counterexample to Pólya's conjecture at $n = 906180359$.[199]

polychoron

An unofficial name (from the Greek *poly* meaning "many" and *choros* meaning "room" or "space") for a four-dimensional **polytope**; it was first advocated by George Olshevsky.

polycube

A **polyhedron** formed by joining unit cubes by their faces. Examples of puzzles that involve polycubes are the **Soma cube** and **Rubik's cube**.

polygon

A plane closed figure whose sides are straight lines. The term *polygon* (from the Greek *poly* for "many" and *gwnos* for "angle") sometimes also refers to the *interior* of the polygon (the open area that this path encloses) or to the **union** of both. A polygon is *simple* if it is described by a single, nonintersecting boundary; otherwise it is said to be *complex*. A simple polygon is called *convex* if it has no internal angles greater than 180°; otherwise it is called *concave*. A polygon is called *regular* if all its sides are of equal length and all its angles are equal (see table, "Regular Polygons"). Any polygon, regular or irregular, has as many angles as it has sides. See **constructible** (for constructible polygons).

Regular Polygons

Name	Sides	Angle (= 180°–360°/sides)
Equilateral **triangle**	3	60°
Square	4	90°
Regular **pentagon**	5	108°
Regular **hexagon**	6	120°
Regular heptagon	7	128.57° (approx.)
Regular **octagon**	8	135°
Regular nonagon	9	140°
Regular **decagon**	10	144°
Regular hectagon	100	176.4°
Regular megagon	10^6	179.99964°
Regular **googolgon**	10^{100}	180° (approx.)

polygonal number

The number of equally spaced dots needed to draw a **polygon**. Polygonal numbers, which are a type of **figurate number**, include square numbers, triangular numbers, and hexagonal numbers.

polyhedron

A three-dimensional object whose faces are all **polygons** and whose edges are shared by exactly two polygons. *Polyhedron* comes from the Greek *poly* for "many" and -*hedron* meaning "base," "seat," or "face." Every polyhedron in three-dimensional space consists of (two-dimensional) faces, (one-dimensional) edges, and (zero-dimensional) vertices. Sometimes the term *polyhedron* is used to apply to figures in more than three dimensions; however, analogs of polyhedra in the **fourth dimension** or higher are also referred to as **polytopes**. Polyhedrons, like polygons, may be *convex* or *nonconvex*. If a line that connects any two points on the surface of a polyhedron is completely inside or on the polyhedron, the figure is convex. Otherwise, it is nonconvex or *concave*. A polygon is *regular* if all of its faces are exactly the same size and shape and if the same number of faces meet at each vertex. There are only five regular convex polyhedra–the **Platonic solids**. Another four regular nonconvex polyhedra are called **Kepler-Poinsot solids**. However, the term *regular polyhedra* is sometimes used to describe only the Platonic solids. A convex polyhedron is said to be *semi-regular* if its faces have a similar arrangement of nonintersecting regular plane convex polygons of two or more different types about each vertex. These solids, of which there are 13 different kinds, are commonly called the **Archimedean solids**. A *dual* of a polyhedron is another polyhedron in which faces and vertices occupy complementary locations. The duals of the Archimedean solids are known as the **Catalan solids**. A **quasi-regular polyhedron** is the solid region interior to two dual regular polyhedra; only two exist: the cuboctahedron and the icoidodecahedron. There are also infinite families of **prisms** and **antiprisms**. In total there are 92 convex polyhedra with regular polygonal faces (and not necessarily equivalent vertices); these are the **Johnson solids**.

The oldest known examples of man-made polyhedra were found on the islands of northeastern Scotland and date back to Neolithic times, between 2000 and 3000 B.C. These stone figures are about 2 inches in diameter and many are carved into rounded forms of regular polyhedra. Examples including cubical, tetrahedral, octahedral, and dodecahedral forms, one which is the dual of the pentagonal prism, are on display in the Museum of Scotland and in Oxford's Ashmolean Museum.

polyiamond

A shape made from identical equilateral **triangles** that have been joined at their edges.

polynomial

An expression in which whole-numbered powers of a variable multiplied by numerical coefficients are added together. The powers of the variable must be positive integers, or zero. An example of a polynomial is $3x^3 + 7x - 2$. A polynomial *equation* has a polynomial expression, or zero, on each side of the equal sign: for example, $4a^2 - 5.6a + 1.7 = 0$ or $5x^2 - 1 = 8x + 2x^4$. The type of polynomial expression or equation is determined by the highest power present of the variable. A **quadratic**, for example, has nothing higher than a squared term. **Cubics**, **quartics**, and **quintics** have maximum powers of three, four, and five, respectively. Mathematicians who have done pioneering work on each of these higher types of polynomial equations tended, for some reason, to have had colorful and star-crossed lives. Niccolo **Tartaglia**, who first solved the cubic, failed miserably as a mathematician for the rest of his life, largely because he spent it trying to discredit Girolamo **Cardano**. Tartaglia

told Cardano his method of solution and swore him to secrecy but Cardano went ahead and published the solution anyway. Cardono himself lived a long unhappy life and his only son was executed for murder. Lodovico Ferrara, Cardano's student, who solved the general quartic, was poisoned, probably by his sister, over an inheritance dispute. Finally, Evariste **Galois**, who showed that the general quintic was unsolvable, died in a duel at the age of 20.

polyomino

A two-dimensional shape made by connecting n squares of the same size along their edges. The polyomino is a generalization of the **domino**, which can be thought of as a polyomino with $n = 2$. Several puzzles involving polyominoes made from three or more squares stuck together became popular around the beginning of the twentieth century. The best-known of these is the Broken Chessboard, presented as problem no. 74 in Henry **Dudeney**'s book *The Canterbury Puzzles* (1907).[87] Dudeney gives an amusing introduction to the problem, first quoting from Hayward's *Life of William the Conqueror* (1613) about an incident in which Prince Henry,

polyomino A complete set of 12 solid pentominos, also known as planar pentacubes. *Kadon Enterprises, Inc., www.gamepuzzles.com*

one of William's sons, smashes a chessboard over the head of his brother, Prince Robert, and then adding his own mathematical appendix. By a curious quirk, Dudeney reveals, the board breaks into 13 pieces, 12 of which are the different possible ways in which five squares can be arranged and one square 2 × 2 piece. The puzzle is to reconstitute the board from these fragments. This is the earliest example of a mathematical recreation that involves pentominoes, a name that wasn't coined until 1953 when Solomon **Golomb** first used it in a talk to the Harvard Math Club. Golomb not only invented the nomenclature for polyominoes but did much of the pioneering research on them; his work was brought to a wide audience and popularized by Martin **Gardner** in his *Scientific American* column, beginning in 1957.[114, 136, 137]

Golomb was particularly interested in the pentomino, for a reason that becomes clear (see table, "Types of Polyomino"). The familiar domino comes in just one configuration as, obviously, does the trivial monomino. A triomino can have two different shapes—three squares in a line or in an L shape. The tetromino (or tetramino) has five distinct arrangements and is used in the popular video game **Tetris**. Thus for $n \leq 4$ the number of distinct pieces is ≤ 5, which restricts the variety of combinations possible. On the other hand, for $n > 5$, the number of different pieces is large, making analysis of problems difficult and games based on such polyominoes difficult and unwieldy. However, pentominoes, for which $n = 5$, come in 12 unique configurations, which is just about an ideal balance between tractability and combinatorial richness, hence the high level of interest in this particular type of polyomino. The problem of tiling an 8 × 8 chessboard with a square hole in the center using pentominoes was first solved in 1935 and found, by computer in 1958, to have exactly 65 solutions. Another standard pentomino puzzle is to arrange the set of 12 possible shapes into rectangles without holes: 3 × 20, 4 × 15, 5 × 12, and 6 × 10. Pentominoes are prominently featured in a subplot of the novel *Imperial Earth* by Arthur C. Clarke.

Hexominoes and heptominoes come in 35 and 108 unique arrangements, respectively. One of the 108 heptomino configurations, however, has a hole—a region that is not tiled with squares but that is unconnected to the exterior of the polyomino—and may or may not be counted as a valid piece depending on the rules of a particular game. All polyominoes made of seven or more squares may contain holes. There is no known algorithm or formula for calculating how many distinct polyominoes of each order there are.

Related to polyominoes are **polyiamond**s (formed from equilateral triangles) and polyhexes (formed from regular hexagons). The three-dimensional analog of polyominoes uses cubes instead of squares; an example is the **Soma cube**.

Types of Polyomino

Name	Number of Unit Squares	Number of Distinct Configurations
Monomino	1	1
Domino	2	1
Triomino	3	2
Tetromino	4	5
Pentomino	5	12
Hexomino	6	35
Heptomino	7	108

polytope

A higher-dimensional analogue of a **polygon** or **polyhedron**. The number of possible regular polytopes depends on the number of dimensions. In two dimensions there

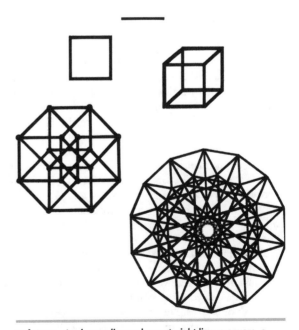

polytope   In descending order: a straight line, a square, a cube, a tesseract (four-dimensional cube), and a regular, multidimensional polytope.

are infinitely many possible regular polygons; in three dimensions there are five possible regular polyhedra; in four dimensions there are six possible regular polytopes; and in each number of dimensions higher than four, there are just three possible regular polytopes, analogous to the three-dimensional tetrahedron, cube, and octahedron. A four-dimensional polytope is also sometimes called a *polyhedroid* or a *polychoron*. Just as a polygon has vertices and edges, and a polyhedron has vertices, edges, and faces, a four-dimensional polytope has vertices, edges, faces, and cells, where a cell is a three-dimensional figure.[75] See also **Boole (Stott), Alicia**.

Poncelet, Jean Victor (1788–1867)

A French mathematician who substantially advanced **projective geometry**. With Brianchon, he proved *Feuerbach's theorem* on the **nine-point circle** in 1820–1821, and also suggested the theorem proved by Jakob **Steiner** and now called the *Poncelot-Steiner theorem* that Euclidean constructions (see **constructible**) can be done with a straightedge alone. As a soldier in Napoleon's army, he was captured and imprisoned in Russia. While in prison from 1813–1814, he organized and wrote down his discoveries, and the result was published as *Traité des Propriétés Projectives des Figures* (Treatise on the properties of the projections of shapes, 1822). To serve as an introduction to this work, he also wrote *Applications D'analyse et de Géométrie* (Applications of analysis and geometry, 2 vols., 1862–1864).

Poncelet's theorem

Given an **ellipse**, and a smaller ellipse entirely inside it, start at a point on the outer ellipse, and, moving clockwise, follow a line that is **tangent** to the inner ellipse until you hit the outer ellipse again. Repeat this over and over again. It may be that this path will never hit the same points on the outer ellipse twice. However, if it *does* close up in a certain number of steps, then something amazing is true: *all* such paths, starting at *any point* on the outer ellipse, close up in the same number of steps. This fact is Poncelet's theorem, also known as *Poncelet's closure theorem,* and is named after the Jean **Poncelet**.

Pony Puzzle

One of Sam **Loyd**'s best known and commercially successful puzzles. It consists of just six pieces of the silhouette of a horse that have to be assembled in the most sensible way. Loyd got the idea for the puzzle from the governor of Philadelphia, Andrew Curtin, when the two were on a steamer returning from Europe and talking about the White Horse of Uffington—a famous chalk figure carved on a hillside in Berkshire, England.

Pony Puzzle   One of Sam Loyd's best known brain-teasers.

postage stamp problems

Mathematical puzzles that involve postage stamps have been around almost as long as postage stamps themselves, the first of which, the Penny Black, was issued by Great Britain on May 6, 1840. Some such puzzles ask what postage amounts can or can't be made with stamps of certain values. Others are based on the ways that a block of stamps can be folded or torn along the perforations.

PUZZLE

One of this type appears as problem no. 285 in Henry **Dudeney**'s *Amusements in Mathematics* (1917).[88] It starts by saying you have just bought 12 stamps in a rectangular block of three rows with four stamps in each. (In Dudeney's diagram, they are labeled 1, 2, 3, 4 across the top row, and so on.) He goes on: "[A] friend asks you to oblige him with four stamps, all joined together—no stamp hanging on by a mere corner. In how many different ways is it possible for you to tear off those four stamps? You see, you can give him 1, 2, 3, 4, or 2, 3, 6, 7, . . . , and so on. Can you count the number of different ways in which those four stamps might be delivered?" This can be thought of as a problem involving tetrominoes, which are a type of **polyomino**.

Solutions begin on page 369.

The postage stamp problem, also known as the *Frobenius problem,* is a long-standing challenge in **number theory** and in **computer science**. Suppose a country issues n different denominations of stamps but allows no more than m stamps to be put on a single letter. The postage stamp problem is to write and implement an algorithm (a stepwise set of rules) that, for any given values of m and

n, computes the greatest consecutive range of postage values, from one on up, and all possible sets of denominations that realize that range. For example, for $n = 4$ and $m = 5$, the stamps with values (1, 4, 12, 21) allow the postage values 1 through 71. If the values of the stamps are constant and not part of the input, algorithms can be, and have been, devised that give a short-cut solution. However, in the general case where the number of stamp values *is* part of the input, the postage stamp problem has been shown to be an **NP-hard problem**, and thus not susceptible to an efficient algorithmic approach.

potato paradox
Fred brings home 100 pounds of potatoes, which (being purely mathematical potatoes) consist of 99 percent water. He then leaves them outside overnight so that they consist of 98 percent water. What is their new weight? The surprising answer is 50 pounds!

potential theory
The study of *harmonic functions*. These functions satisfy **Laplace**'s equation, a certain type of **partial differential equation** that commonly arises in physics in problems to do with gravity and electromagnetism.

power
A word that is almost never used in its correct, original sense any more. Strictly speaking, if we write $8 = 2^3$, then 2 is the **base**, 3 is the **exponent**, and 8 is the power. But almost everyone, including most mathematicians, would say that 3 is the power, and that "power" and "exponent" mean the same thing. The misuse has probably come about from a misunderstanding of statements such "eight is the third power of two."

power law
A type of mathematical pattern in which the frequency of an occurrence of a given size is inversely **proportional** to some **power** (or exponent) of its size. For example, in the case of avalanches or earthquakes, large ones are fairly rare, smaller ones are much more frequent, and in between are cascades of different sizes and frequencies. Power laws define the distribution of catastrophic events in self-organized critical systems (see **self-organization**).

power series
An infinite sum of the form

$$a_0 + a_1 x + a_2 x^2 + a_3 x^3 + \dots,$$

where a is a number of any type and x is the variable. Power series are commonly used to define **functions**. For example, the **sine** function can be written

$$\sin x = x - x^3/3! + x^5/5! - x^7/7! + \dots,$$

where "!" stands for **factorial**. Although the series has infinitely many terms, these get small so quickly that only the first few terms make much of a contribution.

power set
The **set** of all **subsets** of a given set, including the **empty set** and the original set. For example, if the original set is $\{a, b, c\}$ then the power set is $\{\{\varnothing, \{a\}, \{b\}, \{c\}, \{a, b\}, \{a, c\}, \{b, c\}, \{a, b, c\}\}$.

power tower
A way of representing very large numbers in terms of stacks of **exponents**. For example:

$$10^{3^{3^3}} = 10^{3^{27}} = 10^{32541865828329} = \text{enormous}$$

In general, and especially if the number at the top of the tower is fairly large, adding another exponent to the bottom of a tower will make the value of the tower much larger than will increasing the size of the bottom exponent. This leads to the (somewhat counterintuitive) result that to know which of two towers is the larger, you can look at how many exponents are in the tower and know right away which is larger. For example:

$$1.1^{1.1^{1.1^{1,000}}} \text{ is much larger than } 1,000^{1,000^{1,000}}$$

In the case of an infinite power tower of the form

$$x^{x^{x^{\cdots}}},$$

the maximum value that x can take and still cause the tower to converge to a finite value is $e^{(1/e)} = 1.444667\dots$. The minimum value of x that will produce convergence is $1/e^e = 0.065988\dots$.

powerful number
Also known as a *squarefull number*, a positive whole number n such that for every **prime number** p dividing n, p^2 also divides n. Every powerful number can be written as $a^2 b^3$, where a and b are positive integers. The first few powerful numbers are 1, 4, 8, 9, 16, 25, 27, 32, 36, 49, 64, 72, 81, 100, and 108. Pairs of consecutive powerful numbers exist, such as (8,9), (288,289), and (675,676). However, no three consecutive powerful numbers are known and, in 1978, Paul **Erdös** conjectured that none exist.

practical number
A number n such that every positive integer less than n is either a divisor or a sum of distinct divisors of n. The first few practical numbers are 1, 2, 4, 6, 8, 12, 16, 18, 20, and 24. All **perfect numbers** are practical.

prime number

[U]pon looking at prime numbers one has the feeling of being in the presence of one of the inexplicable secrets of creation.

—Don Zagier

An **integer** greater than 1 that is divisible only by 1 and itself. Prime numbers have fascinated mathematicians for centuries, in large part because of how they are distributed. At first sight, their occurrence looks random; yet, on closer inspection, they reveal a subtle order or pattern, that seems to hold deep truths about the nature of mathematics and of the world in which we live. The German-born American mathematician Don Zagier (1951–), in his inaugural lecture at Bonn University, put it this way:

There are two facts about the distribution of prime numbers which I hope to convince you. . . . The first is that despite their simple definition and role as the building blocks of the natural numbers, the prime numbers . . . grow like weeds among the natural numbers, seeming to obey no other law than that of chance, and nobody can predict where the next one will sprout. The second fact is even more astonishing, for it states just the opposite: that the prime numbers exhibit stunning regularity, that there are laws governing their behavior, and that they obey these laws with almost military precision.

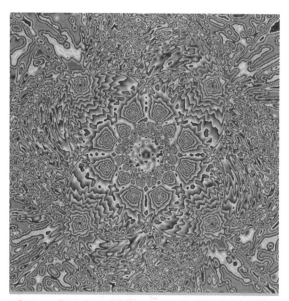

prime number A remarkable pattern generated by the distribution of the first few billion primes. *Jean-Francois Colonna/Ecole Polytechnique*

The prime numbers are 2, 3, 5, 7, 11, 13, 17, 19, 23, 29, 31, 37, 41, 43, 47, 53, 59, ... The *fundamental theorem of arithmetic* declares that the primes are the building blocks of the positive integers: every positive integer is a product of prime numbers in one and only one way, except for the order of the factors. This is the key to their importance: the prime factors of an integer determine its properties. The ancient Greeks proved (c. 300 B.C.) that there are infinitely many primes and that they are irregularly spaced; in fact, there can be arbitrarily large gaps between successive primes. On the other hand, in the nineteenth century it was shown that the number of primes less than or equal to n approaches $n/\log n$, as n gets very large (a result known as the *prime number theorem*), so that a rough estimate for the nth prime is $n \log n$. In his *Disquisitiones Arithmeticae* (1801), Carl **Gauss** wrote: "The problem of distinguishing prime numbers from composite numbers and of resolving the latter into their prime factors is known to be one of the most important and useful in arithmetic. It has engaged the industry and wisdom of ancient and modern geometers to such an extent that it would be superfluous to discuss the problem at length. . . . Further, the dignity of the science itself seems to require that every possible means be explored for the solution of a problem so elegant and so celebrated." The earliest known primality test is the **sieve of Eratosthenes**, which dates from around 240 B.C. However, high-speed computers and fast algorithms are needed to identify large primes. New record-breaking primes tend to be of the variety known as **Mersenne prime**s, since these are the easiest to find. About 6,000 prime numbers are known of which the largest is $2^{20996011} - 1$.

Much remains unknown about the primes. As Martin **Gardner** said:[113] "No branch of **number theory** is more saturated with mystery. . . . Some problems concerning primes are so simple that a child can understand them and yet so deep and far from solved that many mathematicians now suspect they have no solution. Perhaps they are 'undecideable.' Perhaps number theory, like quantum mechanics, has its own uncertainty principle that makes it necessary, in certain areas, to abandon exactness for probabilistic formulations." One of the greatest unsolved problems in mathematics is the **Riemann hypothesis** concerning the distribution of prime numbers. See also **Goldbach's conjecture**, **twin primes**, **Ulam spiral**, and **Ishango bone**.

primitive root

A primitive root for a **prime number** p is one whose powers generate all the nonzero integers **modulo** (or mod) p. For example, 3 is a primitive root modulo 7 since $3 = 3^1$, $2 = 3^2$ mod 7, $6 = 3^3$ mod 7, $4 = 3^4$ mod 7, $5 = 3^5$ mod 7, $1 = 3^6$ mod 7.

prism   A pentagonal prism. *Robert Webb, www.software3d.com; created using Webb's Stella program*

primitive root of unity

The **complex number** z such that $z^n = 1$ but z^k is not equal to 1 for any positive integer k less than n.

primorial

Also known as a *prime factorial*, the product of all **prime numbers** that are less than or equal to a given prime p; it is denoted $p\#$. For example $3\# = 2 \times 3 = 6$, $5\# = 2 \times 3 \times 5 = 30$, and $13\# = 2 \times 3 \times 5 \times 7 \times 11 \times 13 = 30{,}030$.

Prince Rupert's problem

The problem of pushing a **cube** through a hole in another cube of equal or less size; it is named after Prince Rupert (1619–1682), a nephew of England's King Charles I, who won a wager that a hole could be made in one of two equal cubes large enough for the other cube to slide through. The mathematics of cubes passing through cubes was considered by John **Wallis**. Later, in 1816, a solution was published posthumously by the Dutch mathematician Pieter Nieuwland (1764–1794) to the question: What is the largest cube that can be passed through a unit cube (a cube of which each side is one unit long)? Nieuwland answered this by finding the largest square that fits inside a unit cube. When viewed from directly above one apex, a unit cube has the outline of a regular hexagon of side $\sqrt{3}/\sqrt{2}$. The largest square that will go into a cube has a face that can be inscribed within this hexagon; the length of its edge is $\sqrt{6} - \sqrt{2} = 1.03527618$.

prism

A **semi-regular polyhedron** constructed from two **congruent** *n-sided* **polygons** and n **parallelograms**. The word comes from the Greek *prizma*, which relates to cutting or sawing. A *prismoid* resembles a prism but has bases that are **similar** rather than congruent, and sides that are **trapezoids** rather than parallelograms. An example of a prismoid is the **frustum** of a **pyramid**. A *prismatoid* is a polyhedron with all its vertices lying in two parallel planes.

prisoner's dilemma

A problem in **game theory** first described by the Canadian-born Princeton mathematician Albert Tucker (1905–1995) in 1950, while addressing an audience of psychologists at Stanford University, where he was a visiting professor. It runs along these lines: Al and Bob have been arrested for holding up the Anyapolis State Bank and have been put in separate cells. Each cares a lot more about his personal freedom than he does about his accomplice's welfare. A clever prosecutor makes the following offer to each. "You may choose to confess or remain silent. If you confess and your accomplice remains silent, I'll drop all charges against you and use your testimony to ensure that your accomplice does serious time. Likewise, if your accomplice confesses while you remain silent, he'll go free while you do the time. If you both confess I get two convictions, but I'll see to it that you both get early parole. If you both remain silent, I'll have to settle for token sentences on firearms possession charges. If you wish to confess, you must leave a note with the jailer before I come back tomorrow morning." The dilemma faced by the prisoners is that, whatever the other does, each is better off confessing than remaining silent. But the outcome when both confess is worse for each than the outcome if both stay silent!

Tucker's paradox was based on puzzles with a similar structure that had been devised in 1950 by Merrill Flood and Melvin Dresher as part of the Rand Corporation's investigations into game theory (which Rand pursued because of possible applications to global nuclear strategy). Flood and Dresher hadn't published much about their work, but the prisoner's dilemma attracted an enormous amount of attention in subjects as diverse philosophy, biology, sociology, political science, and economics, as well as game theory itself. A common view is that the puzzle illustrates a conflict between individual and group rationality. A group whose members pursue rational self-interest may all end up worse off than a group whose members act contrary to rational self-interest. More generally, if the payoffs aren't assumed to represent self-interest, a group whose members rationally pursue any goals may all meet less success than if they hadn't rationally pursued their goals individually.

probability

A measure of how likely it is that some event will occur, given as a number between 0 (impossible) and 1 (certain). Usually, probability is expressed as a ratio: the number of experimental results that would produce the event divided by the number of experimental results considered possible. For example, the probability of drawing the five of hearts from an ordinary deck of cards is one in fifty-two (1:52).

probability theory

The branch of **mathematics** that deals with the possible outcomes of events and their relative likelihoods. While mathematicians agree on how to calculate the probability of certain events and how to use those

calculations in certain ways, there's plenty of disagreement as to what the numbers actually mean. Probability divides into two main concepts: *aleatory probability*, which represents the likelihood of future events whose occurrence is governed by some random physical phenomenon like tossing dice or spinning a wheel; and *epistemic probability*, which represents our uncertainty of belief about past events that either did or didn't occur, or uncertainty about the causes of future events. An example of the latter is when we say that it's "probable" that a certain suspect committed a crime based on the available evidence. It is an open question whether aleatory probability is reducible to epistemic probability based on our inability to precisely predict every force that might affect the roll of a die, or whether such uncertainties exist in the nature of reality itself, particularly at the level of **quantum mechanics**. One of the earliest mathematical studies on probability was written by Girolamo **Cardano**. Among other important contributions to the development of the subject were those by Blaise **Pascal**, Pierre de **Fermat**, Jakob Bernouilli (see **Bernouilli family**), Joseph **Lagrange**, Pierre **Laplace**, Carl **Gauss**, Siméon **Poisson**, Abraham **de Moivre**, Pafnuty Chebyshev, Andrei Markov (see **Markov chain**), and Andrei **Kolmogorov**.

PUZZLES

The following are two problems in probability. Both are easier to solve if the various possible outcomes are written down in the form of a table.

1. You meet a stranger on the street, and ask how many children he has. He truthfully says two. You ask "Is the older one a girl?" He truthfully says yes. What is the probability that both children are girls? What would the probability be if your second question had been "Is at least one of them a girl?", with the other conditions unchanged?

2. You are in a game of Russian roulette, but this time the gun (a six-shooter revolver) has three bullets in a row in three of the chambers. The barrel is spun only once. Each player then points the gun at his head and pulls the trigger. If he is still alive, the gun is passed to the other player who then points it at his own head and pulls the trigger. The game stops when one player dies. Would you stand a better chance of surviving if you shoot first or second, or does it make any difference?

Solutions begin on page 369.

See also **Buffon's needle, birthday problem, Monty Hall problem,** and **St. Petersburg paradox.**

Proclus Diadochus (c. A.D. 410–485)

The last major Greek philosopher; his *Commentary on Euclid* is the main original source we have on the early history of Greek geometry. He also wrote *Hypotyposis*, which gives a detailed account of the Earth-centered astronomical theories of Hipparchus and Ptolemy.

product

The result of one or more multiplications.

projectile

An object, such as a baseball, spear, or cannonball, that is thrown, fired, or otherwise propelled but that can't propel itself. For centuries, philosophers and mathematicians debated what path a projectile follows under gravity. Galilei Galileo was the first to establish that (in the absence of air resistance) this path is a **parabola**.

projective geometry

The branch of **geometry** that deals with properties of geometric figures that remain unchanged under projection. A mathematical theory of perspective grew out of the studies of Renaissance architects and painters who asked themselves how to best represent a three-dimensional object on a two-dimensional surface. The Greeks had done some early work on perspective, and the great geometer **Pappus of Alexandria** is credited with the first theorem in projective geometry. However, the subject reached mathematical maturity through the efforts first of Girard **Desargues**, and then, much later through the work of Jean **Poncelet** and by Karl von Staudt (1798–1867).

The basic elements of projective geometry are points, lines, and planes. These elements retain their characteristics under projection; for example, the projection of a line is another line, and the point of intersection of two lines is projected into another point that is the intersection of the projections of the two original lines. However, lengths and ratios of lengths are not invariant under projection, nor are angles or the shapes of figures. The concept of parallelism doesn't appear at all in projective geometry; any pair of distinct lines intersects in a point, and if these lines are parallel in the sense of **Euclidean geometry**, then their point of intersection is at infinity. The plane that includes the ideal line, or line at infinity, consisting of all such ideal points, is called the **projective plane**. Two properties that are invariant under projection are the order of three or more points on a line and the harmonic relationship, or cross-ratio, among four points, *A, B, C, D*, that is, *AC/BC : AD/BD*. The most remarkable concept in projective geometry is that of duality. In the plane, the terms *point* and *line* are dual and can be interchanged in any valid statement to yield another valid statement; in space, the terms *plane, line*, and *point* are interchanged with *point, line*, and *plane*, respectively, to yield dual statements. Entire theorems also occur in dual pairs, so that one can be instantly transformed into the other. For example, *Pascal's theorem* (given a hexagon inscribed in a **conic section**, the three pairs of the continuations of opposite sides meet on a straight line) is the dual of *Brianchon's theorem* (given a hexagon circumscribed on a conic section, the lines joining opposite diagonals meet in a single point). In fact, *all* the propositions in projective geometry occur in dual pairs.

projective plane

The surface you would get if you glued the edge of a disk to the edge of a **Möbius band**. This sounds easy to do, since a disk and a Möbius band each have one edge. But the process becomes hopelessly tangled and is, in fact, impossible. The projective plane needs a **fourth dimension**, in addition to the three we live in (up-down, left-right, and back-forth), to be fully realized. The idea of the projective plane arose from the study of perspective by mathematicians and painters in the Renaissance. In trying to represent parallel lines in space on the two-dimensional surface of a painting, it was found useful to introduce the notion of a *line at infinity* on which parallel lines met. The study of the geometry that adds this extra line of ideal points to the ordinary familiar plane came to be known as **projective geometry**, because of its use in studying projections of figures onto different lines. This idea was even more important in three dimensions, since projections are used for representing three-dimensional figures on planes. An interesting property of the projective plane is that any "straight" line on it, followed far enough, comes back to the starting point. (The old arcade version of the game of Asteroids was played on a virtual projective plane: the screen was a disk, and when an asteroid went off one edge of the screen it emerged on the opposite side.) The projective plane is also *nonorientable*, as a result of which any two-dimensional object pushed along a path back to its starting point would be reversed.

prolate

(1) Rounded like an **egg**. (2) Having a polar diameter greater than the equatorial diameter. See also **spheroid**.

pronic number

Also known as a *rectangular* or *oblong number*, a number that is the product of two consecutive integers: 2 (1 × 2), 6 (2 × 3), 12 (3 × 4), 20 (4 × 5), ... The pronic numbers are twice the **triangular numbers**, and represent the lengths that produce the musical intervals: octave (1:2), fifth (2:3), fourth (3:4), major third (4:5) .... Pronic seems to be a misspelling of *promic,* from the Greek *promekes,* for "rectangular" or "oblong"; however, the "n" form goes back at least as far as Leonhard **Euler** who used it in series one, volume fifteen of his *Opera.*

proof

A sequence of statements in which each subsequent statement is derivable from one of the previous statements or from an **axiom** of a **formal system**. The final statement of a proof is usually the **theorem** that one has set out to prove.

proper divisor

See **aliquot part**.

proportional

A variable *a* is said to be *directly proportional* to *b* if *a/b* is a constant. The relationship is written $a \propto b$, which implies that $a = kb$, where *k* is a constant. If *a* is *inversely proportional* to *b*, this is written $a \propto 1/b$.

pseudoprime

A number that passes the test of **Fermat's little theorem** (FLT) for **prime number**s but actually isn't a prime. FLT says that if *p* is prime and *a* is **coprime** to *p*, then $a^{p-1} - 1$ is divisible by *p*. If a number *x* is not prime, *a* is coprime to *x*, and *x* divides $a^{x-1} - 1$, then *x* is called a *pseudoprime to base a.* A number *x* that is a pseudoprime for all values of *a* that are coprime to *x* is called a **Carmichael number**. The smallest pseudoprime in base 2 is 341. This isn't prime because $341 = 11 \times 31$; however, it satisfies FLT: $2^{340} - 1$ is divisible by 341.

pseudosphere

A saddle-shaped surface that is produced by rotating a **tractrix** about its **asymptote**. The name *pseudosphere,* which means "false sphere," is misleading because it suggests something that is spherelike; however, a pseudosphere is almost exactly the opposite of a sphere. Whereas a sphere has a constant positive **curvature** (equal to $+1/r$, where *r* is the radius) at every point on its surface, a pseudosphere has a constant negative curvature (equal to $-1/r$) everywhere. As a result, a sphere has a closed surface and a finite area, while a pseudosphere has an open surface and an infinite area. In fact, although both the two-dimensional plane and a pseudosphere are infinite, the pseudosphere manages to have more room! One way to think of this is that a pseudosphere is more intensely infinite than the plane. Another result of the pseudosphere's negative curvature is that the angles of a triangle drawn on its surface add up to less than 180°. The geometry on the surfaces of both the sphere and the pseudosphere is a two-

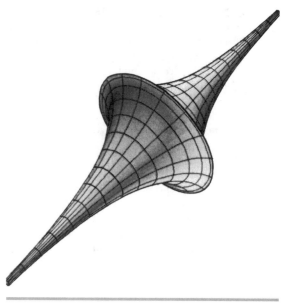

pseudosphere

dimensional **non-Euclidean geometry**–spherical (or elliptical) geometry in the case of the sphere and **hyperbolic geometry** in the case of the pseudosphere. Astronomers currently suspect that the universe we live in may have a hyperbolic geometry and thus have properties analogous to those of a pseudosphere.

Ptolemy's theorem

The sum of the products of the two pairs of opposite sides of a convex cyclic **quadrilateral** (see **cyclic polygon**) is equal to the product of the lengths of the diagonals. The theorem is named after the mathematician, astronomer, and geographer Ptolemy of Alexandria.

pure mathematics

Mathematics for the sake of its internal beauty or logical strength. Compare with **applied mathematics**.

pursuit curve

The path an object takes when chasing after another object in the most effective way. Pursuit curves can arise in a variety of situations, for example when a lion is chasing after a gazelle or a heat-seeking missile is homing in on a moving target. Suppose that four ants are at the corners of a square. They start to crawl clockwise at a constant rate, each moving toward its neighbor. At any instant, they mark the corners of a square. As the ants get closer to

pursuit curve   A set of superimposed snapshots shows the lines of sight at regular intervals of four bugs chasing one another, all moving at the same speed after starting at the corners of a square. *John Sharp*

the original square's center, the new square they define rotates and diminishes in size. In reaching the center, each ant travels on a **logarithmic spiral** with a length equal to the side of the **original** square. Superimposed snapshots of the ants' progress give rise to an intriguing pattern.

puzzle jug

A practical joke in the form of a drinking vessel that, unless the trick is known, spills all over the user when he tries to take a sip. As the jug is tilted, liquid pours out from the pierced decoration surrounding the pot. The puzzle is to learn the secret of how to successfully drink without getting wet. Puzzle jugs are the oldest known **mechanical puzzle**s. Several Phoenician examples are in New York's Metropolitan Museum of Art, and in ninth-century Turkey they were especially popular. Puzzle jugs have been manufactured in Germany, Holland, France, and several other European countries since the end of the thirteenth century. The Exeter puzzle jug, on display in the Royal Albert Museum in Exeter, Devon, was probably made in the Saintonge region of western France, around 1300, and is one of the finest examples of medieval pottery imported to England. Puzzle jugs came back into vogue in the eighteenth and nineteenth centuries and are still manufactured by ceramicists in France, Germany, and Britain. And the secret? That is really quite open, and can be handled by anyone prepared to get to the bottom of it.

puzzle rings

A **mechanical puzzle** consisting of a group of interlocking rings that fit together to form an intricate design; when apart, the rings remain interconnected and pose a challenging puzzle. The puzzle itself has ancient origins in Egypt, though many stories have been told of its use throughout the Middle East. According to one, a person would give the ring to their lover telling them of the magical qualities of the ring. If the ring was ever worn while being amorous to another, it would tell the giver of the ring about the unfaithfulness of the ring wearer. The wearer, wishing to avoid the watchfulness of the ring, might remove it from their finger so that it wouldn't witness an unfaithful act. But once removed, the ring would fall into its separate pieces and so the giver would know their lover had been cheating.

pyramid

A **polyhedron** whose base is a **polygon** and whose other faces are triangles that meet at a common **vertex**, sometimes called the *apex*. A right pyramid has its apex directly above the center of its base. A square-based pyramid, like the pyramids in Egypt, has a square base and four triangular sides. A triangular pyramid, or **tetrahedron**, is made from four triangular sides; if regular, it is one of the **Platonic solids**. The volume of a pyramid is $\frac{1}{3}A_b h$, where A_b is the area of the base and h is the perpendicular height of the apex above the base; the surface area is ps, where p is the base perimeter and s is the slant height.

Pyramidology, which (in its original form) claims links between biblical prophesy and the construction of the Egyptian pyramids, stemmed from the publication in 1859 of *The Great Pyramid: Why Was It Built? And Who Built It?* by

pyramid   The Great Pyramid of Cheops at Giza.

John Taylor, a British amateur mathematician and astronomer who was editor of the *London Observer*. Taylor became convinced that the Great Pyramid of Cheops, in its architectural proportions, embodied various remarkable and deeply meaningful geometric and mathematical properties. Chief among these, Taylor noted, was that the ratio of the perimeter of the pyramid's base to twice its height closely approximates the universal constant **pi**. The said ratio of the Great Pyramid gives a better value for pi than any found in written Egyptian records. The **Rhind papyrus** contains several problems that involve a multistep method for finding the area of a circle from its diameter. This method implies a value of π equal to $^{256}/_{81}$, or 3.1605, which is less than 1% larger than the true value of 3.14159 . . . . The values Taylor used for the Great Pyramid's base and height gave him a value of π correct to two decimal places. Further analysis of the pyramid's dimensions and various manipulations of numbers led him to conclude that the builders of the edifice had used a unit that he called "pyramid inch" (equal to about 1.01 times a standard inch). Twenty-five pyramid inches made a "pyramid cubit," and 10 million pyramid cubits, Taylor pointed out, approximate the length of Earth's polar radius. These and a series of similar calculations provided what Taylor considered to be adequate evidence that the Great Pyramid had been built as a model of Earth. Taylor's fantastic claims may never have become popular had not Charles Piazzi Smyth, the Astronomer Royal of Scotland taken up the cause of pyramidology. He popularized it in Britain, the rest of Europe, and the United States, through a number of books including *Our Inheritance in the Great Pyramid* (1864) and *Life and Work at the Great Pyramid* (1867). (In the churchyard in Sharow, North Yorkshire, Smyth has a pyramidal grave marker.) Others who espoused pyramid numerology have included Helena Blavatsky, the theosophist; Charles Taze Russell, founder of the Watchtower Bible and Tract Society; and Edgar Cayce, the American psychic; not to mention a variety of more recent pseudohistorians who have linked the Great Pyramid to everything from lost civilizations to UFOs. Yet Taylor himself was well aware of other, less dramatic explanations for the pyramid's dimensions, including the possibility that the pyramid had been constructed so that the area of one of its faces would equal the square of its height. The mathematical sophistication required to achieve this is not great and would have resulted in a ratio of the perimeter of the base to twice the height of 3.145—about as close as the approximation of pi used by Smyth in his studies. So, the ratio could have occurred as a completely coincidental byproduct of a design that had nothing to do with the ratio of the circumference of a circle to its diameter. The pyramid is a rich source of the kind of data that Taylor and Smyth worked with, and it would be surprising if they'd been unable to come up with some interesting number combinations given the extent of their efforts.

pyramidal number

The number of dots that may be arranged in a **pyramid** with a regular **polygon** as a base.

Pythagoras of Samos (c. 580–500 B.C.)

A Greek philosopher and mathematician, a native of the Aegean island of Samos, and the founder of a secretive pseudoreligious community in Croton, southern Italy. Pythagoras left no writings and virtually nothing is known about him as an individual, so it is almost impossible to disentangle the beliefs and discoveries of the "Pythagoreans" from those of their leader. To the Pythagoreans, "everything is number" and every number was supposed to be a quantity that could be expressed as the ratio of two integers. We now call such a number a **rational number**. The Pythagoreans used music as an example of the perfection and harmony of numbers that can be expressed as ratios. They showed that pitch could be represented as a simple ratio that came from the length of equally tight strings that could be plucked. Perhaps the most famous of the Pythagoreans' mathematical results is **Pythagoras's theorem**. Then the sky fell in on the Pythagoreans' worldview. Using their very own theorems they showed that not all numbers are rational. Their discovery that the **square root of 2** (the length of the hypotenuse of a triangle with sides 1 and 1) can't be expressed as ratio of two whole numbers was to have been kept a closely guarded a secret, but was later revealed by one of the cult's members.

Pythagoras's lute

The **kite**-shaped figure that forms the enclosing shape for a progression of diminishing **penta**gons and pentagrams, linking the vertices together. The resulting diagram is replete with lines in the **golden ratio**.

Pythagoras's theorem

The square of the length of the **hypotenuse** of a right **triangle** is the sum of the squares of the lengths of the two sides. This is usually expressed as $a^2 + b^2 = c^2$. See also **Pythagorean triplet**.

Pythagorean square puzzle

A deceptively hard assembly puzzle in which a small square piece must be combined with four pieces forming a larger square to make an even larger square.

PUZZLE

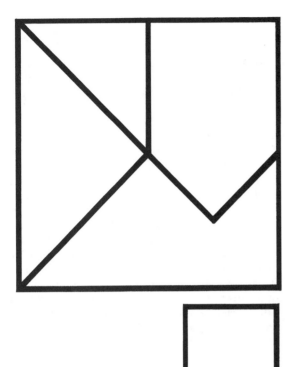

Rearrange the pieces in the Pythagorean square puzzle, including the small square, to make a single larger square.

Solutions begin on page 369.

Pythagorean triangle

A right triangle whose sides are integers. A *primitive Pythagorean triangle* is one whose sides are **coprime**.

Pythagorean triplet

Also called a *Pythagorean triple,* a set of three whole numbers that satisfies **Pythagoras's theorem**, that is, the squares of two of the numbers add up to the square of the third number. Examples include (3, 4, 5), (5, 12, 13), and (7, 24, 25). These are called *primitive triplets* because they have no common divisors. If the members of a primitive triplet are multiplied by the same integer, the result is a new (but not primitive) triplet. In any primitive Pythagorean triplet, one, and only one, of the three numbers must be even (but can't equal 2); the other two numbers are **coprime**. There are infinitely many such triplets, and they are easy to generate using a classic formula, known since ancient times. If the numbers in the triplet are a, b, and c, then: $a = n^2 - m^2$, $b = 2mn$, $c = m^2 + n^2$, where m and n are two integers and m is less than n. Because the **square root of 2** is irrational, there can't be any Pythagorean triplets (a, a, c). However, there are an infinite number of triplets (a, $a + 1$, c), the first three of which (apart from the trivial (0, 1, 1)) are (3, 4, 5), (20, 21, 29), and (119, 120, 169).

There are also an infinite number of Pythagorean quartets (a, b, c, d) such that $a^2 + b^2 + c^2 = d^2$. This is simply the three-dimensional form of Pythagoras's theorem and can be interpreted as the fact that the point in three-dimensions with Cartesian coordinates (a, b, c) lies an integer distance d from the origin. A formula that generates Pythagorean quartets is: $a = m^2$, $b = 2mn$, $c = 2n^2$, $d = (m^2 + 2n^2) = a + c$. Also note that $b^2 = 2ac$. When $m = 1$ and $n = 1$, we get the quartet (1, 2, 2, 3)–the simplest example.

Although there are an infinite number of Pythagorean triplets, **Fermat's last theorem**, which is now know to be true, ensures that there are no triplets for higher powers. See also **Euler's conjecture** and **multigrade**.

QED

Abbreviation for *quod erat demonstrandum* ("that which was to be shown"), used to denote the end of a proof.

quadrangle

A plane figure consisting of four points, each of which is joined to two other points by a line segment. A quadrangle may be **concave** or **convex** depending on whether the line segments do or don't intersect. A convex quadrangle is a **quadrilateral**. The word is from the Latin *quadrangulum* for "four-cornered" and is also used to describe a rectangular area surrounded on all four sides by buildings, or to such buildings themselves.

quadrant

Any one of the four portions of the plane into which the plane is divided by the Cartesian coordinate axes.

quadratic

An expression or an equation that contains the variable squared, but not raised to any higher power. For instance a *quadratic equation* in x contains x^2 but not x^3. Similarly a *quadratic expression,* or a *quadratic form,* contains its variable(s) squared but not raised to any higher power. If there is more than one variable (say, x and y), quadratic can mean that they are multiplied together in pairs (xy) but not in threes (such as x^2y). The graph of a quadratic equation is known as a *quadratic curve;* the curve of the general quadratic equation $y = ax^2 + bx + c$ is a **parabola**.

quadratrix of Hippias

The first curve in recorded history that was not part of a line or a circle, and the first curve known that is not **constructible** in the classical sense; in other words, it can't be drawn using a straightedge and a compass alone, but instead has to be plotted point by point. The quadratrix can be thought of as the intersection of two lines moving with constant velocity: the first line rotates (e.g., counterclockwise) while the second line moves along (say, in the direction of the positive y-axis). It has the Cartesian equation $y = x \, cot(\pi x/2a)$. The quadratrix was discovered by **Hippias of Elis** in about 430 B.C. and was used by him in his work on **trisecting an angle** and **squaring the circle**. In fact, its name refers to its use in turning curvilinear space into a rectangular area.

quadrature

The determination of the area of a geometric figure.

quadric

A surface in three dimensions that is described by equations containing the squares of $x, y,$ and z, but no higher powers of them. Examples of such surfaces include the **sphere**, **ellipsoid**, **cone**, and **cylinder**.

quadrifolium

See **rose curve**.

quadrilateral

A **polygon** that has four sides and four vertices (corners). Quadrilaterals, and polygons in general, may be **convex** or **concave**. A convex quadrilateral may be further classified as a **trapezoid** or a British **trapezium** (one pair of opposite sides are parallel), a trapezium (no sides parallel); an isosceles trapezoid (United States) or an isosceles trapezium (United Kingdom) (two of the opposite sides parallel, the two other sides equal, and the two ends of each parallel side of equal angles); a **parallelogram** (opposite sides are parallel); a **kite** (two

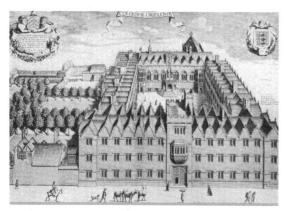

quadrangle The sixteenth-century quadrangle of Oxford University's Oriel College. *Oriel College, Oxford*

adjacent sides of equal length, the other two sides of equal length); a **rhombus** (four sides of equal length); a **rectangle** (each angle is a right angle); or a **square** (four sides of equal length, each angle a right angle). *Quadrangular prisms* and *quadrangular pyramids* are ones whose bases are quadrilateral.

quantifier

In symbolic logic, the *universal quantifier* ∀ indicates "for every" or "for all." For example, ∀ *x A, p(x)* means for all *x* belonging to *A*, the **proposition** *p(x)* is valid. The *existential quantifier* ∃ indicates "there exists." So, for instance, ∃ *x A, p(x)* means there exists at least one *x*, belonging in *A*, for which the proposition *p(x)* is valid.

quantum field theory

The study of force fields such as the electromagnetic field in the context of **quantum mechanics**, and often special **relativity theory**. The mainstay of modern high-energy physics.

quantum mechanics

The science and mathematics that describe the behavior of nature at the atomic and subatomic level. At the heart of quantum mechanics are two basic concepts: (1) that every small bit of matter or energy can behave as if it were either a particle or a wave; and (2) that certain combinations of properties such as position and velocity, and energy and time, can't be known with arbitrary precision. The latter idea is encapsulated in *Heisenberg's uncertainty principle.* See also **many worlds hypothesis.**

quartic

A **polynomial** or polynomial equation that contains the fourth power of the variable, but no higher power. Many famous curves are described by such equations, including the **bicorn, Cartesian oval, conchoid, deltoid, devil's curve, folium, kampyle of Eudoxus,** and **limacon of Pascal.**

quartile

The first quartile of a **sequence** of numbers is the number such that one quarter of the numbers in the sequence are less than this number.

quasicrystal

A strange type of solid whose atomic structure is very regular but never quite repeats. Quasicrystalline structures don't have a simple unit cell that can be repeated periodically in all directions to fill space, although they do have local patterns that repeat almost periodically. They also have local rotational symmetries, such as those of a **pentagon**, that can't exist in ordinary crystals. Prior to the discovery of quasicrystals, it was thought that five-fold crystal symmetry was impossible, because there are no space-filling periodic tilings of this kind. The best known examples of quasicrystals resemble **Penrose tiling**s, which use repeated copies of two different rhombi to cover an infinite plane in intricate, interlocking patterns. In fact, some quasicrystals can be sliced in such a way that the atoms on the surface follow the exact pattern of the Penrose tiling.[293]

quasiperiodic

Refers to a form of motion that is regular but never exactly repeating. Quasiperiodic motion is always composed of multiple but simpler **periodic** motions. In the general case for motion that is the sum of simpler periodic motions, if there exists a length of time that evenly divides the frequencies of the underlying motions, then the composite motion will also be periodic; however, if no such length of time exists, then the motion will be quasiperiodic.

quasiregular polyhedron

A **polyhedron** that consists of two sets of regular **polygons**, *m*-sided and *n*-sided respectively, and is constructed so that each polygon in one set is surrounded by members of the other set. There are three convex quasiregular solids: the *cuboctahedron* ($m = 3$, $n = 4$), the *icosidodecahedron* ($m = 3$, $n = 5$), and the **octahedron** ($m = n = 3$). In each case four faces meet at each vertex in the cyclic order (*m, n, m, n*). Because of this, these polyhedra have some special properties, one of which is that their edges form a system of **great circles**. The edges of the octahedron form three squares; the edges of the cuboctahedron form four hexagons, and the edges of the icosidodecahedron form six decagons. Among the nonconvex polyhedra are two examples of type (*m, n, m, n*): the *dodecadodecahedron* ($m = 5$, $n = {}^5\!/_2$) and the *great icosidodecahedron* ($m = 3$, $n = {}^5\!/_2$), which can be made by truncating the **Kepler-Poinsot polyhedra** at their edge midpoints. There are also three nonconvex examples of type (*m, n, m, n, m, n*): the *small triambic icosidodecahedron* ($m = 3$, $n = {}^5\!/_2$), the *triambic dodecadodecahedron* ($m = {}^5\!/_3$, $n = 5$), and the *great triambic icosidodecahedron* ($m = 3$, $n = 5$). Finally, there is a group of nine *hemihedra*, in which some faces pass through the polyhedron's center. These hemifaces each cut a sphere into two hemispheres.

quaternion

An ordered **set** of four numbers. Quaternions, first introduced by William **Hamilton**, can also be written in the

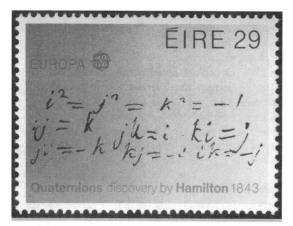

quaternion An Irish stamp showing the quaternion equations in Hamilton's own hand.

form $a + bi + cj + dk$, where a, b, c, and d are real numbers and i, j, and k are imaginary numbers, which is similar to that of **complex numbers**. Whereas complex numbers can be represented by points of a two-dimensional plane, quaternions can be viewed as points in the **fourth dimension**. For a while, quaternions were very influential: they were taught in many mathematics departments in the United States in the late 1800s, and were a mandatory topic of study at Dublin, where Hamilton ran the observatory. But then they were driven out by the **vector** notation of William Gibbs and Oliver Heaviside. Had quaternions come along later, when theoretical physicists were trying to understand patterns among subatomic particles, they may have found a place in modern science; after all, the unit quaternions form the group SU(2), which is perfect for studying spin-$\frac{1}{2}$ particles. But the way things turned out, quaternions had fallen from favor by the twentieth century and Wolfgang Pauli used 2×2 complex matrices instead to describe the generators of SU(2).

queens puzzle

A famous **chess** problem that asks in how many ways eight queens can be placed on a chessboard so that no two attack each other. The generalized problem, to find how many ways n queens can be placed on an $n \times n$ board so that no two attack each other, was first posed by Franz Nauck in 1850. In 1874 Günther and Glaisher described methods for solving this problem based on **determinants**. The number of distinct solutions, not counting rotations and reflections, for board sizes ranging from 1×1 to 10×10 is 1, 0, 0, 1, 2, 1, 6, 12, 46, and 92, respectively. The 6×6 puzzle, for which there is a solitary

unique solution, was sold for one penny in Victorian London in the form of a wooden board with 36 holes into which pins were placed.

quine

Before we put the motion "that the motion be now put," should we not first put the motion "that the motion 'that the motion be now put' be now put?"
—Chairman of the meeting of the Society of Logicians

A term named by Douglas **Hofstadter** after the Harvard logician Willard van Orman Quine. It can be used either as a noun or a verb. (1) Quine (noun). A computer program that produces an exact copy of itself (or, alternatively, that prints its own listing.) This means that when the program is run, it must duplicate (or print out) precisely those instructions that the programmer wrote as part of the program, including the instructions that do the copying (or printing) and the data used in the copying (or printing.) A respectable quine—one that doesn't cheat—is not allowed to do anything as underhand or trivial as seeking the source file on the disk, opening it, and copying (or printing) its contents. Although writing a quine is not always easy, and in fact may seem impossible, it can always be done in any programming language that is Turing complete (see **Turing machine**), which includes every programming language actually in use. (2) Quine (verb). To write a sentence fragment a first time, and then to write it a second time, but with quotation marks around it. For example, if we quine "say," we get "say 'say' "). Thus, if we quine "quine", we get "quine 'quine,' " so that the sentence "quine 'quine' " is a quine. In this linguistic analogy, the verb "to quine," plays the role of the code, and "quine" in quotation marks plays the role of the data.

quintic

A **polynomial** or polynomial equation that contains the fifth power of the variable, but no higher power. Niels **Abel** and Evariste **Galois** independently proved that although there exist formulas for the general solution of **quadratic**, **cubic**, and **quartic** equations, no such formula exists for quintic equations.

quipu

A recording device generally associated with the Incas, who ruled Peru before the Spanish conquest. Quipu consisted of a number of color-coded cords; knots of various kinds were tied on these cords to represent a variety of information. An important use of quipu was to record numbers for use in trade, keeping accounts, and calendars, but the knotted strings might also have served as mnemonics for important historical events, astronomical data, and mythology (the Incas had no written language). Some evidence supports the idea that the knots and cords followed in a decimal system. Similar devices were used by several other Indian tribes and also described in Chinese and Persian documents from the fifth and sixth centuries B.C., and are still used by shepherds in the Andes for keeping accounts of their herds.

quotient

The number of times that one number can be divided exactly into another.

R

radian

A unit of angular measurement such that there are 2π radians in a complete circle. One radian = $180/\pi$ degrees. One radian is approximately $57.3°$.

radical

The symbol that indicates a **root**, $\sqrt[n]{\ }$. It seems to have been first used in 1525 by Christoff Rudolff (1499–1545) in his *Die Coss*.

radical axis

The **locus** of points of equal power with respect to two **circle**s. The *radical center* of three circles is the common point of intersection of the radical axes of each pair of circles.

radius

The distance from the center of a **circle** to its circumference, or from the center of a regular **polygon** to any one of its vertices. The *radius of curvature, r,* at any point of a curve is $r = 1/\kappa$, where κ is the **curvature**.

radix

See **base**.

railroad problems

See **shunting puzzles**.

Ramanujan, Srinivasa Aaiyangar (1887–1920)

An extraordinary, largely self-taught, Indian mathematician who, in the most unorthodox way, made significant contributions to **number theory**, including the subject of **elliptic function**s, **continued fraction**s, and **infinite series**. During much of his early work, he was unaware that he was rediscovering results that had taken other mathematicians centuries to achieve. But even in cases where he arrived at conclusions already known, he'd often travel an original route, and, in many cases, almost purely by intuition. Ramanujan was employed in a lowly clerk's position in Madras, when, in 1913, he wrote letters to three eminent mathematicians in England describing some of his results. Two of the three letters were returned unopened. However, G. H. **Hardy** recognized Ramanu-

jan's abilities and arranged for him to come to Cambridge. Because of his lack of formal training, Ramanujan sometimes failed to distinguish between formal proof and apparent truth based on intuition or numerical evidence. His extraordinary innate familiarity with numbers was revealed by an incident recalled by Hardy:[150] "I remember once going to see him when he was lying ill at Putney. I had ridden in taxi cab number 1729 and remarked that the number seemed to me rather a dull one, and that I hoped it was not an unfavorable omen. 'No,' he replied, 'it is a very interesting number; it is the smallest number expressible as the sum of two cubes in two different ways [$1729 = 1^3 + 12^3 = 9^3 + 10^3$].' "

Unfortunately, Ramanujan's health deteriorated rapidly in England, perhaps due to the unfamiliar climate and food, and to the isolation which Ramanujan felt as the sole Indian and a devout Hindu in a culture which was alien to him. Ramanujan was sent home to recuperate in 1919, but tragically died the following year at the age of only 32. Although he published some of his results in journals, much of his work and conclusions have only

Ramanujan, Srinivasa Aaiyangar The enigmatic mathematician on an Indian commemorative stamp.

come to light more recently, from a scrutiny of his disorganized but fascinating notebooks.[184]

Ramsey theory

A branch of mathematics that asks questions such as: Can order always be found in what appears to be disorder? If so, how much can be found and how big a chunk of disorder is needed to find a particular amount of order in it? Ramsey theory is named after the English mathematician Frank P. Ramsey (1904–1930) who started the field in 1928 while wrestling with a problem in logic. (Frank's one-year-younger brother, Arthur, served as Archbishop of Canterbury from 1961 to 1974.) His life was cut short at the age of 26, following a bout of jaundice. Ramsey suspected that if a system was big enough, even if it seemed to be disorderly to an arbitrary degree, it was bound to contain pockets of order from which information about the system could be gleaned.

random

Without cause; not **compressible**; obeying the statistics of a fair coin toss.

random number

A number generated by a process that is fundamentally nondeterministic and unpredictable. Computer-generated "random numbers," which are calculated through a deterministic process, cannot, by definition, be **random**. Given knowledge of the **algorithm** used to create the numbers and its internal state, it's possible to predict all of the numbers returned by subsequent calls to the algorithm. For this reason the numbers produced by computer-based "random number generators" are often referred to as *pseudorandom numbers*. In the case of genuinely random numbers, knowledge of one number or an arbitrarily long sequence of numbers offers no clue of the next number to be generated. Humans are among the worst random number gen-

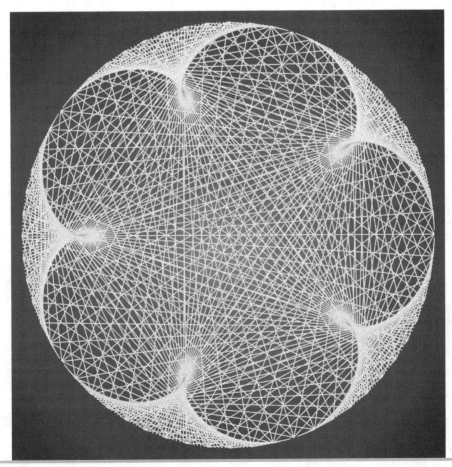

ranunculoid   A ranunculoid curve spun by thread on a computer loom. *Jos Leys, www.josleys.com*

erators. Ask someone to pick a number "at random" between 1 and 20, and the number they're most likely to choose is 17. *Psychologically random numbers,* such as 17, are usually odd and don't end in 5, so that they frequently tend to be **prime numbers**. See also **Chaitin's constant**.

random walk

A process in which the position of a particle changes by discrete steps of fixed length, and the direction of each step is chosen randomly. Random walks have interesting mathematical properties that vary greatly depending on the number of dimensions in which the walk takes place and whether it is confined to a **lattice**. For a random walk in one dimension there are only two directions to choose from. Imagine a drunken person wandering on the **number line** who starts at 0, and then moves left or right (+/−1) with probability ½. The probability that the walker will eventually return to his starting point is 1; in other words, it is certain to happen. The same is true for a random walk in the plane, moving on the integer lattice points, with probability ¼ in each of the coordinate directions: the probability of ending up back at the starting point is 1. However, the situation changes in three dimensions. Suppose a drunken fly moves randomly from one point to another in a three-dimensional lattice with a probability of 1 in 6 of arriving at any of the six adjacent lattice points on each move. No matter how long the fly roams, it has only a 0.34054 . . . probability of ever getting back to where it started. Probabilists say that random walks on the line and plane are *recurrent,* whereas random walks in three dimensions or more are *transient.* Effectively, this is because there is so much more "space" in three or more dimensions. The numbers giving the probability of eventually returning to the starting point are known as *random walk constants.* The random thermal perturbations in a liquid are responsible for a random walk phenomenon known as *Brownian motion,* and the collisions of molecules in a gas are a random walk responsible for *diffusion.*

range

(1) The set of possible values in which a **function**'s output can be. See also **codomain**. (2) The set of all points on a **line segment**.

rank

(1) Any of the rows of squares running crosswise to the files on a playing board in **chess** or checkers. (2) The rank of a **matrix** is equal to the dimension of the largest submatrix that can be obtained by deleting rows and columns of the parent matrix and that has a nonzero determinant. See also **tensor**.

ranunculoid

An **epicycloid** with five cusps ($n = 5$), named after the buttercup genus *Ranunculus.*

ratio

A **rational number** of the form *a/b* where *a* is called the *numerator* and *b* is called the *denominator.* It may be written with a colon (:), as a fraction, or with the word *to.*

rational number

A number that can be written as an ordinary **fraction**–a ratio, *a/b,* of two integers, *a* and *b,* where *b* isn't zero–or as a decimal expansion that either stops (like 4.58) or is periodic (like 1.315315 . . .). Other examples include 1, 1.2, 385.66, and ⅓. Rational numbers are countable, which means that, although there are infinitely many of them, they can always be put in a definite order, from smallest to largest, and can thus be counted. They also form what's called a *densely ordered set;* in other words, between any two rationals there always sits another one–in fact infinitely many others. The rational numbers are a subset of the **real numbers**; real numbers that aren't rational are called, rationally enough, **irrational numbers**. Although rationals are dense on the real **number line**, in the sense that any **open** set contains a rational, they're pretty sparse in comparison with the irrationals. One way to think of this is that the **infinity** of rationals (which, strangely enough, is exactly the same size as the infinity of whole numbers) is smaller than the infinity of irrational numbers. Another way to grasp the scarcity versus density issue, is to realize that the rationals can be covered by a set whose "length" is arbitrarily small. In other words, given a string of any positive length, no matter how short, it will still be long enough to cover all the rationals. In mathematical parlance, the rationals are a **measure** zero set. The irrationals, by contrast, are a measure one set. This difference in measure means that the rationals and irrationals are quite different even though a rational can always be found between any two irrationals, and an irrational exists between any two rationals.

raven paradox

A **paradox** put forward by the German logician Carl Hempel (1905–1997) in the 1940s to highlight a situation where the logic of **induction** seems to fly in the face of intuition. According to the *principle of induction,* the more that a theory is supported by observation, the greater the probability that the theory is true. Consider, said Hempel, the theory that all ravens are black. After each observation of a black raven, our belief in the theory "all ravens are black" increases. But here's the rub. The statement "all ravens are black" is logically equivalent to the statement "all nonblack things are nonravens." The observation of a white swan is consistent with this statement. A white swan is a nonblack thing, and when we examine it, we observe that it is a nonraven. So by the principle of induction, observing a white swan should increase our belief that all ravens are black!

Many solutions have been offered to this enigma. The American logician Nelson Goodman (1906–1998) suggested imposing restrictions to our reasoning, such as never considering an instance as support for "All P are Q" if it would also support "No P are Q." Others have questioned the *principle of equivalence*. Perhaps seeing a white swan should strengthen our belief in the theory "all nonblack things are nonravens," without increasing our conviction that "all ravens are black." Yet others have argued that our intuition is flawed. Observing a white swan really does increase the probability that all ravens are black! After all, if you were shown all the non-black things in existence, and you noticed that none was a raven, then you could properly conclude that all ravens were black. The example only seems counterintuitive because the set of nonblack things is vastly larger than the set of ravens. Thus observing one more non-black thing that is not a raven should make a tiny difference to our degree of belief in the proposition compared to the difference made by observing one more raven that is black.

A way to sidestep the paradox is by using **Bayes's theorem**. According to this the probability of a hypothesis H must be multiplied by the ratio:

$$\frac{\text{probability of observing } X \text{ if } H \text{ is true}}{\text{probability of observing } X}$$

If a swan is picked at random, the probability of it being white is independent of the colors of ravens. The numerator in the above ratio will equal the denominator, the ratio will equal one, and the probability will remain unchanged. Seeing a white swan doesn't affect our belief about whether all ravens are black. If a nonblack thing is chosen at random, and a white swan is shown, then the numerator will be bigger than the denominator by a tiny amount. Seeing the white swan will only slightly increase our belief that all ravens are black. We'd have to see almost every nonblack thing in the universe (and see that they're all nonravens) before our belief in "all ravens are black" would increase appreciably. In both cases, these results are in line with intuition.

ray

A straight path of points that begins at one point and continues in one direction.

real number

Any number that can be represented as a decimal, possibly infinitely long and nonrepeating. Real numbers stand in one-to-one correspondence with the points on a continuous line, known as the *real number line*, that stretches from zero to infinity in both directions. The set of real numbers contains the set of all **rational numbers** and the set of all **irrational numbers**. The name "real number" is a retronym, coined by René **Descartes** in response to the concept of **imaginary numbers**. Number systems that are even more general than the real numbers include the **complex numbers** and, of much more recent discovery, **hyperreal numbers** and **surreal numbers**.

realm

A term advocated for a three-dimensional version of the two-dimensional **plane**.

reciprocal

One over a given number; for example, the reciprocal of 4 is ¼.

Recorde, Robert (c. 1510–1558)

A Welsh physician and mathematician, born in Tenby, Pembrokeshire, and trained at Oxford and Cambridge, who held various positions, including master of the mint in Bristol and later in Ireland, and wrote a number of influential math textbooks. These books formed a complete course and were written in English, rather than the usual Latin or Greek, so that they could be read by anyone. In one of them Recorde introduces the "=" sign for "equals."

rectangle

A **quadrilateral** whose interior angles are all 90°. If all of its sides are the same length, it is a **square**. The smallest square that can be cut into $m \times n$ rectangles, such that all m and n are different integers, is the 11×11 square, and the **tiling** uses five rectangles. The smallest rectangle that can be cut into $m \times n$ rectangles, such that all m and n are different integers, is the 9×13 rectangle; this tiling also uses five rectangles.

rectangular coordinates

See **Cartesian coordinates**.

rectangular hyperbola

See **hyperbola**.

recursion

See **recursion**.

No, seriously. Recursion is a process that wraps back on itself and feeds the output of a process or function back in as the input. Using some sort of *recurrence relation,* an entire class of objects can be built up from a few initial values and a small number of rules. The **Fibonacci sequence**, for example, is defined recursively, as are many **fractal** figures. *Self-recursion,* of which the first two lines of this entry are an example, leads to an endless feedback loop. One, somewhat disturbing, notion of reality is that we live in a *recursive universe* in which nature resembles an infinite nest of Russian dolls. One day, we will simulate the big bang in one of our supercomputers. Inside this artificial universe (rather like an immense *Star Trek* "holodeck" simulation) will evolve new star systems and new life forms. They too will evolve intelligence one day and invent computers. And they too may one day simulate the big bang inside one of their supercomputers. This chain of existence will continue as long as we continue to run our simulation. Our universe will continue to exist as long as our parent universes' continue to run their simulations. And therein lies madness.

recursive function

Strictly speaking, a **function** that is *computable;* however, in the usual sense of the word, a function is said to be recursive if its definition makes reference to itself (see **recursion**). For example, **factorial** can be defined as $x! = x(x-1)!$ with the base case of 1! equal to 1. See also **self-referential sentence**.

recursively enumerable set

A potentially infinite **set** whose members can be enumerated by a **universal computer**; however, a universal computer may not be able to determine that something is not a member of a recursively enumerable set. The *halting set,* a concept related to the **halting problem**, is recursively enumerable but not recursive.

reductio ad absurdum

"Reduction to the absurd"; the process of demonstrating that an idea is probably false by first assuming its truth, and then showing how that truth leads to conclusions that can't possibly be true. In *A Mathematician's Apology* (1941),[151] G. H. **Hardy** said: "Reductio ad absurdum, which Euclid loved so much, is one of a mathematician's finest weapons. It is a far finer gambit than any chess play: a chess player may offer the sacrifice of a pawn or even a piece, but a mathematician offers the game."

reductionism

The idea that nature can be understood by taking it apart. In other words, knowing the lowest-level details of how things work (at, say, the level of subatomic physics)

reveals how higher-level phenomena come about. This is a bottom-up way of looking at the universe, and is the exact opposite of **holism**.

redundancy

The existence of repetitive patterns or structures. In an important sense, redundancy refers to *order* in a **complex system** since order is defined as the existence of structures that maintain themselves over time. In **information theory**, redundancy refers to repetition in patterns of messages in a communication channel. If the message contains these redundancies, they can be compressed further; for example, a message containing a series of 250 ones, could be compressed into a command that effectively says "and then repeat one 250 times," instead of writing out all 250 ones.

reentrant angle

An inward-pointing angle of a **concave** polygon.

reflection

A way of transforming a shape in the same way that a mirror does. The reflection of a shape in a mirror line is an identical shape that has been flipped over. When an object is placed a certain distance in front of a mirror, its image in the mirror appears the same distance away from the edge of the mirror. Likewise, all the points on a shape and all the points on its image are the same distance away from the mirror line.

reflex angle

An angle between 180° and 360°.

reflexible

Having a plane of mirror symmetry. Compare with **chiral**.

regular polygon

A **polygon** in which all the sides are equal and all the angles are equal.

regular polyhedron

A polyhedron in which every face and vertex figure is regular. There are nine regular polyhedra: the five **Platonic solid**s and the four **Kepler-Poinsot solid**s. However, others are sometimes allowed, depending on the definition of polyhedron.

relativity theory

The physical and mathematical theory due to Albert Einstein (1879–1955) that revolutionized our understanding of **space**, **time**, and gravity. In it, space and time are seen as a unified and inseparable whole—the four-dimensional continuum of **space-time**. The curvature of space-time due to the presence of matter becomes the extraordinary new explanation and interpretation of gravity. Einstein's theory of relativity was published by him in two great parts. Special relativity, published in 1905, deals exclusively with *inertial* frames of reference, that is, reference frames that don't accelerate with respect to one another. Its two central premises are that the laws of physics are the same in all reference frames and that the speed of light (in a vacuum) is constant in all reference frames. General relativity, published in 1915, centers on the *equivalence principle,* the idea that acceleration and gravity are equivalent. See also **non-Euclidean space**.

renormalization
A mathematical technique for looking at a physical system at different levels of magnification.

repdigit
A number composed of repetition of a single digit in a given base, generally taken as base 10 unless otherwise specified. For example, the **beast number** 666 is a (base-10) repdigit.

representation theory
A theory that seeks to understand an abstract algebraic system such as a **group** by obtaining it in a more concrete

way as a *permutation group* or as a group of matrixes (see **matrix**).

rep-tile
A repetitive **tiling**: a shape with the property that it tiles a larger version of itself, using identical copies of itself. A simple example is a square because four copies of any square tile a larger square. Any triangle also is a rep-tile, because four copies of it tile a larger version of this triangle. Rep-tiles that require n tiles to build a larger version of themselves are said to be rep-n; thus a square is rep-4. Since any of these larger replicas can be combined to give an even larger, second-generation copy, a rep-n tile is also rep-n^2, rep-n^3, and so on. Often tiles have several rep-numbers. If a tile is rep-n and rep-m, it is also rep-mn, since replicas can be built with n tiles, then combined, m at a time, to give a yet larger version.

The set of rep-tiles is a subset of the set of *irreptiles*. An irreptile is any shape that tiles a larger version of itself using either differently sized or identical copies of itself. The problem to find all irreptiles in the Euclidean plane has been studied but not yet completely solved. A

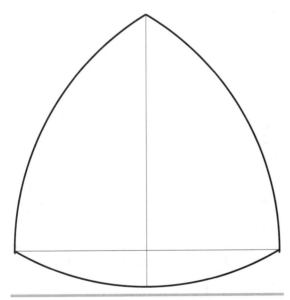

Reuleaux triangle   © *Jan Wassenaar, www.2dcurves.com*

Reutersvärd, Oscar   A Swedish stamp depicting one of Reutersvärd's impossible figures.

re-lated set of problems is to find for each irreptile the minimum number of smaller copies needed to tile the original shape; in many cases it is difficult to prove such a minimality. The name "rep-tile" was coined by Simon **Golomb**.

rep-unit

A number whose digits are all units; the rep-unit (repeated unit) with n digits is denoted R_n. For example, $R_1 = 1$, $R_2 = 11$, $R_3 = 111$, and $R_n = (10^n - 1)/9$. R_n divides R_m whenever n divides m. No rep-unit can be a square, but it is not known if one can be a cube. *Rep-unit primes* are rep-units that are **prime numbers**. The only known rep-unit primes are R_2 (11), R_{19}, R_{23}, R_{317}, and R_{1031}, though R_{49081} and R_{86453} are suspected primes.

resultant

A **vector** that is the sum of a given set of vectors.

Reuleaux polytope

A **convex** body in the plane or in higher dimensions that, like the **Reuleaux triangle**, consists of pieces of round spheres, each centered at one of the corners of the convex body.

Reuleaux triangle

The simplest noncircular **curve of constant width**; also known as the *Reuleaux wheel*, it is named after the German engineer and mathematician Franz Reuleaux (1829–1905). Although it was known to earlier mathematicians, Reuleaux was the first to show its constant-width properties. To form a Reuleaux triangle, take the three points at the corners of an equilateral triangle and connect each pair of points by a circular arc centered at the remaining point. The ratio of the circumference to the width of the triangle is, remarkably, **pi**. By rotating the centroid of a Reuleaux triangle appropriately, the figure can be made to trace out a square, perfect except for slightly rounded corners. This idea has formed the basis of a drill that will carve out squares, first patented by Harry Watts in 1914. Bits for square, pentagonal, hexagonal, and octagonal holes are still sold by the Watts Brothers Tool Works in Wilmerding, Pennsylvania. The actual drill bit for the square is a Reuleaux triangle made concave in three spots to allow for unobstructed corner-cutting and the discharge of shavings. The Reuleaux triangle may also form the shape of the piston in a rotary, or Wankel, engine, in which gasoline burns in crescent-shaped chambers, turning a rotating piston that drives an axle through its center.

Reutersvärd, Oscar (1915–)

A Swedish artist who pioneered the creation and design of **impossible figure**s. His work in this area goes back to one day in 1934 when, as a young student in Stockholm, he started doodling in the margins of a textbook during a long lecture. Reutersvärd's doodle began with an outline of a perfect six-pointed star. Once the star was complete he added cubes around the star, nestled into the spaces between the points. He soon realized that what he'd drawn was paradoxical: something that couldn't be built in the real world. A different version of this figure, independently created by Roger **Penrose**, would later be called the **Penrose triangle**. Thus began a lifelong fascination with such objects, which later included work on the impossible staircase (a design he sketched in 1950 while on a cross-country train ride), which is known as the **Penrose stairway**, and the discovery of the **tribar illusion**. In the early 1980s the Swedish government honored Reutersvärd's achievements by issuing a set of three stamps depicting impossible figures, including a version of his 1934 weird cubes design.

Rhind papyrus

A papyrus scroll, 33 cm high and 565 cm wide, found in a tomb in Thebes, that is the most valuable source of information we have about Egyptian mathematics. The

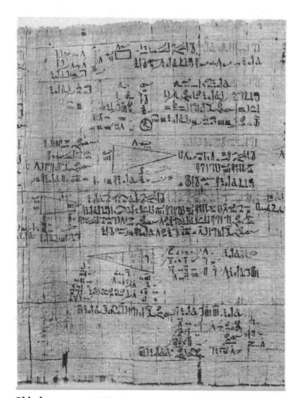

Rhind papyrus *British Museum*

scroll was bought at a market in Luxor in 1858 by a 25-year-old Scotsman, Henry Rhind, who went to Egypt for health reasons and became interested in archeology. After his early death at the age of 30, the scroll found its way to the British Museum in London in 1864 and has remained there ever since. It is often referred to as the *Rhind mathematical papyrus,* or RMP for short. The hieroglyphs on the papyrus were deciphered in 1842, while the Babylonian clay-tablet cuneiform writing was deciphered later in the nineteenth century. The text begins by stating that the scribe "Ahmes" is writing it (in about 1600 B.C., and he is thus the earliest named individual in the history of mathematics) but that he has copied it from "ancient writings," which probably go back to at least 2000 B.C. Although there is some strictly practical mathematics on the papyrus, including calculations needed for surveying, building, and accounting, some of which involve **Egyptian fractions**, many of the problems in the RMP take the form of arithmetic puzzles. One of these is: Seven houses contain seven cats. Each cat kills seven mice. Each mouse had eaten seven ears of grain. Each ear of grain would have produced seven hekats of wheat. What is the total of all of these? This is very similar to the **St. Ives problem**.

There are also four lesser documents preserving Egyptian arithmetic: the Moscow papyrus and the Berlin papyrus (named for the places they are kept), the Kahun papyrus (named for where it was found), and the Leather Roll (named for its composition). The Moscow papyrus is sometimes called the Golenischev papyrus after the Russian V. S. Golenischev, who purchased it in 1893 from two Egyptian brothers who found it in a tomb at Deir el-Bahri. It measures 8 cm high and 540 cm wide and contains 25 problems and their solutions. The most unusual are the tenth, which seems to give the area of the surface of a hemisphere or perhaps a cylinder, and the fourteenth, which gives the formula for the volume of the frustum of a **pyramid**.

rhombus

A **quadrilateral** in which both pairs of opposite sides are parallel and all sides are the same length, that is, an equilateral **parallelogram**. A rhombus is also sometimes called a *rhomb* or a diamond. A rhombus whose acute angles are 45° is called a *lozenge*. The diagonals p and q of a rhombus are perpendicular and satisfy the relationship $p^2 + q^2 = 4a^2$. The area of a rhombus is given by $A = \frac{1}{2} pq$.

Richard's paradox

See **Berry's paradox**.

Riemann, (Georg Friedrich) Bernhard (1826–1866)

A German mathematician who was the first person to provide a thorough treatment of **non-Euclidean geome-**

try and to see how it might be applied in physics; he thus helped pave the way for the general **relativity theory**. Among several profound aspects of mathematics now named after him are the **Riemann hypothesis** and the related **Riemann zeta function**. His father, a Lutheran pastor, encouraged him to study theology at Göttingen. But even as a child Riemann had shown a tremendous aptitude for mathematics and, in 1847, he persuaded his father to let him go to Berlin to learn mathematics from the likes of Karl **Jacobi**, Peter **Dirichlet**, and Jakob **Steiner**. Two years later, he returned to Göttingen to study for his Ph.D. and begin his climb up the professorial ladder. In 1854, his inaugural lecture, "Concerning the hypotheses which underlie geometry," covered a breathtaking array of topics, including a workable definition of the curvature of space and how it could be measured, the first description of **elliptical geometry**, and, most important of all, the extension of geometry into more than three dimensions with the aid of algebra.

Riemann hypothesis

If I were to awaken after having slept for a thousand years, my first question would be: Has the Riemann Hypothesis been proven?

–David Hilbert

The most important open question in **number theory** and, possibly, in the whole of mathematics. A $1 million prize has been offered by the Clay Mathematics Institute for a proof. The hypothesis was first formulated by Bernhard **Riemann** in 1859, was included in David **Hilbert**'s list of challenging problems for twentieth-century mathematicians, and is widely believed to be true. Yet a proof remains tantalizingly out of reach. What the Riemann hypothesis says is that the nontrivial zeros of the **Riemann zeta function** all have real parts equal to ½. In plain language, the hypothesis asserts that there is an underlying order, akin to musical harmonics, in the way **prime numbers** are distributed. It's known that for any given number n there are approximately $n/\log n$ prime numbers that are less than n. The formula is not exact: sometimes it is a little high and sometimes it is a little low. Riemann looked at these deviations and found that they contain periodicities. His hypothesis quantifies and formalizes this discovery, positing that the zeros of the zeta function can be regarded as the harmonic frequencies in the distribution of primes. If the Riemann hypothesis turns out to be true, what do these harmonics in the "music" of the primes mean? Remarkably, it is been found by the English physicist Michael Berry and his colleagues that there is a deep connection between the harmonics–the Riemann zeros–and the allowable energy states of physical systems that are on the border between the quantum world (see **quantum**

mechanics) and the everyday world of classical physics. The Riemann harmonics, or "magic numbers," behave exactly like the energy levels in quantum systems that classically would be chaotic. This deep connection between number theory and the physics of the real universe, if upheld, is utterly astonishing. If the Riemann hypothesis is proved true, it could open an entirely new window on the nature of reality and the relationship between the abstract world of mathematics and the behavior of matter and energy. On the other hand, if it is disproven, there will be an even deeper mystery to explore: How can the Riemann zeta function so convincingly mimic a quantum system without actually being one?[40]

Riemann integral

The kind of **integral** familiar from **calculus** texts and normally used by scientists and engineers, in which the process of **integration** of function on an interval amounts to finding the area under the curve.

Riemann sphere

A topological sphere consisting of the **complex plane** and the point at infinity; an example of a **Riemann surface**.

Riemann surface

Also known as a *complex curve*, a complex **manifold** with one complex dimension. The Riemann surface is a conformal structure (see **conformal mapping**).

Riemann zeta function ζ(s)

We may—paraphrasing the famous sentence of George Orwell—say that "all mathematics is beautiful, yet some is more beautiful than the other." But the most beautiful in all mathematics is the zeta function. There is no doubt about it.
    —Krzysztof Maslanka, Polish cosmologist

One of the most profound and mysterious objects in modern mathematics; from it has sprung the **Riemann hypothesis** and all that this conjecture seems to imply. The Riemann zeta function is closely tied to the distribution of **prime numbers**. It is an extension of the *Euler zeta function*, first studied by Leonhard **Euler**, which is the sum

$$\zeta(s) = 1 + 1/2^s + 1/3^s + 1/4^s + \ldots = \sum_{n=1}^{\infty} 1/n^s.$$

Euler found that this function is linked to the occurrence of prime numbers by the following fundamental relationship:

$$\zeta(s) = 1 + 1/2^s + 1/3^s + 1/4^s + \ldots = 2^n/(2^n - 1)$$
$$\times 3^n/(3^n - 1) \times 5^n/(5^n - 1) \times 7^n/(7^n - 1) \times \ldots$$
$$= \prod_p 1/(1 - p^{-s})$$

The Riemann zeta function extends the definition of Euler's zeta function to all **complex numbers**.

Riemannian geometry

See **elliptical geometry**.

right

A *right angle* is an angle of 90°. A *right triangle* is a triangle that contains a right angle.

ring

(1) Another name for an **annulus**. (2) A number system in which addition, subtraction, and multiplication are always defined and the **associative** and **distributive** laws are valid. Compare with **field**.

Rithmomachia

A medieval, chesslike board game for two players that is based on the number theories of Pythagoras and Boethius. Rithmomachia or "battle of numbers" (*rithmo*, "arithmetic, numbers"; *machia* "battle") dates back to about A.D. 1150 although the earliest publication of the rules was by Jean de Boissiere in the sixteenth century. Used as an educational tool (the only game allowed in medieval schools and universities) and as an intellectual exercise, it enjoyed a last wave of popularity during the Renaissance before the early Scientific Revolution led to its disappearance.

The game is played on an 8 × 16 board. Each player starts with either 24 black or white pieces: 8 circles, 8 triangles, 7 squares, and 1 pyramid. Each of these has a *number*, which is how many places it can move. A square can move 4 spaces, a triangle 3, a circle 1, and the pyramid as many spaces as the player chooses. Opponent pieces can be captured in a variety of ways: *siege capture* (surrounding the opponent piece on all four sides; *meeting capture* (attacking a piece with the same type of piece); *assault capture* (achieved if the piece's number times the number of spaces it moved lands it next to an opposing piece that equals the product); *ambuscade* (achieved if two pieces of a player that are on either sides of a piece sum to equal the opponent's piece). There are also a number of different ways to win, including: *de corpore* (players agree on a number of pieces to be captured); *de bonis* (players agree on a number value target); *de lite* (the winner is determined by the sum of the pieces as well as the number of digits on all those pieces); *victoria magna* (if there is a common difference between the pieces a player has captured, or the squares of three consecutive integers are captured; or there is a difference of 2, 4, and 6 in even pieces, or 3, 5, and 7 in odd pieces).

river-crossing problem

A puzzle in which a variety of objects and living things, some of them mutually incompatible, must be conveyed in small groups from one side of a river to another without any loss along the way. The earliest known examples are in *Propositiones ad Acuendos Juvenes* (Propositions for sharpening youths), which is generally attributed to Abbott **Alcuin**. They are: the problem of three jealous husbands (each of whom won't let another man be alone with his wife), the problem of the two adults and two children where the children weigh half as much as the adults (and the boat has a limited weight capacity), and the problem of the wolf, the goat, and the cabbage. In the last case, the difficulty is that only one item can be ferried across at once but, if left unattended, the sheep will eat the cabbage and the wolf will eat the sheep. The solution, which involves a stratagem common to all these types of problems, is to bring back to the starting bank of the river an item that has already been taken across. In this case, the sheep must be taken across first, followed by either the cabbage or the wolf, but then the sheep must be brought back before the next item is taken across to avoid the sheep becoming either a diner or a dinner.

These medieval puzzles were considered and elaborated on by Niccoló **Tartaglia**, Luca **Pacioli**, and Claude-Gaspar **Bachet**, and even more so by later mathematicians such as Edouard **Lucas** and Gaston **Tarry**. Ways of complicating river-crossing problems include adding more people and objects, using a bigger boat, and inserting an island in the river. The reader may care to try the problem of the missionaries and the cannibals.

PUZZLE

Three missionaries and three cannibals must cross a river in a boat that holds a maximum of two people. If the cannibals outnumber the missionaries, on either side of the river, the missionaries are in trouble. Each missionary and each cannibal is capable of rowing the boat. How can all six get across the river safely?

Solutions begin on page 369.

Robinson, Abraham (1918–1974)

A German-born mathematician who founded **nonstandard analysis** and did important work in a wide diversity of fields from aerodynamics to mathematical logic. A modern counterpart to Gottfried **Leibniz** in his range of interests and the significance of his research on **infinitesimals**, Robinson (he changed his name from Robinsohn) taught at various universities in Israel, England, Canada, and the United States.

Rolle's theorem

Suppose a continuous function (see **continuity**) crosses the x-axis at two points a and b and is differentiable at all points between a and b; that is, it has a **tangent** at all points on the curve between a and b. Then there's at least one point between a and b where the **derivative** is 0, and the tangent is parallel to the x-axis.

Roman numerals

A **number system** in which each symbol represents a fixed value regardless of its position; this differs from the place-value system of **Arabic numerals**. The earliest form of the Roman system was, however, decimal. In this primitive version a series of I's represented any number from 1 to 9, and a new symbol was introduced for each higher power of 10: X for 10, C for 100, and M for 1,000. The symbols V, L, and D, which stand for 5, 50, and 500, are thought to have been introduced by the Etruscans. A common remark is that multiplication and division using Roman numerals is so awkward that it is totally impractical. However, an article by James G. Kennedy in *The American Mathematical Monthly* in 1980 gives algorithms for these operations that are actually more straightforward in the Roman system than in the Arabic. In multiplication the first step is to rewrite the numbers in a simple place-value notation. Seven columns are set up, headed by the symbols M, D, C, L, X, V, and I, and tallies are marked in each column corresponding to the number of times that symbol appears in the multiplicand. For example, if the multiplicand is XIII (13), one tally is marked in the X column and three tallies are marked in the I column. The multiplier is written in the same way. The multiplication itself is done by forming partial products according to two simple rules. In most cases the partial product given by any one tally in the multiplier is simply the set of tallies that represents the multiplicand, shifted to the left an appropriate number of columns. If the multiplier digit is I, the multiplicand is not shifted at all; the multiplicand is shifted one place to the left for V, two places for X, three places for L, and so on. The second rule is applied only when one Etruscan character is multiplied by another. In such cases the tallies representing the multiplicand digit are written twice in the appropriately shifted column and an additional tally is written one column to the right. Once a partial product has been formed for every tally in the multiplier, the tallies in each column are accumulated and replaced by the Roman symbol at the head of the column, giving the final answer. Only a slight change in the method is needed for Roman numerals in "subtractive notation," where 10 is written as IX, and so on. If all this sounds not quite so simple, the method for multiplying Arabic numbers is just as

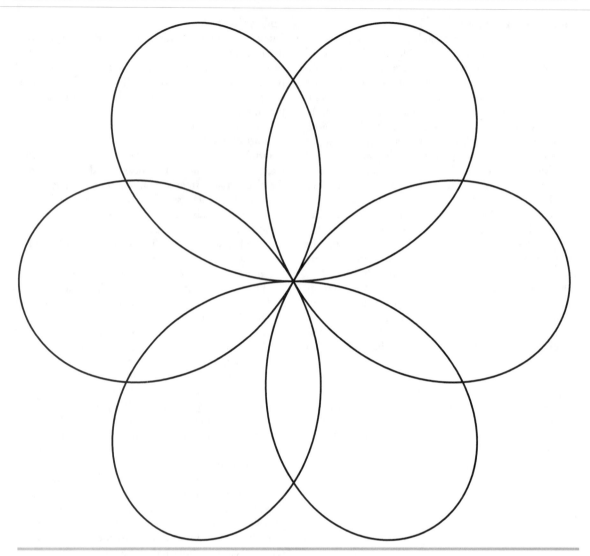

rose curve   © Jan Wassenaar, www.2dcurves.com

involved if they are written in explicit form. Furthermore, Arabic operations require a multiplication table giving the 100 products of all the possible pairs of Arabic digits. No comparable table is needed with Roman numerals, where all arithmetical operations can be defined in terms of shifting rules, addition, and subtraction.

rooks problem

To find the maximum number of rooks that can be placed on an $n \times n$ **chess**board such that no rook attacks another. Since each rook attacks all squares in the rank and file upon which it rests, this number is n: the rooks may be placed along the main diagonal. The total number of ways of placing n nonattacking rooks is n factorial ($n!$).

root

(1) A number used to build up another number by repeated multiplication. For example, since $2 \times 2 \times 2 = 8$, two is said to be the third root or cube root of eight. (2) A solution of an equation. For example, 3 is a root of the

Rubik's cube *Peter Knoppers, www.buttonius.com*

equation $x^2 = 9$. A root is also called a *zero of a function* because it is a value that will make the function zero ($x = 3$ will make the function $f(x) = x^2 - 9$ zero). The word comes from the Indo-European *werad*, which originally meant the roots of a plant but was later generalized to mean the origins or beginnings of something, whether it was physical or mental.

root of unity
A solution of the equation $x^n = 1$, where n is a positive integer.

rope around the earth puzzle
Imagine a rope that fits snugly all the way around Earth like a ring on a person's finger. Now imagine the rope is made just 1 meter longer and lifted uniformly off the surface until it is once again taught. What will its height be above the surface? This puzzle, or one very like it, appeared in a students' book on Euclid written in 1702 by

the English clergyman, mathematician, and natural philosopher William Whiston (1667–1752). The answer in the form just given is remarkable: about 16 cm. It comes simply from the formula for the circumference of a circle. If the extra radius of the rope is r and Earth's radius is R, then

$$2\pi(R + r) = 2\pi R + 100$$
$$\text{so that } r = 100/2\pi \cong 15.9.$$

Rosamund's bower
A legendary **maze**, located in Woodstock Park, Oxfordshire, whose purported site is marked today by a well and fountain. It was supposedly intended to conceal Rosamund Clifford, the mistress of King Henry II (1133–1189), from the queen, Eleanor of Aquitaine. Legend has it that about 1176, Eleanor managed to solve the maze and confronted Rosamund with the choice of a dagger or poison; she drank the poison and Henry

never smiled again. Historically, Henry had imprisoned Eleanor for fomenting rebellion by her sons, and Rosamund was his acknowledged mistress. Rosamund probably spent her last days at a nunnery in Godstow, near Oxford. The legend of the bower dates from the fourteenth century and her murder is a later addition. In the nineteenth century, many puzzle collections had a maze called Rosamund's bower.

rose curve

A curve that has the shape of a flower with petals; it was named *rhodonea* ("rose") by the Italian mathematician Guido Grandi in the 1720s. It is given by the polar equation

$$r = a \sin (n\theta).$$

If n is odd the rose has n petals, if n is even the rose has $2n$ petals, and if $n = 2$ the rose becomes the **quadrifolium**. If n is an *irrational* number, then there are an infinite number of petals.

rotation

A **transformation** in which a figure turns through a specific angle about a fixed point, called the center of rotation. The center of rotation may be inside or outside the figure that is being transformed. If the figure is turned anticlockwise, the rotation is considered positive, while a negative rotation turns the figure clockwise.

rotor

A **convex** figure that can be rotated inside a **polygon** (or **polyhedron**) while always touching every side (or face). The least area rotor in a square is the **Reuleaux triangle**. The least area rotor in an equilateral triangle is a lens with two 60° arcs of circles and a radius equal to the triangle's altitude. There exist nonspherical rotors for the tetrahedron, octahedron, and cube, but not for the dodecahedron and icosahedron. See also **curve of constant width**.

roulette

(1) The curve traced by a fixed point on a closed **convex** curve as that curve rolls without slipping along a second curve. (2) A gambling game in which players bet on which slot of a rotating disk a small ball will come to rest in. On August 18, 1913, on an unbiased roulette wheel at Monte Carlo, evens came up 26 times in a row. The probability of this occurring is 1 in 136,823,184.

round

In **topology**, the terms *circle* and *sphere* refer to topological objects and not geometric ones, so that the surface of

an egg shape is a sphere. A *round sphere* is, topologically speaking, not a tautology, but a sphere with constant **curvature**; that is, a sphere in the sense of geometry.

rounding

Replacing a number by another number having fewer **significant digits** or, for integer numbers, fewer value-carrying (nonzero digits). For example, 386.804 may be rounded successively to 386.80, 386.8, 387, 390, and 400. Rounding may be carried out in two ways: by *rounding down*, which is equivalent to truncation, and by *rounding up* the last digit to be retained by one unit. See also **banker's rounding**.

round-off error

The error accumulated during a calculation due to rounding intermediate results. See also **banker's rounding**.

Rubik's cube

A $3 \times 3 \times 3$ cube in which the 26 subcubes on the outside are internally hinged in such a way that rotation (by a quarter turn in either direction or a half turn) is possible in any plane of cubes. Each of the six sides is painted a distinct color, and the goal of the puzzle is to return the cube to a state in which each side has a single color after it has been randomized by repeated rotations. Invented in 1974 by the Hungarian Ernö Rubik, patented in 1975, and put on the market in Hungary in 1977, it went on to sell some 100 million copies worldwide over the next decade. Since there are over 43 million trillion different arrangements of the small cubes, only one of which corresponds to the desired goal, to solve Rubik's cube in a time significantly less than the current age of the universe (let alone the world record, which stands at around 20 seconds) calls for some kind of methodical approach. Algorithms exist for solving a cube from an arbitrary initial position, but they are not necessarily optimal (i.e., requiring a minimum number of turns).[270, 297]

Rucker, Rudy (Rudolf von Bitter) (1946–)

An American mathematician best known for his entertaining popular mathematics, science, and science fiction books, including *Infinity and the Mind*,[275] *The Fourth Dimension*,[273] and *Mind Tools*.[274] Rucker has a doctorate in mathematical logic from Rutgers and teaches at San José State University. His great-great-great-grandfather was the famous German philosopher Georg Hegel.

ruled surface

A surface that is built up from an infinite number of perfectly straight lines. A **cylinder**, for example, is a ruled surface of parallel straight lines. A **cone** is a ruled surface of straight lines that meet at the apex of the cone. Also known as *scrolls,* ruled surfaces have been studied for centuries by geometers such as the Jesuits Roger Boscovich and Andre Tacquet as well as by their famous students, including Gaspar **Monge** and Phillippe de Lahire. Examples that stand out because of both their striking shape and their relative ease of construction include **hyperboloids**, **helicoids** and **Möbius bands**. Most ruled surfaces, however, are so complicated that, before the computer age, they were almost impossible to construct.

ruler-and-compass construction
See **Mascheroni construction**.

Russell, Bertrand Arthur William (1872–1970)
A British philosopher, mathematician, and logician who rose to prominence with his first major work, *The Principles of Mathematics* (1902), in which he attempted to remove mathematics from the realm of abstract philosophical notions and to give it a precise scientific framework. Russell then collaborated for eight years with the British philosopher and mathematician Alfred North Whitehead (1861–1947) to produce the monumental work *Principia Mathematica* (3 volumes, 1910–1913). This work showed that mathematics can be stated in terms of the concepts of general **logic**, such as class and membership in a class. It became a masterpiece of rational thought. Russell and Whitehead proved that numbers can be defined as classes of a certain type, and in the process they developed logic concepts and a logic notation that established symbolic logic as an important specialization within the field of philosophy. In his next major work, *The Problems of Philosophy* (1912), Russell borrowed from the fields of sociology, psychology, physics, and mathematics to refute the tenets of idealism, the dominant philosophical school of the period, which held that all objects and experiences are the product of the intellect. Russell, a realist, believed that objects perceived by the senses have an inherent reality independent of the mind.

Russell condemned both sides in World War I, and for his uncompromising stand he was fined, imprisoned, and deprived of his teaching post at Cambridge. In prison he wrote *Introduction to Mathematical Philosophy* (1919). After the war he visited the Soviet Union, and in his book *Practice and Theory of Bolshevism* (1920) he expressed his disappointment with the form of socialism practiced there. He felt that the methods used to achieve a Communist system were intolerable and that the results obtained were not worth the price paid. Russell taught at Beijing University during 1921 and 1922, and in the United States

from 1938 to 1944, though he was barred from teaching at the College of the City of New York (now City College of the City University of New York) by the state supreme court because of his attacks on religion and his advocacy of sexual freedom. Russell returned to England in 1944 and was reinstated as a fellow of Trinity College. Although he abandoned pacifism to support the Allied cause in World War II, he became an ardent opponent of nuclear weapons. Russell received the 1950 Nobel Prize for Literature and was cited as "the champion of humanity and freedom of thought." He led a movement in the late 1950s advocating unilateral nuclear disarmament by Britain, and at the age of 89 he was imprisoned after an antinuclear demonstration.[276]

Russell's paradox
A **paradox** uncovered by Bertrand **Russell** in 1901 that forced a reformulation of **set theory**. One version of Russell's paradox, known as the barber paradox, considers a town with a male barber who, every day, shaves every man who doesn't shave himself, and no one else. Does the barber shave himself? The scenario as described requires that the barber shave himself if and only if he does not! Russell's paradox, in its original form considers the set of all sets that aren't members of themselves. Most sets, it would seem, aren't members of themselves—for example, the set of elephants is not an elephant—and so could be said to be "run-of-the-mill." However, some "self-swallowing" sets *do* contain themselves as members, such as the set of all sets, or the set of all things except Julius Caesar, and so on. Clearly, every set is either run-of-the-mill or self-swallowing, and no set can be both. But then, asked Russell, what about the set S of all sets that aren't members of themselves? Somehow, S is neither a member of itself nor not a member of itself. Russell discovered this strange situation while studying a foundational work in symbolic logic by Gottlob **Frege**. After he described it, set theory had to be reformulated axiomatically in a way that avoided such problems. Russell himself, together with Alfred North Whitehead (1861–1947), developed a comprehensive system of types in *Principia Mathematica*. Although this system does avoid troublesome paradoxes and allows for the construction of all of mathematics, it never became widely accepted. Instead, the most common version of axiomatic set theory in use today is the *Zermelo-Fraenkel set theory,* which avoids the notion of types and restricts the universe of sets to those that can be built up from given sets using certain axioms. Russell's paradox underlies the proof of **Gödel's incompleteness theorem** as well as Alan **Turing**'s proof of the undecidability of the **halting problem**.

Russian multiplication

Multiplication by repeated doubling, also known as *peasant multiplication*. For example to multiply 17 by 13, double the 17 and halve the 13 rounding down to the next whole number where necessary; then add the doubles that correspond to an odd number in the other column.

So	17×13
doubled and halved	34×6
doubled and halved	68×3
doubled and halved	136×1

adding up the numbers in the first column that correspond to an odd number in the second $(17 + 68 + 136) = 221 = 17 \times 13$.

Saccheri, Giovanni Girolamo (1667–1733)

A Jesuit priest, philosopher, and mathematician who did early work on **non-Euclidean geometry**, although he didn't see it as such. His *Euclides ab Omni Naevo Vindicatus* (1733) was actually an attempt to prove Euclid's **parallel postulate** but ended up laying the groundwork for both **hyperbolic geometry** and **elliptical geometry**.

saddle

A type of surface that is neither a peak nor a valley but still has a zero gradient. Saddle points are situated such that moving in one direction takes one uphill, while moving in another direction would be downhill. A *saddle function* is a **function** $f(x, y)$ of two vectors x and y (which typically lie in different **vector spaces**) that is **concave** up in x and concave down in y. See also **pseudosphere**.

St. Ives problem

A well-known and simple puzzle in arithmetic set by Mother Goose in the rhyme "As I Was Going to St. Ives":

> As I was going to St. Ives, I met a man with seven wives. Every wife had seven sacks, and every sack had seven cats, every cat had seven kittens. Kittens, cats, sacks, and wives, how many were going to St. Ives?

The various individuals and items make up a **geometric sequence**: 7 wives + 7^2 sacks + 7^3 cats + 7^4 kittens = 7 + 49 + 343 + 2,401. This gives a total of 2,800 (or 2,801 if the narrator also happens to be a wife). It's been said that the real answer to this rhyme is 1, since "I *met* a man with seven wives" means the 2,800 were going in the *opposite* direction!

A similar problem, and solution, is contained in the **Rhind papyrus** by Ahmose, written about 1650 B.C. (some of which is copied from an older document of about 1800 B.C.). Here the geometric series has one more power of 7: 7 houses + 49 cats + 343 mice + 2,401 ears of grain + 16,807 hekats (a measure) of grains, giving a total of 19,607.

St. Petersburg paradox

A strange state of affairs that arises from a game proposed by Nikolaus (I) Bernoulli (see **Bernoulli family**) in 1713. It is named after the fact that a treatise on the paradox was written by Nikolaus' cousin, Daniel, and published (1738) in the *Commentaries of the Imperial Academy of Science of St. Petersburg*. The game goes as follows. You toss a coin. If it shows heads, you win whatever is in the pot and the game is over. If it shows tails, the pot is doubled and you get to toss the coin again. If the coin shows heads on the first toss you win $2; if it shows tails, you toss again. If the coin now shows heads you win $4, and so on. After n tosses you get 2^n if heads appear for the first time. The only catch is you have to pay to play the game. How much should you be willing to pay? Classical decision theory says that you should be willing to pay any amount up to the expected prize, the value of which is obtained by multiplying all the possible prizes by the probability that they are obtained and adding the resulting numbers. The chance of winning $2 is ½ (heads on the first toss); the chance of winning $4 is ¼ (tails followed by heads); the chance of winning $8 is ⅛ (tails followed by tails followed by heads); and so on. Since the expected payoff of each possible consequence is $1 ($2 × ½, $4 × ¼, etc.) and there are an infinite number of them, the total expected payoff is an infinite sum of money. A rational gambler would enter a game if and only if the price of entry was less than the expected value. In the St. Petersburg game, any finite price of entry is smaller than the expected value of the game. Thus, the rational gambler would play no matter how large the entry price was! But there's clearly something wrong with this. Most people would offer between $5 and $20 on the grounds that the chance of winning more than $4 is only 25% and the odds of winning a fortune are very small. And therein lies the paradox: If the expected payoff is infinite, why is no one willing to pay a huge amount to play?

The classical solution to this mystery, provided by Daniel Bernoulli and another Swiss mathematician, Gabriel Cremer, goes beyond probability theory to touch areas of psychology and economics. Bernoulli and Cremer pointed out that a given amount of money isn't always of the same use to its owner. For example, to a millionaire $1 is nothing, whereas to a beggar it can mean not going hungry. In a similar way, the utility of $2 million is not twice the utility of $1 million. Thus, the important quantity in the St. Petersburg game is the *expected utility* of the game (the utility of the prize multiplied by its probability) which is far less than the *expected prize*. This explanation forms the theoretical basis of the insurance business. The existence of a **utility function** means that most people prefer, for example, having $98 in cash to gambling in a lottery where they could win $70 or $130 each with a chance of 50%, even though the lottery has the higher expected prize of $100. The difference of $2 is the premium most of us would be willing to pay for insurance. That many people pay for insurance to avoid any risk, yet at the same time spend money on lottery tickets in order to take a risk of a different kind, is another paradox, which is still waiting to be explained.

salient

At a *salient point,* two branches of a curve meet and stop, and have different **tangents**. A *salient angle* is an outward-pointing angle of a **polygon**; compare with **reentrant angle**.

salinon

A figure formed from four connected semicircles. The word *salinon* is Greek for "salt cellar," which the figure resembles. In his *Book of Lemmas,* **Archimedes** proved that the salinon has an area equal to the circle having the line segment joining the top and bottom points as its diameter. See also **arbelos**.

Sallows, Lee C. F. (1944–)

A British electronics engineer and puzzle enthusiast at the University of Nijmegen in the Netherlands. Among his many accomplishments in recreational mathematics, he devised the first **self-enumerating sentence** (published in 1982), introduced **alphamagic square**s in 1986, coined the word *golygon* in 1990, invented reflexicons (minimal self-enumerating phrases) in 1992, demonstrated the parallelogram theorem for 3 × 3 **magic squares** in 1997, and discovered **geometric magic squares** in 2001.

scalar

A quantity specified by a single number or value (as opposed to a vector, matrix, or array) that contains multiple values. Examples of scalars include mass, volume, and temperature. A *scalar field* is an arrangement of scalar values distributed in a space.

scalene triangle

A **triangle** whose sides are all unequal.

schizophrenic number

An informal name for an **irrational number** that displays such persistent patterns in its decimal expansion, that it

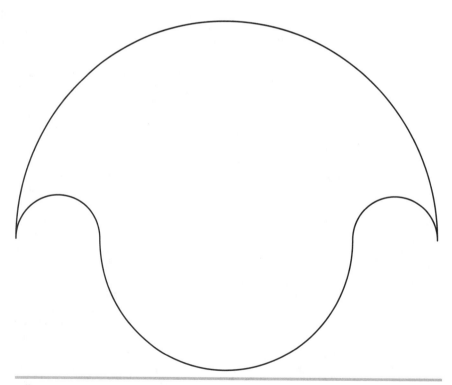

salinon   © *Jan Wassenaar, www.2dcurves.com*

has the appearance of a **rational number**. A schizophrenic number can be obtained as follows. For any positive integer n let $f(n)$ denote the integer given by the recurrence $f(n) = 10 f(n-1) + n$ with the initial value $f(0) = 0$. Thus, $f(1) = 1, f(2) = 12, f(3) = 123$, and so on. The square roots of $f(n)$ for odd integers n give rise to a curious mixture appearing to be rational for periods, and then disintegrating into irrationality. This is illustrated by the first 500 digits of $\sqrt{f(49)}$:

1111111111111111111111111.11111111111111111111111   0860
55555555555555555555555555555555555555555555555   2730541
66666666666666666666666666666666666666666666   0296260347
2222222222222222222222222222222222222222   0426563940928819
44444444444444444444444444444444   38775551250401171874
9999999999999999999999999999   80824968771148630533841
66666666666666666666   59871857386214406386555598958
333333333333333333   08434604076276082069402770996093749
99999999999999   06422275875559830666394303215874565979
222222222   18634920167911808330818449 . . . .

The repeating strings become progressively shorter and the scrabbled strings become larger until eventually the repeating strings disappear. However, by increasing n we can forestall the disappearance of the repeating strings as long as we like. The repeating digits are always 1, 5, 6, 2, 4, 9, 6, 3, 9, 2, ...

Schläfli, Ludwig (1814–1895)

A German mathematician whose work centered on geometry, arithmetic, and the theory of functions. He made an important contribution to **non-Euclidean geometry** when he proposed that spherical three-dimensional space could be thought of as the surface of a **hypersphere** in Euclidean four-dimensional space. Schläfli started out as a schoolteacher and amateur mathematician. He was also an expert linguist and spoke many languages, including Sanskrit. In 1843 he served as a translator for the great mathematicians Jakob **Steiner**, Karl **Jacobi**, and Peter **Dirichlet** during their visit to Rome and learned a great deal from them. Ten years later he became professor of mathematics at Bern. However, his true importance was only appreciated following the publication of his magnum opus *Theory of Continuous Manifolds* in 1901, several years after his death.

Schläfli symbol

A notation, devised by Ludwig **Schläfli**, which describes the number of edges of each **polygon** meeting at a **vertex** of a regular or semi-regular **tessellation** or solid. For a **Platonic solid**, it is written $\{p, q\}$, where p is the number of sides each face has, and q is the number of faces that touch at each vertex.

schoolgirls problem

A problem in **combinatorics** posed by the Rev. Thomas **Kirkman** in a letter in 1850 following a paper he wrote on the same subject in 1847:[188] A school mistress has fifteen girl pupils and she wishes to take them on a daily walk. The girls are to walk in five rows of three girls each. It is required that no two girls should walk in the same row more than once per week. Can this be done?

In fact, provided n is divisible by 3, we can ask the more general question about n schoolgirls walking for $(n-1)/2$ days so that no girl walks with any other girl in the same triplet more than once. Solutions for $n = 9, 15$, and 27 were given in 1850 and much work was done on the problem thereafter. The general problem of how many triads can be made out of **n** symbols, so that no pair of symbols is comprised more than once among them gave rise to the study of Steiner triple systems. However, Jakob **Steiner** had little to do with them and they should rightfully be named after Kirkman. They are important in the modern theory of **combinatorics**.

Schröder's reversible staircase

A classic example of an **ambiguous figure**, first drawn by Schröder in 1858. Not to be confused with the **Penrose stairway**.

Schubert, Hermann Cäsar Hannibal (1848–1911)

A German mathematician who worked mainly in enumerative geometry—the parts of algebraic geometry that

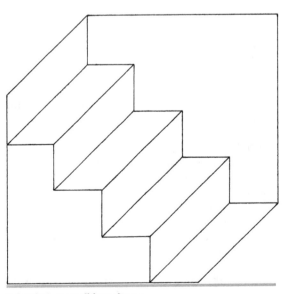

Schröder's reversible staircase

involve a finite number of solutions. He also wrote extensively on recreational math.

Schuh, Frederick (1875–1966)

A Dutch mathematician who wrote many textbooks and a number of books on recreational mathematics, including *The Master Book of Mathematical Recreations* (1943), the English edition of which appeared in 1968 (Dover Publications).[291] Schuh was professor of mathematics at the Technische Hoogeschool at Delft (1907–1909 and 1916–1945) and professor of mathematics at Groningen (1909–1916).

scientific notation

A number of the form $a \times 10^n$, where n is an integer, positive or negative, and a is a **real number** larger than or equal to 1, but less than 10. Scientific notation provides a compact way of writing **large number**s.

scintillating grid illusion

See **Hermann grid illusion**.

score

A group of 20 items. The word comes from the Old Norse *skor* for a heavy mark used to indicate a string of 20 smaller marks; *skor*, in turn, is descended from the Indo-European *sker*, for "cutting" or "slicing." From about 1400, *score* was also the word for a record or an amount due—the total of the score marks on a tally. It became a common word for the total of a tradesman's or innkeeper's account. So, *to settle the score* originally meant just to pay one's bill. But it acquired the figurative sense of taking revenge on somebody, and that's usually what is meant by the expression now. The more general meaning of score, as a tally, is used daily when the results of sports competitions are reported.

search space

A characterization of every possible solution to a problem instance.

secant

A straight line that meets a curve in two or more points.

second

One-sixtieth of a **minute** in both time and angle. Literally, the second division of the hour or the circle, the minute being the first; from the Latin *secundus*. In the System International d'Unites (SI units) one second is defined as the duration of 9,192,631,770 periods of radiation corresponding to the transition between two hyperfine levels of cesium-133 in a ground state at a temperature of 0°K (Kelvin).

secretary problem

See **sultan's dowry**.

sector

Part of a **circle** bounded by two radii and the included **arc**.

segment

Part of a **circle** bounded by a **chord** and the **arc** subtending the chord.

self-enumerating sentence

Also known as an *autogram,* a **self-referential sentence** whose text consists solely of the enumeration of its letter content. The answer to the question whether such a sentence exists in English was given by Lee **Sallows**, and was first published in *Scientific American* in January 1982:

> Only the fool would take trouble to verify that his sentence was composed of ten a's, three b's, four c's, four d's, forty-six e's, sixteen f's, four g's, thirteen h's, fifteen i's, two k's, nine l's, four m's, twenty-five n's, twenty-four o's, five p's, sixteen r's, forty-one s's, thirty-seven t's, ten u's, eight v's, eight w's, four x's, eleven y's, twenty-seven commas, twenty-three apostrophes, seven hyphens and, last but not least, a single !

This remarkable sentence counts not only its own letters, but also its punctuation marks, although it fails to enumerate three letters of the alphabet (j, q, and z). Sallows went on to devise a "pangram machine"–a computer purposely built to search for sentences of this type. Among its many successes is:

> This pangram lists four a's, one b, one c, two d's, twenty-nine e's, eight f's, three g's, five h's, eleven i's, one j, one k, three l's, two m's, twenty-two n's, fifteen o's, two p's, one q, seven r's, twenty-six s's, nineteen t's, four u's, five v's, nine w's, two x's, four y's, and one z.

self-intersecting

A *self-intersecting polygon* is a **polygon** with edges that cross other edges. A *self-intersecting polyhedron* is a **polyhedron** with faces that cross other faces.

self-organization

A process in a **complex system** whereby new emergent structures, patterns, and properties arise without being externally imposed on the system. Not controlled by a centralized, hierarchical "command and control" center, self-organization is usually distributed throughout a system. It requires a complex, **nonlinear system** under appropriate conditions, variously described as "far-from-equilibrium," critical values of control parameters leading to "bifurcation," or the "edge of chaos." First investigated in the 1960s in physical systems by Ilya Prigogine and his followers, as well as the Synergetics School founded by Hermann Haken, self-organization is now studied mainly through computer simulations (see **cellular automaton**), Boolean networks, and other phenomena of **artificial life**. However, self-organization is now recognized as a crucial way of understanding emergent, collective behavior in a large variety of systems including the economy, the brain and nervous system, the immune system, and ecosystems. The buildup of system order via self-organization is now conceived as a primary tendency of complex systems in contrast to the past emphasis on the degrading of order in association with the principle of entropy (Second Law of Thermodynamics). However, rather than denying entropy, self-organization can be understood as a way that entropy increases in complex, nonlinear systems.

self-organized criticality
A mathematical theory that describes how systems composed of many interacting parts can tune themselves toward dynamical behavior that is critical in the sense that it is neither stable nor unstable but at a region near a **phase transition**. See also **edge of chaos** and **self-organization**.

self-referential sentence
A sentence that refers to itself and nothing else. Here are some examples:

This statement is short.

This sentence has five words.

The last word of this sentence is "wrong."

"Pentasyllabic" is pentasyllabic.

How long is the answer to this question? Ten letters.

Some self-referential statements take the form of jokes. For example:

The two rules for success are: Never tell them everything you know.

There are three kinds of people in the world: those who can count and those who can't.

Finally, some take the form of maxims, as in this case by Thomas Macaulay (1800–1859):

Nothing is so useless as a general maxim.

See also **Hofstadter's law**.

self-similarity
The property an object has when a part of itself looks the same or similar to the whole. Many objects in the real world, such as coastlines, are statistically self-similar: parts of them show the same statistical properties at many scales. Self-similarity is a defining characteristic of **fractal**s.

semigroup
A **set** together with a method of combining elements, such as addition or multiplication, to get new ones, which satisfies only some of the properties required to get a **group**. In particular, a semigroup need not have an **identity** element and elements need not have **inverse**s.

semi-magic square
A square array of n numbers such that sum of the n numbers in any row or column is a constant (known as the magic sum). See also **magic square**.

semi-regular polyhedron
A **polyhedron** that consists of two or more types of **regular polygon**s, all of whose vertices are identical. This category includes the **Archimedean solid**s, **prism**s and **antiprism**s, and the nonconvex uniform polyhedra (see **nonconvex uniform polyhedron**).

Senet
A popular two-player board game in ancient Egypt, enjoyed by both commoners and nobility, that may be an ancestor of modern **backgammon**. The rules are not known, though about 40 sets have been found in tombs, some in very good condition, together with paintings of games on tomb walls, dating back to the reign of Hesy (c. 2686–2613 B.C.). Senet, or the "game of passing," was played on a rectangular board consisting of three rows of 10 squares called "houses" that represented good or bad fortune. The board could be a grid drawn on a smooth surface or an elaborate box of wood and other precious materials. A perfectly preserved traveling version of Senet was found in Tutankhamen's tomb. The pieces, called *ibau* ("dancers" in Egyptian), varied in number from five to ten per player–five and seven being commonest. Cone-shaped pieces were pitted against reel-shaped pieces. The object was to get one's pieces on the board, then around the board in an S-shaped pattern, and finally off again at the far end. Strategy was mixed with chance (as it is in backgammon), introduced by the throw of four, two-sided sticks (as depicted in the Hesy painting) or, in later times, of knucklebones. Later depictions of the game, in the New Kingdom period, often showed just one player in competition–the opponent being a spirit from the afterlife. This has been interpreted as a change

Senet A modern version of the ancient Egyptian board game. *Fundex Games Ltd.*

in the significance of Senet, from a simple amusement to a symbolic representation of the deceased's journey through the underworld. See also **nine men's morris**.

sensitivity
The tendency of a system, which may be chaotic (see **chaos**) to change dramatically with only small **perturbation**s.

sequence
An ordered list of values that may be finite or infinite in length. Among the different types of sequence are **arithmetic sequence**s, **geometric sequence**s, and the **harmonic sequence**.

series
A sum of all or some of the terms of a **sequence**. A series may or may not *converge* to a particular value as more and more terms are included. A series is said to be *absolutely convergent* if the sum of the absolute values of the terms converges; in this case the series converges no matter how the terms of the sum are arranged. Series that are *conditionally convergent* only converge for some arrangements of the terms and, even then, converge to different values for different arrangements.

serpentine
A curve named and studied by Isaac **Newton** in 1701 and contained in his classification of **cubic curve**s. It had been studied earlier by **de L'Hôpital** and Christiaan

Huygens in 1692. The curve is given by the Cartesian equation

$$y(x) = abx/(x^2 - a^2).$$

set
A finite or infinite collection of objects known as *elements*. Sets are one of the most basic and important concepts in mathematics. An example of a finite set is the set of whole numbers from 1 to 58; an example of an infinite set is the set of all the **rational number**s. Two sets are equal if, and only if, they contain the same objects. Standard notation uses braces around the list of elements, as in: {red, green, blue}. If A and B are two sets and every x in A is also contained in B, then A is said to be a *subset* of B. Every set has as subsets itself, known as the *improper subset*, and the **empty set**. The *union* of a collection of sets $S = \{S_1, S_2, S_3, \ldots\}$ is the set of all elements contained in at least one of the sets $S_1, S_2, S_3, \ldots$. The *intersection* of a collection of sets $T = \{T_1, T_2, T_3, \ldots\}$ is the set of all elements contained in all of the sets. The union and intersection of sets, say A_1 and A_2, is denoted $A_1 \cup A_2$ and $A_1 \cap A_2$, respectively. The set of all subsets of X is called its *power set* and is denoted 2^X or $P(X)$. See also **set theory**, **Venn diagram**, and **Russell's paradox**.

set of all sets
See **Russell's paradox**.

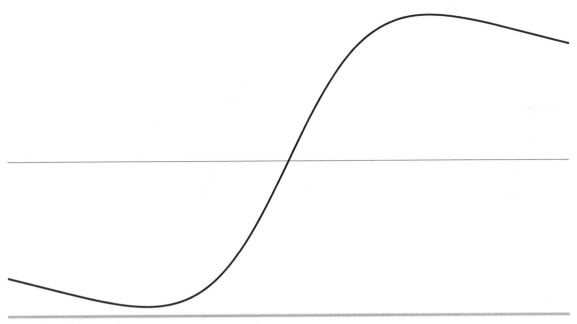

serpentine The serpentine curve. © *Jan Wassenaar, www.2dcurves.com*

set theory

A branch of **mathematics** created by Georg **Cantor** at the end of the nineteenth century. Initially controversial, set theory has come to play a foundational role in modern mathematics, in that it is used to justify assumptions made concerning the existence of mathematical objects (such as numbers or functions) and their properties. Formal versions of set theory also figure centrally in specifying a theoretical ideal of mathematical rigor in proofs. Cantor's basic discovery was that if we define two sets A and B to have the same number of members (the same *cardinality*), then there is a way of pairing off members of A exhaustively with members of B. The appearance around the turn of the century of set-theoretical paradoxes, such as **Russell's paradox**, prompted the formulation in 1908 by Ernst Zermelo of an axiomatic theory of sets. The axioms for set theory now most often studied and used are those called the *Zermelo-Fraenkel axioms,* usually together with the **axiom of choice**. The Zermelo-Fraenkel axioms are commonly abbreviated to ZF, or ZFC if the axiom of choice is included. An important feature of ZFC is that every object that it deals with is a set. In particular, every element of a set is itself a set. Other familiar mathematical objects, such as numbers, must be subsequently defined in terms of sets.

seven

A lucky number in the eyes of many people and one that has been given much spiritual significance. The early religious and cultural use of the seven-day week almost certainly stems from the fact that the Moon goes through its four phases in a bit over 28 days, which divides nicely into seven days per phase. There are seven moving objects in the sky visible to the naked eye (the Sun, Moon, Mercury, Venus, Mars, Jupiter, and Saturn), seven seas, seven orders of architecture, seven deadly sins, seven liberal arts and sciences, and seven dwarfs. The seventh son of a seventh son is supposed to be born gifted (Donny Osmond was such a person). In the Bible, there were seven years of famine and seven years of plenty, and seven years were taken to construct King Solomon's Temple. The Pythagoreans were especially intrigued by the number as it is the sum of three and four, which are the number of sides of a triangle and a square—shapes of enormous importance to the sect. These links with Solomon's Temple and the Pythagoreans help explain the importance of seven in freemasonry. Seven is the smallest positive integer whose reciprocal has a pattern of more than one repeating digit: $1/7 = 0.142857142857\ldots$ and is the smallest number for which the digit sequence of $1/n$ is of length $n - 1$ (the longest such a sequence can be). The next such numbers are 17, 19, 23, 29, 47, 59, 61, 97,

109, 113, ... Other curios: the citrus soda 7-UP, created in 1929, was so called because the original containers were 7 ounces and "up" was the direction of the bubbles, and seven is the maximum number of times you can fold any sheet of paper (try it!).

seventeen

The number most often picked in response to the request "Pick a random number from 1 to 20." Seventeen is a *Fermat prime* (a **prime number** of the form $2^{2n} + 1$, where n is a positive integer), the exponent of a *Mersenne prime* (a prime p for which $2^p - 1$ is prime), and the only prime that is the sum of four consecutive primes ($2 + 3 + 5 + 7$). Seventeen is also the smallest number for which the sum of the digits of its cube is equal to the number: $17^3 = 4913$, $4 + 9 + 1 + 3 = 17$, and the smallest number that can be written as $a^2 + b^3$ in two different ways: $17 = 3^2 + 2^3 = 4^2 + 1^3$. The pair $(8, 9)$, whose sum is 17, is the only pair of consecutive numbers where one is a square and the other is a cube (a result proved by Leonhard **Euler**.) There are 17 planar crystallographic groups, called **wallpaper groups**. The minimum number of faces on a convex **polyhedron** that has only one stable face is 17. (A stable face is one that the figure can rest on without falling over; most polygons have more than one such face.) Seventeen is also the answer to the following problem: At a party where any two people have previously met each other in one of three other places, what is the least number of people who must be at the party to guarantee that there is at least one group of three people who have met each other before in the same place?

sexagesimal

Of, relating to, or based on the number 60. Sexagesimal refers especially to the **number system** with base 60. The Babylonians began using such a scheme around the beginning of the second millennium B.C. in what was the first example of a **place-value system**. Our degree of 60 minutes, minute of 60 seconds (in both time and angle measure), and hour of 60 minutes hark back to this ancient method of numeration. Why the Babylonians counted using sexagesimal isn't known, but 60 certainly has more factors than any other number of comparable size.

Shannon, Claude Elwood (1916–2001)

An American mathematician and a pioneer of **information theory**. Shannon was the first to realize that any sort of message can be transmitted as a series of 0's and 1's, regardless of whether it consists of words, numbers, pictures, or sound. In his master's thesis, he explained how electrical switches could represent **binary** digits—a 1 when the switch is on and 0 when it is off. He also used **Boolean algebra** to show that complex operations could be carried out automatically on these electrical circuits, thus manipulating the data they were storing. It was in one of his papers, "A Mathematical Theory of Communication" published in 1948, that the word "bit" (short for binary digit) was used for the first time. In fact the framework and terminology for information theory he developed remains standard today. Shannon was driven by curiosity, and in his own words he "just wondered how things were put together." Among his inventions were rocket-powered Frisbees, motorized Pogo sticks, a device that could solve the **Rubik's cube** puzzle, and a juggling machine. (He could ride a unicycle while juggling three balls.) He was involved in pioneering **artificial intelligence** research, which included building the electromechanical mouse called "Theseus" that could navigate a metal **maze** using magnetic signals. Shannon also built a **chess**-playing computer, many years before IBM's Deep Blue, that played well against the world champion of the time, Mikhail Botvinnik (the computer lost only after 42 moves).

shell curve

See **Dürer's shell curve**.

shuffle

How many shuffles does it take to **randomize** a deck of cards—in other words, to mix up the cards about as thoroughly as dropping them all on the table and stirring them around for several minutes (the author's usual method). The answer depends on the kind of shuffle used. The beginner's *overhand shuffle*, for example, is a really bad way to mix cards: about 2,500 such shuffles are need to randomize a deck of 52 cards. A magician's *perfect shuffle*, on the other hand, in which the cards are cut exactly in half and then perfectly interlaced, *never* produces randomization (see below). One of the most effective ways to get a random deck is the *riffle shuffle* in which the deck is cut in half and imperfectly interlaced by dropping cards one by one from either half of the deck with a probability proportional to the current sizes of the deck splits. In 1992 Persi Diaconis (then at Harvard) and David Bayer demonstrated that, starting with a completely ordered deck, it takes *seven* riffle shuffles to produce randomization.[27] Any more than this and there's no significant increase in the randomness; any less and the shuffle is far from random. In fact, not only are five or six riffles not enough to randomize, there are some configurations of cards that are impossible to reach in this number of shuffles! To understand this, suppose the starting order of the cards is marked 1 to 52, top to bottom. After one shuffle, only configurations with two or fewer rising sequences are possible. A rising sequence is a maximal increasing sequential ordering of cards that appear in the deck (with other cards possibly interspersed) as it is run through from top to bottom. For instance, in an eight-card deck, 12345678 is the ordered deck and it has one rising sequence. After one shuffle, 16237845 is a possible configuration, and there are two rising sequences (the underlined numerals form one, the nonunderlined numerals form the other). Clearly the rising sequences are formed when the deck is cut before the cards are interleaved in the shuffle. After two shuffles, there can be at most four rising sequences, since each of the two rising sequences from the first shuffle has a chance of being cut in the second. This pattern continues: the number of rising sequences can at most double during each shuffle. After five shuffles, there are at most 32 rising sequences. But the *reversed* deck, numbered 52 down to 1, has 52 rising sequences. Thus, this is one (of many) arrangements that are unattainable in five riffle shuffles. Interestingly, Diaconis and other researchers have also found that decks can undergo sudden changes in their degree of randomness; after six riffle shuffles, a deck is still visibly ordered, but this order vanishes one shuffle later.

Perfect shuffles do the exact opposite of randomizing: they preserve order at every stage. There are two kinds of perfect shuffles. The *out-shuffle* is one in which the top card stays on top; the *in-shuffle* is one in which the top card moves to the second position of the deck. Amazingly, eight perfect out-shuffles restore the deck to its original order! Magicians use combinations of out and in shuffles to perform a variety of baffling tricks and to control the position of any given card in a deck. How could you make the top card (call it position 0) go to position *n*? Easy: write *n* in **binary** (base 2), read the 0's and 1's from left to right, perform an out-shuffle for a 0 and an in-shuffle for a 1, and, as if by magic, the top card will have materialized at position *n*.

shunting puzzles

Railroad modelers, especially those with limited space available for their layouts, often enjoy setting up track that allows interesting shunting problems to be tried out and solved. The most famous mathematical puzzle of this type, called the *railroad shunting puzzle*, comes in a number of variations, but basically the problem is that there are two trains (*A* and *B* in the diagram) facing each other on a single line with just one short siding, which will only hold one item of rolling stock at a time. In order to enable the two trains to pass each other and to continue their journey, a series of movements using the siding is required. First time around it's quite a brain-teaser, which probably explains why railroad companies all over the world took the more costly but easier way out and built passing sidings!

Siegel's paradox

If a fixed fraction *x* of a given amount of money *P* is lost, and then the same fraction *x* of the remaining amount is gained, the result is less than the original and equal to the final amount if a fraction *x* is first gained, then lost.

Sierpinski, Waclaw Franciszek (1882–1969)

A Polish mathematician who made outstanding contributions to **set theory**, which included research on the **axiom of choice** and the **continuum hypothesis**, **number theory**, and **topology**. Two well-known fractals, the **Sierpinski carpet** and the **Sierpinski gasket**, are named after him.

Sierpinski carpet

A fractal, named after Waclaw **Sierpinski**, that is derived from a square by cutting it into nine equal squares with a 3×3 grid, removing the central piece, and then applying the same procedure ad infinitum to the remaining eight squares. It is one of two generalizations of the Cantor set to two dimensions; the other is the **Cantor dust**. The carpet's **Hausdorff dimension** is log 8/log 3 = 1.8928. . . .

Sierpinski gasket

A fractal, also known as the *Sierpinski triangle* or *Sierpinski sieve* after its inventor Waclaw **Sierpinski**. It is produced by the following set of rules: (1) start with any triangle in a plane; (2) shrink the triangle by ½, make three copies, and translate them so that each triangle touches the two other triangles at a corner; (3) repeat step 2 ad infinitum. The gasket can also be made by starting with **Pascal's**

shunting puzzles How can two trains, traveling in opposite directions on a single track, get past each other by using a siding that can only accommodate a single item of rolling stock at a time?

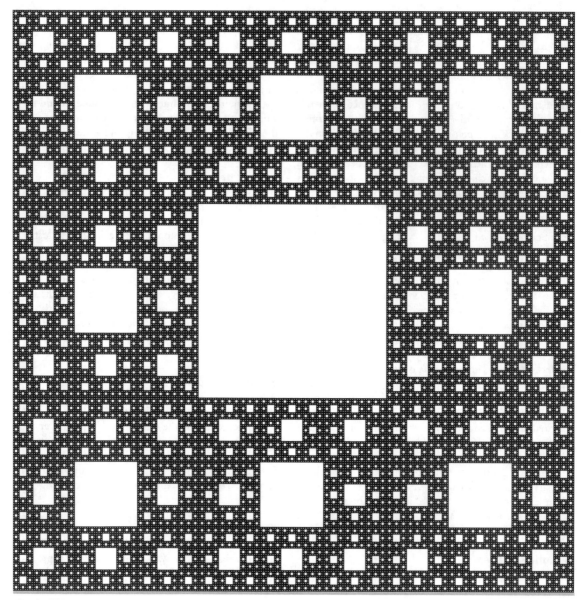

Sierpinski carpet

triangle, then coloring the even numbers white and the odd numbers black. Most curiously, it can be generated by a game of chance. Begin with three points, labeled 1, 2, and 3, and any starting point, S. Then select randomly 1, 2, or 3, using a die or some other method. Each random number defines a new point halfway between the latest point and the labeled point that the random number indicates. When the game has gone on long enough

the pattern produced is the Sierpinski gasket. The gasket has a **Hausdorff dimension** of $\log 3/\log 2 = 1.585\ldots$, which follows from the fact that it is a union of three copies of itself, each scaled by a factor of $\frac{1}{2}$. Adding rounded corners to the defining curve gives a nonintersecting curve that traverses the gasket from one corner to another and which Benoit **Mandelbrot** called the *Sierpinski arrowhead*.

Sierpinski gasket

Sierpinski number

A positive, odd integer k such that k times $2^n + 1$ is *never* a **prime number** for any value of n. In 1960 Waclaw **Sierpinski** showed that there were infinitely many such numbers (though he didn't give a specific example). This is a strange result. Why should it be that while the vast majority of expressions of the form m times $2^n + 1$ eventually produce a prime, some don't? For now, mathematicians are focused on a more manageable problem posed by Sierpinski: What is the smallest Sierpinski number? In 1962, John Selfridge discovered the smallest known Sierpinski number, $k = 78,557$. The next largest is 271,129. Is there a smaller Sierpinski number? No one yet knows. However, to establish that 78,557 is really the smallest, it would be sufficient to find a prime of the form $k(2^n + 1)$ for every value of k less than 78,557. In early 2001, there were only 17 candidate values of k left to check: 4,847; 5,359; 10,223; 19,249; 21,181; 22,699; 24,737; 27,653; 28,433; 33,661; 44,131; 46,157; 54,767; 55,459; 65,567; 67,607; and 69,109. In March 2002, Louis Helm of the University of Michigan and David Norris of the University of Illinois started a project called "Seventeen or Bust," the goal of which is to harness the computing power of a worldwide network of hundreds of personal computers to check for primes among the remaining candidates. The team's effort have so far eliminated six candidates–5,359; 44,131; 46,157; 54,767; 65,567; and 69,109. Despite this encouraging start, it may take as long as a decade, with many additional participants, to check the eleven remaining candidates.

sieve of Eratosthenes

The most efficient way to find all of the smallest **prime numbers**. First described by **Eratosthenes of Cyrene**, it involves making a list of all the integers less than or equal to n (and greater than one), then striking out the multiples of all primes less than or equal to the square root of n. The numbers that are left are the primes. For example, to find all the primes less than or equal to 30, we list the numbers from 2 to 30:

2, 3, 4, 5, 6, 7, 8, 9, 10, 11, 12, 13, 14, 15, 16, 17, 18, 19, 20, 21, 22, 23, 24, 25, 26, 27, 28, 29, 30

The first number, 2, is prime, so we keep it and strike out all of its multiples, leaving

2, 3, 5, 7, 9, 11, 13, 15, 17, 19, 21, 23, 25, 27, 29

The next number left, 3, is prime, so again we retain it and delete all of its multiples, leaving

2, 3, 5, 7, 11, 13, 17, 19, 23, 25, 29

Now we do the same thing for 5, another prime, to give

2, 3, 5, 7, 11, 13, 17, 19, 23, 29

The next number, 7, is larger than the square root of 30, so all of the numbers left are primes.

significant digits

The digits that define a numerical value. The significant digits of a given number begin with the first nonzero integer digit or, if this number is less than unity, with the first (zero or nonzero) decimal digit. They end with the final (zero or nonzero) decimal digit; the final zero or zeros of an integer may or may not be significant.

similar

Having the same shape but not necessarily the same size. Two triangles are similar if they have equal angles and their corresponding sides, say, a_1, b_1, c_1 and a_2, b_2, c_2, have a common ratio, r: $a_1/a_2 = b_1/b_2 = c_1/c_2$. In general, a *similarity* is a transformation under which the distance between any corresponding pair of points changes by the same factor.

simple group

A **group** that has no nontrivial proper normal **subgroups**. Simple groups are important because they can be thought of as the blocks out of which other groups can be built. Much activity has been expended in the classification of all finite simple groups.

simplex

The n-dimensional generalization of the **triangle** and the **tetrahedron**; in other words, a **polytope** in n dimensions with $n + 1$ vertices.

simply connected

The condition of a geometrical object if it consists of one piece and doesn't have any holes or "handles." For example, a line, disk, and sphere are simply connected, but a torus (doughnut) and teapot are not. See also **connected**.

simulation

Experimentation in the space of theories, or a combination of experimentation and theorization. Some numerical simulations are programs that represent a model for how nature works. Usually, the outcome of a simulation is as much a surprise as the outcome of a natural event, due to the richness and uncertainty of computation.

sine

The **trigonometric function** of an angle of a right-angled triangle other than the right angle that is equal to the length of the side adjacent to the angle divided by the length of the hypotenuse. The curve of $y = \sin x$ is called the sine curve, or *sinusoid*.

Singmaster, David

A professor at the school of computing, information systems, and mathematics at South Bank University, London, who is one of the world's leading compilers and historians of mathematical puzzles.

singularity

(1) A point at which the **derivative** does not exist for a given **function** but every neighborhood of which contains points for which the derivative exists. (2) A point in **space-time** at which gravitational forces cause matter to have infinite density and infinitesimal volume, and space and time to become infinitely distorted.

six

The smallest **perfect number**, the number of faces of a **cube**, and the number of sides of a **hexagon**. There are six players on a volleyball team, six kinds of chessmen, and six types of quark (not including antiquarks). A touchdown in American football earns six points and a hit across the boundary rope of a cricket field without bouncing scores six runs. Long ago people indicated a number by pointing to a part of their body; this is echoed in the New Guinea word for six, which is the same as that for "wrist."

six circles theorem

Given a triangle and a circle inside it that touches two of the triangle's sides, draw a second circle touching another two sides and touching the first circle. Draw a third circle touching two sides and the second circle, and so on. This chain ends with the sixth circle, which will touch the first.

sixty
See **sexagesimal**.

skeletal division
A long division in which most or all of the digits are replaced by symbols (usually asterisks) to form a **cryptarithm**.

skew lines
Also known as *crossing lines*, lines that lie in different planes and do not intersect one another.

Skewes' number
A famous **large number**, commonly given as $10^{10^{10^{34}}}$, that was first derived in 1933 by the South African mathematician Samuel Skewes in a proof involving **prime numbers**.[299] G. H. **Hardy** once described Skewes' number as "the largest number which has ever served any definite purpose in mathematics," though it has long since lost that distinction. Skewes' numbers—there are actually two of them—came about from a study of the frequency with which **prime numbers** occur. Gauss's well-known estimate of the number of prime numbers less than or equal to n, pi(n), is the integral from $u = 0$ to $u = n$ of $1/(\log u)$; this integral is called Li(n). In 1914 the English mathematician John Littlewood proved that pi(x) – Li(x) assumes both positive and negative values infinitely often. For all values of n up to 10^{22}, which is as far as computations have gone so far, Li(n) has turned out to be an overestimate. But Littlewood's result showed that above some value of n it becomes an underestimate, then at an even higher value of n it becomes an overestimate again, and so on. This is where Skewes' number comes in. Skewes showed that, if the **Riemann hypothesis** is true, the first crossing can't be greater than $e^{e^{e^{79}}}$. This is called the first or *Riemann true Skewes' number*. Converted to base 10, the value can be approximated as $10^{10^{10^{34}}}$, or more accurately as $10^{10^{8.852142 \times 10^{33}}}$ or $10^{10^{8852142197543270606106100452735038.55}}$. In 1987, the Dutch mathematician Herman te Riele[330] reduced dramatically the upper bound of the first crossing to $e^{e^{27/4}}$, or approximately 8.185×10^{370}, while John **Conway** and Richard Guy[68] have made the contradictory claim that the lower bound is 10^{1167}. In any event, Skewes' number is now only of historical interest. Skewes also defined the limit if the Riemann hypothesis is false: $10^{10^{10^{1000}}}$. This is known as the second Skewes' number.

slide rule
A calculating device consisting of two sliding logarithmic scales.

sliding-piece puzzle
A type of *sequential-movement puzzle,* within the larger category of **mechanical puzzles**, that involves sliding one piece at a time into a single vacant opening in order to advance toward the solution—a certain orderly arrangement of the pieces. The best known is Loyd's **Fifteen Puzzle**.

Slocum, Jerry
An American historian of and writer on **mechanical puzzles**.

slope
"Rise over run." For a straight line in the plane, the *slope* is the **tangent** of the angle it forms with the positive x-axis. For a curve, the slope is, by definition, the slope of the *tangent* line. Therefore, if the slope is constant, a line is straight.

Slothouber-Graatsma puzzle
A **packing** puzzle in which six $1 \times 2 \times 2$ blocks and three $1 \times 1 \times 1$ blocks must be fitted together to make a $3 \times 3 \times 3$ cube. There is only one solution. A similar but much more difficult puzzle, named after its inventor, John **Conway**, calls for packing three $1 \times 1 \times 3$ blocks, one $1 \times 2 \times 2$ block, one $2 \times 2 \times 2$ block, and thirteen $1 \times 2 \times 4$ blocks into a $5 \times 5 \times 5$ box.

Smith number
A **composite number**, the sum of whose digits equals the sum of the digits of its prime factors (its **factors** that are **prime numbers**). The name stems from a phone call in 1984 by the mathematician Albert Wilansky to his brother-in-law Smith, during which Wilansky noticed that the phone number, 493-7775, obeyed the condition just mentioned. Specifically:

$$4,937,775 = 3 \times 5 \times 5 \times 65,837$$
$$4 + 9 + 3 + 7 + 7 + 7 + 5 = 3 + 5 + 5 + 6 + 5 + 8 + 3 + 7$$

Trivially, all **prime numbers** have this property, so they are excluded. The first few Smith numbers are: 4, 22, 27, 58, 85, 94, 121, 166, 202, 265, 274, 319, 346, ... In 1987, Wayne McDaniel proved that there are infinitely many Smiths.

smooth
(1) Infinitely differentiable; possessing infinitely many **derivatives**. For example, $\sin(x)$ is a smooth function, while $|x|^3$ is not. More complicated mathematical objects such as **manifolds** are called smooth if they are defined or described by smooth functions. (2) Continuously differentiable (see **continuity**); possessing a continuous **tangent** or derivative.

Smullyan, Raymond (1919–)

An American mathematical logician, puzzle-maker, and magician, who has taught in various colleges but is best known for his books on recreational mathematics. Among these are *What Is the Name of This Book?*, *The Lady or the Tiger*, *The Tao is Silent*, and *This Book Needs No Title: A Budget of Living Paradoxes*.[304–311] Smullyan is an inventive maker of logic paradoxes and a pioneer of chess problems that involve "retrograde analysis," in which the object is to deduce the past history of a game from some given present position.

snow

It is often said that no two snowflakes are alike. While this is hard to prove, individual samples can be captured on a chilled glass microscope slide and preserved with artist's spray fixative. All are six-sided and the more ornate kind, called *dendritic snowflakes,* form when the air temperature is between −12°C and −16°C (10°F and 3°F). Typical snowflakes fall at a rate of a meter or two per second; assuming 1.5 m/s and a cloud base of 3,000 m (roughly the height of nimbostratus clouds) gives a descent time of 20 minutes. One of the great urban legends is that the Inuit have *n* words for "snow," where *n* is a large number. This story may have started in 1911 when anthropologist Franz Boaz casually mentioned that the Inuit—he called them "Eskimos," using the derogatory term of a tribe to the south of them for eaters of raw meat—had four different words for snow. With each succeeding reference in textbooks and the popular press the number grew to as many as 400 words. A problem with trying to pin down exactly how many Inuit words there are for snow and/or ice, or for anything else, is that the various dialects of Inuit are polysynthetic, which means that words can effectively be made up on the spot by concatenating various particles to the root word. For example, the suffix *-tluk,* for "bad," might be added to *kaniktshaq,* for "snow," to give *kaniktshartluk,* "bad snow." This can give rise to any number of snow terms, from *akelrorak* ("newly drifting snow") to *mitailak* ("soft snow over an opening in an ice floe").

snowball prime

Also known as a *right-truncatable prime,* a **prime number** whose digits can be chopped off, one by one, from the right-hand side, yet still leave a prime number. This means that even if you stop writing before you finish the number, you will still have written a prime. The largest snowball prime is 73,939,133 (7, 73, 739, . . . , 73,939,133 are all prime).

snowflake curve

See **Koch snowflake**.

soap film

See **bubbles**.

Soddy circle

A solution to the three-circle form of the **Apollonius problem** in which each of the given circles is tangent to (just touches) the other two.[312] There are two Soddy circles: the *outer Soddy circle,* which surrounds the three given circles, and the *inner Soddy circle,* which is interior to them. The inner Soddy circle is the solution to the **four coins problem**.

Soddy's formula

If four circles *A, B, C,* and *D,* of radii r_1, r_2, r_3, and r_4, are drawn so that they do not overlap but each touches the other three, and if we let $b_1 = 1/r_1$, etc., then

$$(b_1 + b_2 + b_3 + b_4)^2 = 2(b_1^2 + b_2^2 + b_3^2 + b_4^2).$$

solid

Of or relating to three-dimensional geometric figures or bodies.

solid angle

An angle formed by three or more planes intersecting at a common point. Solid angles are measured in **steradian**s.

solid geometry

The geometry of three-dimensional space.

solidus

The slanted line in a fraction such as *a/b* dividing the **numerator** from the **denominator**.

solitaire

See **peg solitaire**.

solitary number

A number that is not one of a pair of **amicable numbers**. Examples include all **prime number**s, all integer powers or primes, and other numbers such as 9, 16, 18, 52, and 160.

soliton

A solitary wave that can travel for long distances without changing its shape or losing energy. Mathematically, a soliton is a solution to a **partial differential equation** that is localized in some directions but not localized in time, and which does not change its shape. The first soliton to be described was a water wave seen by the engineer and shipbuilder John Scott Russell (1808–1882) in 1834 when he was riding by the Grand Union Canal at Hermiston, Glasgow. He observed that when a canal boat stopped, its bow wave continued onward as a well-defined elevation of the water at constant speed. (Another account says Scott Russell observed it on the Glasgow and Ardrossan Canal when a horse bolted with a light canal boat in tow.) The phenomenon was largely forgotten until the 1960s, when the American physicist Martin Kruskal rediscovered it and called it a soliton wave. On July 12, 1995, a viaduct at Hermiston was renamed the John Scott Russell viaduct, with Kruskal unveiling plaques and attempting to re-create a soliton wave. A classic soliton occurs on the River Severn in England, which starts at the head of the triangular Bristol Channel and narrows rapidly upstream. When the tide comes in, it is greatly compressed and produces the Severn Bore, a tidal wave up to 6 feet (2 m) high which rushes about 20 miles (32 km) up the river to Gloucester at a speed of up to 10 mph (16 km/hr). It is strongest at the spring tides, i.e. at full and new moons, and is now popular with surfers and canoeists.

solution

A value that satisfies the requirements of an equation. See also **lost solutions**.

Soma cube

A mathematical puzzle devised by Piet **Hein** in 1936 during a lecture on quantum mechanics by Werner Heisenberg in which the great German physicist was describing a space sliced into cubes. In a moment of genius, Hein grasped that the result of combining all seven of the irregular shapes that can be made from no more than four unit cubes joined at their faces is a single larger (3 × 3 × 3) cube. The Soma cube was first brought to popular attention by Martin **Gardner** in his "Mathematical Games" column in *Scientific American* in 1958. All 240 possible solutions were first identified by John **Conway** and Mickael Guy in 1961. The pieces can also be used to make a variety of other interesting three-dimensional

Soma cube   How the pieces of the cube fit together. *Mr. Puzzle Australia, www.mrpuzzle.com.au*

Soma cube   One of many animals into which the cube pieces can be arranged. *Mr. Puzzle Australia, www.mrpuzzle.com.au*

shapes so that the Soma cube is often regarded as a three-dimensional analog of **tangrams**. It is possible that the puzzle is named after the fictitious drug "soma" in Aldous Huxley's novel *Brave New World*. See also **poly-omino**.[112]

Sophie Germain prime

Any **prime number** p such that $2p + 1$ is also prime; the smallest examples are 2, 3, 5, 11, 23, 29, 41, 53, 83, 89, 113, and 131. Around 1825 Sophie **Germain** proved that the first case of **Fermat's last theorem** (FLT) is true for such primes. Soon after, Adrien-Marie Legendre began to generalize this by showing the first case of FLT also holds for odd primes p such that $kp + 1$ is prime, $k = 4, 8, 10, 14,$ and 16. In 1991 Fee and Granville extended this to $k < 100$, where k is not a multiple of three. Many similar results were also shown, but now that FLT has been proven correct, they are of less interest.

soroban

See **abacus**.

space

(1) The three-dimensional theater in which things as we know them can exist or in which events can take place. In the Einsteinian worldview, space and **time** are united inextricably in a **space-time** continuum and there is also the possibility of **higher dimensions**. (See also **fourth dimension**.) (2) In mathematics, there are additionally many other types of space, most of them too abstract to imagine or to describe accurately in a few sentences. Generally, a mathematical space is a **set** of points with additional features. In a *topological space* (see **topology**) every point has a collection of neighborhoods to which it belongs. In an *affine space,* which is a generalization of the familiar concepts of a straight line, a plane, and ordinary three-dimensional space, a defining feature is the ability

to fix a point and a set of coordinate axes through it so that every point in the space can be represented as a "tuple," or ordered set, of coordinates. Other examples of mathematical spaces include *vector spaces, measure spaces,* and *metric spaces.*

space-filling curve
A curve that passes through every point of a finite region (such as a unit square or unit cube) of an *n*-dimensional space, where $n \geq 2$. A well-known example is the **Peano curve**.

space-time
The inseparable four-dimensional **manifold**, or combination, which **space** and **time** are considered to form in the special and general theories of relativity (see **relativity theory**). A point in space-time is known as an *event*. Each event has four coordinates (x, y, z, t). Just as the x, y, z coordinates of a point depend on the axes being used, so distances and time intervals, (which are invariant in Newtonian physics) may depend (in relativistic physics) on the reference frame of an observer; this can lead to bizarre effects such as *length contraction* and *time dilation.* A *space-time interval* between two events is the invariant quantity analogous to distance in **Euclidean space**. The space-time interval *s* along a curve is defined by the quantity

$$ds^2 = dx^2 + dy^2 + dz^2 - c^2 dt^2,$$

where *c* is the speed of light. A basic assumption of relativity theory is that coordinate transformations leave intervals invariant. However, note that whereas distances are always positive, intervals may be positive, zero, or negative. Events with a space-time interval of zero are separated by the propagation of a light signal. Events with a positive space-time interval are in each other's future or past, and the value of the interval defines the proper time measured by an observer traveling between them.

special function
A **function**, often named after the person who introduced it, that has a particular use in physics or some branch of mathematics. Examples include *Bessel functions, Lagrange polynomials,* **beta functions**, **gamma functions**, and **hypergeometric functions**.

special relativity
See **relativity theory**.

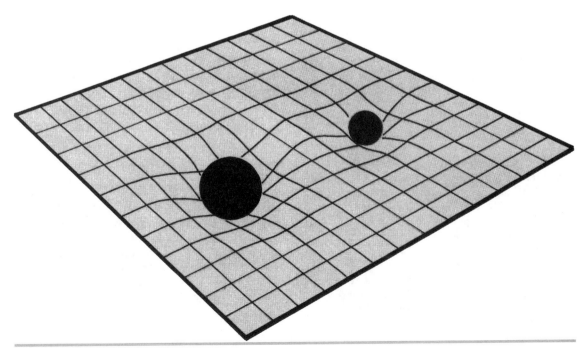

space-time   Gravity seen as a curvature of the fabric of space and time.

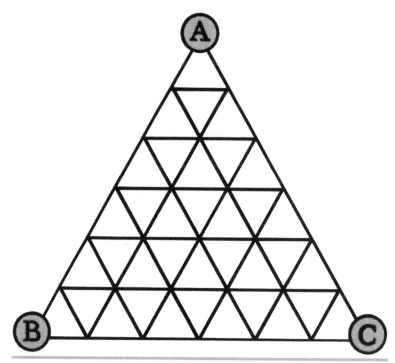

Sperner's lemma   A triangular lattice that can be used for demonstrating Sperner's lemma. To understand how, see the accompanying entry.

spectrum

(1) In **quantum mechanics**, the set of allowable energy levels of a particle or system. It is directly related to bright or dark lines in a spectrum of light produced by a prism. (2) In mathematics, the set of **eigenvalues** of a linear transformation. By historical coincidence, it is equivalent to the notion of a spectrum in quantum mechanics.

Sperner's lemma

Take a triangle ABC, labeled counterclockwise, and subdivide it into lots of smaller triangles in any arbitrary way. Then label all the new vertices as follows: (1) vertices along AB may be labeled either A or B, but not C; (2) vertices along BC may be labeled either B or C, but not A; (3) vertices along CA may be labeled either C or A, but not B; (4) vertices inside triangle ABC may be labeled A or B or C. Now shade in every small triangle that has three different labels. Use two different shadings to distinguish the triangles that have been labeled counterclockwise (i.e., in the same sense as triangle ABC) from the triangles that have been labeled clockwise (i.e., in the sense opposite to that of triangle ABC). Then

there will be exactly one more counterclockwise triangle than clockwise triangles. In particular, the number of shaded triangles will be odd. This is Sperner's lemma, named after its discoverer, the German mathematician Emanuel Sperner (1905–1980). Sperner's lemma is equivalent to the **Brouwer fixed-point theorem**; a version of it holds in all dimensions.

sphere

Roughly speaking, a ball-shaped object. In everyday usage a sphere is often considered to be solid; mathematicians call this the *interior* of the sphere. In mathematics, a sphere is a **quadric** consisting only of a surface and is therefore hollow. More precisely, a sphere is the set of all points in three-dimensional **Euclidean space** that lie at distance r, the radius, from a fixed point. In **analytical geometry** a sphere with center (x_0, y_0, z_0) and radius r is the set of all points (x, y, z) such that

$$(x - x_0)^2 + (y - y_0)^2 + (z - z_0)^2 = r^2.$$

A sphere can also be defined as the **surface of revolution** formed by rotating a circle about its diameter. If the circle is replaced by an **ellipse**, the shape becomes a **spheroid**. The surface area of a sphere is $4\pi r^2$ and its volume is $4\pi r^3/3$. The sphere has the smallest surface area among all surfaces enclosing a given volume and it encloses the largest volume among all closed surfaces with a given surface area. In nature, **bubbles** and water drops tend to form spheres because surface tension always tries to minimize surface area. The circumscribed cylinder for a given sphere has a volume which is 3/2 times the volume of the sphere. If a spherical egg were cut up by an egg-slicer with evenly spaced wires, the bands between the cuts (on the surface of the sphere) would have exactly the same area. Spheres can be generalized to other dimensions. For any natural number n, an n-sphere is the set of points in $(n + 1)$-dimensional **Euclidean space** that lie at distance r from a fixed point of that space. A 2-sphere is therefore an ordinary sphere, while a 1-sphere is a **circle** and a 0-sphere is a pair of points. An n-sphere for which $n = 3$ or more is often called a **hypersphere**.

sphere packing
See **packing**, **Kepler's conjecture**, and **cannonball problem**.

spherical geometry
See **elliptical geometry**.

sphericon
A curious and mathematically delightful three-dimensional object made from a right double-**cone**—two identical, 90° cones joined base to base—and an added twist. To create a sphericon, a right double-cone is sliced along a plane that includes both vertices. The resulting cross section is a square, which enables one of the halves to be rotated through a right angle and the two halves to be glued back together without any overlap. This final twist enables the sphericon to roll in an unusual way. An ordinary cone placed on a flat surface rolls around in circles. A double-cone can roll in a clockwise circle or a counterclockwise one. A sphericon, in contrast, performs a controlled wiggle, with first one conical sector in contact with the flat surface, then the other. Two sphericons placed next to each other can roll on each other's surfaces. Four sphericons arranged in a square block can all roll around one another simultaneously. And eight sphericons can fit on the surface of one sphericon so that any one of the outer solids can roll on the surface of the central one. The sphericon was first found by the En-glishman Colin Roberts in 1969, while he was still in school. In 1999 he brought his discovery to the attention of Ian **Stewart** who subsequently wrote about the new object in his "Mathematical Recreations" column in *Scientific American*.[322]

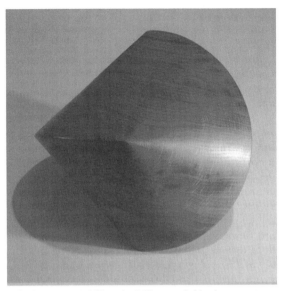

sphericon   A model in oak of this remarkable shape. *Paul and Colin Roberts*

spheroid
A surface in three dimensions obtained by rotating an **ellipse** about one of its principal axes. If the ellipse is rotated about its major axis, the surface is called a *prolate spheroid* (similar to the shape of a rugby ball). If the minor axis is chosen, the surface is called an *oblate spheroid* (similar to the shape of Earth). The **sphere** is a special case of the spheroid in which the generating ellipse is a **circle**. A spheroid is a special case of an **ellipsoid** where two of the three major axes are equal.

Sphinx riddle
In Greek mythology, the Sphinx sat outside Thebes and asked this riddle of all travelers who passed by. If the traveler failed to solve the riddle, then the Sphinx killed him/her. And if the traveler answered the riddle correctly, then the Sphinx would destroy herself. The riddle: what goes on four legs in the morning, on two legs at noon, and on three legs in the evening? Oedipus solved the riddle, and the Sphinx destroyed herself. The solution: a man, who crawls on all fours as a baby, walks on two legs as an adult, and walks with a cane in old age. Of course morning, noon, and night are metaphors for the times in a person's life. Such metaphors are common in riddles. There were two Thebes; apparently the Thebes in this myth was the one in Greece, and the Sphinx was different from the one that stands at Giza, in Egypt.

spider-and-fly problem

A puzzle that was originally posed in the *Weekly Dispatch*, an English newspaper, on June 14, 1903, by Henry **Dudeney**, and that appears as one of the problems in *The Canterbury Puzzles* (1907).[87] A simple but elegant exercise in **geodesic**s, it is Dudeney's best-known brain-teaser. In a cuboidal (shoebox-shaped) room measuring 30′ × 12′ × 12′, a spider is in the middle of one 12′ × 12′ wall, 1 foot away from the ceiling. A fly is in the middle of the opposite wall 1 foot away from the floor. If the fly remains stationary, what is the shortest total distance (the geodesic) the spider must crawl along the walls, ceiling, and floor in order to get to the fly? The answer, 40′, can be obtained by flattening out the walls. Note that this distance is shorter than the 42′ the spider would have to travel if it first crawled along the wall to the floor, then across the floor, then up 1′ to get to the fly. A twist to the problem can be obtained by a spider that suspends himself from a strand of cobweb and thus takes a shortcut by not being forced to remain glued to a surface of the room. If the spider attaches a strand of cobweb to the wall at his starting position and lowers himself down to the floor (thus not crawling a single inch), he can then walk across the length of the room (30′) and ascend a single foot, thus reaching his prey after a total crawl of 31′ (although the total distance traveled is of course 42′). If the spider is not proficient with fastening strands to vertical walls, he must first ascend 1′ to the ceiling, from where it can lower himself to the floor, traverse the length of the room, and climb one foot to get to the fly, for a total distance crawled of 32′.

Dudeney and Sam **Loyd** offered several versions of the problem in a rectangular room. In 1926 Dudeney gave a version on a cylindrical glass with the source and the target on opposite sides.

spinor

A mathematical object similar to a **vector**, but which changes sign when rotated through 360°. Spinors were invented by Wolfgang Pauli and Paul **Dirac** (they were named by Paul Ehrenfest) to represent the spin of a subatomic particle. In the early 1930s, Dirac, Piet **Hein**, and others at the Niels Bohr Institute created games such as **Tangloids** to teach and model the calculus of spinors.

spiral

A curve that turns around some central point, getting progressively closer to it or progressively farther from it, depending on which way the curve is followed. Among the best known types are the **Archimedean spiral**, the **logarithmic spiral**, the **circle involute**, and the **lituus**. Like their space-curve cousin the **helix**, all spirals are asymmetric and each come in two forms that are mirror reflections of one another.

spirograph curve

See **roulette**.

Sprague-Grundy theory

A theory of certain classes of games called *impartial games*, discovered independently by Roland Percival Sprague (in 1936) and Patrick Michael Grundy (in 1939)

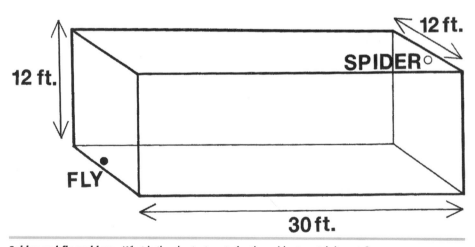

Spider-and-fly problem What is the shortest route for the spider to catch its prey?

and originally applied to **Nim**. In simple terms, they showed that one could take any impartial game and analyze it in terms of Nim heaps that could grow or decrease in size. The theory was developed further by E. R. Berlekamp, John **Conway**, and others, and presented comprehensively in the books *Winning Ways for Your Mathematical Plays* and *On Numbers*. Sprague-Grundy theory has been applied to other combinatorial games, including kayles.

Sprouts

A pencil-and-paper game invented by John **Conway** and Michael S. Paterson at Cambridge University in 1967. Sprouts is played by two players, starting with a few dots (called *spots*) drawn on a sheet of paper. To make a move, a player draws a curve between two spots or a loop from a spot to itself; the curve may not cross any other curve. The player marks a new spot on the curve, dividing it in two. Each spot can have at most three curves connected to it. The player who makes the last move wins. Sprouts has been studied from the perspectives of **graph** theory and of **topology**. It can be proven that a game started with n spots will last at least $2n$ moves and at most $3n - 1$ moves. By enumerating all possible moves, one can show that the first player is guaranteed a win in games involving three, four, or five spots, while the second player can always win a game that starts with one, two, or six spots. Following a 1990 computer analysis of the game at Bell Labs out to 11 spots, David Applegate, Guy Jacobsen, and Daniel Sleator conjectured that the first player has a winning strategy when the number of spots divided by six leaves a remainder of three, four, or five. Sprouts was featured in the plot of the first part of the science fiction novel *Macroscope* by Piers Anthony.

square

(1) A **quadrilateral** with four equal sides that meet at right angles. (2) To square something is to multiply it by itself. To take the *square root* is the reverse process. A *square number* is the square of a whole number (see also **figurate number**). A well-known formula for the *difference of squares* is $a^2 - b^2 = (a + b)(a - b)$. This formula enables some otherwise difficult calculations to be done easily in the head. For example, $43 \times 37 = (40 + 3)(40 - 3) = 40^2 - 3^2 = 1600 - 9 = 1591$.

square free

An integer that is not divisible by a perfect square, n^2, for $n > 1$.

square pyramid problem

See **cannonball problem**.

square root of 2 ($\sqrt{2}$)

The first number shown to be what is now known as an **irrational number** (a number that can't be written in the form a/b, where both a and b are integers). This discovery was made by **Pythagoras** or, at any rate, by the Pythagorean group that he founded. The square root of 2 is the length of the hypotenuse (longest side) of a right **triangle** whose other two sides are each one unit long. A *reductio ad absurdum* proof that $\sqrt{2}$ is irrational is straightforward. Suppose that $\sqrt{2}$ is rational, in other words that $\sqrt{2} = a/b$, where a and b are coprime integers (that is, they have no common factors other than 1) and $b > 0$. It follows that $a^2/b^2 = 2$, so that $a^2 = 2b^2$. Since a^2 is even (because it has a factor of 2), a must be even, so that $a = 2c$, say. Therefore, $(2c)^2 = 2b^2$, or $2c^2 = b^2$, so b must also be even. Thus, in a/b, both a and b are even. But we started out by assuming that we'd reduced the fraction to its lowest terms. So there is a contradiction and therefore $\sqrt{2}$ must not be irrational. This type of proof can be generalized to show that any root of any natural number is either a natural number or irrational.

As a **continued fraction**, $\sqrt{2}$ can be written $1 + 1/(2 + 1/(2 + 1/(2 + \ldots)))$, which yields the series of rational approximations: 1/1, 3/2, 7/5, 17/12, 41/29, 99/70, 239/169, ... Multiplying each numerator (number on the top) by its denominator (number on the bottom) gives the series 1; 6; 35; 204; 1,189; 6,930; 40,391; 235,416; ... which follows the pattern: $A_n = 6A_{n-1} - A_{n-2}$. Squaring each of these numbers gives 1; 36; 1,225; 41,616; 1,413,721; 48,024,900; 1,631,432,881; ... each of which is also a **triangular number**. The numbers in this sequence are the only numbers that are both square and triangular.

squarefull number

See **powerful number**.

squaring the circle

The problem to find a construction (see **constructible**), using only a straightedge and compass, that would give a square of the same area as a given circle. It is now known that this is impossible because, as the German mathematician Ferdinand von Lindemann showed in 1882, it would amount to finding a **polynomial** expression for **pi**, which can't be done as π is a **transcendental number**. (All the coordinates of all points that can be constructed with ruler and compass are algebraic numbers.) This minor inconvenience, however, hasn't prevented some amateur mathematicians from continuing to claim they have found proofs that the circle can be squared. Some of the most bizarre attempts have involved proposing a different and rational value for π. In 1897, the Indiana

State Legislature came within a hair's breadth of introducing a bill to set the value of π equal to 3.2! In fact, long before π was finally proved to be transcendental, most learned societies had conjectured that a proof was impossible and had stopped considering circle-squaring arguments sent to them.

squaring the square

The problem of how to tile a square with *integral squares* (squares of integral side-length). Of course squaring the square is a trivial task unless additional conditions are set. The most studied restriction is the *perfect squared square:* a square such that each of the smaller squares has a different size. The name was coined in humorous analogy with **squaring the circle** and is first recorded as being studied by R. L. Brooks, C. A. B. Smith, A. H. Stone, and W. T. Tutte at Cambridge University. The first perfect squared square was found by Roland Sprague in 1939. If such a **tiling** is enlarged so that the formerly smallest tile becomes as big as the original square, it becomes clear that the whole plane can be tiled with integral squares, each having a different size. It is still an unsolved problem, however, whether the plane can be tiled with a set of integral square tiles such that each natural number is used exactly once as the size of a tile. A *simple squared square* is one where no subset of the squares forms a rectangle. The smallest simple perfect squared square was discovered by A. J. W. Duijvestin using a computer search. His tiling uses 21 squares, and has been proved to be minimal. Other possible conditions that lead to interesting results are nowhere-neat squared squares and no-touch squared squares. Developments leading to squaring the square can be traced back to 1902 and the first appearance of Henry **Dudeney**'s Lady Isabel's casket, later published as problem #40 in *The Canterbury Puzzles*.[87]

standard deviation

A measure of the spread of a **set** of data. For a **Gaussian** distribution, the standard deviation hints at the width of the tails of the distribution **function**.

Stanhope, Earl

A line of English earls, several of whom were notable mathematicians and polymaths. The third earl, Charles Stanhope (1753–1816), invented the process of stereotyping and, in 1777, devised the first mechanical logical calculator. The fourth earl, Philip Henry Stanhope (1781–1855) was a mathematician who applied his knowledge to mapmaking and **maze** design. From 1818 to 1830 he planted a maze according to a basic plan laid down by the second earl (1714–1786) that was probably the first to incorporate islands that defeat the familiar "hand-on-

wall" method. (There are islands in the famous Hampton Court maze, yet the hand-on-wall method solves it.) His designs often had a number of islands, or just isolated lengths of hedge.

Star of David

Also known as a *hexagram*, the six-pointed star obtained by extending the sides of a regular **hexa**gon to the points of intersection.

star of Lakshmi

An eight-pointed star design often used in architecture, particularly as a tiling or other decoration on the floor of a room that has four- or eight-fold symmetry. A notable example is in the **octagonal** central lobby in the Houses of Parliament in London. It was used by Hindus to symbolize *Ashtalakshmi*, the eight forms of wealth.

state space

See **phase space**.

stationary point

A point on the graph of a **function** where the **tangent** to the graph is parallel to the x-axis or, equivalently, where the **derivative** of the function is 0. There are four kinds of stationary points: (1) a *local minimum*, where the derivative of the function changes from negative to positive; (2) a *local maximum*, where the derivative changes from positive to negative; (3) a *rising point of inflection*, where the derivative is positive on both sides of the stationary point; and (4) a *falling point of inflection*, where the derivative is negative on both sides of the stationary point.

statistical mechanics

The study of statistical and thermal properties of physical materials and their idealized mathematical models.

statistics

The study of ways that lots of data can be represented using a few numbers and the study of how such numbers can be chosen and used to draw reasonable conclusions about the data. The word *statistics* comes from the Latin *statis* for "political state"; one of the main tasks of the subject involves analyzing facts and figures about governments, resources, and populations. Although a powerful tool, statistics is open to abuse, both intentional and unintentional. Benjamin Disraeli (1804–1881) may have gone a little over the top when he said, "There are lies, damned lies, and statistics," but Scottish author Andrew Lang (1844–1912) could have been describing many a politician when he remarked, "He uses statistics as a drunken man uses lamp posts—for support rather than

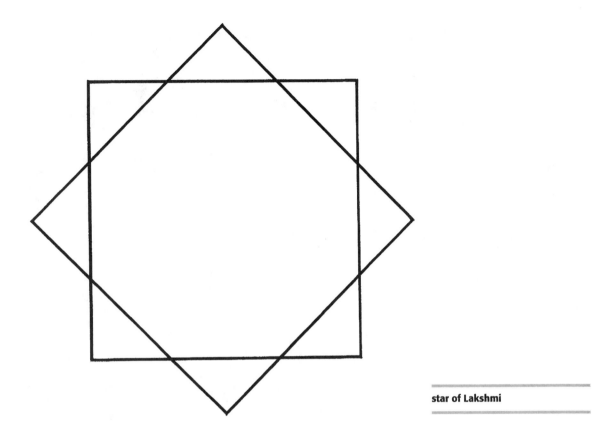

star of Lakshmi

illumination." The branch of statistics most commonly used in recreational mathematics is **probability theory.**

Steiner, Jakob (1796–1863)

A Swiss mathematician, considered by many to be the greatest geometer since **Apollonius** of Perga. Largely self-taught, he became a professor at the University of Berlin, and was a pioneer in the field of **projective geometry**. He had an important influence on his students, including Bernhard **Riemann**.

Steiner-Lehmus theorem

Any triangle that has two equal angle bisectors (each measured from a vertex to the opposite sides) is an **isosceles** triangle. In 1840, a Berlin professor Ludolph Lehmus wondered if this statement is true, given that it is the inverse of the already proven rule: If a triangle is isosceles then two of its internal bisectors are equal. He put the problem to Jakob **Steiner** who was quickly able to show its validity. Shortly after, Lehmus himself found a neater proof and it is has since become a favorite pas-

time of geometry hobbyists to search for still simpler proofs of the theorem.

Steinhaus, Hugo Dyonizy (1887–1972)

A Polish mathematician who was an influential member of the *Lvov school,* based at the Jan Kazimierz University in Lvov, which also included Stefan **Banach** and which focused on problems in functional **analysis**, real functions, and probability in the 1920s and '30s. Early on, Steinhaus's work revolved around applications of the **Lebesque** measure and integral. In 1923 he published the first rigorous account of the theory of tossing coins based on **measure theory**, and in 1925 was the first to define and discuss the concept of strategy in **game theory**. During World War II, as a Jew he was compelled to hide from persecution by the Nazis, yet continued his mathematical work despite great hardship. In 1944, Steinhaus proposed the problem of dividing a cake into *n* pieces so that it is proportional and envy free (see **cake-cutting**). He is also well known as the author of the widely read *Mathematical Snapshots*.[316]

stellation The final stellation of the icosahedron. *Robert Webb, www.software3d.com; created using Webb's Stella program*

stellation

(1) The process of constructing a new **polyhedron** by extending the face planes of a given polyhedron past their edges. (2) The new polyhedron thus obtained. Starting with the **icosahedron**, for example, there are 59 possible stellations, including the *great icosahedron,* which is one of the **Kepler-Poinsot solids.**

steradian

The international (SI) unit in which **solid angle**s are measured. The steradian (sr) is a solid angle with its **vertex** in the center of a **sphere** and which cuts off from the spherical surface an area equal to the square of the radius of the sphere. The full sphere represents a spherical angle of 4π steradians.

Stewart, Ian (1945–)

A British mathematician at Warwick University who has written many books and articles on recreational math

and popular science. From 1990 to 2001 he wrote the "Mathematical Recreations" column in *Scientific American.* His books include *Flatterland: Like Flatland Only More So* (a modern version of Edwin Abbott's classic *Flatland: A Romance of Many Dimensions*); *Does God Play Dice: The New Mathematics of Chaos,* and *Evolving the Alien* (with Jack Cohen).[317, 318, 321, 323]

Stewart toroid

A **polygon** with a hole through it that meets some or all of the criteria specified by Bonnie Madison Stewart in his 1980 book *Adventures among the Toroids.* These criteria are as follows: all faces must be regular; faces that meet mustn't lie in the same plane; the polygon must be quasi-convex; the hole through the polygon must change its **genus**; and, the faces aren't allowed to intersect with themselves or each other. Stewart toroids, pierced relatives of the familiar Platonic solids, combine dazzling complexity with attractive symmetry.

stellation   A stellation of the small stellated truncated dodecahedron, a uniform polyhedron. *Robert Webb, www.software3d.com; created using Webb's Stella program*

stochastic
Something that is **random**.

stochastic process
A **dynamical system** with **random** fluctuations at each iteration or that is influenced by random noise. A random variable that, at each stage in time, depends on its previous values and on further random choices. For example, the price of a stock is often modeled as a stochastic process.

Stomachion
See **Loculus of Archimedes**.

straight
Having no deviations. A *straight line* is usually simply called a **line**. A *straight angle,* or *flat angle,* is exactly 180°.

strange attractor
See **chaotic attractor**.

strange loop
A phenomenon in which, whenever movement is made upward or downward through the levels of some hierarchical system, the system unexpectedly arrives back where it started. Douglas **Hofstadter** has used the strange loop as a paradigm in which to interpret paradoxes in logic, such as **Grelling's paradox** and **Russell's paradox**, and has called a system in which a strange loop appears a *tangled hierarchy.*

strategy
In **game theory**, a policy for playing a game. A strategy is a complete recipe for how a player should act in a game under all circumstances. A policy may employ **random-ness**, in which case it is referred to as a **mixed strategy**.

Stewart toroid A drilled truncated dodecahedron. *Robert Webb, www.software3d.com; created using Webb's Stella program*

string

Any sequence of letters, numbers, digits, **bit**s, or symbols.

string puzzle

The origins of puzzles using string, a form of *disentanglement puzzle,* are lost in antiquity but one of the earliest stories involving this type of problem is that of the **Gordian knot.** Today many kinds of string puzzles are available, most of which involve removing a ring from loops of string that are threaded through a solid (typically wooden) body. See also **mechanical puzzle**s and **knot**.

string theory

An important theory in modern physics in which the fundamental particles in nature are thought of as the "musical notes" or excitation modes of vibrating elementary strings. These strings have the shortest mean-

ingful physical length, the so-called *Planck length* (equal to about 10^{-33} cm), and no thickness. Even more bizarre, for the theory to make sense, the universe must have nine space dimensions and one time dimension, for a total of 10 dimensions. This idea of a 10-dimensional universe first appeared in the **Kaluza-Klein theory.** We're familiar with time and three of the space dimensions: the other six together are known as **Calabi-Yau space**s. In string theory, as in the case of a stringed instrument, the string must be stretched under tension in order to vibrate. This tension is fantastically high—equivalent to a weight of about 10^{39} tons. String theories are classified according to whether or not the strings are required to be closed loops, and whether or not the particle spectrum includes *fermions.* In order to include fermions in string theory, there must be a special kind of symmetry called *supersymmetry,* which means that for every *boson* (a particle, of integral spin, that transmits a force) there is a corresponding fermion (a particle, of half-integral spin, that makes up matter). So supersymmetry relates the particles that transmit forces to the particles that make up matter. Supersymmetric partners to currently known particles have not been observed in particle experiments, but theorists believe this is because supersymmetric particles are too massive to be produced by present-day high-energy accelerators. Particle accelerators could be on the verge of finding evidence for high energy supersymmetry in the next decade. Evidence for supersymmetry at high energy would be compelling evidence that string theory was a good mathematical model for nature at the smallest distance scales. In string theory, all of the properties of elementary particles—charge, mass, spin, and so forth—come from the vibration of the string. The easiest of these to understand is mass. The more frantic the vibration, the more energy; and since mass and energy are the same thing, higher mass comes from faster vibration.

strobogrammatic prime

A **prime number** that remains unchanged when rotated through 180°. An example is 619, which looks the same when read upside-down. To be strobogrammatic, a prime cannot contain digits other than 0, 1, and 8, which have a horizontal line of symmetry (ignoring font variations), and 6 and 9, which are vertical reflections of each other. An *invertible prime* is one that yields a different prime when the digits are inverted. Of course, these definitions are not taken seriously by mathematicians!

strophoid

A looping curve, first studied in 1670 by Isaac Barrow (1630–1677), the first Lucasian Professor of Mathematics at Cambridge and the immediate predecessor of Isaac **Newton** in this job. The strophoid, which is a special case of the general **cissoid**, was named by Enrico Montucci in 1846 from the Latin for "twisted belt shape." The elaborate rules for drawing a strophoid are as follows. Let C be a curve, let O be a fixed point called the pole, and let O' be a second fixed point. Let P and P' be points on a line through O meeting C at Q such that $P'Q = QP = QO'$. The locus of P and P' is the strophoid of C with respect to the pole O and fixed point O'. If the curve C is a straight line, the pole P is not on C, and the second point O' is on C, the resulting strophoid is called an *oblique strophoid*. If these same conditions apply except that O' is the point where the perpendicular from O to C cuts C, then the strophoid produced is called a *right strophoid*. On the other hand if C is a circle, O is the center of the circle, and O' a point on its circumference, then the strophoid that results is known as *Freeth's nephroid*. The French mathematician Gilles Roberval (1602–1675) found the

strophoid in a different way—as the result of planes cutting a **cone**. When the plane rotates (about the tangent at its vertex) the collection of foci of the obtained conics gives the strophoid.

subgroup

A subset of a **group** that is a group under the same operation.

sublime number

A number such that both the *sum* of its divisors and the *number* of its divisors are **perfect numbers**. The smallest sublime number is 12. There are 6 divisors of 12—1, 2, 3, 4, 6, and 12—the sum of which is 28. Both 6 and 28 are perfect. The second sublime number begins 60,865..., ends...91,264, and has a total of 76 digits! It is not known if there are larger even sublime numbers, nor if there are any odd sublime numbers.

subset

A **set** whose members are members of another set; a set contained within another set.

substitution cipher

A **cipher** that replaces each plaintext (original message) symbol with a ciphertext (coded text) symbol. The receiver decodes using the inverse substitution. A simple example is the **Caesar cipher**.

subtraction

The **binary operation** of finding the difference between two quantities or numbers.

sultan's dowry

A sticky problem in **probability** that first came to light in Martin **Gardner**'s "Mathematical Recreations" column in the February 1960 issue of *Scientific American*. Gardner's original version has become known as the secretary problem. In the exactly equivalent form called the sultan's dowry problem, a sultan has granted a commoner the chance to marry one of his hundred daughters. The commoner will be shown the daughters one at a time and will be told each daughter's dowry. The commoner has only one chance to accept or reject each daughter; he can't go back and choose one that he has previously rejected. The sultan's catch is that the commoner may only marry the daughter with the highest dowry. What is the commoner's best strategy, assuming that he knows nothing in advance about the way the dowries are distributed?

Many mathematicians have tackled this question and numerous papers have been written on the subject. It has even spawned its own area of study within the field of management science. The consensus among those who have worked on the problem is that the commoner's best strategy is to let a certain fraction of the daughters pass and then choose the next one who has a dowry higher than any of the ones seen up to that point. The exact number to skip is determined by the condition that the odds that the highest dowry has already been seen is just greater than the odds that it remains to be seen *and that if it is seen it will be picked*. This amounts to finding the smallest x such that:

$$x/n > x/n \times [1/(x + 1) + \ldots + 1/(n - 1)]$$

Substituting $n = 100$ leads to the conclusion that the commoner should wait until he has seen 37 of the daughters, then pick the first daughter with a dowry that is bigger than any that have already been revealed. With this strategy, his odds of choosing the daughter with the highest dowry are surprisingly high: about 37%.[224]

supercomputer

A computer that, in its day, is faster than any contemporary conventional computer. Supercomputers are typically used for enormous number-crunching tasks such as weather-forecasting, simulations of fusion experiments or galaxy evolution, design of cars and planes, and code-cracking. At the start of 2004, the world's most powerful computer, by a long way, is the Earth Simulator in Japan, which is capable of a maximum of 40,960 gigaflops, that is, 40.96 trillion "floating point operations" per second. The top 500 supercomputers are listed at the following Web site: http://www.top500.org.

superegg

The **surface of revolution** of a **superellipse** given by the formula $|x/a|^{2.5} + |y/b|^{2.5} = 1$, where $a/b = 4/3$. The superegg was named by Piet **Hein** and singled out by him because of an unusual property: stood on either end it has a peculiar and surprising stability. Supereggs, made of metal, wood, and other materials, were sold as novelties in the 1960s; small, solid-steel ones were marketed as an "executive toy." The world's largest superegg, made of steel and aluminum and weighing 1 ton, was placed outside Kelvin Hall in Glasgow in 1971 to honor Hein's appearance there as a speaker.

superellipse

A **Lamé curve**, described by the formula $|x/a|^n + |y/b|^n = 1$, for which $n > 2$. Superellipses have a form partway between an **ellipse** and a rounded rectangle (or, if $a = b$, partway between a circle and a rounded square). The Danish poet and architect Piet **Hein** decided that the superellipse with $n = 5/2$ and $a/b = 6/5$ is the most pleas-

superegg

ing to the eye. This so-called *Piet Hein ellipse* was quickly adopted as the basic motif for planning an open space at the center of Stockholm and was also incorporated into Scandinavian designs for office tables, desks, beds, and even roundabouts in roads. The **surface of revolution** of a superellipse is a *superellipsoid*, one special form of which has been nicknamed the **superegg**.

superfactorial

A **function** based on the **factorial** that produces **large numbers** very quickly. It is a recent creation that has not yet entered the mathematical mainstream, and has been defined in a couple of different ways. In 1995, in his book *Keys to Infinity*, Clifford Pickover gave the superfactorial as

$$n\$ = \underbrace{n!^{n!^{\cdot^{\cdot^{\cdot^{n!}}}}}}_{n! \text{ terms}}$$

In the same year, Sloane and Plouffe offered an alternative definition:

$$n\$ = \prod_{i}^{n} i!$$

Pickover's superfactorial grows with extraordinary speed. The first two terms are 1 and 4, but the third term 3\$, already has more digits than could be written down on this page. Sloan and Plouffe's factorial grows sedately by comparison—the first few values are 1, 1, 2, 12, 288, and 345,690—and is related to **Bell number**s.

supersymmetry

A theory in physics that postulates a counterintuitive symmetric relationship between fermions, which are particles such as electrons that obey the Pauli exclusion principle and thus cannot occupy the same quantum state, and bosons, which are particles such as photons that can coexist in the same state.

supertetrahedral number

A type of **figurate number** in four dimensions (see **fourth dimension**). Supertetrahedral numbers are obtained by piling up the **tetrahedral number**s 1, 4, 10, 20, 35, etc., as in:

$$
\begin{aligned}
1 &= 1 \\
1 + 4 &= 5 \\
1 + 4 + 10 &= 15 \\
1 + 4 + 10 + 20 &= 35 \\
1 + 4 + 10 + 20 + 35 &= 70
\end{aligned}
$$

etc.

supplementary angles

Two angles that add up to 180°.

surd

A **radical** that expresses an **irrational number**. Surds may be *quadratic* (e.g., $\sqrt{2}$), *cubic* (e.g., $\sqrt[3]{2}$), *quartic* (e.g., $\sqrt[4]{2}$), and so on. (The term is sometimes used as a synonym for irrational number.) A *pure surd*, or entire surd, contains no **rational number**; that, is, all its factors or terms are surds (e.g., $\sqrt{2} + \sqrt{3}$). A *mixed surd* contains at least one rational number (e.g., $2 + \sqrt{3}$ or $3\sqrt{2}$).

surface

In mathematics, any object that locally (if you zoom in close enough to it) looks like a piece of a flat **plane**. A **sphere**, a **torus**, a **pseudosphere**, and a **Klein bottle** are examples of different types of surfaces.

surface of revolution

A surface produced by rotating a line or a curve about some axis. For example, a **sphere** is the surface of revolution generated when a circle spins about its diameter.

surreal number

A member of a mind-boggling vast class of numbers that includes all of the **real numbers**, all of Georg Cantor's infinite **ordinal number**s (different kinds of infinity), a set of **infinitesimal**s (infinitely small numbers) produced from these ordinals, and strange numbers that previously lived outside the known realm of mathematics. Each real number, it turns out, is surrounded by a "cloud" of surreals that lie closer to it than do any other real numbers. One of these surreal clouds occupies the curious space between zero and the smallest real number greater than zero and is made up of the infinitesimals.

Surreal numbers were invented or discovered (depending on your philosophy) by John **Conway** to help with his analysis of certain kinds of games. The idea came to him after watching the British **Go** champion playing in the mathematics department at Cambridge. Conway noticed that endgames in Go tend to break up into a sum of games, and that some positions behave like numbers. He then found that, in the case of infinite games, some positions behaved like a new kind of number–the surreals. The name "surreal" was introduced by Donald **Knuth** in his 1974 book *Surreal Numbers: How Two Ex-Students Turned on to Pure Mathematics and Found Total Happiness.*[191] This novelette is notable as being the only instance where a major mathematical idea has been first presented in a work of fiction. Conway went on to describe the surreal numbers and their use in analyzing games in his 1976 book *On Numbers and Games.*[67] The surreals are similar to the **hyperreal number**s, but they are constructed in a very different way and the class of surreals is larger and contains the hyperreals as a subset.

swastika

An ancient ideogram, signs of which have been found in the Euphrates-Tigris Valley, and in some areas of the Indus Valley, dating back 3,000 years; it became commonly used

swastika

around 1000 B.C., possibly first in ancient Troy, in the northwest of modern Turkey. The swastika is an irregular icosagon (a 20-sided **polygon**), which in Arabic and Indian culture originally represented good luck. Of course, in more recent times, it was adopted as the symbol of the Nazi Party in Hitler's Germany and thus came to stand for anti-Semitism.

syllogism

An argument composed of three parts—a major premise, a minor premise, and a conclusion. For example: All men are mortal (major premise). Socrates is a man (minor premise). Therefore, Socrates is mortal (conclusion). The syllogism forms the basis of Aristotle's system of logic, which went unchallenged for over 2,000 years. Aristotle believed that by setting out any argument in syllogistic form, it should be possible to avoid fallacies. However, Bertrand **Russell** discovered several formal errors in the doctrine of syllogism.

Sylvester, James Joseph (1814–1897)

An English mathematician and lawyer who, in 1850, first used the term *matrix* in mathematics and gave it its present meaning of a rectangular array of numbers from which **determinant**s may be formed. Together with Arthur **Cayley**, he founded the *theory of invariants*. Hotheaded and vociferous in opposition of anti-Semitism, he was thrown out of the University of London for threatening another student with a table knife. Later he studied at Cambridge, emerging as Second Wrangler but without a degree (though in 1871 he earned an MA) because, as a Jew, he refused to accept the articles of the Church of England. After a spell working as an actuary and a barrister, and also taking private pupils (one of whom was Flor-ence Nightingale), he met Cayley, with whom he forged a lifelong friendship and collaboration, and returned to mathematics. He became professor of mathematics at the Royal Military Academy at Woolwich (1855–1870) and at the newly established Johns Hopkins University at Baltimore (1877–1883), founding the *American Journal of Mathematics,* before accepting the Savilian chair at Oxford (1883–1894). Remarkably, especially in the field of mathematics, he produced an extraordinary flood of ideas well into his old age. At 82, he worked out the theory of compound partitions. He also published on the roots of **quintic** equations and on **number theory**. His partnership with Cayley worked perfectly, since Cayley supplied the rigor which the brilliantly creative Sylvester lacked.

Sylvester's problem of collinear points

A problem posed in 1893 by James **Sylvester** who wrote: "Prove that it is not possible to arrange any finite number of real points so that a right line through every two of them shall pass through a third, unless they all lie in the same right line." No correct proof was forthcoming at the time, but the problem was revived by Paul **Erdös** in 1943 and correctly solved by T. Grünwald in 1944.

symmedian

Reflection of a **median** of a triangle about the corresponding angle **bisector**.

symmetric group

The **group** of all **permutation**s of a finite set. The symmetric group of a **set** of size n is denoted S_n and has $n!$ elements.

symmetry

An intrinsic property of a mathematical object that allows it to remain unchanged under certain types of **transformation**, such as rotation, reflection, or more abstract operations. The mathematical study of symmetry is systematized and formalized in the extremely powerful subject known as **group** theory. Symmetries and apparent symmetries in the laws of nature have played a part in the construction of physical theories since the time of Galileo and Newton. The most familiar symmetries are spatial or geometric ones. In a **snow**flake, for example, the presence of a symmetrical pattern can be detected at a glance. One of the most remarkable developments of the past half century has been the emergence of symmetry as a central theme of subatomic physics. This came about through a series of subtle evolutions in the concept of symmetry itself. Many researchers believe that this evolutionary process has not come to an end, and that further meaning of the concept of symmetry, with perhaps new mathematical structures, will develop in the coming years.

symmetry group

The **group** formed by the set of all rigid motions (translations, rotations, reflections, etc.) of Euclidean space that map all points of a subset F into F.

system

Something that can be studied as a whole. Systems may consist of subsystems that are interesting in their own right. Or they may exist in an environment that consists of other similar systems. Systems are generally understood to have an internal state, inputs from an environment, and methods for manipulating the environment or themselves. Since cause and effect can flow in both directions of a system and environment, interesting systems often possess **feedback**, which is self-referential in the strongest case.

Szilassi polyhedron

A **toroid**al heptahedron (seven-sided **polyhedron**) first described in 1977 by the Hungarian mathematician Lajos Szilassi. It has 7 faces, 14 vertices, 21 edges, and 1 hole. The Szilassi polyhedron is the **dual** of the **Császár polyhedron** and, like it, shares with the **tetrahedron** the property that each of its faces touches all the other faces. Whereas a tetrahedron demonstrates that four colors are necessary for a map on a surface topologically equivalent to a sphere, the Szilassi and Császár polyhedra show that seven colors are necessary for a map on a surface topologically equivalent to a **torus**.

tachyon

A hypothetical particle that travels faster than the speed of light. Tachyons were first proposed in prerelativistic times by the physicist Arnold Sommerfeld and named in the 1960s by Gerald Feinberg from the Greek *tachys* meaning "swift." By extension of this terminology, particles that travel slower than light are called *tardyons* (or *bradyons* in more modern usage) and particles, such as photons, that travel exactly at the speed of light are called *luxons.* The existence of tachyons is allowed by the mathematics of special **relativity theory**, one of the basic equations of which is

$$E = m/\sqrt{(1 - v^2/c^2)}$$

where E is the mass-energy of a particle, m its rest mass, v its velocity, and c the speed of light. This shows that for tardyons (particles of ordinary matter), E increases as v increases and becomes infinite when $v = c$, thus preventing an initially slower-than-light particle from being accelerated up to the speed of light and beyond. What about a particle for which v is always greater than c? In this case, $v^2/c^2 > 1$, so that the denominator in the equation above is an **imaginary number**—the square root of a negative real number. If m has a real value, E is imaginary, which is hard for physicists to swallow because E is a measurable quantity. If m takes an imaginary value, however (because one imaginary number divided by another is real), then E is real. Tachyons are allowed, therefore, providing (a) they never cross from one side of the light-barrier to the other, and (b) they have an imaginary rest mass (which is physically more acceptable, since the rest mass of an object that never stops isn't directly measurable). Tachyons would slow down if they lost energy, and accelerate if they gained energy. This leads to a problem in the case of charged tachyons because charged particles that move faster than the speed of light in the surrounding medium give off energy in the form of Cherenkov radiation. Charged tachyons would continuously lose energy, even in a vacuum, through Cherenkov emission. This would cause them to gain speed, thus lose energy at an even greater rate, thus accelerate even more, and so on, leading to a runaway reaction and the release of an arbitrarily large amount of energy.

More worrisome, as the physicist Gregory Benford and his colleagues first pointed out in their 1970 paper

"The Tachyonic Antitelephone," tachyons seem to lead to a **time travel** paradox because of their ability to send messages into the past. Suppose Alice on Earth and Boole on a planet circling around Sirius can communicate using what has been called a tachyon "antitelephone." They agree in advance that when Boole receives a message from Alice, he will reply immediately. Alice promises to send a message to Boole at noon her time, if and only if she has not received a message from Boole by 10 A.M. The snag is that both messages, being superluminal, travel back in time. If Alice sends her message at noon, Boole's reply could reach her before 10 A.M. "Then," as Benford and colleagues wrote, "the exchange of messages will take place if and only if it does not take place. . . ." Perhaps not surprisingly, despite numerous searches, no tachyon detection has so far been confirmed. The same is true of another hypothetical faster-than-light particle called a dybbuk (Hebrew for a "roving spirit"), which would have imaginary mass, energy, and momentum. Dybbuks, proposed by Raymond Fox of the Israel Institute of Technology, have properties even stranger than those of tachyons yet, interestingly, they avoid the **causality** problem that affects their superluminal cousins.

TacTix

A two-player game of strategy, devised by Piet **Hein**, that is essentially a two-dimensional version of **Nim**. Though the game is nontrivial, the first player can always win, at least in the 5 × 5 matrix version by choosing the center piece and symmetrically mirroring the second player's moves. On a 5 × 5 grid, players alternate taking away as many contiguous pieces as desired from a single row or column.

Tafl game

A type of board game in which the contest is between two forces of unequal number or strength. The earliest form of Tafl (old Norse for "table"), known as **Hnefa-Tafl** ("king's table"), originated in Scandinavia before A.D. 400 and was then exported by the Vikings to Greenland, Iceland (where it is mentioned in the *Grettis Saga* dating back to A.D. 1300), Ireland, England, Wales, and as far east as the Ukraine. Several boards unearthed in both Viking and Anglo-Saxon contexts, including the board found at the Gokstad ship burial, have had

Hnefa-Tafl on one side and **nine men's morris** on the other. Later variants of Tafl include Tabula (the medieval ancestor of **backgammon**, introduced from the French as *Quatre* and thus *Kvatru-Tafl*), **fox and geese** (*Ref-Skak,* "fox chess", *Hala-Tafl* or *Freys-Tafl*), **three men's morris** (*Hræ-Tafl,* "Quick-Tafl"), and **nine men's morris**.

Tait, Peter Guthrie (1831–1901)

A Scottish scientist and mathematician who carried out the world's first systematic investigation of **knot** theory. Early in his career he formed a friendship with William **Hamilton** and became fascinated in the application of Hamilton's **quaternion**s to problems in physics. In 1857, he also took an interest in Hermann Helmholtz's theories on the behavior of vortex rings, and began experimenting with smoke rings and their interactions. These experiments greatly impressed William Thomson (Lord Kelvin) who saw in them a possible way (wrong, as we now know) to explain atomic structure and the buildup of different elements. This idea, in turn, led Tait, Thomson, and James Maxwell to do seminal work on knot theory, since the basic building blocks in Thomson's vortex atom model were rings knotted in three dimensions. Without any rigorous theory, which would have been well beyond nineteenth-century mathematics, Tait began to classify knots using his geometric intuition. By 1877 he had classified all knots with seven crossings. He then went on to consider the coloring of **graph**s and put forward a hypothesis (see **Tait's conjecture**) that, if true (which it wasn't), would have proved the **four-color map problem**. Among his many other accomplishments, Tait wrote a classic paper on the trajectory of golf balls (1896). This was a subject close to his heart because the third of his four sons was Frederick Gutherie Tait, the leading amateur golfer in 1893 and winner of the Open Golf Championship in 1896 and 1898.

Tait's conjecture

A hypothesis put forward by Peter **Tait** in 1884, which says that every **polyhedron** has a **Hamilton** circuit through its vertices. In other words, it is possible to travel around all the edges of a polyhedron, passing through each vertex (corner) exactly once and arriving back at the starting point. If true, Tait's conjecture would have provided an immediate proof of the **four color map problem**. However, in 1946, the British mathematician William Tutte (1917–2002), whose work at Bletchley Park on cracking the German FISH cipher played an important role in World War II, found a counterexample to the conjecture in the form of a polygon with 25 faces, 69 edges, and 46 vertices.

tally

To count or keep score. In the past this was often done by making marks on a stick; the word comes from the Latin *talea* meaning "one who cuts," which is also the root of *tailor.* The oldest known tally stick is thought to be the **Lebombo bone** dating back about 37,000 years. Until around 1828 British tax records were kept on wooden tally sticks. When the system was finally abandoned, the government was left with a mountain of wood which, in 1834, it decided to dispose of by having a giant bonfire. So successful was that blaze that it also burned down the parliament buildings. What Guy Fawkes had failed to do with dynamite, the Exchequer did with tally sticks!

tangent

(1) A straight line that touches a given curve exactly once, at a given point. (2) In a right triangle, if one of the angles is θ, then the tangent of θ is the ratio of lengths of the side opposite θ to the side next to θ. See also **trigonometric function**.

tangle

A system in which a **strange loop** appears.

tangled graph

A **graph** in 3-dimensional space; equivalently, a graph drawn in the plane so that when edges cross, one edge goes over the other.

Tangloids

A mathematical game for two players devised by Piet **Hein** to model the calculus of **spinor**s. Two flat blocks of wood each pierced with three tiny holes are joined with three parallel strings. Each player holds one of the blocks of wood. The first player holds one block of wood still, while the other player rotates the other block of wood around any axis for 4π radians (two full revolutions). Then the first player tries to untangle the strings without rotating either piece of wood. Only translations (sliding the pieces) are allowed. Afterward, the players reverse roles; whoever can untangle the strings fastest is the winner.

tangrams

A puzzle of Chinese origin, the objective of which is to form given shapes using a set of seven pieces (five triangles of various sizes, one square, and one parallelogram) that come from slicing up a square. The produced shape has to contain all the pieces, which mustn't overlap. Tangrams became popular in England around the middle of the nineteenth century, having been brought back by sailors from Hong Kong. It received a further boost when Lewis **Carroll** used the pieces to create illustrations of the characters in the Alice books. The origin of the name

tangrams The seven tangrams. *Kadon Enterprises, Inc., www.gamepuzzles.com*

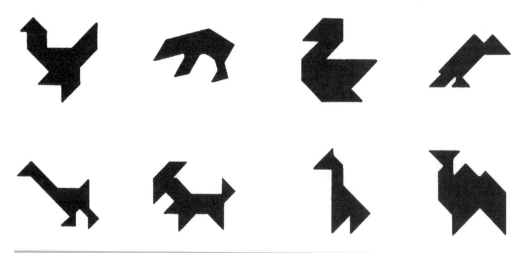

tangrams A tangram menagerie built up from the seven simple shapes.

isn't certain. One theory is that it comes from the Cantonese word for chin. A second is that it stems from a mispronunciation of a Chinese term that the sailors used for the ladies of the night from whom they learned the game! A third suggestion is that it is from the archaic Chinese root for seven, which still persists in the Tanabata festival held on July 7 in Japan. The first definitive appearance of tangrams in the Far East seems to be in Japan in the early eighteenth century; by around 1805 it had become a fad in both China and Europe. The **loculus of Archimedes** is a similar game and it has been suggested, though without evidence, that this was a direct precursor of tangrams, having been transmitted to the East via Arab sources.[93, 265]

Tarry, Gaston (1843–1913)

A French civil servant and amateur mathematician who spent the whole of his working career in Algeria. He published numerous articles on geometry, **number theory**, and **magic squares** from 1882 until his death, and is best known for his contribution to Euler's **thirty-six officers problem**. He also published an algorithm for exploring **mazes** that is named after him.

Tarski, Alfred (1902–1983)

A Polish-American mathematician who, along with Aristotle, Friedrich **Frege**, and Kurt **Gödel** is considered one of the greatest logicians of all time and certainly the most prolific: his collected works, excluding books, runs to 2,500 pages. Tarski made important contributions in many areas of mathematics, including **set theory**, **topology**, algebraic **logic**, and metamathematics. His most important contribution to logic is the *semantic method*–a technique that allows a more exacting study of formal scientific languages.

Tartaglia, Niccoló Fontana (1499–1557)

An Italian mathematician who, along with Girolamo **Cardano**, discovered the algebraic solution of the **cubic**. He was also a well-known inventor of mathematical recreations. He devised many arithmetical problems, and contributed especially to **measuring and weighing puzzles** and to **river-crossing problem**s. Although his real name was Niccoló Fontana, he is always referred to as Tartaglia, "the stammerer."

tautochrone problem

Find the curve down which an object can slide from any point to the bottom (accelerated by gravity and ignoring friction), always in the same length of time. *Tautochrone* comes from the Greek *tauto* for "the same" (which also gives us *tautology*) and *chronos* for "time." The solution, first found by Christiaan **Huygens** and published in his

Horologium oscillatorium (1673), is a **cycloid**. Thus, if you were to upturn a cycloid, in the manner of an inverted arch, and then release a marble from any point on it, it would take exactly the same time to reach the bottom, no matter where on the curve you started. Huygens used his discovery to design a more accurate pendulum–one with curved jaws from the point of support that forced the string to follow the right curve no matter how large or small the swing. The cycloid's unique property is mentioned in the following passage from Herman Melville's *Moby Dick:* "[The try-pot] is also a place for profound mathematical meditation. It was in the left-hand try-pot of the *Pequod*, with the soapstone diligently circling round me, that I was first indirectly struck by the remarkable fact, that in geometry all bodies gliding along a cycloid, my soapstone, for example, will descend from any point in precisely the same time." The cycloid is also the curve that answers the **brachistochrone problem**.

tautology

In mathematics, a logical statement in which the conclusion is equivalent to the premise. According to the school of thought known as *logicism,* all of mathematics is derived from **logic** and is thus inherently tautological. Tautology is also the needless, pointless, meaningless, and unwarranted repetition of words and phrases that mean the same thing. Examples are to be found in the previous sentence and the next one. Have a slice of pizza (*pizza* is Italian for "slice") and spin the **roulette** wheel (*roulette* is French for "little wheel").

ten

The **base** of our familiar **number system**, which stems directly from the fact that we have 10 fingers on which to count. Ten is the only **triangular number** that is a sum of consecutive odd squares ($10 = 1^2 + 3^2$) and the only composite integer such that all of its positive integer divisors other than 1 are of the form $x^2 + 1$ ($2 = 1^2 + 1$, $5 = 2^2 + 1$, $10 = 3^2 + 1$). Strange but true: the life span of a taste bud is 10 days.

tensor

A generalization of the concept of a **vector**. A *scalar* is a tensor of rank zero and a vector is a tensor of rank one. There are tensors of rank two, three, and so on, used mainly in manipulating and transforming sets of equations within and between different coordinate systems. A tensor of order n has n^2 components and is often best thought of as an array of values that can be manipulated like a **matrix**. To illustrate a tensor of rank two, imagine a plane surface area with a force acting on it. The total effect depends on two things: the magnitude and direction of the force, and the size of the area and its orientation. In fact this latter property can be represented uniquely by a vector of magnitude proportional to the size of the area, and in a direction normal to the area. So the effect of the force upon the surface depends on two vectors and is equivalent to a tensor of rank two. Tensors were used by Einstein in deriving his law of gravitation in general **relativity theory**.

tera-

Prefix for 10^{15}, from the Greek *pentakis* for "five times."

terminating decimal

A decimal fraction that comes to an end, as in 0.3 or 0.7194. All terminating decimals are **rational numbers**, but not all rational numbers have terminating decimal expansions. For example, $\frac{1}{3}$ is rational, but its decimal expansion goes on forever.

ternary

(1) Having the **base** three. (2) Involving three **variables**. From the Latin *ternarius* ("three each").

tessellation

See **tiling**.

tesseract

The four-dimensional analogue of a cube, also known as a 4-space **hypercube** or an 8-cell. The name, possibly coined by Charles **Hinton**, comes from *tesser*, meaning "four," and *aktis*, meaning "ray," thus "four rays." Just as a cube is obtained by "thickening" a square in the third dimension, which can be imagined as stacking infinitely many infinitely thin sheets of paper, a tesseract is a cube thickened in the **fourth dimension**. We can't imagine this because we can't think four dimensionally but it is possible to appreciate that just as perspectives of cubes can be drawn on a 2-d surface, so real cubes can serve as perspectives of tesseracts. A square drawn inside a larger square with the vertices connected by lines is one way to provide a perspective of a cube. Similarly, a hypercube is sometimes portrayed as a small cube within a larger cube with lines drawn from the vertices of the smaller cube to the vertices of the larger cube. This kind of representation is a bit misleading, however, and reveals very little of the nature of a tesseract. It doesn't show, for instance, how a tesseract can be subdivided into smaller 4-d blocks in the same way that a cube can be divided into smaller cubes, or a square into smaller squares. A more useful way to think of a tesseract is as a folding, in the fourth dimension, of a 3-d net of eight cubes, just as a cube is a folding in the third dimension of a 2-d net of six squares. Start with a stack of four cubes, with four more cubes

arranged in a cross around the second cube from the top. A tesseract is made by folding (in the fourth dimension) so that the top face of the cube at the top of the stack merges with the bottom face of the bottom cube, and so that the adjacent edges of the cubes in the cross join. (See table, "A Comparison of the Square, Cube, and Tesseract.")

A Comparison of the Square, Cube, and Tesseract

	Vertices	Edges	Squares	Cubes
Square	4	4	1	—
Cube	8	12	6	1
Tesseract	16	32	24	8

A tesseract is bounded by eight hyperplanes, each of which intersects it to form a cube. Two cubes, and so three squares, intersect at each edge. There are three cubes meeting at every vertex, the vertex polyhedron of which is a regular tetrahedron, leading to the **Schläfli symbol** {4,3,3}. The distance between opposite corners of a hypercube is twice the length of a side—much tidier than the corresponding values of $\sqrt{2}$ for a square and $\sqrt{3}$ for a cube.

If a cube is hung from one of its vertices and sliced horizontally through its center, the result is a hexagon. What if the same is done to a tesseract? The slice will yield a 3-dimensional object—but what kind? The answer is an octahedron. By analogy with the slice of the 3-cube, the slice of the 4-cube must cut every "face." The number of faces of a 4-cube is eight and the only regular eight-sided solid is an octahedron.

Tesseracts turn up in both art and literature. Salvador Dali's *Christus Hypercubus* shows Christ being crucified on a tesseract. In Robert Heinlein's short story "And He Built a Crooked House" (1940), a house built as a three-dimensional projection of a tesseract collapses, or folds up, to become a real tesseract—with unusual consequences for the person trapped inside. The tesseract is also mentioned in Madeleine l'Engles children's fantasy *A Wrinkle in Time* as a way of introducing the concept of higher dimensions.[161, 338]

tetragon

The less familiar name for what's normally called a **quadrilateral**.

tetrahedral number

A number that can be made by considering a tetrahedral pattern of beads in three dimensions. For example, if a tri-

tetrahedron   Rosie's pyramid puzzle, a vintage game that calls for four pieces to be assembled into a tetrahedron. *Sue & Brian Young/Mr. Puzzle Australia, www.mrpuzzle.com.au*

angle of beads is made with three beads to a side, and on top of this is placed a triangle with two beads to a side, and on top of that a triangle with one bead to a side, the result is a tetrahedron of beads. In this case the total number of beads is (third triangular number) + (second triangular number) + (first triangular number) = 6 + 3 + 1 = 10. In general the nth tetrahedral number is equal to the sum of the first n triangular numbers. This is the same as the fourth number from the left in the $(n + 3)$th row of **Pascal's triangle**. We can use the **binomial** formula for numbers in Pascal's triangle to show that the nth tetrahedral number is $^{n+2}C_3$, or $(n + 2)(n + 1)n/6$. The only numbers that are both tetrahedral and square are 4 ($= 2^2 = T_2$) and 19,600 ($= 140^2 = T_{48}$).

tetrahedron

A four-sided **polyhedron**. A regular tetrahedron, one of the **Platonic solid**s, is a regular three-sided pyramid in which the base-edges and side-edges are of equal length. The projection of a regular tetrahedron can be an equilateral triangle or a square. The centers of the faces of a tetrahedron form another tetrahedron.

tetraktys

The sum $1 + 2 + 3 + 4 = 10$ (the fourth **triangular number**), held in reverence by the Pythagoreans (see **Pythagoras of Samos**). The tetraktus, or "holy fourfoldness," was taken to represent the four elements: fire, water, air, and earth.

Tetris

A video and computer game, invented in 1985 by the Russian Alexey Pajitnov, that has become one of the most widely played games of all time. In 2002, computer scientists Erik Demaine, Susan Hohenberger, and David Liben-Nowell of the Massachusetts Institute of Technology (MIT) analyzed the game to determine its computational complexity and found it to be an **NP-hard problem** (one that is immune to simple solution and instead demands exhaustive analysis to work out the best way to be completed). Many people first played Tetris on the Nintendo Gameboy handheld console but it has since become available for virtually every personal computer–based device. The game gives the player the task of creating complete lines from a series of regularly shaped blocks–tetrominos, which are a type of **polyomino**–that advance steadily

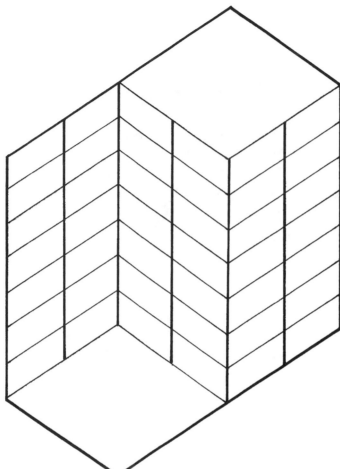

Thiery figure

down a narrow grid. The blocks can be spun to make them fit together better and complete lines. The game gets faster as levels are completed, making it harder to spin and fit blocks together fast enough to form lines. The MIT team found that, subject to certain conditions, Tetris has much in common with some of the knottiest mathematical conundrums, including the **traveling salesman problem**. Because Tetris is NP-hard, there is no easy way to maximize a score for the game, even when the sequence of blocks is known in advance.

tetromino

A four-square **polyomino**.

theorem

A major mathematical proposition that has been proved correct. More precisely, a statement in a **formal system** for which there exists a **proof**. See also **conjecture** and **lemma**.

theory of everything

A unified theory of all fundamental forces and interactions in nature; a grand unified theory that also includes gravity or general relativity (see **relativity theory**).

Thiery figure

A classic **ambiguous figure** devised by the psychologist A. Thiery in the late nineteenth century.

thirteen

The unluckiest of numbers if you happen to be superstitious. This belief has a couple of historical roots. According to biblical tradition, there were 13 people at Christ's Last Supper, and Christ was crucified on a Friday the thirteenth. Further back in time, Alexander the Great decided he wanted to be the thirteenth god alongside the 12 that already stood for each month of the year, so he had a thirteenth statue built on the place of his capital. His death shortly after gave the number a bad name. Many buildings don't have a floor labeled 13 and many hotels will have a room numbered 12A instead of 13. There is even a name for a morbid fear of 13: triskaidekaphobia.

Fresh disasters involving the number hardly help triskaidekaphobics overcome their affliction. The most notorious of these involved the Apollo 13 Moon mission, which was launched on April 11, 1970 (the sum of 4, 11, and 70 equals 85, the digital sum of which is 13), from Pad 39 (3×13) at 13:13 local time, and suffered an explosion on April 13. (The astronauts did, however, make it home safely, which could be considered good luck.) There is always at least one Friday the thirteenth in each year; in some years there are two, and rarely three (e.g., 1998 and 2009). There were 13 original U.S. colonies (hence the 13 stripes on the American flag) and 13 signers of the Declaration of Independence. In Japan, the numbers 4 and 9 are considered unlucky, not because 13 can be represented as sum of these two perfect squares but because of their pronunciation. In Japanese, four is *shi,* which is pronounced the same as the word for death; nine is *ku,* which sounds like the word for torture. And speaking of torture, it was not unusual in times past for bakers to come in for stiff punishment if they shortchanged their customers. In ancient Egypt, someone found selling light loaves might end up with his ear nailed to a doorpost, while in medieval Britain the punishment was likely to be a spell in the pillory. This led to the custom of adding a thirteenth loaf to every batch of 12 to be on the safe side, and hence the expression "a baker's dozen."

Mathematically, the reverse of the square of 13 is the same as the square of the reverse of 13: $13^2 = 169$; the reverse of 169 is 961 and the reverse of 13 is 31; $31^2 = 961$. Thirteen is the smallest **prime number** that can be expressed as the sum of the squares of two prime numbers: $13 = 2^2 + 3^2$. Also the sum of all prime numbers up to 13 ($2 + 3 + 5 + 7 + 11 + 13$) is equal to the thirteenth prime number (41), and this is the largest such number. On the anagrammatical front is this nice equation:

$$\text{ELEVEN} + \text{TWO} = \text{TWELVE} + \text{ONE}$$

See also **dollar.**

thirty colored cubes puzzle

A game devised in 1921 by Percy **MacMahon,** which was marketed under the name "Mayblox." It is played with 30 cubes that have all possible permutations of six different colors on their faces. A number of different games can be played with the blocks. One is to choose a cube at random and then choose seven other cubes to make a $2 \times 2 \times 2$ cube with the same arrangement of colors for its faces as the first chosen cube. Each face of the

$2 \times 2 \times 2$ cube has to be a single color and the interior faces have to match in color. After building one such cube, only one other with the same properties can by made from the remaining 22 cubes—the mirror image of the first.

thirty-six officers problem

Arrange 36 officers in a 6×6 square so that one officer from each of six regiments appears in each row and one from each of six ranks appears in each column. This problem, first posed by Leonhard **Euler** in 1779, is equivalent to finding two mutually orthogonal **Latin square**s of order six. Euler correctly conjectured that there was no solution; the search for a proof led to important developments in **combinatorics.**

Thompson, D'Arcy Wentworth (1860–1948)

A Scottish naturalist and polymath. Thompson held a professorial chair at St. Andrews and Dundee, Scotland, for the amazing period of 64 years, a record for tenure unlikely ever to be broken. Although he wrote more than 300 scientific articles and books, his reputation is based primarily on his efforts to reduce biological phenomena to mathematics in his magnum opus *On Growth and Form* (1917).[332] In this book, full of marvelous sketches of such things as nautilus shells and honeycombs, Thompson claimed that much about animals and plants could be understood by the laws of physics, as mirrored in the

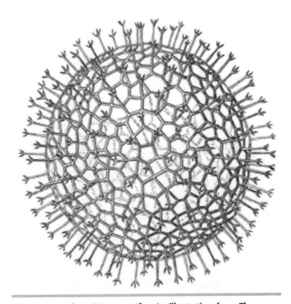

Thompson, D'Arcy Wentworth An illustration from Thompson's *On Growth and Form:* the shell of the radiolarian Aulastrum triceros.

structures and patterns of mathematics. His most novel idea was to show how mathematical functions could be applied to the shape of one organism to continuously transform it into other, physically similar organisms. One memorable example is the squeezing and stretching of a rectangular Cartesian grid that transforms the fish species *Scarus sp.* to the species *Pomacanthus*. Thompson used the same principle to transform skulls of baboons into those of other primates, and to show how corresponding bones like the shoulder blade are related in different species. Doubtless, if he were alive today, he would be heavily into "morphing"—the digital technique that allows a computer to do exactly the same kind of transformation of one object to another, but vastly more efficiently. Thompson acquired a local reputation as a mild eccentric; indeed, older inhabitants of St. Andrews still recall seeing him strolling about town with a parrot on his shoulder.

thousand

A number that, linguistically, comes from an extension of "hundred" and has roots in the Germanic *teue* and *hundt*. *Teue* refers to a thickening or swelling, and *hundt* is the source of our present-day hundred. A thousand, then, literally means a swollen or large hundred. The root *teue* is the basis of such common words today as thigh, thumb, tumor, and tuber. One thousand is the answer to the question: If you were to spell out numbers, how far would you have to go until you would find the letter "A"? While Americans may say "Thanks a million" to express gratitude, Norwegians offer "Thanks a thousand" ("*tusen takk*").

three

> *1 + 1 = 3, for large values of 1.*
> —Anonymous

The number of dimensions of space in which we live; three is also the smallest odd **prime number**, the second **triangular number**, and member of the **Fibonacci sequence**. Three is often the number of repetitions in jokes and children's stories (for example, the tale of the *Three Little Pigs* and of the number of chairs, beds, and porridge-eating bears in *Goldilocks*), because it is the minimum number needed to establish a pattern (such as a regular tempo) or to convey the impression of an ongoing sequence or succession. In the Christian tradition, three plays a crucial role: Christ represents one third of the Trinity (the Father, Son, and Holy Spirit), was visited by the three wise men and, 33 years later, when Peter disowned him three times, rose on the third day after the crucifixion, having died at 3 P.M. Other things that come

in threes: musketeers, primary colors, wishes, blind mice, bad luck, and London buses. See also **Triangular Lodge** and **tinner's rabbits**.

three-hat problem
See **hat problem**.

three men's morris

An old game played on a 3×3 board that is thought to be a direct ancestor of **tic-tac-toe**; it is known by many other names, including nine holes, and is related to six men's morris and **nine men's morris**. The game involves two differently colored sets of four pieces (one set for each player). Players take turns placing pieces on intersection points, and the first person to place three along a line wins the game. The earliest known board for three men's morris was found on the roof of the temple in Kurna, Egypt, dating back almost three and a half thousand years. Its earliest known appearance in literature is in Ovid's *Ars Amatoria*. The Chinese are believed to have played it under the name *Luk tsut K'i* during the time of Confucius (c. 500 B.C.). Boards for three men's morris dating back to the thirteenth century can be found carved into the cloister seats at the cathedrals at Canterbury, Gloucester, Norwich, and Salisbury, and at Westminster Abbey.

three-body problem

The problem of determining the future positions and velocities of three gravitational bodies. The problem was proved unsolvable in the general case by Henri **Poincaré**, which foreshadowed the importance of **chaos**. Although no analytical solutions are possible in the worst case, a numerical solution is sometimes sufficient for many tasks.

Thue-Morse constant

Also known as the *parity constant*, the number defined as follows: Take a string of 1's and 0's and follow it by its complement (the same string with 1's switched to 0's and vice versa) to give a string twice as long. Repeat this process forever (starting with 0 as the initial string) to get the sequence 01101001100101101001010110 . . . . Make this a binary fraction, $0.0110100110010110 \ldots_2$, and rewrite it in base 10. The resulting **transcendental number**, 0.41245403364 . . . , is the Thue-Morse constant.

tic-tac-toe

Also known as noughts and crosses, and spelled in a variety of ways (such as ticktacktoe), this well-known pastime is not as ancient as popularly believed, though it certainly has roots in older games. The earliest clear description of the rules, but without a name, comes from Charles **Babbage** around 1820. Later on, Babbage started to call the game tit tat to and by slight variants on this, gave the first detailed analysis of it, and designed the first robot to play it! He speculated that "[T]he machine might consist of the figures of two children playing against each other, accompanied by a lamb and a cock. That the child who won the game might clap his hands whilst the cock was crowing, after which, that the child who was beaten might cry and wring his hands whilst the lamb was bleating."

His plan was to exhibit the machine in London to raise money for his more serious projects, including the fabulous Analytical Engine. However, on hearing that similar devices, such as a mechanical composer of Latin verse, had flopped financially he abandoned the scheme.

Two expert players of tic-tac-toe will always draw. In other words, there is no winning strategy against an opponent who knows the game well. Nor is it difficult to become unbeatable. Despite the fact that there are $9 \times 8 \times 7 \times 6 \times 5$ ($= 15,120$) possible layouts of noughts and crosses for the first five moves alone, these reduce to just a few basic patterns of play and counterplay that ensure the canny contestant never loses. The game is only interesting when at least one novice is involved. The strongest opening play is in a corner because an unwary opponent will be trapped unless her countermove is in the center. By the same token, a center opening must be countered by a corner response or the first player has an easy win. Of course, a master player will not only never lose but will learn his opponent's weaknesses and exploit them in the most devastating way.

The game also becomes more interesting when played on a larger board and/or in more dimensions. An exception to this is the $3 \times 3 \times 3$ cube which gives an easy win to the first player; in fact, it cannot end in a draw because the first player has 14 plays and it is impossible to make all of these without scoring. The $4 \times 4 \times 4$ cube is much more interesting and is sold commercially as the well-known Score Four game. For the ambitious, tic-tac-toe can be played in four dimensions on a **tesseract** by sectioning it into two-dimensional squares. Older, similar, and mathematically more interesting relatives of tic-tac-toe include **nine men's morris** and **Ovid's game.**

tie knots

How many ways can you tie a tie? For many years there were just three styles of **knot**: the Four-in-Hand, the Windsor, and the Half-Windsor. Then the Pratt was introduced to the world on the cover of the *New York Times* in 1989. Intrigued that only one new knot had been added to the tie-tying repertoire in more than half a century, two researchers from the University of Cambridge's physics department, Thomas Fink and Yong Mao, decided to see how many tie knots were actually possible. To this end, they applied **random walk** theory–a technique useful for describing movement which, although unpredictable in detail, reveals large-scale patterns. Such patterns, the researchers realized, were essential to a successfully accomplished tie-knot. For example, if the end of the tie is moved to the right, its next move can't be to the right again–it has to be either to the left or to the center. This means that each move made in tying a tie can only be followed by one of two alternatives. Fink and Mao found that the simplest possible knot involves just three moves. They went on to discover 85 possible tie-knots, including the four popular knots, six new knots that they consider aesthetically pleasing, and two complicated nine-move knots.[100]

tiling

Also called a *tesselation,* a collection of smaller shapes that precisely covers a larger shape, without any gaps or overlaps. Usually, the shape to be tiled is a flat plane but other shapes and three-dimensional objects can be tiled, too. In a game that involves tiling, certain conditions are applied; for example, all the tiles may have to be identical, or they may all have to be squares but every one of a different size. It's been known for some time that all simple regular tilings in the plane belong to one of the 17 plane symmetry groups known as **wallpaper group**s. All 17 of these patterns are known to exist in the **Alhambra** palace. This doesn't exhaust the apparently simple problem of tiling the plane: adding extra constraints or removing the requirement for regularity leads to a large number of interesting problems. These include *alternating tilings,* for examples of squares or dominoes, such that two tiles have a side or a part of a side in common, or *colored tilings,* in which no two adjacent tiles have the same color. Colored tilings are also called *colored maps.* The most famous problem relating to colored tilings is the **four-color map problem,** which has been solved. Other problems involve *n*-tesselations, in which each tile has an integral area and for each natural number *n* there is exactly one tile with area *n*. See also **Penrose tiling, rep-tile, rectangle, squaring the square, dissection,** and **packing.**

time

Time is a great teacher. Unfortunately, it kills all its pupils.

–Hector Berlioz (1803–1869)

One of the most familiar and yet mysterious properties of the universe. The "flow" of time is one of the strongest impressions we have, yet it may simply be

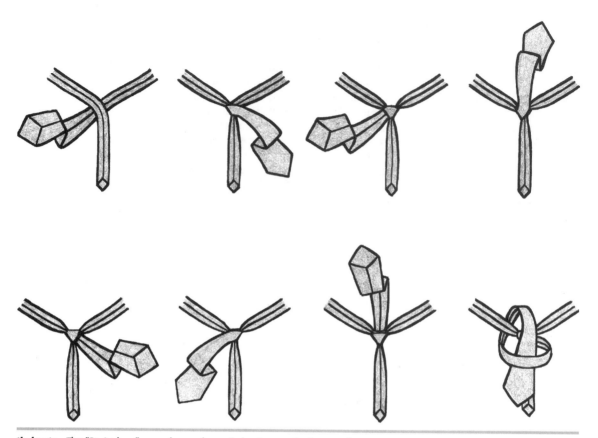

tie knots The "St. Andrew": one of several new tie knots recently discovered. *Thomas Fink*

an illusion or a product of the conscious mind. The very notion that time somehow moves leads to a logical paradox because, as the Australian philosopher J. J. C. Smart asked: "In what units is the rate of time flow to be measured? Seconds per ____?" John Dunne in his classic book *An Experiment with Time,* argued that the human mind has the ability to rove back and forth along the time line so that precognition is a physical possibility. However his theory involves an infinite regress of time and of the observer that is philosophically hard to swallow. All the same, it may be that the apparent movement from past to present to future has less to do with the universe at large than it has to do with our individual subjective experience. In some way, still to be fathomed, time, consciousness, free will, and the individual are intimately entwined. In physics, by contrast, time is treated no differently (with one important exception, noted below) than space. It is simply another **dimension**—another axis, or extension, of physical reality. Just as the various spatial dimensions pre-

vent everything from happening at a single point, so time prevents everything from happening all at once. As one wag put it, "Time is just one damned thing after another!" In Einstein's **relativity theory**, time is effectively "spatialized" so that, instead of speaking of an absolute three-dimensional space and a separate one-dimensional time, there is a four-dimensional **space-time** continuum. So closely related are time and space in relativity theory that time can be converted into space and vice versa. In particular, different observers may not agree on the distance or the duration between any two events in space-time, but they will always agree on the *space-time interval*. If the two points events occur at (t, x, y, z) and $(t + dt, x + dx, y + dy, z + dz)$, then the (constant) space-time interval between them is given by

$$s^2 = c^2(t_2^2 - t_1^2) - (x_2^2 - x_1^2) - (y_2^2 - y_1^2) - (z_2^2 - z_1^2).$$

But the time of relativity, like that of classical physics, remains reversible.

tiling A computer-generated periodic tiling. *Jos Leys, www.josleys.com*

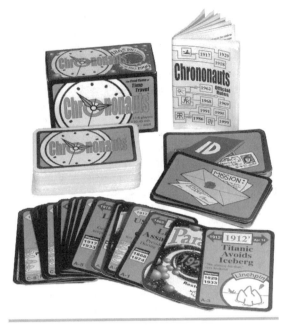

time travel The game of Chrononauts, in which players must resolve a series of paradoxes that threaten the time line. *Looney Labs, www.LooneyLabs.com*

time complexity

A **function** that describes the amount of time required for a program to run on a computer to perform a particular task. The function is parameterized by the length of the program's input.

time dilation

See **relativity theory**.

time-reversible

A property of **dynamical system**s that can be run unambiguously both forward and backward in time. The **Lorenz system**, for example, is time-reversible.

time travel

Can we travel through **time**? Of course, we do it all the time! But can we do it at a different rate than normal? Again, the answer is "yes" because of the phenomenon known as *time dilation* in Einstein's **relativity theory**. However, time dilation enables, even in principle, only a limited kind of leap into the future—one from which we cannot return to the present. Genuine time travel is the ability to jump forward or backward through time at a rate other than that of the ordinary progression of events or that enabled by the relativistic time dilation effect.

The possibility of traveling through time poses such a threat to **causality** and opens the door to so many disturbing paradoxes that many scientists feel inclined to dismiss it out of hand. However, it has been a favorite theme of science fiction since the 1880s. In *The Time Machine* (1895), H. G. Wells gives a pleasant preamble about the nature of the **fourth dimension** before whisking his hero 802,000 years into the future. Says the Time Traveller (we never learn his real name), "[A]ny real body must have extension in four directions: it must have Length, Breadth, Thickness, and–Duration. There are really four dimensions, three of which we call the three planes of Space, and a fourth, Time. There is, however, a tendency to draw an unreal distinction between the former three dimensions and the latter, because it happens that our consciousness moves intermittently in one direction along the latter from the beginning to the end of our lives."

Unfortunately for would-be chrononauts (an early version of *The Time Machine* was called *The Chronic Astronauts*), Wells is not specific about how his time traveling device works, though we know that "Parts were of nickel, parts of ivory, parts had certainly been filed or sawn out of rock crystal." In more recent times, physicists, speculating on some of the more esoteric byways of relativity and **quantum mechanics**, have been a little more forthcoming about how time travel might be achieved in practice. These speculations have variously involved wormholes (shortcuts outside of normal space and time), faster-than-light particles known as **tachyon**s, and unusual cosmological models, such as the *Gödel universe,* which allow movement to any point in the future or the past. Let us leave aside the practical aspects, however, and focus on the logic of breaking the time barrier.

The various time travel possibilities dealt with in science fiction fall into two broad categories. In the first the timeline, from deepest past to darkest future, is frozen and immutable, like a filmstrip. Any time traveling that takes place is constrained by this preordained structure—effectively, already written into the narrative of the world (the "block universe" of Einsteinian physics)—and is thus prevented from leading to paradoxes. In one variant of this scenario, the *Novikov self-consistency principle* applies. Named after Igor Novikov, an astrophysicist at Copenhagen University, this asserts that any attempt at time travel that would lead to a paradox, such as the **grandfather paradox**, is bound to fail even if the cause of failure is an extremely improbable event. In other words, try as you might to introduce a contradiction into the timeline, like killing yourself or one of your ancestors in the past, circumstances will always conspire to prevent you. An excellent example of this type of universe is found in Robert L. Forward's novel *Timemaster.* Another variant on the fixed timeline concept is that any event that appears to have caused a paradox has, in fact, created a new timeline. The old timeline remains unaltered, and the time traveler becomes part of a new temporal branch line. One difficulty with this arrangement is that it might violate the principle of conservation of mass-energy, unless the mechanics of time travel demand that mass-energy be exchanged in precise balance between past and future at the moment of travel. However, the concept of branching universes and alternative histories is not outrageous in physics where the **many worlds hypothesis** and of Feynmann's sum-over-histories are routinely debated.

The second main type of time travel entertained in science fiction assumes that the timeline is flexible and changeable. This can lead to all sorts of mind-boggling difficulties and contradictions. A way to offset some of these problems is to stipulate that the timeline is very resistant to change. In the extreme case, as writer Larry Niven has argued, it may be a fundamental rule that in any universe where time travel is allowed, no actual time machine is ever invented. The English physicist and mathematician Stephen Hawking put this idea on a more formal footing with his *chronology protection conjecture.* On the other hand if the timeline is presumed to be easily changed, paradoxes threaten to spring up at every turn. One of the most remarkable of these is the *closed causal curve paradox* in which, it seems, something can be gotten for nothing. Samuel Mines summarized the plot of his 1946 short story as follows:

> A scientist builds a time machine, goes 500 years into the future. He finds a statue of himself commemorating the first time traveler. He brings it back to his own time and it is subsequently set up in his honor. You see the catch here? It had to be set up in his own time so that it would be there waiting for him when he went into the future to find it. He had to go into the future to bring it back so it could be set up in his own time. Somewhere a piece of the cycle is missing. When was the statue made?

Closed loops in time can also conjure knowledge out of thin air. A man builds a time machine and travels into the past to give the plans for the device to his younger self who then builds the machine, travels into the past, and so on. Where did the plans originate? A curious thing about time loops is that they have no easily discernible future or past because all the events taking place in them affect one another in a circular way. Time loops also put a question mark over free will. What happens if the younger man, given the time machine plans by his older self, decides not to build the device? *Can* he make that choice given that, in some sense, he has already built it? Perhaps the apparent absence of time travelers and time machines in the real world is a sign that we do not have to worry about such issues—at least, for the present.

tinner's rabbits

A name that has emerged recently to describe a pattern of three rabbits or hares that has been found in many parts of the world, including England and Wales, mainland Europe, China, and Russia. It occurs, for example, on the medieval roof bosses of some churches in Devon and Cornwall and is thought to be connected with the local tin-mining industry. One theory suggests the following link: Tin is alloyed with copper to make bronze, copper came from Cyprus (the words *Cyprus* and *copper* have the same root), Cyprus is the island of the goddess Venus or Aphrodite (she was born there), rabbits are symbols of Venus. Three intertwined fishes are a common Christian symbol, so the three rabbits may also have stood for the Trinity.

Tippee Top

Also known as a *Tippy Top,* a type of top, patented in Britain in 1953, that consists of a peg with a ball-shaped body. If the top is spun quickly on the rounded body, with the peg pointing upward, it flips itself over and spins on its peg. Crucial to this counterintuitive behavior is the shape of the body, which is smooth and spheroidal with no sharp point. After release, the top, like tops of every description, begins to show precession, that is, its axis of rotation moves in a small circle. After a while, the contact point of the top with the table no longer coincides with the rotation axis but instead moves to other points of the top's head. Due to frictional forces and precession, the top seeks a more stable position, which it finds by flipping on to its stalk. See also **celt**.

Titchener illusion

Also known as the *Ebbinghaus size illusion,* a well-known **distortion illusion**. Two circles are surrounded by either six big circles or six small circles. Despite appearances, the two center circles are exactly the same size.

Toeplitz matrix

A **matrix** in which all the elements are the same along any **diagonal** that slopes from northwest to southeast.

topological group

Also called a *continuous group,* a **set** that has both the structure of a **group** and of a **topological space** in such a way that the operations defining the group structure give continuous maps in the topological structure. Many groups of matrices (see **matrix**) give topological groups.

topological dimension

An integer that defines the number of coordinates needed to specify a given point of an object of set X. A single point, therefore, has a topological dimension equal to zero; a curve has dimension one, a surface has dimension two, and so on.

topological space

A type of generalized mathematical space in which the idea of closeness, or limits, is described in terms of relationships between **sets** rather than in terms of distance. Every topological space consists of: (1) a set of points; (2) a class of subsets defined axiomatically as open sets; and (3) the set operations of union and intersection.

topology

The study of those properties of mathematical objects that remain unaffected by smooth deformations, such as stretching and squeezing, but that don't involve tearing.

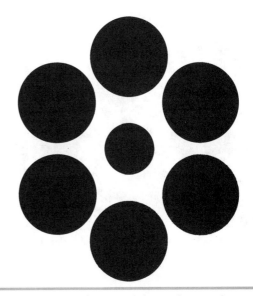

Titchener illusion   The center circles are the same size.

The word comes from the Greek *topos* for "place," and was introduced into English by Solomon Lefschetz in the late 1920s. A topologist has been described as someone who doesn't know the difference between a doughnut and a coffee cup. Substitute "care about" for "know" and this becomes more accurate. Imagine a donut made of soft clay. A potter can easily shape this into a cup with a handle without removing or creating any new holes. Both shapes, in topology, are said to be genus 1–objects with a single hole. A sphere, by contrast, is genus 0 (no holes), while an eyeglass frame, with the lenses removed, is genus 2. For more on topologically intriguing structures, see **Möbius band** and **Klein bottle**.

torus Inside the torus of an experimental nuclear fusion reactor. *Joint European Torus*

torus

A doughnut, bagel, or inner-tube shape; the word comes from the Latin for "bulge" and was first used to describe the molding around the base of a column. One way to think of a torus is as a **surface of revolution** obtained by rotating a circle around an axis that lies in the plane of the circle but doesn't intersect the circle. In the general case, where the shape being so rotated is any closed plane curve, the resulting surface is called a *toroid*. Although the usual torus in three-dimensional space is shaped like a doughnut, the concept of the torus is extremely useful in higher dimensional space as well.

tour

A sequence of moves by a **chess** piece on a chessboard in which each square of the board is visited exactly once. See also **knight's tour** and **magic tour**.

Tower of Brahma

A romantic legend manufactured by Edouard **Lucas** as an accompaniment to the popular game he invented, the **Tower of Hanoi**. According to the tale of the Tower of Brahma, in the Indian city of Benares, beneath a dome that marked the center of the world, is to be found a brass plate in which are set three diamond needles, "each a cubit high and as thick as the body of a bee." Brahma placed 64 disks of pure gold on one of these needles at the time of Creation. Each disk is a different size, and each is placed so that it rests on top of another disk of greater size, with the largest resting on the brass plate at the bottom and the smallest at the top. Within the temple are priests whose job it is to transfer all the gold disks from their original needle to one of the others, without ever moving more than one disk at a time. No priest can ever place any disk on top of a smaller one, or anywhere

Tower of Hanoi   A vintage version of the game called Pyramids, manufactured by Knapp Electric, Inc., of New York. *Sue & Brian Young/Mr. Puzzle Australia, www.mrpuzzle.com.au*

else except on one of the needles. When the task is done, and all 64 disks have been successfully transferred to another needle, "tower, temple, and Brahmins alike will crumble into dust, and with a thunder-clap the world will vanish." The prediction (thunder-clap aside) seems fairly safe given that the number of steps required to transfer all the disks is $2^{64} - 1$, which is approximately 1.8447×10^{19}. Assuming one second per move, this would take about five times longer than the current age of the universe! Interestingly, $2^{64} - 1$ is also the answer to the **wheat and chessboard problem**.

Tower of Hanoi

A game invented by Edouard **Lucas** and sold as a toy in 1883. Early versions of it carried the name "Prof. Claus" of the College of "Li-Sou-Stain," but these were quickly discovered to be anagrams for "Prof. Lucas" of the College of "Saint Louis." The game, in its usual form, consists of three pegs on one of which are eight disks, stacked from largest to smallest. The problem is to transfer the tower to either of the vacant pegs in the fewest possible moves, by moving one disk at a time and never placing

any disk on top of a smaller one. The minimum number of moves turns out to be $2^n - 1$, where n is the number of disks; this equals 255 in the case of eight disks. The original toy came with a description saying that it was a small version of the great **Tower of Brahma**.[257]

T-puzzle

A surprisingly difficult puzzle, given that there are only four pieces; it dates back to the start of the twentieth century. Photocopy and cut out the four pieces shown in the figure on the following page, and then try to arrange them to make the symmetric capital T. You are allowed to rotate the pieces as you wish and even turn them over, but they mustn't overlap in the final letter. In fact *two* different symmetric capital T letters can be made from the pieces. Also, two other symmetric shapes can be formed from the set, including an isosceles trapezoid. Can you find all of these?

trace

The sum of the terms along the **main diagonal** of a **matrix**.

T-puzzle The pieces of the T-puzzle. *Kadon Enterprises, Inc., www.puzzlegames.com*

tractrix

A curve, sometimes called the *trajectory curve* or *equitangential curve,* that is the answer to a question asked by the Frenchman Claude Perrault (1613–1688). Perrault is not a giant in the annals of mathematics; in fact, he trained as a doctor and gained a minor reputation as an architect and an anatomist before dying in unusual style as a result of an infection he caught while dissecting a camel. His greatest claim to fame, aside from his connection with

the tractrix, is that he was the brother of the author of "Cinderella" and "Puss-in-Boots."

In 1676, at about the time Gottfried **Leibniz** was doing groundbreaking work on the calculus, Perrault placed his pocket watch on the middle of a table, pulled the end of its chain along the edge of the table, and asked: What is the shape of the curve traced by the watch? The first known solution was given in a letter to a friend in 1693 by Christiaan **Huygens**, who also coined the name "tractrix" from the Latin *tractus* for something that is pulled along. (The corresponding German name is *hundkurve,* or "hound curve," which makes sense if you imagine the path a dog might follow on its leash as its master walks away.) The tractrix can also be found by taking the **involute** of a **catenary**. (Imagine a horizontal bar held at the vertex of the catenary and the point of contact marked as *P.* When the bar is rolled against the catenary without slipping, the path of *P* is a tractrix.) It is described by the parametric equations: $x = 1/\cosh(t)$, $y = t - \tanh(t)$. The **surface of revolution** of the tractrix is the **pseudosphere,** which is the classic model for **hyperbolic geometry** and one possible three-dimensional analog for the shape of the four-dimensional space-time in which we live.

tractrix © *Jan Wassenaar, www.2dcurves.com*

trajectory

(1) The path of a **projectile** or other moving body through space. (2) A curve that intersects all curves of a given family at the same angles; if the intersection is at a right angle, this is an *orthogonal trajectory*. (3) The path through **phase space** taken by a system.

transcendental number

A number that can't be expressed as the root of a **polynomial** equation with integer coefficients. Transcendental numbers are one of the two types of **irrational number**, the other being **algebraic numbers**. Their existence was proved in 1844 by the French mathematician Joseph Liouville (1809–1882). Although transcendentals make up the vast majority of real numbers, it is often surprisingly hard, and may even be impossible, to tell whether a certain number is transcendental or algebraic. For example it is known that both **pi** (π) and e are transcendental and also that at least one of $\pi + e$ and $\pi \times e$ must be transcendental, but it is not known which. It is also known that e^{π} is transcendental. This follows from the *Gelfond-Schneider theorem,* which says that if a and b are algebraic, a is not 0 or 1, and b is not rational, then a^b is transcendental. Using Euler's formula, $e^{i\pi} = -1$, and taking both sides to the power $-i$ gives $(-1)^{-i} = (e^{i\pi})^{-i} = e^{\pi}$. Since the theorem tells us that the left-hand side is transcendental, it follows that the right-hand side is too. (It also follows that $e \times \pi$ and $e + \pi$ are not both algebraic, because if they were, then the equation $x^2 + x(e + \pi) + e\pi = 0$ would have roots e and π, making both numbers algebraic.) But although it is known that e^{π} is transcendental, the status of e^e, π^e, and π^{π} remains uncertain.

transfinite number

Any of the infinite **ordinal numbers** first described by Georg **Cantor**.

transformation

In geometry, a change to an object due to a process such as **rotation**, **reflection**, enlargement, or **translation**. In algebra, a transformation is the action of a **function**; in other words, what happens when there is a **one-to-one** mapping between sets of objects.

translation

Any **transformation** that takes the form of a constant offset with no rotation or distortion.

transpose

An operation that flips a **matrix** about the **main diagonal**.

transposition cipher

A **cipher** that encodes a message by reordering the plaintext. The receiver decodes the message using the inverse transposition. A simple kind of transposition cipher writes the message in a rectangle by rows, for example:

Asimplekin
doftranspo
sitionciph
erwritesth
emessagein
toarectang
lebyrowsan
dreadsitou
tbycolumns

and reads it by columns:

Adsee tldts oirmo erbif tweab eymti rsrya cproi serdo lanta cosle ncegt wiuks iseas tmipp tinao nnohh ngnus.

This type of cipher can be made more difficult to crack by permuting the rows and columns. See also **substitution cipher**.

transversal

A line that cuts across parallel lines, intersecting each of them.

trapezoid

A **quadrilateral** with one pair of parallel sides; in Britain this shape is known as a **trapezium**. If the parallel sides are of length a and b, and h is the perpendicular distance between them, then the area of the trapezoid is given by $A = \frac{1}{2}(a + b)h$.

trapezium

The American definition of a trapezium is a **quadrilateral** with no parallel sides. The British definition is equivalent to that of a **trapezoid**.

traveling salesman problem

Given a number n of cities, along with the cost of travel between each pair of them, find the cheapest way of visiting all the cities and returning to the starting point. This is equivalent to finding the **Hamilton circuit** of minimum weight in a weighted **complete graph**. Mathematical problems related to the traveling salesman problem (TSP) were treated in the nineteenth century by William **Hamil-**

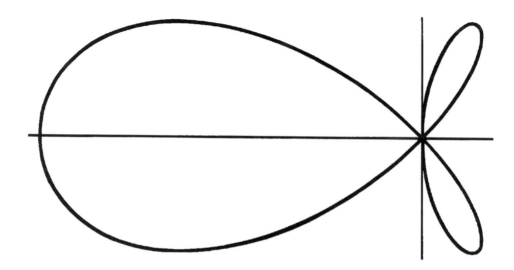

trefoil curve

ton, for example in his **Icosian game**, and by Thomas **Kirkman**. The general form of the TSP appears to be have been first studied by mathematicians in the 1930s, notably by Karl Menger in Vienna and Harvard, and later promoted by Hassler Whitney and Merrill Flood at Princeton. It has become a classic challenge to computer scientists seeking fast **algorithm**s to complex problems. An approximate solution to the TSP for 15,112 cities, towns, and villages in Germany was found in 2001 by Princeton researchers using 110 computer processors and the equivalent of more than five years computer time on a 2 GHz machine.

tree

A **graph** with the property that there is a unique path from any **vertex** to any other vertex traveling along the **edge**s.

trefoil curve

The plane curve given by the equation

$$x^4 + x^2y^2 + y^4 = x(x^2 - y^2).$$

triangle

A three-sided **polygon**. The sum of the interior angles of a triangle is always 180°, unless the triangle is drawn in a **non-Euclidean geometry**. Triangles can be classified either by their angles, as **acute**, **obtuse**, or **right**; or by their sides, as scalene (all different), **isosceles** (two the same), or **equilateral** (all equal).

Triangular Lodge

One of the few triangular buildings in England; it was built by Sir Thomas Tresham in about 1595 at Rushton, Northamptonshire. Tresham was a Catholic (spending some 15 years in prison because of this) and also a mystic numerologist. The whole design of the Lodge is based on the number **three**, which Tresham saw connected with his own surname and as an expression of his faith in the Christian Trinity. The Lodge's ground plan is a perfect **equilateral** triangle, each side 33 feet long–by tradition, the age of Christ at his death. The building has three floors, each floor has three windows, and each window is a three-fold **trefoil**. There are three gables and three gargoyles on each side. Even the central chimney is three-sided. The inscriptions all have 33 letters. Other buildings with a three-sided equilateral theme in Europe include a triangular castle at Gripsholm in Sweden and part of the Chateau de Chantilly in France, which is based on an equilateral plan of gigantic scale.

triangular number

Any number that can be represented by a triangular array of dots: 1, 3, 6, 10, ... The nth triangular number is $n(n + 1)/2$. Every integer is the sum of at most three triangular numbers. Every triangular number is a **perfect number**. If T is a triangular number, $8T + 1$ is a perfect square and $9T + 1$ is another triangular number. The square of the nth triangular number is equal to the sum of the first n cubes. Certain triangular numbers are also squares, but no triangular number can be a third, fourth, or fifth power, nor can one end in 2, 4, 7, or 9.

triangulation
A **tiling** of some object such as a **manifold** by simplices (see **simplex**).

tribar illusion
See **Penrose triangle**.

tricuspoid
See **deltoid**.

trident of Newton
A curve investigated and named "trident" by Isaac **Newton** as part of his systematic study of **cubic** equations. René **Descartes** also studied it and it is sometimes called the parabola of Descartes, although it isn't a parabola. It has the Cartesian equation

$$xy = cx^3 + dx^2 + ex + f.$$

trifolium
See **rose curve**.

trigonometric curve
Any of the curves produced when a **trigonometric function** is graphed.

trigonometric function
Any of the **functions** *sine* (sin), *cosine* (cos), *tangent* (tan), *secant* (sec), *cosecant* (cosec), or *cotangent* (cot), or their inverses, sin^{-1}, etc, which deal with certain proportions in right **triangles**. For example, the sine of an angle θ, sin θ, in a right triangle, is equal to the side opposite the angle divided by the hypotenuse (the longest side). Similarly cos is adjacent over hypotenuse and tan is opposite over adjacent. Sec, cosec, and cot are the multiplicative inverses of cos, sin, and tan, respectively: sec θ = 1/cos θ, and so forth. These are not the same as the inverse functions cos^{-1}, sin^{-1}, and tan^{-1}, which are also known as arccos, arcsin, and arctan. Graphs of trigonometric functions produce **trigonometric curves**.

trigonometry
The branch of **mathematics** that deals with the relationships between the sides and the angles of **triangles** and the calculations based on them, particularly the **trigonometric functions**. Sherlock Holmes relies on a little trigonometry to solve a 250-year-old mystery known as the Musgrave Ritual (in a short story of the same name)—an enigmatic series of clues that refers to the shadow of

an elm tree when the sun is just visible at the top of a nearby oak to point toward buried treasure. The great detective recalls to Watson his conversation with Reginald Musgrave:

> "Have you any old elms?" . . .
>
> "There used to be a very old one over yonder, but it was struck by lightening ten years ago, and we cut down the stump."
>
> "You can see where it used to be?"
>
> "Oh, yes." . . .
>
> "I suppose it is impossible to find out how high the elm was?"
>
> "I can give it to you at once. It was sixty-four feet. . . . When my old tutor used to give me an exercise in trigonometry, it always took the shape of measuring heights." . . .
>
> I went with Musgrave to his study and whittled myself this peg, to which I tied this long string with a knot at each yard. Then I took two lengths of a fishing-rod, which came to just six feet. . . . The sun was just grazing the top of the oak. I fastened the rod on end, marked out the direction of the shadow. . . . It was nine feet in length. Of course, the calculation was now a simple one. If a rod of six feet threw a shadow of nine, a tree of sixty-four feet would throw one of ninety-six. . . . I measured out the distance . . . and I thrust a peg into the spot.

trillion
In American and general usage, a million million—1,000,000,000,000 or 10^{12}. A European trillion is a million times larger than this, or 10^{18}. Counting one number every second, 24 hours a day, it would take 31,688 years to reach one (American) trillion. The first trillion-dollar lawsuit ($116 trillion) was filed in August 2002 by 600 family members against a company run by Osama bin Laden's family, Saudi Arabian princes, and Sudan. See also **large number**.

trimorphic number
See **automorphic number**.

trinomial
An algebraic expression consisting of three terms.

triomino
Also called a *tromino*, a three-square **polyomino**.

triple
A multiple of **three**. A *triple integral* is one in which the integrand is integrated three times. See also **Pythagorean triplet**.

trisecting an angle

Whereas **bisecting an angle** could hardly be simpler, splitting an angle in three equal parts with compass and straightedge alone is impossible, except in a few special cases such as when the angle happens to be 90°. Trisecting an arbitrary angle can be done if you cheat by using a measuring ruler instead of a plain straightedge, or even if you draw just two marks on the straightedge, but not if you play by the rules and the straightedge is completely blank. The Greeks put a huge effort into the problem but couldn't crack it. In fact, the question of whether trisection could ever be done in the general case remained open until 1837, when it was finally shown to be impossible by Pierre Wantzel, a 23-year-old French mathematician. Why is it impossible? Wantzel showed that the two problems of trisecting an angle and of solving a cubic equation are equivalent. Moreover, he showed that only a very few cubic equations can be solved using the straightedge-and-compass method. He thus deduced that most angles cannot be trisected.

trisector theorem

See **Morley's miracle.**

trisectrix

A general name for curves that can be used in **trisecting an angle.** The name "trisectrix," on its own, is often applied specifically to the **limaçon of Pascal.** Other famous trisectrix curves include the **Maclaurin trisectrix** and the **conchoid** of Nicomedes.

triskaidekaphobia

See **thirteen.**

trochoid

The curve formed by the path of a point on the extension of a radius of a circle as it rolls along a curve or line. It is also the curve formed by the path of a point on a perpendicular to a straight line as the straight line rolls along the convex side of a base curve. By the first definition, the trochoid is derived from the **cycloid**; by the second definition it is derived from the **involute.** See also **roulette.**

truel

A three-cornered gunfight or its logical equivalent. Imagine a truel between Arnie, Bullseye, and Clint, who are standing at the corners of an equilateral triangle. All know that Arnie's chance of hitting a target is 0.3 and Clint's is 0.5, while Bullseye never misses. They have to fire at their choice of target in the order Arnie then Bulls-

eye then Clint until only one man is left. A man who's been hit is out of the fight and can no longer be shot at. What should Arnie's strategy be?

Truels, like this one, have become a significant topic in **game theory** because they're analogs of various real-life situations, from rivalry among animals to competition between television networks. Small changes in the rules can lead to strikingly different, sometimes counterintuitive outcomes. Different firing rules are possible: sequential in fixed order (players fire one at a time in a predetermined, repeating sequence), sequential in random order (the first player to fire and each subsequent player is chosen at random from among the survivors), or simultaneous (all surviving players fire at the same time in every round). In certain truels, a participant is allowed to shoot at the ground rather than try to eliminate an opponent (an optimal strategy if the firing order is fixed and each player has only one bullet and is a perfect shot). If the first shooter misses on purpose, he eliminates himself as a threat, and the other two fight it out, leaving two survivors in the end. Any other course of action would lead to the first shooter's own demise, with only one survivor. Even if the players have an unlimited supply of bullets, the truel may still end with more than one survivor because no player wants to be the first to shoot. Indeed, under the fixed firing order rule, no player has an incentive to eliminate another player. Only in the case of simultaneous firing is there a chance that nobody will survive.

Most of the mathematical research on truels concerns the relationship between a player's marksmanship (probability of hitting a target) and his or her survival probability. It's possible to show, for example, that better marksmanship can hurt in many situations. In a sequential truel in which contendors aren't allowed to shoot in the air, a player maximizes his probability of survival by firing at the opponent against whom he'd less prefer to fight in a duel—regardless of what the other players do. If his shot misses, it makes no difference who the target was. If the shot hits the target, the shooter is better off because his opponent in the next duel is weaker. Thus, the first shooter fires at the opponent who's the better marksman. In general, depending on the marksmanship values, the survival probabilities of the truelists could end up in any order, including one that is the reverse order of shooting skill. Optimal play can be very sensitive to slight changes in the rules, such as the number of rounds of play allowed. On the other hand, some factors are fairly constant: the disadvantage of being the best marksman, the weakness of pacts, the possibility that an endless supply of ammunition may stabilize rather than undermine cooperation, and the deterrent effect of an indefinite number of rounds of play (which can prevent players from trying to get the last shot). Some of these findings are counterintuitive, even paradoxical. An understanding of them might well dampen the desire of aggressive players to score quick but temporary wins, rendering them more cautious. In particular, contemplating the consequences of a long, drawn-out conflict, truelists may come to realize that their own actions, while immediately beneficial, may trigger forces that ultimately lead to their own destruction.

truncate
To slice off a corner of a **polyhedron** around a **vertex**.

truncatable prime
A **prime number** n that remains a prime when digits are deleted from it one at a time. For example 410,256,793 is a truncatable prime because each number created by the removal of the digit underlined produces a new prime: 41$\underline{0}$,256,793; 41,256,7$\underline{9}$3; 4,1$\underline{2}$5,673; 4$\underline{1}$5,673; 45,67$\underline{3}$; 4,$\underline{5}$67; $\underline{4}$67; $\underline{6}$7; 7. It is conjectured that there are infinitely many of these primes. If the digits from a prime can be deleted only from the right, to leave a prime, then n is called a *right truncatable prime*. If they can be deleted only from the left, to leave a prime, then n is a *left truncatable prime*. The list of primes from which *any* digit can be deleted at each step to leave a prime is very short indeed, because it demands that each digit be a prime and also that no digit occurs twice. Only these numbers satisfy this requirement: 2, 3, 5, 7, 23, 37, 53, and 73.

Tschirnhaus's cubic
A curve with the Cartesian equation $3ay^2 = x(x - a)^2$.

Turing, Alan Mathison (1912–1954)
An English mathematician considered to be one of the fathers of modern digital computing. At an early age, Turing showed signs of the genius and eccentricity that became hallmarks of his adult personality. He taught himself to read in three weeks and made a habit of stopping at street corners to read the serial numbers of traffic lights. Later, he became a near-Olympic-class runner and ran long distances with an alarm clock tied to his waist to time himself.

At Cambridge, Turing studied under G. H. **Hardy** and got involved with problems that David **Hilbert** and Kurt **Gödel** had proposed to do with completeness and decidability in mathematics. In 1936, he introduced the idea of what became known as **Turing machines**—formal devices capable of solving any conceivable mathe-

Turing, Alan Mathison

matical problem that could be represented by an **algorithm**. However, the Turing machine was only a theoretical possibility at that time and not a working implementation. It would remain for later researchers to solve the various practical difficulties required to make the computer a reality. Turing also showed that there were mathematical problems that a Turing machine could never solve. One of these is the **halting problem**. While his proof was published after that of Alonzo **Church**, Turing's work is more accessible and intuitive. During World War II, Turing was a major player at Bletchley Park, near present-day Milton-Keynes (a town built after the war), in the successful efforts to crack the Nazi Enigma ciphers. While serving at Bletchley Park (1939–1944), he stayed at the Crown Inn, Shenley Brook End, and somewhere near here he buried two silver bars, carefully recording the site with respect to local landmarks. When he returned to recover them, the area had been rebuilt and all his landmarks were gone. Despite several attempts with metal detectors, he never recovered them and no one else is known to have found them. The Crown is now a private house and the area where he buried the bars is a housing estate.

Turing's interest in computing continued after the war, when he worked at the National Physical Laboratory on the development of a stored-program computer (the ACE or Automatic Computing Engine). In 1948 he moved to the University of Manchester, where the first stored-program digital computer ran later that year (see

Babbage, Charles for a photo of this machine). In 1950, in the article "Computing Machinery and Intelligence," Turing tackled the problem of **artificial intelligence**, and proposed an experiment now known as the **Turing test**.

In 1952 his lover helped a compatriot to break into Turing's house and commit larceny. Turing went to the police to report the crime. As a result of the police investigation, he was charged with homosexuality (formerly a crime), offered no defense, and was convicted. Following the well-publicized trial, he was given a choice between incarceration and libido-reducing hormone injections. He chose the latter, which lasted for a year and had side effects including the development of breasts during that period. In 1954 he died of poisoning after eating a cyanide-laced apple. Most (though not his mother) believed that his death was intentional, and the death was ruled a suicide. According to one urban legend the Apple company's logo is symbolic of this event: an apple with two bites (or possibly bytes) out of it and rainbow colors that code for homosexuality. See also **Church-Turing thesis**.[168]

Turing machine

An abstract model of computer execution and storage introduced in 1936 by Alan **Turing** to give a mathematically precise definition of **algorithm**. A Turing machine can be thought of as a black box that carries out a calculation of some kind on an input number. If the calculation reaches a conclusion, or halts, then an output number is returned. Otherwise, the machine theoretically carries on forever (see **halting problem**). There are an infinite number of Turing machines, as there are an infinite number of calculations that can be done with a finite list of rules. A Turing machine that can simulate any other Turing machine is called a *universal Turing machine* or a **universal computer**. The concept of Turing machines is still widely used in theoretical computer science, especially in **complexity theory** and the theory of computation.

Turing test

A proposed way of deciding if a machine has human-level intelligence. First described by Alan **Turing** in 1950, it goes like this: A human judge engages in a natural language conversation with other parties; if the judge can't reliably tell whether the other party is human or machine, then the machine is said to pass the test. It is assumed that both the human and the machine try to appear human. The origin of the test is a party game in which guests try to guess the gender of a person in another room by writing a series of questions on notes and reading the answers sent back. In Turing's original proposal, the human participants had to pretend to be the other gender, and the test was limited to a 5-minute conversation. These features are nowadays not considered to be essential and are generally not included in the specification of the Turing test. Turing proposed the test in order to replace the emotionally charged and, for him, meaningless question "Can machines think?" with a more well-defined one. Turing predicted that machines would eventually be able to pass the test. In fact, he estimated that by the year 2000, machines with 10^9 bits (about 119MB) of memory would be able to fool 30% of human judges during a 5-minute test. He also predicted that people would then no longer consider the phrase "thinking machine" contradictory.

It has been argued that the Turing test can't serve as a valid definition of **artificial intelligence** for at least two reasons: (1) A machine passing the Turing test might be able to simulate human conversational behavior, but this could be much weaker than true intelligence. The machine might just follow some cleverly devised rules. (2) A machine might well be intelligent without being able to chat like a human. Simple conversational programs, such as ELIZA, have fooled people into believing they are talking to another human being; however, such limited successes don't amount to passing the Turing test. Most obviously, the human party in the conversation has no reason to suspect he is talking to anything other than a human, whereas in a real Turing test the questioner is actively trying to determine the nature of the entity he is chatting with. The Loebner Prize is an annual competition to determine the best Turing test competitors. See also **Chinese room**.

twelve

A number heavily used for grouping things (inches, hours, 12-packs), partly because it can be divided evenly in several different ways (by 2, 3, 4, and 6) and partly because there are roughly 12 cycles of the Moon for every one of the Sun. The Latin *duodecim* (two + ten) for 12 forms the root of dodecagon (originally duodecagon), meaning a 12-sided shape, and duodenum, the first part of the intestine that is about 12 inches long. Contracted and modified over the years, *duodecim* became "dozen." Multiples of 12 have also been used by many cultures for various units and measures. A "shock" was 60 or five dozen (a dozen for each finger on one hand), and many cultures had a "great hundred" of 120 or 10 dozen (a dozen for each finger on both hands). The Romans used a fraction system based on 12 and the smallest part, an *uncil*, became our word for "ounce." The French emperor Charlemagne established a monetary system that had a base of 12 and 20, the remnants of which persist. Until 1970, the English pound sterling consisted of 20 shillings, and each shilling contained 12 pence. In 1944, The Duodecimal Society was formed in New York with the purpose of proposing a switch to base 12 for all scientific work. There are 12 signs of the zodiac and there were 12 apostles of Christ. Twelve is the smallest **abundant number**, a **Harshad number**, and a semiperfect number (because $12 = 1 + 2 + 3 + 6$; see **perfect number**). See also **gross**.

twelve-color map problem

If on a plane or sphere each country has at the most one colony, requiring the same color as its parent country, at most 12 different colors are needed to distinguish the political regions on a map. The problem of determining if this is true or not remains open. See also **four-color map problem**.

twenty

Many early cultures, including some in Europe; the Mayans of Central America; and the Ainu, the indigenous people of the Japanese islands, used a **base** of 20 for counting. The base 20 system was retained until about 1970 in the British monetary system, in which there were 20 shillings to the pound. A group of 20 is often called a **score**. Two of the five Platonic solids involve 20: the **icosahedron** has 20 triangular faces and the **dodecahedron** has 20 vertices. Twenty is a **Harshad number**, a semiperfect number (see **perfect number**), and a **practical number**. It is also a tetratrahedral number—the sum of consecutive triangular numbers ($1 + 3 + 6 + 10$).

twin primes

Pairs of **prime numbers** that differ by two, the first of which are 3 and 5, 5 and 7, 11 and 13, and 17 and 19. The largest example known, as of February 2003, is a pair of 51,090-digit primes discovered by Yves Gallot and Daniel Papp, with the value $33,218,925 \times 2^{169690} \pm 1$. Other than the first, all twin primes have the form $\{6n - 1, 6n + 1\}$; also, the integers n and $n + 2$ form twin primes if and only if $4[(n - 1)! + 1] \equiv -n \pmod{n(n + 2)}$. In 1919 the Norwegian mathematician Viggo Brun (1885–1978) showed that the sum of the reciprocals of the twin primes converges to a sum now known as *Brun's constant*:

$$(1/3 + 1/5) + (1/5 + 1/7) + (1/11 + 1/13) + (1/17 + 1/19) + \ldots$$

In 1994, by calculating the twin primes up to 10^{14} (and discovering the infamous Pentium bug in the process), Thomas Nicely of Lynchburg College estimated Brun's constant to be 1.902160578. According to the (unsolved) *twin-prime conjecture* there are infinitely many twin primes. The twin-prime conjecture generalizes to prime pairs that differ by any even number n, and generalizes even further to certain finite patterns of numbers separated by specified even differences. For example, the following triplets of primes all fit the pattern k, $k + 2$, and $k + 6$: 5, 7, and 11; 11, 13, and 17; 17, 19, and 23; 41, 43, and 47. It is believed that for any such pattern not outlawed by divisibility considerations there are infinitely many examples. (The pattern k, $k + 2$, and $k + 4$ has only one solution in primes, 3, 5, and 7, because any larger such triplet would contain a number divisible by 3.) Quartets of the form k, $k + 2$, $k + 6$, and $k + 8$ (the smallest example is 5, 7, 11, and 13) are thought to be infinite. For some patterns no example is known, or only one.

twins paradox
See **relativity theory**.

twisted cubic

A curve in three-dimensional space or projective space whose points are given by $(x(t), y(t), z(t))$ for a parameter t and where x, y, z are **polynomials** of at most degree 3.

two

The first even number and the only even **prime number**. The word comes from the Greek *dyo* and the Latin *duo* through the Old English *twa*. Early languages often had both feminine and masculine forms for two and so there are a lot of diverse roots related to "two-ness." Many "two" words use the Greek root *bi* such as biannual, binary, biscuit, and biceps. Others come from the Old English *twa*, such as between, twilight, twist, and twin. From *duo* we get dual, duet, dubious (of two minds), duplex (two layers), and double. The Latin *di* gives us diploma (two papers) and dihedral. The earlier Greek *dyo* produces dyad, composed of two parts. Two is the only positive real number that gives the same result when added to itself as when multiplied by itself. It is conjectured that 2 is the only even integer that cannot be written as the sum of two primes (see **Goldbach conjecture**) and it has recently been proven that 2 is the largest value of n for which the equation $x^n + y^n = z^n$ has nonzero integer solutions (see **Fermat's last theorem**). Two is the base of the **binary** number system.

two-dimensional world

Life in three dimensions is familiar and there is a huge body of literature on the **fourth dimension**. But what would a universe of just two dimensions be like? The first and the most charming book on the subject is Edwin Abbott's *Flatland: A Romance of Many Dimensions*

(1884).[1] This was followed by Charles **Hinton**'s lengthier *An Episode of Flatland* (1907)[272] in which the 2-d world is not a plane, as in Abbott's yarn, but the rim of a large circular world called Astria. Hinton was the first to explore in some depth what science and technology might be like in two dimensions; in fact, an earlier pamphlet of his called "A Plane World," (reprinted in *Scientific Romances* in 1884) may have helped inspire Abbott's novel. Hinton's speculations were taken much further by Alexander **Dewdney** in *Planiverse* (1984).[81]

Ulam, Stanislaw Marcin (1909–1984)

A Polish-born American mathematician and physicist who solved the problem of how to initiate fusion in the hydrogen bomb and also devised the **Monte Carlo method** of solving mathematical problems using statistical sampling. He first came to the United States in 1935 following an invitation from John **von Neumann**. One morning in 1946 an event happened that changed Ulam's life, as colleague Gian-Carlo Rota recalled:

> Ulam, a newly appointed professor at the University of Southern California, awoke to find himself unable to speak. A few hours later, he underwent a dangerous surgical operation after the diagnosis of encephalitis. . . . In time, however, some changes in his personality became obvious to those who knew him . . . [H]is ideas, which he spouted out at odd intervals, were fascinating beyond anything I have witnessed before or since. However, he seemed to

studiously avoid going into details . . . . [H]e came to lean on his unimpaired imagination for his ideas, and on . . . others for technical support . . . . A crippling technical weakness coupled with an extraordinarily creative imagination is the drama of Stan Ulam.

Ulam spiral

A remarkable geometric pattern accidentally found among the **prime numbers** by Stanislaw **Ulam**;[314] it is also known as the *Prime Spiral*. During a boring meeting one day in 1963, Ulam drew a square, marked the number 1 at the center, and then wrote the increasing whole numbers as a spiral that wound its way out to the edge of the paper. He then circled all the prime numbers and was immediately struck by how they tended to fall on diagonal lines radiating from the central 1. In Ulam's words the arrangement of primes "appears to exhibit a strongly nonrandom appearance." Ulam rushed home and

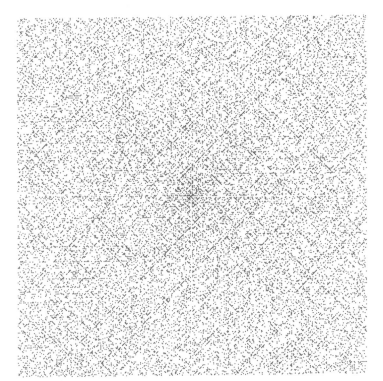

Ulam spiral   Patterns amid the primes.

expanded the spiral to cover a much larger portion of the number sequence. The strange pattern persisted. Primes had a tendency to occur in clusters and all clusters tended to make a beautiful image that couldn't be predicted. With the help of computers this pattern can now be explored almost indefinitely and it reveals a wonderfully rich combination of symmetry and surprise—very reminiscent of some **fractal**s.

The Ulam spiral should perhaps be known as the "Clarke spiral" in view of the fact that Arthur C. Clarke described the phenomenon in his novel *The City and the Stars* (1956, ch. 6, p. 54),[63] predating Ulam's discovery by several years. Clarke wrote:

Jeserac sat motionless within a whirlpool of numbers. The first thousand primes.... Jeserac was no mathematician, though sometimes he liked to believe he was. All he could do was to search among the infinite array of primes for special relationships and rules which more talented men might incorporate in general laws. He could find how numbers behaved, but he could not explain why. It was his pleasure to hack his way through the arithmetical jungle, and sometimes he discovered wonders that more skillful explorers had missed. He set up the matrix of all possible integers, and started his computer stringing the primes across its surface as beads might be arranged at the intersections of a mesh.

Ulam's conjecture
See **Collatz problem**.

uncountable set
A **set** of numbers that can't be put in a definite order from smallest to largest and so can't be counted. All uncountable sets are infinite, but not all infinite sets are uncountable. The best known uncountable set is the set of all **real number**s. By contrast the set of all **natural number**s, which represents the "smallest" type of **infinity**, is countable.

undulating number
An integer whose digits, in a given **base**, alternate—that is, one written in the form *ababab*..., where *a* and *b* are digits. For example, 434,343 and 101,010,101 are undulating numbers.

unduloid
A member of a family curves that is formed by films or liquid drops suspended between certain boundaries. Examples of unduloids are seen on a spider web when viewed through a microscope. They consist of blobs of viscous liquid that make up the sticky part of the web and

are mostly gathered into a lemon shape. The family of unduloids includes shapes ranging from very thin to almost spherical, depending on the diameter of the thread and the volume of liquid in the blob. The shape of the curve is a result of the equality of pressure throughout the blob, which means that the total **curvature** at all points on the surface must be the same. The total curvature is the sum as the curvatures in two planes at right angles, and so varies from one blob to the next. A common property of all unduloids, however, is that they have a constant nonzero main curvature.

unexpected hanging
A remarkable logical **paradox** that appears to have begun circulating by word of mouth in the 1940s, often in the form of a puzzle about a man condemned to be hanged. A judge, with a reputation for reliability, tells a prisoner on Saturday that he will be hung on one of the next seven days but that he will not know which day until he is informed on the morning of the execution. Back in his cell, the prisoner reasons that the judge must be wrong. The hanging cannot be left until Saturday, because the prisoner would certainly know, if this day dawned, that it was his last. But if Saturday is eliminated, the hanging cannot take place on Friday either, because if the prisoner survived Thursday he would know that the hanging was scheduled for the next day. By the same argument, Thursday can be crossed off, then Wednesday, and so forth, all the way back to Sunday. But with every other day ruled out for a possible unexpected hanging, the hangman cannot arrive on Sunday without the prisoner knowing in advance. Thus, the condemned man reasons, the sentence can't be carried out as the judged decreed. But then Wednesday morning comes around and, with it, the hangman—unexpectedly! The judge was right after all and something was awry with the prisoner's seemingly impeccable logic. But what? More than half a century of attack by numerous logicians and mathematicians has failed to produce a resolution that is universally accepted. The paradox seems to stem from the fact that whereas the judge knows beyond doubt that his words are true (the hanging will occur on a day unknown in advance to the prisoner), the prisoner does not have this same degree of certainty. Even if the prisoner is alive on Saturday morning, can he be certain that the hangman will arrive?[61, 260]

uniform polyhedron
A **polyhedron** in which each face is regular and each **vertex** is equivalently arranged. Uniform polyhedra include the **Platonic solid**s, the **Archimedean solid**s, the **prism**s and **antiprism**s, and the nonconvex uniform polyhedra (see **nonconvex uniform polyhedron**).

uniform polyhedron The great icosicosidodecahedron, a uniform polyhedron. *Robert Webb, www.software3d.com; created using Webb's Stella program*

unilluminable room problem
Imagine an L-shaped room in which Amy is standing near one corner holding a match. If Bob stands round the corner, he can see light from the match because a light ray can bounce off the two opposite walls. This is true wherever Bob stands in the room: the whole room is illuminated by the one match. Would this be true for a room of *any* shape, or is there at least one room that is *so* complicated that there's somewhere inside it that light from the match never reaches. This problem was first asked by Ernst Strauss in the 1950s. Nobody knew the answer until 1995, when George Tokarsky of the University of Alberta showed that the answer is "yes," there *is* a room that is not completely illuminable. His published floor plan showed a room with 26 sides–the smallest such room currently known. But a mystery remains. The room Tokarsky found contains one particular place where the match can be held which leaves part of the room dark. But if the match is moved slightly, the whole room is lit

up again. Is there a room so fiendishly complicated that *wherever* the match is held there are some places that its light can never reach? For the moment we remain in the dark.[320, 334]

unique number
The constant U_n that results if a number A_n consisting of n consecutive digits, in ascending order, is subtracted from the number A_n' obtained by reversing the digits of A_n. For example, a three-digit number 345, if subtracted from its reverse 543, yields a difference of 198. Any other three-digit number subtracted from its reverse gives the same difference. Thus $U_3 = 198$. Similarly for a number with four consecutive digits, the unique number $U_4 = 3,087$. The first ten unique numbers are: $U_1 = 0$, $U_2 = 9$, $U_3 = 198$, $U_4 = 3,087$, $U_5 = 41,976$, $U_6 = 530,865$, $U_7 = 6,419,754$, $U_8 = 75,308,643$, $U_9 = 864,197,532$, and $U_{10} = 9,753,086,421$. Unique numbers are related to **Kaprekar numbers**, K_n, by the formula

unilluminable room problem No single light source can light up every corner of a room with this shape.

$$U_n + U'_n = K_n + K'_n.$$

For example, when $n = 4$, $K_4 = 6,174$, $K_4' = 4,716$, $U_4 = 3,087$, $U_4' = 7,803$, and

$$3,087 + 7,803 = 10,890 = 6,174 + 4,716.$$

unimodal sequence
A **sequence** that first increases and then decreases.

unimodular matrix
A square **matrix** whose **determinant** is 1.

union
The **set** of all elements that belong to at least one of two or more given sets. It is denoted by the symbol ∪.

unit circle
A circle with radius 1.

unit cube
A cube with edge length 1. A *unit square* has a side length of 1.

unit fraction
A fraction whose **numerator** (number on top) is 1.

universal approximation
Having the ability to approximate any **function** to an arbitrary degree of accuracy. **Neural networks** are universal approximators.

universal computation
Capable of computing anything that can in principle be computed; being equivalent in computing power to a **Turing machine** or the **lambda calculus**.

universal computer
A computer that is capable of **universal computation**, which means that given a description of any other computer or program and some data, it can perfectly emulate this second computer or program. Strictly speaking, home PCs are not universal computers because they have only a finite amount of memory. However, in practice, this is usually ignored.

Universal Library
A library that contains not just one copy of every book that has ever been printed but one copy of every book that it is possible to print. A version of such a fantastic place is described by Jorge Luis **Borges** in his melancholic short story "Library of Babel" from *The Garden of Forking Paths* (1941). It begins: "The universe (which others call the Library) is composed of an indefinite, perhaps infinite number of hexagonal galleries." Each gallery is identical to all the others and contains 800 books identical in format. "[E]ach book contains four hundred ten pages; each page, forty lines, each line, approximately eighty black letters. . . ." There are 25 symbols—22 letters, the comma, the period, and the space. Because the library contains every possible combination of these

Universal Library   Inside the library that goes on forever.
Joseph Formoso

symbols it contains, in addition to vast tracts of gibberish, every truth, falsehood, idea, novel, thought, and description of events, past and future, that are possible. It contains, writes Borges,

All–the detailed history of the future, the autobiographies of the archangels, the faithful catalog of the Library, thousands and thousands of false catalogs, the proof of the falsity of those catalogs, a proof of the falsity of the *true* catalog, the gnostic gospel of Basilides, the commentary upon that gospel, the commentary on the commentary of that gospel, the true story of your death, the translation of every book into every language, the interpolations of every book into all books, the treatise Bede could have written (but did not) on the mythology of the Saxon people, the lost books of Tacitus.

The name "Universal Library" was first used by the German philosopher and science fiction writer Kurt Lasswitz (1848–1910) as the title of a short story published in 1901. He, in turn, borrowed the concept from the German psychologist **Theodor Fechner**. But to get to the root of speculation about listing all possible combinations of words and meanings, we have to go back much further, to Ramon Lully (1235–1315), a Spaniard who was a missionary and a mystic philosopher and who had an idea that later became known as Lully's Great Art. His idea was simply this: if one property of a thing is chosen, say the color of blood, and all the possibilities for that property are listed–blood is green, blood is yellow, etc.–then one of them must be true. One list alone might not be enough to point out the truth. However, other lists could be made that would eliminate some of

the possible colors. Done in the right way, the one and only true answer should emerge. Lully even tried to build a device that used a series of concentric rings to bring different combinations of words into alignment. Eventually, the idea came down to Fechner who ruminated on the idea of permuting all combinations of letters to express all possible statements and concepts. There are, however, two barriers to this dream of ultimate truth. The first is that there isn't enough matter or space in the universe to represent all the different ways that book-length sequences of letters can expressed. Second, even if there were, it would take an all-seeing, all-knowing intelligence to sort the rare grain of meaningful wheat from the vast quantities of vapid chaff. See also **monkeys and typewriters**.

universal set
The **set** that contains all elements capable of being accepted to the problem. Also known as the *universe*, as in the **universe of discourse**, it is usually denoted *U*.

universe of discourse
The part of the world under discussion; more precisely, the **set** of all objects presumed or hypothesized to exist for some specific purpose. Objects may be concrete (e.g., a specific carbon atom, Confucius, the Sun) or abstract (e.g., the number 2, the set of all integers, the concept of justice). Objects may be primitive or composite (e.g., a circuit that consists of many subcircuits). Objects may even be fictional (e.g., a unicorn, Sherlock Holmes). The universe of discourse is a familiar concept in **logic**, linguistics, and mathematics.

unknown
A quantity, denoted by a letter, that is to be found by solving one or more equations.

untouchable number
A number that is not the sum of the **aliquot part**s of any other number. The first few untouchable numbers are 2, 5, 52, 88.

up-arrow notation
See **Knuth's up-arrow notation**.

upside-down picture
A picture (or figure) that, when inverted, looks the same or changes into the picture of a different subject. Possibly the most remarkable examples of upside-down art were the cartoons drawn by Gustave Verbeek for the *Sunday New York Herald* in the early 1900s. The first part of the cartoon is read normally; then the newspaper is turned

upside-down picture An upside-down figure from China.
Sue & Brian Young/Mr. Puzzle Australia, www.mrpuzzle.com.au

through 180° and the second part read from the same boxes in reverse order. As if by magic, Little Lady Lovekins transforms into Old Man Muffaroo, a giant fish

becomes a giant bird, and a pouncing tiger turns into a tiger buried under a pile of stones.

Ussher, James (1581–1656)

An Irish clergyman, born in Fishamble Street, Dublin, who studied at Trinity College, Dublin, and later became a fellow there. He entered the Church and eventually became Archbishop of Armagh. In 1650 he published his famous assertion that the Creation had taken place "upon the entrance of the night preceding" Sunday, October 23, 4004 B.C.

utility function

The lesson of the **St. Petersburg paradox** is that people do not play games as if they are maximizing the expected monetary value they receive. However, certain rationality assumptions about the way people behave, known as the *von Neumann and Morganstern axioms,* imply that people *do* act as though they are maximizing *something.* This "something" is often referred to as a utility function.

vampire number

A natural number x that can be factorized as $y \times z$ in such a way that the number of occurrences of a particular digit in the representation of x in a given **base** (say 10) appears the same number of times in the representations in that same base of y and z together. For example, 2,187 is a vampire number since $2,187 = 21 \times 87$; similarly 136,948 is a vampire because $136,948 = 146 \times 938$. Vampire numbers are a whimsical idea that was introduced by Clifford Pickover in 1995.[253]

van der Pol, Balthazar (1889–1959)

A Dutch electrical engineer who began the modern experimental study of **dynamical system**s in the 1920s and '30s. Van der Pol discovered that electrical circuits employing vacuum tubes display stable oscillations, now called limit cycles, but that when these circuits are driven with a signal whose frequency is near that of the limit cycle, the periodic response shifts its frequency to that of the driving signal. The resulting waveform, however, can be quite complicated and contain a rich structure of harmonics and subharmonics. In 1927, van der Pol and his colleague van der Mark reported that an "irregular noise" was heard at certain driving frequencies between the natural entrainment frequencies. It's now clear that, without realizing it, they had described one of the first experimental instances of **chaos**.

vanishment puzzle

A **mechanical puzzle** in which the total area of a collection of pieces, or the number of items in a picture, appear to change following some manipulation. A well-known puzzle, first seen in 1868, involves an 8×8 square that is divided up into two triangles and two trapezoids. The pieces are reassembled into an oblong shape that measures 5×13. Where did the extra bit of area come from? The answer is that final shape is not perfectly rectangular but instead has a narrow gap that runs the length of one of the diagonals. Most famous of all vanishment puzzles is **Get off the Earth**.

variable

An **unknown** that has no fixed quantitative value.

vector

A quantity that is specified by a number, indicating size or "magnitude," and a direction; for example, "80 kilo-

meters per hour, heading due south." More generally, a vector is any element of a **vector space** and is also a type of **tensor**. A vector is usually shown in a graph or other diagram as an arrow whose length and direction represent the vector's magnitude and direction, respectively. In n-dimensional space, it is easy to deal with vectors algebraically in the form of n-tuples, which are ordered lists (one-dimensional arrays) of n components.

vector space

Also known as a *linear space,* the most fundamental concept in **linear algebra**. It is a generalization of the set of all geometric **vectors** and is used throughout modern mathematics. Like the concepts of **group**, **ring**, and **field**, that of a vector space is entirely abstract.

Venn diagram

A simple way of representing **sets** and subsets, which makes use of overlapping circles. Venn diagrams are named after the Englishman John Venn (1834–1923), a fellow of Cambridge University. Venn was a cleric in the Anglican Church, an authority on what was then called "moral science," the compiler of a massive index of all Cambridge alumni, and a rather mundane mathematician who worked in logic and probability theory. The diagrams he used for representing **syllogism**s appear to have been first called "Venn diagrams" by Clarence Irving in his book *A Survey of Symbolic Logic* in 1918. However, Venn was lucky to be so immortalized. Both Gottfried **Leibniz** and Leonhard **Euler** used very similar forms of representation many years earlier.

vertex

The point where two sides of a closed figure, or two sides of an angle, meet; otherwise known as a *corner.* A cube, for example, has eight vertices.

vertex figure

The **polygon** that appears if a **polyhedron** is **truncate**d at a **vertex**. The vertex figure of a cube, for example, is an equilateral triangle. To ensure consistency, the truncation may be done at the midpoints of the edges.

vertical-horizontal illusion

Vertical lines looks considerably longer than horizontal ones of the same length. For centuries, it has been known

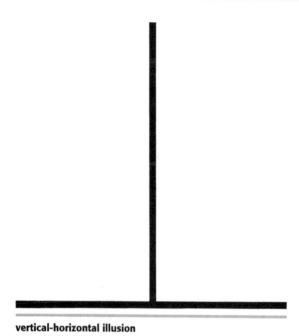

vertical-horizontal illusion

that the height of trees and buildings is perceived to be greater than the horizontal distance between them. This may be because we live in gravity. Rotate the book and you will see that the vertical line in the diagram appears to shrink in relation to what was originally the horizontal line. See also **distortion illusion**.

Vesica Piscis
Literally "fish's bladder," the almond shape formed when two identical circles overlap so that the outside edge of each just touches the center of the other. The Vesica Piscis appears as the Christian fish symbol and also frequently in medieval art and architecture. Most significantly, it provides the template for the pointed Gothic **arch**.

Vickrey auction
A silent auction in which the object being sold goes to the *second-highest bidder*. In a normal silent auction prospective buyers place their bids in sealed envelopes and the highest bid wins. This makes it risky for the seller because if the object is highly valuable and all buyers think they are the only ones who recognize this fact, they may offer much less than they think the object is actually worth. However, a Vickrey auction induces people to bid truthfully. Why? Because when all other bids are fixed and unknown to a given bidder, that bidder's optimal strategy is to bid what she thinks the object is worth. Suppose Alice places a bid for an antique vase. Let V be the amount that Alice thinks the vase is actually worth, and

B the bid that she actually makes. Let M be the maximum of all other bids. If M is more than V, then Alice should set her bid B less than or equal to V, so that she does not get the vase for more than she thinks it is worth. If M is less than V, then Alice should set $B = V$, because if she bids any less, she will not get the vase any cheaper, and she may lose it altogether.

vigesimal
Of, relating to, or based on the number 20; the term comes from the Latin *vigesimus* for "twentieth." Mayan arithmetic, which took account of all the toes as well as the fingers, used a vigesimal system. In place of the multiples of 10 used in the decimal system: 1; 10; 100; 1,000; 10,000, . . . , the Mayans dealt in multiples of 20: 1; 20; 400; 8,000; 160,000; . . . .

vinculum
The bar that is placed over repeating decimal fractions to indicate the portion of the pattern that repeats. In the original Latin, *vinculum* referred to a small cord for binding the hands or feet. The symbol was once used in the same way that parentheses and brackets are now used to bind together a group of numbers or symbols. Originally the line was placed under the items to be grouped. What today might be written $7(3x + 4)$ the early users of the vinculum would write $\underline{3x + 4}$ 7. Sometimes the horizontal fraction bar is called a vinculum as it binds the numerator and denominator into a single value.

Vinogradov's theorem
Every sufficiently large odd number can be expressed as the sum of three **prime number**s. The theorem was named after the Russian mathematician Ivan Vinogradov (1891–1983) who proved it in 1937. This is a partial solution of the **Goldbach conjecture** and is related to **Waring's conjecture**.

Viviani's curve
The space curve that marks the intersection of the cylinder $(x - a)^2 + y^2 = a^2$ and the sphere $x^2 + y^2 + z^2 = a^2$. It is given by the parametric equations:

$$x = a(1 + \cos t)$$
$$y = a \sin t$$
$$z = 2a \sin(\tfrac{1}{2} t)$$

Viviani's theorem
For a given point inside an equilateral **triangle**, the sum of the perpendicular distances from the point to the sides is equal to the height of the triangle. If the point is outside the triangle, the relationship still holds if one or more of the perpendiculars is treated as a negative value.

Volvox fractal   *Jos Leys, www.josleys.com*

Viviani's theorem generalizes to a regular *n*-sided **polygon**: the sum of the perpendicular distances from an interior point to the *n* sides being *n* times the **apothem** of the figure. The theorem is named for Vincenzo Viviani (1622–1703), a pupil of Galileo and Torricelli, who is also remembered for a reconstruction of a book on the **conic section**s of **Apollonius** and for finding a way of **trisecting an angle** through the use of an equilateral hyperbola.

volume

The measure of space occupied by a solid body.

Volvox fractal

A **fractal** that is similar in appearance to *Volvox*—a unicellular life form that lives in spherical colonies with thousands of members.

von Neumann, John (1903–1957)

Young man, in mathematics you don't understand things, you just get used to them.

A Hungarian-American mathematician who made important contributions to **set theory**, computer science, economics, and **quantum mechanics**. He received a Ph.D. in mathematics from the University of Budapest and later he worked at the Institute for Advanced Study in Princeton. *Theory of Games and Economic Behavior*,[230] which he coauthored with Oskar Morgenstern in 1944, is considered a seminal work in the field of **game theory**. Von Neumann devised the von Neumann architecture used in all modern computers and studied cellular automata (see **cellular automaton**) in order to construct the first examples of self-replicating automata, now known as **von Neumann machines**. Von Neumann had a mind of great

Voronoi diagram Top (left to right): Start from a grid of points. Draw lines that limit areas closest to the points of the grid. Bottom (left to right): Color the areas in some fashion. An infinite variety of patterns is possible. *Jos Leys, www.josleys.com*

ingenuity, nearly total recall of what he'd learned, immense arrogance, and a great love of jokes and humor.[259]

von Neumann machine

(1) A model for a computing machine that uses a single storage structure to hold both the set of instructions on how to perform the computation and the data required or generated by the computation. John **von Neumann** helped to create the model as an example of a general-purpose computing machine. By treating the instructions in the same way as the data, the machine could easily change the instructions. In other words the machine was reprogrammable. (2) A self-replicating machine. In principle, if a machine (e.g., an industrial

robot) could be given enough capability, raw material, and instructions, then that robot could make an exact physical copy of itself. The copy would need to be programmed in order to do anything. If both robots were reprogrammable, then the original robot could be instructed to copy its program to the new robot. Both robots would now have the capability of building copies of themselves. Since such a machine is capable of reproduction, it could arguably qualify as a simple form of life.

Voronoi diagram

Also known as a *Dirichlet tesselation,* a partitioning of space into cells, each of which consists of the points closer to one particular object than to any others. More specifically, in two dimensions a Voronoi diagram consists of breaking up a plane containing n points into n convex **polygon**s in such a way that each polygon contains exactly one point and every point in a given polygon is closer to its central point than to any other. Voronoi diagrams, their boundaries (known as *medial axes*), and their **dual**s (called *Delaunay triangulations*) have been reinvented, given different names, generalized, studied, and applied many times over in many different fields. Voronoi diagrams tend to be involved in situations where a space should be partitioned into "spheres of influence"; examples include models of crystal and cell growth and protein molecule volume analysis.

vulgar fraction
See **common fraction.**

walking in the rain problem

The question is whether you get wetter by walking or running a given distance through rain that is falling at a constant rate. An early appearance of this problem was in Bagley's *Paradox Pie* (1944). A simple answer is that moving faster is better. If the rain falls vertically and the density of water in the air is assumed constant, then, no matter what your speed, you'll sweep out the same volume and will always get the same amount of water hitting your front. However, running rather than walking will reduce the amount of water landing on your head.

Wallis, John (1616–1703)

The most influential English mathematician before Isaac **Newton** and an important contributor to the origins of **calculus**. He was a skilled linguist, was one of the first to proclaim in public Harvey's discovery of the circulation of the blood, and had an extraordinary memory for figures. His *Arithmetica Infinitorum* was described as "the most stimulating mathematical work so far published in England" and introduced the symbol ∞ for infinity (see also **aleph**). It contained the germs of the differential calculus, and it suggested to Newton, who was delighted by it, the **binomial theorem**.

Wallis formula

See **pi**.

wallpaper group

Also known as a *crystallographic group*, a distinct way to tile the plane that repeats indefinitely in two dimensions; that is, a collection of two-dimensional symmetric patterns on a plane surface, containing two nonparallel **translation**s (see **periodic tiling**). There are only 17 kinds of these patterns, known as isometries (see **isometry**), each uniquely identified by its translation and rotation symmetries, as discovered in the late nineteenth century by E. S. **Fedorov** and, independently, by the German A. M. Schoenflies and the Englishman William Barlow. Thirteen of the isometries include some kind of rotational symmetry, while four do not; twelve show rectangular symmetries, while five involve hexagonal symmetries. Every two-dimensional repetitive pattern in wallpaper, textiles, brickwork, or the arrangement of atoms in a plane of a crystal is just a minor variation on one of these 17 patterns.

Wari

See **Mancala**.

Waring's conjecture

A hypothesis given, without proof, by the English mathematician Edward Waring (1734–1798) in his *Meditationes algebraicae* (1770). It states that for every number k, there is another number s such that every natural number can be represented as the sum of s kth powers. For example, every natural number can be written as a sum of 4 squares, 9 cubes and so on. Waring's conjecture was first proven in full by David **Hilbert** in 1909.

weak inequality

An **inequality** that permits the equality case. For example, a is less than or equal to ($\leq$) b.

Weierstrass, Karl Wilhelm Theodor (1815–1897)

A German mathematician who is considered the father of modern **analysis**. Compelled by his father to study law, he instead spent four years at the University of Bonn, fencing, drinking, and reading math. He left under a cloud and

wallpaper group   A pattern made from one of the wallpaper groups. *Jos Leys, www.josleys.com*

wheat and chessboard problem **349**

ended up teaching in secondary schools for many years. In 1854 he published a paper, written 14 years earlier when he was fresh out of college, in *Crelle's Journal*, on *Abelian functions* which completed work that Niels **Abel** and Karl **Jacobi** had begun. Its importance was immediately recognized and Weierstrass was appointed a professor at the Royal Polytechnic School and a lecturer at the University of Berlin. He went on to give the first rigorous definitions of **limit**, **derivative**, differentiability, and **convergence**, and investigated under what conditions a power series will converge.

Weierstrass's nondifferentiable function

The earliest known example of a *pathological function*—a **function** that gives rise to a **pathological curve**. It was investigated by Karl **Weierstrass**, but had been first discovered by Bernhard **Riemann**, and is defined as:

$$f(x) = \sum_{k=1}^{\infty} \frac{\sin(\pi k^2 x)}{\pi k^2}$$

The Weierstrass function is everywhere continuous (see **continuity**) but nowhere differentiable; in other words, no **tangent** exists to its curve at any point. Constructed from an infinite sum of trigonometric functions, it is the densely nested oscillating structure that makes the definition of a tangent line impossible.

weighing puzzles

See **measuring and weighing puzzles**.

weird number

See **abundant number**.

Wessel, Caspar (1745–1818)

A Norwegian surveyor whose mathematical fame rests on a single paper, published in 1799, that gave the first geometrical interpretation of **complex numbers**. His priority in this discovery, however, went unrecognized for many years. Thus what should really be called a *Wessel diagram* is known instead as an **Argand diagram** after the man whose work on the same subject, published in 1806, first came to the attention of the mathematical world. Wessel's paper, by contrast, wasn't noticed by the mathematical community until 1895 when the Danish mathematician Sophus Juel drew attention to it and, in the same year, Sophus **Lie** republished Wessel's paper. Astonishingly, Wessel's remarkable work was not translated into English until 1999—its bicentenary!

Weyl, Hermann Klaus Hugo (1885–1955)

A German mathematician (known as "Peter" to his close friends) whose work involved symmetry theory, **topology**, and **non-Euclidean geometry**. Weyl studied under David Hilbert at Göttingen. Then, as a colleague of Albert Einstein at Zurich 1913, he got involved with **relativity theory** and came to believe (erroneously) that he had found a way to unite gravity and electromagnetism. From 1923 to 1938 he concentrated on **group** theory and made some important contributions to **quantum mechanics**. As the Nazi tide swept over Europe, Weyl came to the United States and spent the rest of his career at the Institute for Advanced Studies at Princeton. He said: "My work always tried to unite the truth with the beautiful, but when I had to choose one or the other, I usually chose the beautiful." See also **beauty and mathematics**.

wff

A well-formed formula.

what color was the bear?

A hunter walks one mile due south, then one mile due east, then one mile north and arrives back at his starting point. He shoots a bear. What color was it? Such a trip is possible if the hunter starts from one of the geographical poles, then circumnavigates the sides of a spherical triangle (see **elliptical geometry**). Since there are no bears in Antarctica, the trip is assumed to have taken place in the Arctic where there are polar bears and thus the answer is "white." Versions of this problem began to appear in the 1940s. A closer examination reveals that there are many more points on the globe, other than an exact pole, from which the hunter could have begun his trek. One example is any point (of which there are an infinite number) on a circle drawn at a distance of slightly more than $1 + \frac{1}{2}\pi$ mile (about 1.16 miles) from a pole—"slightly more" because of Earth's curvature. But this is not all. The hunter could also satisfy the conditions by starting at points closer to the pole, so that the walk east would carry him exactly twice around the pole, or three times, and so on. Of course, the bear would still be white (except that polar bears don't live that far north!).

wheat and chessboard problem

According to one myth, **chess** was invented by Grand Vizier Sissa Ben Dahir and then given to King Shirham of India. The king was so pleased that he offered his subject a great reward in gold, but the wily vizier said that he would be happy merely to have some wheat: one grain for the first square of the chessboard, two grains for the second square, four for the third, and so on, doubling each time. The king thought this was a very modest request, granted it, and asked for a bag of wheat to be brought in. However the bag was emptied by the twentieth square. The king asked for another bag, but then realized that this entire bag was needed for the next square. In 20 more squares, as many bags would have been exhausted as there

Whitney's umbrella *Takashi Nishimura*

were grains in the first bag! The number of grains on the sixty-fourth square would have been 2^{63}, and the total for the whole board $2^{64} - 1 = 18,446,744,073,709,551,615$. This is more wheat than in the entire whole world; in fact, it would fill a building 40 km long, 40 km wide, and 300 meters tall. See also **Tower of Brahma**.

Whitney's umbrella

A strange-looking, whimsically named geometrical object first studied by Hassler Whitney in the 1940s. It can be pictured as a self-intersecting rectangle in three dimensions. A *pinch point*, also known as a *Whitney singularity* or a *branch point* occurs at the top endpoint of the segment of self-intersection: every neighborhood of the pinch point intersects itself.

Wiener, Norbert (1894–1964)

An American mathematician who established the subject of **cybernetics**. As a precocious youngster, Wiener hopped from subject to subject at college, finally earning a Ph.D. in mathematics from Harvard at 19, before embarking on an even more erratic early career that took him into a variety of activities, including journalism. Having settled upon mathematical research, obtaining a post at the Massachusetts Institute of Technology in 1919, he nevertheless continued to range across fields, from random processes, including ergodic theory (concerned with the onset of chaos in a system), to integral equations, quantum mechanics, and potential theory. Wartime work that involved applying statistical methods to control and communication engineering, led to him extending these studies into control and communication in complex electronic systems and in animals, especially humans—the science of cybernetics.

Wiles, Andrew (1953–)

The English mathematician who, in 1994, finally proved **Fermat's last theorem**. Wiles studied at Oxford (B.A. 1974) and Cambridge (Ph.D. 1977) and has held posts at Cambridge, Oxford, and Princeton. From the mid-1980s his work was focused on proving a proposition known as the *Shimura-Taniyama conjecture*, since from this, it had been shown, Fermat's last theorem would follow. In 1993 he gave a series of lectures at Cambridge University ending on June 23, 1993. At the end of the final lecture he announced he had a proof of Fermat's last theorem. However, when the results were written up for publication, a subtle error was found. Wiles worked hard for about a year, helped in particular by a colleague, R. Taylor, and by September 19, 1994, having almost given up, he decided to have one last try. As he recalled, "suddenly, totally unexpectedly, I had this incredible revelation. It was the most important moment of my working life. . . . [I]t was so indescribably beautiful, it was so simple and so elegant . . . [that] I just stared in disbelief for twenty minutes, then during the day I walked round the department. I'd keep coming back to my desk to see it was still there—it was still there."

In 1994 Wiles was appointed Eugene Higgins Professor of Mathematics at Princeton. His paper that proves Fermat's last theorem is called "Modular elliptic curves and Fermat's Last Theorem" and appeared in the *Annals of Mathematics* in 1995.[296]

Wilson's theorem

Any number p is a **prime number** if, and only if, $(p - 1)!$ + 1 is divisible by p. We can easily check this for some small numbers: $(2 - 1)! + 1 = 2$, which is divisible by 2; $(5 - 1)! + 1 = 25$, which is divisible by 5; $(9 - 1)! + 1 = 40,321$, which is not divisible by 9. The theorem is named for Sir John Wilson (1741–1793), who came across it (but left no formal proof) while he was a student at Peterhouse College, Cambridge. Wilson went on to become a judge and seems to have done little else in mathematics. The theorem was first published and named after Wilson by Edward Waring (1734–1798) around 1770. However, it is now clear that the result was known to Gottfried **Leibniz** and perhaps, much earlier to Ibn al-Haytham (965–1040). The first known proof was provided by Joseph **Lagrange**.

winding number

The number of times a **closed** curve in the plane passes around a given point in the counterclockwise direction.

wine

The creation of a fine wine may be a complex business, but it is not mathematically intractable according to a Californian winemaker and businessman, Leo McCloskey, who runs a consulting firm. McCloskey's formula takes into account color, fragrance, and flavor and attempts to predict wine quality two to three years in advance. The formula stems from a comparison of key chemical characteristics for a range of wines with what experts made of the final product. Putting this information into a database, McCloskey has claimed, enables the math of wine flavor to be derived. More than 50,000 wines were analyzed for the database but, in each case, only a handful of the 400 to 500 constituent chemicals were considered crucial flags to future color, taste, and bouquet. The proportion of tannins and phenols is particularly crucial in determining a wine's future, thus reducing the number of essential flags for an accurate mathematical model to just 10 or 20. With this system, McCloskey believes, he can predict what a wine will be like before it is bottled, giving winemakers the opportunity to modify the flavor if necessary. However, whether any wine formula can effectively take into account the subjective impressions of the end user remains to be seen. See also **bottle sizes.**

Witch of Agnesi curve
See **Agnesi.**

word puzzles

1. What positive integer, when spelled out, has a Scrabble score equal to that integer?
2. What is the only English word that ends in "mt?"
3. Only four words in the English language end in "-dous." What are they?
4. What is the shortest complete sentence in the English language?
5. Name the only word (not including proper names) in the English language that has two "i"'s together.
6. Name a one-syllable word that becomes a three-syllable word by adding one letter to the end of it.
7. What word begins and ends with "und?"

Solutions begin on page 369.

word trivia

• The hardest word to define briefly is thought to be the word *mamihlapinatopai* from the Fuegian language spoken by the natives of the Andaman Islands. Its simplest definition is "two people looking at each other without speaking hoping that the other will offer to do something which both parties desire but neither are willing to do."

• No word in the English language rhymes with orange, silver, purple, or month.

• The only 15-letter word that can be written without repeating a letter is *uncopyrightable.*

• The combination "ough" can be pronounced in nine different ways. The following sentence contains them all: "A rough-coated, dough-faced, thoughtful ploughman strode through the streets of Scarborough; after falling into a slough, he coughed and hiccoughed."

• *Facetious* and *abstemious* contain all the vowels in the correct order, as does *arsenious,* meaning "containing arsenic."

• The longest word in the English language, according to the Oxford English Dictionary, is *pneumonoultramicroscopicsilicovolcanoconiosis.* The only other word with the same amount of letters is its plural: *pneumonoultramicroscopicsilicovolcanoconioses.* It means an infection of the lungs.

• The longest English word with one vowel is *strengths.*

• The "sixth sick sheik's sixth sheep's sick" is said to be the toughest tongue twister in the English language.

• Richard Millhouse Nixon was the first U.S. president whose name contains all the letters from the word *criminal.*

• In Scotland, a new game was invented. It was called Gentlemen Only Ladies Forbidden and thus the word GOLF entered the vocabulary.

• The longest English words that don't use any of vowels a, e, i, o, or u are "rhythm" and "syzygy."

• The verb *cleave* is one of many English words with two synonyms which are antonyms of each other: adhere and separate.

• There is a seven letter word in the English language that contains ten words without rearranging any of its letters, *therein:* the, there, he, in, rein, her, here, ere, therein, herein.

• The longest English word that consists entirely of consonants is *crwth,* which is from the fourteenth century and means crowd.

• The word *trivia* comes from the Latin *tri-* + *via,* and means "three streets." This is because in ancient times, at an intersection of three streets in Rome, they would have a type of kiosk where ancillary information was listed. You might be interested in it, you might not, hence they were bits of "trivia."

worldline

The path of an object through **space-time.** On a *Minkowski diagram,* in which the three dimensions of space are represented by the horizontal axis and time is represented by the vertical axis, worldlines appear as wiggly curves extending from the past into the future. *My*

Worldline is the title of the autobiography of George Gamow, the Ukrainian-American physicist.

Wundt illusion

A **distortion illusion** devised by the German "father of experimental psychology," and one-time assistant to the physicist Hermann von Helmholtz at Heidelberg, Wilhelm Wundt (1832–1920). In the figure, two horizontal lines are both straight, although they look as if they bow in at the middle. The distortion is induced by crooked lines on the background, as in **Orbison's illusion**. The simplest of all distortion illusions—the **vertical-horizontal illusion**—was also discovered by Wundt.

Wythoff's game

A variation on the game of **Nim** suggested by W. A. Wythoff in 1907. It is played with two heaps of counters in which a player may take any number from either heap or the same number from both. The player who takes the last counter wins.

x

The most frequently used symbol for an unknown in an expression or equation.

x-axis

The horizontal axis of a two-dimensional plot in **Cartesian coordinates**.

X-pentomino

A pentomino, a five-square **polyomino**, in the shape of the letter X.

Y

yard

A unit of distance equal to 3 feet (36 inches) or 0.9144 meters. The name originates with the old German word *gazdaz* for a staff or stick, which could be used for measurement. This changed to *gierd* in Old English and eventually to *yard*. The yard-arm of a sailing ship—a tapered spar used to support a square sail—harkens back to the earlier meaning. In France, the equivalent of the yard measure is called a "verge," from the Latin *virga* for a stick or rod.

y-axis

The vertical axis of a two-dimensional plot in **Cartesian coordinates**.

yin-yang symbol

See **Great Monad**.

yocto-/yotta-

The prefix yocto-, for 10^{-24}, derives from the Greek *oktakis* ("eight times"). The prefix yotta-, for 10^{24}, comes from the same source.

Z

Zeller's formula

A formula invented by the German clergyman Christian Zeller (1824–1899) for figuring out the day of the week of a given date, without the need of tables. Let

J be the century-number,

K the last two digits in the year,

e the residue, which remains when J is divided by 4,

m the number of the month,

q the day of the month,

h the number of the day of the week.

Then h is the remainder that results when

$$h = q + 26(m + 1)/10 + K + K/4 - 2e$$

is divided by 7. For the formula to work, January and February have to be taken as months 13 and 14 of the preceding year. For example, Frederick the Great was born on 24 January 1712, so $J = 17$, $e = 1$, $K = 11$ (not 12 because of the special way of numbering the months), $m = 13$, and $q = 24$. Plugging these values into the formula, we get

$$24 + 26(13 + 1)/10 + 11 + 11/4 - 2 \times 1$$
$$= 71 = 70 \times 1 + 1.$$

The remainder h is 1, so Frederick the Great was born on the first day of the week, Sunday.

Zenodorus (c. 180 B.C.)

The Greek mathematician and philosopher who proposed in *On Isometric Figures* that the circle has the maximum area of all isoperimetric figures in a plane, and that the sphere has the maximum volume of all bodies with equal surface.

Zeno's paradoxes

A series of **paradox**es posed by the philosopher Zeno of Elea (c. 490–c. 425 B.C.). Little is known about Zeno's life. He was born in Elea (now Lucania) in southern Italy and was a friend and student of Parmenides. None of his writings survive but he is known to have written a book, which **Proclus** says contained 40 paradoxes. Four of these, which all concern motion, have had a profound influence on the development of mathematics. They are described in Aristotle's great work *Physics* and are called the Dichotomy, Achilles (and the Tortoise), the Arrow, and the Stadium.

The Dichotomy argues that "there is no motion because that which is moved must arrive at the middle of its course before it arrives at the end." In order to traverse a line segment it's necessary to reach the halfway point, but this requires first reaching the quarter-way point, which first requires reaching the eighth-way point, and so on without end. Hence motion can never begin. This problem isn't alleviated by the well-known infinite sum $\frac{1}{2} + \frac{1}{4} + \frac{1}{8} + \ldots = 1$ because Zeno is effectively insisting that the sum be tackled in the reverse direction. What is the first term in such a series?

Zeno's paradox of Achilles is told by Aristotle in this way: "The slower when running will never be overtaken by the quicker; for that which is pursuing must first reach the point from which that which is fleeing started, so that the slower must necessarily always be some distance ahead." Thus, Achilles, however fast he runs, will never catch the plodding Tortoise who started first. And yet, of course, in the real world, faster things do overtake slower ones. So how is the paradox to be solved? The German set theorist Adolf Frankel (1891–1965) is one of many modern mathematicians (Bertrand Russell is another) who have pointed out that 2,000 years of attempted explanations have not cleared away the mysteries of Zeno's paradoxes: "Although they have often been dismissed as logical nonsense, many attempts have also been made to dispose of them by means of mathematical theorems, such as the theory of convergent series or the theory of sets. In the end, however, the difficulties inherent in his arguments have always come back with a vengeance, for the human mind is so constructed that it can look at a continuum in two ways that are not quite reconcilable."

zepto-/zetta-

The prefix zepto-, for 10^{-21}, derives from the Greek *heptakis* ("seven times"). The prefix zetta-, for 10^{21}, comes from the same source.

zero

The **integer**, denoted 0, which, when used as a counting number, indicates that no objects are present. It is the only integer that is neither negative nor positive. Zero is both a **number** and a **numeral**. The number zero is the size of the **empty set** but it is not the empty set itself, nor is it the same thing as **nothing**. The numeral or digit zero is used in positional **number system**s, where the position of a digit signifies its value, with successive positions having higher values, and the digit zero is used to skip a position. The earliest roots of the numeral zero stretch back 5,000 years to the Sumerians in Mesopotamia, who inserted a slanted double wedge between cuneiform characters for numbers, written positionally, to indicate a number's absence. The symbol changed over time as positional notation made its way to India, via the Greeks (in whose own culture zero made a late and only occasional appearance). Our word *zero* derives from the Hindi *sunya* for "void" or "emptiness," through the Arabic *sifr* (which also gives us *cipher*), and the Italian *zevero*. As a number in its own right, aside from its use as a position marker, zero took a much longer time to become established, and even now is not equal in status to other numbers: division by zero is not allowed.

zero divisors

Nonzero elements of a **ring** whose product is 0.

zero of a function

See **root**.

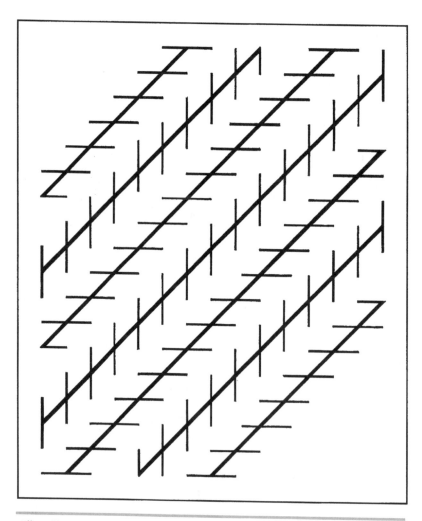

Zöllner illusion

zero-sum game
In **game theory**, a game in which a win for one player results in an equal but opposite loss for the other players.

zeta function
A **function** that has certain properties and is calculated as an infinite sum of negative powers. The most commonly encountered zeta function is the **Riemann zeta function**.

zigzag
A general word for a type of curve that consists of several straight lines joined at points. A zigzag usually goes alternately from side to side.

Zöllner illusion
A line **distortion illusion** first published by the astronomer Johann Zöllner in 1860. The diagonal lines, although parallel, appear not to be. The illusion was one of a series specifically designed to cause errors in optical equipment of that time. They did cause errors, and also great concern among scientists over the validity of all human observations. See also **Poggendorff illusion**.

zombie
A hypothetical being that behaves like us and may share our functional organization and even, perhaps, our neurophysiological makeup, but lacks consciousness or any form of subjective awareness. The concept is used in discussions of **artificial intelligence**.

zone
The portion of a **sphere** between two parallel planes.

zonohedron
A **polyhedron** in which the faces are all **parallelograms** or parallel-sided. The faces of a zonohedron can be grouped into *zones*—encircling bands of faces which share a common edge direction (and length).

References

1. Abbott, Edwin A. *Flatland: A Romance of Many Dimensions.* London: Seely and Co., 1884. Reprint, Mineola, N.Y.: Dover, 1992.

2. Abraham, R. M. *Diversions and Pastimes.* London: Constable & Co., 1933. Reprinted, 1964.

3. Ackermann, A. S. E. *Scientific Paradoxes and Problems.* London, 1925.

4. Adams, Douglas. *Life, the Universe and Everything.* New York: Harmony Books, 1982.

5. Adams, Douglas. *The Hitchhiker's Guide to the Galaxy.* New York: Ballantine, 1995.

6. Ahrens, W. *Mathematische Unterhaltungen und Spiele.* Leipzig, Germany: Teubner, 1910.

7. Ainley, Stephen. *Mathematical Puzzles.* London: Bell, 1977.

8. Allais, M. "Le comportement de l'homme rationnel devant le risque: Critique des postulats et axiomes de l'école américaine." *Econometrica,* 21: 503–546 (1953).

9. Amthor, A., and B. Krumbiegel. "Das Problema bovinum des Archimedes." *Zeitschrift für Mathematik und Physik,* 25: 121–171 (1880).

10. Andrews, William S. *Magic Squares and Cubes,* 2nd ed. Mineola, N.Y.: Dover, 1960.

11. Apéry, R. "Irrationalité de $\zeta(2)$ et $\zeta(3)$." *Astérisque,* 61: 11–13 (1979).

12. Appel, K., and W. Haken. "The Solution of the Four-Color Map Problem." *Scientific American,* 237: 108–121 (1977).

13. ApSimon, Hugh. *Mathematical Byways in Ayling, Beeling, and Ceiling.* New York: Oxford University Press, 1984.

14. Argand, R. *Essai sur une manière de représenter les quantités imaginaires dans les constructions géométriques.* Paris: Albert Blanchard, 1971. Reprint of the 2nd ed., published by G. J. Hoel in 1874. First published in 1806 in Paris.

15. Arno, Steven. "A Note on Perrin Pseudoprimes." *Mathematics of Computation,* 56(193): 371–376 (1991).

16. Arrow, Kenneth. *Social Choice and Individual Values.* New York: John Wiley & Sons, 1951.

17. Ascher, Marcia. *Ethnomathematics: A Multicultural View of Mathematical Ideas.* Pacific Grove, Calif.: Brooks/Cole Publishing Co., 1991.

18. Avedon, Elliot M., and Brian Sutton-Smith. *The Study of Games.* New York: John Wiley & Sons, 1971.

19. Averbach, Bonnie, and Orin Chein. *Mathematics: Problem Solving through Recreational Mathematics.* New York: W. H. Freeman, 1980.

20. Bagley, William A. *Puzzle Pie: A Unique Collection of Scientific Paradoxes, Posers and Oddities.* London: Vawser and Wiles, 1944.

21. Bailey, D. H., P. B. Borwein, and S. Plouffe. "On the Rapid Computation of Various Polylogarithmic Constants." *Mathematics of Computation,* 66(218): 903–913 (1997).

22. Bakst, Aaron. *Mathematical Puzzles and Pastimes.* New York: Van Nostrand, 1954.

23. Bakst, Aaron. *Mathematics: Its Magic and Mystery,* 3rd ed. Princeton, N.J.: Van Nostrand, 1967.

24. Ball, Walter William Rouse. *Mathematical Recreations and Problems.* London: Macmillan, 1892. (Ball, W. W. Rouse and H. S. M. Coxeter. *Mathematical Recreations and Essays,* 13th ed. Mineola, N.Y.: Dover, 1987.)

25. Banchoff, Thomas F. *Beyond the Third Dimension: Geometry, Computer Graphics, and Higher Dimensions.* New York: W. H. Freeman, 1990.

26. Barnsley, Michael. *Fractals Everywhere,* 2nd edition. San Francisco: Morgan Kaufmann, 1993.

27. Bayer, D., and P. Diaconis. "Trailing the Dovetail Shuffle to Its Lair." *Annals of Applied Probability,* 2(2): 294–313 (1992).

28. Beasley, John D. *The Ins and Outs of Peg Solitaire.* New York: Oxford University Press, 1985.

29. Beckmann, P. *A History of Pi,* 3rd ed. New York: Dorset Press, 1989.

30. Beiler, Albert H. *Recreations in the Theory of Numbers.* Mineola, N.Y.: Dover, 1964.

31. Bell, A. H. " 'Cattle Problem.' By Archimedes 251 B.C." *American Mathematical Monthly,* 2: 140 (1895).

32. Bell, E. T. *Men of Mathematics.* New York: Simon & Schuster, 1937.

33. Bell, Robbie, and Michael Cornelius. *Board Games Round the World: A Resource Book for Mathematical Investigations.* New York: Cambridge University Press, 1990.

34. Benford, F. "The Law of Anomalous Numbers."

Proceedings of the American Philosophical Society, 78: 551–572 (1938).

35. Bergholt, E. *Complete Handbook to the Game of Solitaire on the English Board of Thirty-Three Holes.* Routledge: London, 1920.

36. Berlekamp, Elwyn R., John Horton Conway, and Richard K. Guy. *Winning Ways for Your Mathematical Plays,* 2 vols. New York: Academic Press, 1982.

37. Berloquin, Pierre. *Le Jardin du Sphinx.* Paris: Dunod, 1981.

38. Berrondo, Marie. *Mathematical Games.* Englewood Cliffs, N.J.: Prentice Hall, 1983.

39. Besicovitch, A. S. "On Kakeya's Problem and a Similar One." *Mathematische Zeitschrift,* 27: 312–320, (1928).

40. Berry, Michael. "Quantum Physics on the Edge of Chaos." *New Scientist,* 116: 44–47 (1987).

41. Birtwistle, Claude. *The Calculator Puzzle Book.* New York: Bell Publishing Co., 1978.

42. Blatner, David. *The Joy of Pi.* New York: Walker, 1997.

43. Blocksma, Mary. *Reading the Numbers: A Survival Guide to the Measurements, Numbers, and Sizes Encountered in Everyday Life.* New York: Viking Penguin, 1989.

44. Blyth, Will. *Match-Stick Magic, Puzzles, Games and Conjuring Tricks.* London: C. A. Pearson Ltd., 1921.

45. Bogoshi, J., Naidoo, K., and Webb, J. "The Oldest Mathematical Artifact." *Mathematical Gazette,* 71: 458 (1987).

46. Bond, Raymond T. *Famous Stories of Code and Cipher.* New York: Rinehart & Co., 1947.

47. Bondi, Hermann. "The Rigid Body Dynamics of Unidirectional Spin." *Proceedings of the Royal Society,* A405: 265–274, 1986.

48. Bord, Janet. *Mazes and Labyrinths of the World.* London: Latimer, 1976.

49. Borel, Emile. *Algebre et Calcul des Probabilites,* vol. 184. Paris: Comptes Rendus Academie des Sciences, 1927.

50. Bouton, Charles L. "Nim, a Game with a Complete Mathematical Theory." *Annals of Mathematics,* Series 2. 3: 35–39, (1901–1902).

51. Brams, S. J., and A. D. Taylor. "An Envy-Free Cake Division Protocol." *American Mathematical Monthly,* 102: 9–19 (1995).

52. Brams, S. J., and A. D. Taylor. *Fair Division: From Cake-Cutting to Dispute Resolution.* New York: Cambridge University Press, 1996.

53. Brandreth, Gyles. *Numberplay.* New York: Rawson, 1984.

54. Branges, L., de, "A Proof of the Bieberbach Conjecture." *Acta Mathematica,* 154: 137–152 (1985).

55. Burger, Dionys. *Sphereland: A Fantasy about Curved Spaces and an Expanding Universe.* New York: Barnes & Noble, 1983.

56. Cadwell, J. H. *Topics in Recreational Mathematics.* New York: Cambridge University Press, 1966.

57. Carroll, Lewis. *Mathematical Recreations of Lewis Carroll.* 2 vols. Mineola, N.Y.: Dover, 1958.

58. Carroll, Lewis. *Symbolic Logic and the Game of Logic.* 2 vols. Mineola, N.Y.: Dover, 1958.

59. Carroll, Lewis. *Pillow Problems and A Tangles Tale.* New York: Dover, 1958.

60. Chaitin, G. J. "The Berry Paradox." *Complexity,* 1: 26–30 (1995).

61. Chow, T. Y. "The Surprise Examination or Unexpected Hanging Paradox." *American Mathematical Monthly,* 105: 41–51 (1998).

62. Church, Alonzo. *Introduction to Mathematical Logic.* Princeton, N.J.: Princeton University Press, 1956.

63. Clarke, Arthur C. *The City and the Stars.* New York: Harcourt, Brace, 1956.

64. Coffin, Stewart T. *The Puzzling World of Polyhedral Dissections.* New York: Oxford University Press, 1990.

65. Collins, A. Frederick. *Fun with Figures.* New York: D. Appleton & Co., 1928.

66. Conway, J. H. "Mrs. Perkins's Quilt." *Proceedings of the Cambridge Philosophical Society,* 60: 363–368, 1964.

67. Conway, John Horton. *On Numbers and Games.* New York: Academic Press, 1976.

68. Conway, John Horton, and Richard K. Guy. *The Book of Numbers.* New York: Springer-Verlag, 1996.

69. Conway, J. H., and S. P. Norton. "Monstrous Moonshine." *Bulletin London Mathematical Society,* 11: 308–339 (1979).

70. Conway, J. H., and N. J. A. Sloane. "The Monster Group and its 196884-Dimensional Space" and "A Monster Lie Algebra?" Chs. 29–30 in *Sphere Packings Lattices, and Groups,* 2nd ed. New York: Springer-Verlag, pp. 554–571, 1993.

71. Cooper, Necia Grant (ed.). *From Cardinals to Chaos: Reflections on the Life and Legacy of Stanislaw Ulam.* Cambridge: Cambridge University Press, 1989.

72. Coxeter, H. S. M. *Non-Euclidean Geometry.* Toronto: The University of Toronto Press, 1942.

73. Coxeter, H. S. M., M. S. Longuet-Higgins, and J. C. P. Miller. "Uniform Polyhedra." *Philosophical Transactions of the Royal Society of London,* Series, *A* 246: 401–450, 1954.

74. Coxeter, H. S. M. *Introduction to Geometry,* 2nd ed. New York: John Wiley & Sons, 1969.

75. Coxeter, H. S. M. *Regular Polytopes,* 3rd ed. New York: Dover, 1973.

76. Coxeter, H. S. M., P. DuVal, H. T. Flather, and J. F. Petrie. *The Fifty-Nine Icosahedra.* Toronto: University of

Toronto Press, 1938. Reprint, New York: Springer-Verlag, 1982.

77. Coxeter, H. S. M., M. Emmer, R. Penrose, and M. L. Teuber, (eds). *M. C. Escher: Art and Science.* New York: North-Holland, 1986.

78. Cromwell, Peter R. *Polyhedra.* Cambridge: Cambridge University Press, 1997.

79. Császár, Ákos. "A polyhedron without diagonals." *Acta Univ Szegendiensis, Acta Scientia Math,* 13: 140–142, 1949.

80. Cundy, H. Martyn, and A. P. Rollett. *Mathematical Models,* 3rd ed. Oxford: Clarendon Press, 1961; Diss, England: Tarquin Publications, 1981.

81. Dewdney, A. K. *The Planiverse: Computer Contact with a Two-dimensional World.* New York: Simon & Schuster, 1984.

82. Dickson, Leonard E. *History of the Theory of Numbers.* 3 vol. London: Chelsea Pub. Co., 1919–23. Reprint, Providence, R.I.: American Mathematical Society, 1999.

83. Domoryad, A. P. *Mathematical Games and Pastimes.* Elmsford, N.Y.: Pergamon Press, 1963.

84. Dötzel, Gunter. "A Function to End All Functions." *Algorithm: Recreational Programming 2.4,* 16–17 (1991).

85. Doxiadis, Apostolos K. *Uncle Petros and Goldbach's Conjecture.* New York: Bloomsbury, 2000.

86. Dresner, Simon (ed.). *Science World Book of Brain Teasers.* New York: Scholastic Book Services, 1962.

87. Dudeney, H. E. *The Canterbury Puzzles.* London: Nelson, 1907. Reprinted Mineola, N.Y.: Dover, 1958.

88. Dudeney, H. E. *Amusements in Mathematics.* New York: Dover, 1917. Reprinted Mineola, N.Y.: Dover, 1958.

89. Dudeney, Henry E. *The World's Best Word Puzzles.* London: Daily News, 1925.

90. Dudeney, H. E. *536 Puzzles and Curious Problems,* ed. by Martin Gardner. New York: Charles Scribner, 1967.

91. Edgeworth, R., Dalton, B. J., and Parnell, T. "The Pitch Drop Experiment." *European Journal of Physics,* 5: 198–200 (1984).

92. Eiss, Harry Edwin. *Dictionary of Mathematical Games, Puzzles, and Amusements.* Westport, Conn.: Greenwood Press, 1988.

93. Elffers, J. *Tangram—The Ancient Chinese Puzzle.* New York: Harry N. Abrams, 1979.

94. Escher, M. C. *The Graphic Work of M. C. Escher,* New York: Ballantine, 1971.

95. Ewing, John, and Kosniowski, Czes. *Puzzle It Out: Cubes, Groups and Puzzles.* New York: Cambridge University Press, 1982.

96. Fadiman, Clifton. *Fantasia Mathematica: Being a set of stories, together with a group of oddments and diversions, all drawn from the universe of mathematics.* New York: Simon and Schuster, 1958.

97. Fadiman, Clifton. *The Mathematical Magpie: Being more stories, mainly transcendental, plus subsets of essays, rhymes, music, anecdotes, epigrams and other prime oddments and diversions, rational or irrational, all derived from the infinite domain of mathematics.* New York: Simon and Schuster, 1962.

Federico, P. J. "The Melancholy Octahedron." *Mathematics Magazine,* 45: 30–36, 1972.

98. Filipiak, Anthony S. *100 Puzzles: How to Make and Solve Them.* New York: A. S. Barnes, 1942.

99. Filipiak, Anthony S. *Mathematical Puzzles and Other Brain Twisters.* New York: A. S. Barnes, 1964.

100. Fink, Thomas M., and Yong Mao Mao. "Designing Tie Knots Using Random Walks." *Nature,* 398: 31 (1999).

101. Fisher, Adrian. Arian Fisher Maze Design Web site: http://www.mazemaker.com.

102. Fisher, John (ed.). *The Magic of Lewis Carroll.* New York: Simon and Schuster, 1973.

103. Foster, James E. *Mathematics as Diversion.* Astoria, IL: Fulton County Press, 1978.

104. Fraser, James. "A new visual illusion of direction." *British Journal of Psychology,* 2: 307–337 (1908).

105. Frederickson, Greg N. *Dissections: Plane and Fancy.* Cambridge: Cambridge University Press, 1997.

106. Frey, Alexander H., Jr., and David Singmaster. *Handbook of Cubik Math.* Hillside, N.J.: Enslow, 1982.

107. Gamow, G., and M. Stern. *Puzzle-Math.* New York: Viking, 1958.

108. Gardiner, M. *Mathematical Puzzling.* New York: Oxford University Press, 1987.

109. Gardner, Martin. *Mathematics, Magic and Mystery.* Mineola, N.Y.: Dover, 1956.

110. Gardner, M. *Mathematical Puzzles and Diversions.* New York: Simon and Schuster, 1959.

111. Gardner, Martin, ed. *The Mathematical Puzzles of Sam Loyd,* 2 vols. Mineola, N.Y.: Dover, 1959.

112. Gardner, M. *The Second Scientific American Book of Mathematical Puzzles & Diversions.* New York: Simon and Schuster, 1961.

113. Gardner, Martin. "The Remarkable Lore of the Prime Numbers," *Scientific American,* March 1964.

114. Gardner, M. *New Mathematical Diversions from Scientific American.* New York: Simon and Schuster, 1966.

115. Gardner, M. "Mathematical Games." *Scientific American,* 237, 18–28, Nov. 1977.

116. Gardner, Martin. *Mathematical Magic Show: More Puzzles, Games, Diversions, Illusions and Other Mathematical Sleights-of-Mind from Scientific American.* New York: Vintage, 1978.

117. Gardner, Martin. *The Ambidextrous Universe.* New York: Charles Scribner's Sons, 1978.

118. Gardner, Martin. *Mathematical Circus: More Games, Puzzles, Paradoxes, and Other Mathematical*

Entertainments from Scientific American. New York: Alfred A. Knopf, 1979.

119. Gardner, Martin. *Science Fiction Puzzle Tales.* New York: Clarkson N. Potter, 1981.

120. Gardner, Martin. *Wheels, Life, and other Mathematical Amusements.* New York: W. H. Freeman, 1983.

121. Gardner, Martin. *Martin Gardner's Sixth Book of Mathematical Diversions from Scientific American.* Chicago: University of Chicago Press, 1983.

122. Gardner, Martin. *The Magic Numbers of Dr. Matrix.* Buffalo, N.Y.: Prometheus Books, 1985.

123. Gardner, M. *Knotted Doughnuts and Other Mathematical Entertainments.* New York: W. H. Freeman, 1986.

124. Gardner, Martin. *Riddles of the Sphinx And Other Mathematical Puzzle Tales.* Washington, DC: Mathematical Association of America, 1987.

125. Gardner, Martin. *Time Travel and Other Mathematical Bewilderments.* New York: W. H. Freeman, 1987.

126. Gardner, Martin. *Hexaflexagons and Other Mathematical Diversions.* Chicago: University of Chicago Press, 1988.

127. Gardner, Martin. *Penrose Tiles and Trapdoor Ciphers . . . and the Return of Dr. Matrix.* New York: W. H. Freeman, 1989.

128. Gardner, Martin. *Mathematical Carnival.* Washington, D.C.: Mathematical Association of America, 1989.

129. Gardner, Martin. *Mathematical Magic Show.* Washington, D.C.: Mathematical Association of America, 1990.

130. Gardner, M. *Fractal Music, Hypercards, and More Mathematical Recreations from Scientific American Magazine.* New York: W. H. Freeman, 1992.

131. Gardner, Martin (introduction and notes). *The Annotated Alice: Alice's Adventures in Wonderland and Through the Looking Glass.* New York: Random House, 1998.

132. Gasser, Ralph. "Solving Nine Men's Morris." *Computational Intelligence,* 12: 24–41 (1996).

133. Gerver, J. L. "On Moving a Sofa Around a Corner." *Geometriae Dedicata,* 42: 267–283 (1992).

134. Gillam, B. "Geometrical Illusions." *Scientific American,* 242: 102–111 (Jan. 1980).

135. Gleick, James. *Chaos: Making a New Science.* New York: Penguin, 1988.

136. Golomb, S. W. "Checker Boards and Polyominoes." *American Mathematical Monthly,* 61: 675–682 (1954).

137. Golomb, Solomon W. *Polyominoes: Puzzles, Patterns, Problems, and Packings.* New York: Charles Scribner's, 1965. Reprint, Princeton, N.J.: Princeton University Press, 1995.

138. Gomme, Alice Bertha. *Traditional Games of England, Scotland and Ireland.* 2 vols. London: Thames and Hudson, 1894–1898.

139. Gotze, Heinz. *Castel del Monte, Geometric Marvel of the Middle Ages.* New York: Prestel, 1998.

140. Greenberg, Marvin J. *Euclidean and non-Euclidean Geometries: Development and History.* San Francisco: W. H. Freeman, 1974.

141. Greenblatt, M. H. *Mathematical Entertainments.* London: George Allen and Unwin, 1968.

142. Gregory, R. L. "Analogue Transactions with Adelbert Ames." *Perception,* 16: 277–282 (1987).

143. Grunbaum, B., and G. C. Shephard. *Tilings and Patterns.* New York: W. H. Freeman, 1987.

144. Guy, Richard K. *Unsolved Problems in Number Theory,* 2nd ed. New York: Springer-Verlag, 1994.

145. Guy, Richard K., and Robert E. Woodrow (eds). *The Lighter Side of Mathematics: Proceedings of the Eugène Strens Memorial Conference on Recreational Mathematics and its History.* Washington, D.C.: Mathematical Association of America, 1994.

146. Haeckel, Ernst. Kunstformen der Natur (Art Forms in Nature). Leipzig, Wien: Bibliographisches Institut 1899–1904. Reprinted Mineola, N.Y.: Dover, 1974.

147. Haldeman-Julius, E. *Problems, Puzzles and Brain-Teasers.* Girard, KS: Appeal to Reason, 1937.

148. Hall, Trevor H. *Old Conjuring Books.* London: Duckworth, 1972.

149. Hankins, Thomas L. *Sir William Rowan Hamilton.* Baltimore: Johns Hopkins University Press, 1980.

150. Hardy, G. H. *Ramanujan: Twelve Lectures on Subjects Suggested by His Life and Work.* London: Cambridge University Press, 1940.

151. Hardy, G. H. *A Mathematician's Apology.* London: Cambridge University Press, 1940.

152. Hargrave, Catherine Perry. *A History of Playing Cards and a Bibliography of Cards and Gaming.* New York: Houghton Mifflin, 1930. Reprint, New York: Dover, 1966.

153. Harmer, G. P., and D. Abbott. "Losing Strategies Can Win by Parrondo's Paradox." *Nature,* 402: 864 (1999).

154. Hartley, Miles C. *Patterns of Polyhedra.* Ann Arbor, MI: Edwards Brothers, 1957.

155. Heafford, Philip. *The Math Entertainer.* New York: Vintage Press, 1983.

156. Heinlein, Robert. "And He Built a Crooked House," in Isaac Asimov (ed.) *Where Do We Go From Here?* New York: Doubleday, 1971.

157. Heinzelin, J., de, "Ishango." *Scientific American,* 206(6): 105–116 (Jun. 1962).

158. Held, Richard (ed.). *Image, Object and Illusion.* San Francisco: W. H. Freeman, 1974.

159. Henderson, Linda Dalrymple. *The Fourth Dimension and Non-Euclidean Geometry in Modern Art.* Princeton, N.J.: Princeton University Press, 1983.

160. Hendricks, John R. "Magic Tesseracts and N-Dimensional Magic Hypercubes." *Journal of Recreational Mathematics.* 6(3) (Summer 1973).

161. Hendricks, John R. "Black and White Vertices of a Hypercube." *Journal of Recreational Mathematics,* 11(4), 1978–1979.

162. Hendricks, John R. "Magic Hypercubes." Winnipeg, Canada: self-published pamphlet, 1988.

163. Hendricks, John R. "Images from the Fourth Dimension." Winnipeg, Canada: self-published pamphlet, 1993.

164. Hill, T. P. "The First Digit Phenomenon." *American Scientist,* 86: 358–363 (1998).

165. Hilton, Peter, and Jean Pedersen. *Build Your Own Polyhedra.* New York: Addison Wesley, 1988.

166. Hinton, Charles Howard. *A New Era of Thought.* London: Swan Sonnenschein, 1888.

167. Hinton, Charles Howard. *The Fourth Dimension.* London: Allen & Unwin, 1904.

168. Hodges, Andrew, and Douglas Hofstadter. *Alan Turing: The Enigma.* New York: Walker & Co., 2000.

169. Hoffman, Paul. *Archimedes' Revenge: The Joys and Perils of Mathematics.* New York: W. W. Norton, 1988.

170. Hoffman, Paul. *The Man Who Loved Only Numbers: The Story of Paul Erdos and the Search for Mathematical Truth.* New York: Hyperion, 1998.

171. Hoffmann, Professor (Angelo Louis). *Puzzles Old and New.* London: Frederick Warne, 1893. Reprint, L. Edward Hordern, 1993.

172. Hofstadter, Douglas R. *Gödel, Escher, Bach: An Eternal Golden Braid.* New York: Basic Books, 1999.

173. Hogben, Lancelot. *Mathematics for the Million.* New York: W. W. Norton, 1933.

174. Holden, Alan. *Shapes, Spaces and Symmetry.* New York: Columbia University Press, 1971. Reprint, Mineola, N.Y.: Dover, 1991.

175. Hordern, Edward. *Sliding Piece Puzzles.* New York: Oxford University Press, 1986.

176. Hovis, R. Corby, and Helge Kragh. "P. A. M. Dirac and the Beauty of Physics." *Scientific American,* 268(5): 104–109, May 1993.

177. Ingham, A. E. "On Two Conjectures in the Theory of Numbers." *American Journal of Mathematics,* 64: 313–319 (1942).

178. Ittelson, W. H., and F. P. Kilpatrick. "Experiments in Perception." *Scientific American,* 185(2), 50–55, August 1951.

179. Jackson, John. *Rational Amusement for Winter Evenings.* London: Longmans & Co., 1821.

180. Jayne, Caroline Furness. *String Figures and How to Make Them: A Study of Cat's-Cradle in Many Lands.* New York: Dover, 1975.

181. Jones, John Winter. *Riddles, Charades, and Conundrums.* London, 1821.

182. Jones, Samuel Isaac. *Mathematical Nuts for Lovers of Mathematics.* Nashville, 1932.

183. Kadesch, Robert R. *Math Menagerie.* New York: Harper & Row, 1970.

184. Kanigal, Robert. *The Man Who Knew Infinity: A Life of the Genius Ramanujan.* New York: Scribner, 1991.

185. Kapur, J. N. *The Fascinating World of Mathematics.* 3 vols. New Delhi, India: Mathematical Sciences Trust Society, 1989.

186. Kasner, Edward, and James R. Newman, *Mathematics and the Imagination.* New York: Simon and Schuster, 1940.

187. King, T. *The Best 100 Puzzles Solved and Answered.* Slough, England: W. Foulsham & Co., 1927.

188. Kirkman, T. P. "On a Problem in Combinatorics." *Cambridge and Dublin Math. J.,* 2: 191–204 (1847).

189. Klarner, David A., ed. *The Mathematical Gardner.* Boston: Prindle, Weber and Schmidt, 1981.

190. Klarner, David A., ed. *Mathematical Recreations: A Collection in Honor of Martin Gardner.* New York: Dover, 1998.

191. Knuth, Donald. *Surreal Numbers: How Two Ex-Students Turned On to Pure Mathematics and Found Total Happiness.* Reading, Mass.: Addison-Wesley, 1974.

192. Körner, T. W. *The Pleasures of Counting.* Cambridge: Cambridge University Press, 1997.

193. Kraitchik, Maurice. *Mathematical Recreations.* New York: Norton, 1942.

194. Krasnikov, S. V. "Causality Violations and Paradoxes." *Physical Review D,* 55(6): 3427–30 (1997).

195. Langley, E. M. "Problem 644." *Mathematical Gazette,* 11: 173, 1922.

196. Langman, H. *Play Mathematics.* New York: Hafner, 1962.

197. Lausmann, Raymond F. *Fun With Figures.* New York: McGraw-Hill, 1965.

198. Lavine, S. *Understanding the Infinite.* Cambridge, Mass.: Harvard University Press, 1994.

199. Lehman, R. S. "On Liouville's Function." *Mathematics of Computation,* 14: 311–320, 1960.

200. Licks, H. E. *Recreations in Mathematics.* New York: D. Van Nostrand, 1921.

201. Lindgren, Harry. *Geometric Dissections.* New York: Van Nostrand Reinhold, 1964.

202. Lindgren, Harry. *Recreational Problems in Geometrical Dissections and How to Solve Them.* New York: Dover Publications, 1972.

203. Lines, Malcolm. *Think of a Number.* New York: Adam Hilger, 1990.

204. Littlewood, J. E. *A Mathematician's Miscellany.* London: Methuen and Co., 1953.

205. Livio, Mario. *The Golden Ratio: The Story of Phi, the World's Most Astonishing Number.* New York: Broadway Books, 2002.

206. Locher J. L., ed. *M. C. Esher: His Life and Complete Graphic Work.* New York: Abradale Press, 1982.

207. Loyd, Sam. *Sam Loyd's Cyclopedia of 5000 Puzzles, Tricks, and Conundrums.* 1914. Reprint, New York: Pinnacle, 1976.

208. Loyd, Sam. *Sam Loyd's Picture Puzzles with Answers.* Brooklyn: S. Loyd, 1924. Reprint, Ontario, Canada: Algrove, 2000.

209. Lucas, Edouard. *Récréations Mathématiques,* 4 vols. Paris: Gauthier-Villars et fils, 1881–1894. Reprinted by Paris: A. Blanchard, 1960–1974.

210. MacMahon, Alexander: *New Mathematical Pastimes.* Cambridge: Cambridge University Press, 1921.

211. Madachy, Joseph S. *Mathematics on Vacation.* New York: Charles Scribner's, 1966. Republished as *Madachy's Mathematical Recreations.* New York: Dover, 1979.

212. Manning, H. P. *Geometry of Four Dimensions.* New York: Dover, 1956.

213. Manning, H. P. *The Fourth Dimension Simply Explained.* New York: Dover, 1960.

214. Maor, E. *To Infinity and Beyond: A Cultural History of the Infinite.* Boston, Mass.: Birkhäuser, 1987.

215. Maor, E. *e: The Story of a Number.* Princeton, N.J.: Princeton University Press, 1994.

216. Martin, G. "Absolutely abnormal numbers." *American Mathematical Monthly,* 108: 746–754, 2001.

217. Matijasevic, Yu. V. "Solution of the Tenth Problem of Hilbert." *Matematikai Lapok,* 21: 83–87 (1970).

218. Matijasevic, Yu. V. *Hilbert's Tenth Problem.* Cambridge, Mass.: MIT Press, 1993.

219. Matthews, W. H. *Mazes & Labyrinths: A General Account of Their History and Development.* London: Longmans, Green and Co., 1922. Reprint, New York: Dover, 1970.

220. Maxwell, E. A. *Fallacies in Mathematics.* New York: Cambridge University Press, 1959.

221. Meeus, J., and P. J. Torbijn. *Polycubes.* Paris: Cedic, 1977.

222. Melzak, Z. A. *Companion to Concrete Mathematics.* 2 vols. New York: John Wiley, 1973, 1976.

223. Mercer, J. W., et al. "Solutions to Langley's Adventitious Angles Problem." *Mathematical Gazette,* 11: 321–323, 1923.

224. Mosteller, F. *Fifty Challenging Problems in Probability with Solutions,* Reading, Mass.: Addison-Wesley, 1965, #47; "Mathematical Plums," edited by Ross Honsberger, pp. 104–110.

225. Mott-Smith, Geoffrey. *Mathematical Puzzles for Beginners and Enthusiasts,* 2nd rev. ed. New York: Dover, 1954.

226. Moyer, Ann E. *The Philosophers' Game: Rithmomachia in Medieval and Renaissance Europe.* Ann Arbor, Mich.: University of Michigan Press, 2001.

227. Murray, H. J. R. *History of Chess.* Oxford: Clarendon Press, 1913.

228. Murray, H. J. R. *A History of Board-Games Other than Chess.* Oxford: Clarendon Press, 1952.

229. Nasar, Sylvia. *A Beautiful Mind: A Biography of John Forbes Nash, Jr.* New York: Simon & Schuster, 1998.

230. Neumann, J. von, and Morgenstern, O. *Theory of Games and Economic Behavior.* New York: Wiley, 1964.

231. Neville, E. H. *The Fourth Dimension.* Cambridge: Cambridge University Press, 1921.

232. Newcomb, S. "Note on the Frequency of the Use of Digits in Natural Numbers." *American Journal of Mathematics,* 4: 39–40, 1881.

233. Newing, Angela. "The Life and Work of H. E. Dudeney." *Mathematical Spectrum,* 21: 37–44 (1988–1989).

234. Northrop, Eugene P. *Riddles in Mathematics. A Book of Paradoxes.* Princeton, N.J.: Van Nostrand, 1944.

235. O'Beirne, T. H. *Puzzles and Paradoxes: Fascinating Excursions in Recreational Mathematics.* Mineola, N.Y.: Dover, 1984.

236. Ogilvy, C. Stanley. *Tomorrow's Math: Unsolved Problems for the Amateur,* 2nd ed. New York: Oxford University Press, 1972.

237. Olivastro, Dominic. *Ancient Puzzles.* New York: Bantam, 1993.

238. Panofsky, Erwin. *The Life and Art of Albrecht Durer,* Princeton, N.J.: Princeton University Press, 1955.

239. Pardon, G. F. *Parlour Pastimes: A Repertoire of Acting Charades, Fire-Side Games, Enigmas, Riddles, Etc.* London, 1868.

240. Pennick, Nigel. *Mazes and Labyrinths.* London: Robert Hale, 1990.

241. Penrose, L. S., and R. Penrose. "Impossible Objects: A Special Type of Illusion," *British Journal of Psychology,* 49: 31, 1958.

242. Penrose Roger. *The Emperor's New Mind.* New York: Oxford University Press, 1989.

243. Péter, Rózsa. *Playing with Infinity: Mathematical Explorations and Excursions.* Mineola, N.Y.: Dover, 1976.

244. Peterson, Ivars. *The Mathematical Tourist: Snapshots of Modern Mathematics.* New York: W. H. Freeman, 1988.

245. Peterson, Ivars. *Islands of Truth: A Mathematical Mystery Cruise.* New York: W. H. Freeman, 1990.

246. Peterson, Ivars. *The Jungles of Randomness: A Mathematical Safari.* New York: Wiley, 1997.

247. Peterson, Ivars. *The Mathematical Tourist: New and Updated Snapshots of Modern Mathematics.* New York: W. H. Freeman, 1998.

248. Peterson, Ivars. *Mathematical Treks: From Surreal Numbers to Magic Circles.* Washington, D.C.: Mathematical Association of America, 2001.

249. Phillips, Hubert. *Journey to Nowhere: A Discursive Autobiography.* London: Macgibbon & Kee, 1960.

250. Phillips, Hubert. *My Best Puzzles in Mathematics.* Mineola, N.Y.: Dover, 1961.

251. Phillips, Hubert. *My Best Puzzles in Logic and Reasoning.* Mineola, N.Y.: Dover, 1961.

252. Picciotto, Henri. *Pentomino Activities.* Mountain View, Calif.: Creative Publications, 1986.

253. Pickover, Clifford. *Keys to Infinity.* New York: W. H. Freeman, 1995.

254. Pickover, Clifford. *Surfing Through Hyperspace.* New York: Oxford University Press, 1999.

255. Pickover, Clifford A. *Wonders of Numbers: Adventures in Mathematics, Mind, and Meaning.* Oxford: Oxford University Press, 2001.

256. Pickover, C. A. *The Zen of Magic Squares, Circles, and Stars: An Exhibition of Surprising Structures across Dimensions.* Princeton, N.J.: Princeton University Press, 2002.

257. Poole, D. G. "The Towers and Triangles of Professor Claus (or, Pascal Knows Hanoi)." *Mathematics Magazine,* 67: 323–344 (1994).

258. Poundstone, William. *The Recursive Universe.* New York: William Morrow, 1984.

259. Poundstone, William. *Prisoner's Dilemma: John Von Neumann, Game Theory and the Puzzle of the Bomb.* New York: Doubleday, 1992.

260. Quine, W. V. O. "On a So-Called Paradox." *Mind,* 62: 65–67 (1953).

261. Rabinowitz, Stanley. *Index to Mathematical Problems 1980–1984.* Westford, Mass.: Mathpro Press, 1992.

262. Rademacher, Hans, and Otto Toeplitz. *The Enjoyment of Mathematics: Selections from Mathematics for the Amateur.* Princeton, N.J.: Princeton University Press, 1957.

263. Raimi, R. A. "The Peculiar Distribution of First Digits." *Scientific American,* 221: 109–119 (Dec. 1969).

264. Ransom, William R. *One Hundred Mathematical Curiosities.* Portland, Maine: Weston Walch, 1955.

265. Read, Ronald C. *Tangrams: 330 Puzzles.* Mineola, N.Y.: Dover, 1965.

266. Ribenboim, P. *Catalan's Conjecture.* Boston, Mass.: Academic Press, 1994.

267. Robbin, Tony. *Fourfield: Computers, Art, and the Fourth Dimension.* Boston: Little, Brown, 1992.

268. Robertson, J., and Webb, W. *Cake-cutting Algorithms.* Natick, Mass.: A. K. Peters, 1998.

269. Rossi, Hugo. "Mathematics Is an Edifice, Not a Toolbox." *Notices of the AMS,* 43(10) (Oct. 1996).

270. Rubik, Erno, et al. *Rubik's Cubic Compendium.* New York: Oxford University Press, 1987.

271. Rucker, Rudolf v.B. *Geometry, Relativity, and the Fourth Dimension.* New York: Dover, 1977.

272. Rucker, R. (ed). *Speculations on the Fourth Dimension: Selected Writings of C. H. Hinton.* New York: Dover, 1980.

273. Rucker, Rudy. *The Fourth Dimension.* Boston: Houghton Mifflin, 1984.

274. Rucker, Rudy. *Mind Tools.* Boston: Houghton Mifflin, 1987.

275. Rucker, Rudy. *Infinity and the Mind.* Princeton, N.J.: Princeton University Press, 1995.

276. Russell, Bertrand. *Bertrand Russell Autobiography.* New York: Routledge, 1992.

277. Saaty, T. L., and P. C. Kainen. *The Four-Color Problem: Assaults and Conquest.* New York: Dover, 1986.

278. Sallows, L. C. F. "Alphamagic squares." *Abacus* 4 (No. 1): 28–45, 1986.

279. Sallows, L. C. F. "Alphamagic squares, part II". *Abacus* 4 (No. 2): 20–29, 43, 1987.

280. Sallows, L. C. F. "Alphamagic squares." In *The Lighter Side of Mathematics: Proceedings of the Eugène Strens Memorial Conference on Recreational Mathematics and Its History,* R. K. Guy and R. E. Woodrow, eds. Washington, D.C.: Mathematical Association of America, 1994.

281. Sanford, Vera. *A Short History of Mathematics.* Boston, Mass.: Houghton Mifflin, 1930.

282. Scarne, John. *Scarne's New Complete Guide to Gambling.* New York: Simon and Schuster, 1986.

283. Schaaf, William Leonard. *A Bibliography of Recreational Mathematics.* 4 vols. Washington, D.C.: National Council of Teachers of Mathematics, 1955–78.

284. Schattschneider, Doris. *M. C. Escher: Visions of Symmetry.* New York: W. H. Freeman, 1990.

285. Scheick, William J. "The Fourth Dimension in Wells's Novels of the 1920s." *Criticism,* 20: 167–190, 1978.

286. Schrauf, M., B. Lingelbach, E. Lingelbach, and E. R. Wist. "The Hermann Grid and the Scintillation Effect." *Perception,* 24, suppl. A: 88–89, 1995.

287. Schreiber, P. "A New Hypothesis on Dürer's Enigmatic Polyhedron in His Copper Engraving 'Melencholia I'," *Historia Mathematica,* 26: 369–377, 1999.

288. Schubert, Hermann. *Mathematical Essays and Recreations.* Thomas J. McCormack, trans. Chicago: Open Court, 1899.

289. Schubert, Hermann. *Mathematische Mussestunden.* Walter de Gruyter, 1940. First published 1900? (German).

290. Schubert, Hermann. *Mathematische Mussestunden II.* Berlin: G. J. Goshen'sche, 1909.

291. Schuh, Fred. *The Master Book of Mathematical Recreations.* Mineola, N.Y.: Dover, 1968.

292. Searle, John R. "Minds, Brains and Programs" in *The Brain and Behavioral Sciences,* vol. 3. Cambridge: Cambridge University Press, 1980.

293. Senechal, M. *Quasicrystals and Geometry.* New York: Cambridge University Press, 1995.

294. Shepherd, Walter. *For Amazement Only.* New York: Penguin, 1942. Reprinted as *Mazes and Labyrinths: A Book of Puzzles.* New York: Dover, 1961.

295. Simon, William. *Mathematical Magic.* New York: Charles Scribner's, 1964.

296. Singh, Simon. *Fermat's Enigma: The Epic Quest to Solve the World's Greatest Mathematical Problem.* New York: Walker & Co., 1997.

297. Singmaster, David. *Notes on Rubik's 'Magic Cube.* Hillside, N.J.: Enslow, 1981.

298. Singmaster, David. *Sources in Recreational Mathematics: An Annotated Bibliography,* 7th prelim. ed. Unpublished.

299. Skewes, S. "On the difference $\pi(x) - li(x)$." *Journal of the London Mathematical Society,* 8: 277–283 (1933).

300. Slocum, Jerry. *Compendium of Mechanical Puzzles.* 3rd ed. Slocum and Haubrich, 1977.

301. Slocum, Jerry. *The Tangram Book: The Story of the Chinese Puzzle with Over 2000 Puzzles to Solve.* New York: Sterling, 2003.

302. Smith, David Eugene. *Number Stories of Long Ago.* Washington, D.C.: National Council of Teachers of Mathematics, 1919.

303. Smith, David Eugene. *A Source Book in Mathematics.* 2 vols. New York: McGraw-Hill, 1929.

304. Smullyan, Raymond M. *What Is the Name of This Book? The Riddle of Dracula and Other Logical Puzzles.* Englewood Cliffs, N.J.: Prentice Hall, 1978.

305. Smullyan, Raymond M. *The Chess Mysteries of Sherlock Holmes.* New York: Alfred A. Knopf, 1979.

306. Smullyan, Raymond. *This Book Needs No Title: A Budget of Living Paradoxes.* Englewood Cliffs, N.J.: Prentice-Hall, 1980.

307. Smullyan, Raymond M. *The Lady or the Tiger? And Other Logic Puzzles.* New York: Alfred A. Knopf, 1982.

308. Smullyan, Raymond M. *Alice in Puzzleland.* New York: William Morrow, 1982.

309. Smullyan, Raymond M. *To Mock a Mockingbird.* New York: Alfred A. Knopf, 1985.

310. Smullyan, Raymond M. *Forever Undecided: A Puzzle Guide to Godel.* New York: Alfred A. Knopf, 1987.

311. Smullyan, Raymond. *The Tao is Silent.* New York: HarperCollins, 1992.

312. Soddy, F. "The Kiss Precise." *Nature,* 137: 1021, 1936.

313. Sprague, Roland. *Recreation in Mathematics.* Mineola, N.Y.: Dover, 1963.

314. Stein, M. L., S. M. Ulam, and M. B. Wells. "A Visual Display of Some Properties of the Distribution of Primes." *American Mathematical Monthly,* 71: 516–520, 1964.

315. Steinhaus, H. "Sur la division pragmatique." *Ekonometrika (Supp.),* 17: 315–319, 1949.

316. Steinhaus, Hugo. *Mathematical Snapshots.* 3rd rev. ed. New York: Dover, 1983.

317. Stewart, Ian. *Does God Play Dice: The New Mathematics of Chaos.* Cambridge, Mass.: Basil Blackwell, 1989.

318. Stewart, Ian. *Game, Set, and Math: Enigmas and Conundrums.* Cambridge, Mass.: Basil Blackwell, 1989.

319. Stewart, Ian. "Tales of a Neglected Number." *Scientific American,* 274: 102–103 (Jun. 1996).

320. Stewart, Ian. "Unilluminable Rooms." *Scientific American,* 275(2), 100–103 (Aug. 1996).

321. Stewart, Ian and Cohen, Jack Cohen. *Figments of Reality.* Cambridge: Cambridge University Press, 1997.

322. Stewart, Ian. "Cone with a Twist." *Scientific American,* 281: 116–117 (Oct. 1999).

323. Stewart, Ian. *Flatterland: Like Flatland Only More So.* New York: Perseus, 2001.

324. Stromberg, K. "The Banach-Tarski Paradox." *American Mathematical Monthly,* 86: 3, 1979.

325. Stubbs, A. Duncan. *Miscellaneous Puzzles.* London: Frederick Warne, 1931.

326. Struik, Dirk Jan, ed. *A Source Book in Mathematics 1200–1800.* Princeton, N.J.: Princeton University Press, 1969.

327. Strutt, Joseph. *The Sports and Pastimes of the People of England.* London: J. White, 1801.

328. Stuwe, M. *Plateau's Problem and the Calculus of Variations.* Princeton, N.J.: Princeton University Press, 1989.

329. Swade, Doron. *The Difference Engine: Charles Babbage and the Quest to Build the First Computer.* New York: Viking, 2000.

330. te Riele, H. J. J. "On the Sign of the Difference $\pi(x) - Li(x)$." *Mathematics of Computation,* 48: 323–328 (1987).

331. Thom, R. *Structural Stability and Morphogenesis: An Outline of a General Theory of Models.* Reading, Mass.: Addison-Wesley, 1972.

332. Thompson, d'Arcy W. *On Growth and Form.* Cambridge: Cambridge University Press, 1917. (Reprint, New York: Dover, 1992.)

333. Thorpe, Edward O. *Beat the Dealer,* rev. ed. New York: Random House, 1966.

334. Tokarsky, George W. "Polygonal Rooms Not Illuminable from Every Point." *American Mathematical Monthly,* 102(10): 867–879, 1995.

335. Toole, Betty A. (ed.). *Ada, the Enchantress of Numbers: A Selection from the Letters of Lord Byron's Daughter and Her Description of the First Computer.* Mill Valley, Calif.: Strawberry Press, 1998.

336. Tormey, Alan and Tormey, Judith Farr. "Renaissance Intarsia: The Art of Geometry." *Scientific American,* 247: 136–143, July 1982.

337. Turing, A. M. "On Computable Numbers, with an Application to the Entscheidungsproblem." *Proceedings*

of the London Mathematical Society, Series 2, 42: 230–265, 1937.

338. Turney, P, "Unfolding the Tesseract." *Journal of Recreational Mathematics,* 17(1): 1–16, 1984–1985.

339. Underwood, Dudley. "The First Recreational Mathematics Book." *Journal of Recreational Mathematics,* 3, 164–169, 1970.

340. Van Cleve, James. "Right, Left, and the Fourth Dimension." *The Philosophical Review,* 96: 33–68 (1987).

341. Vardi, I. "Archimedes' Cattle Problem." *American Mathematical Monthly,* 105: 305–319 (1998).

342. Vardi, Ilan. *Computational Recreations in Mathematica.* Redwood City, Calif.: Addison-Wesley, 1991.

343. Wagon, S. *The Banach-Tarski Paradox.* New York: Cambridge University Press, 1993.

344. Waller, Mary D. *Chladni Figures: A Study in Symmetry.* London: G. Bell & Sons, 1961.

345. Watson, G. N. "The Problem of the Square Pyramid." *Messenger of Mathematics,* 48: 1–22 (1918–1919).

346. Weisstein, Eric W. *The CRC Concise Encyclopedia of Mathematics.* Boca Raton, Fla.: CRC Press, 1998.

347. Wells, David Graham. *Hidden Connections, Double Meanings: A Mathematical Exploration.* Cambridge: Cambridge University Press, 1988.

348. Wells, David. *The Penguin Dictionary of Curious and Interesting Geometry.* London: Penguin Books, 1991.

349. Wells, David. *The Penguin Book of Curious and Interesting Mathematics.* New York: Penguin, 1997.

350. Wells, Kenneth. *Wooden Puzzles and Games.* New York: Sterling Publications, 1983.

351. Weyl, Hermann. *Symmetry.* Princeton, N.J.: Princeton University Press, 1952.

352. White, Alain C. *Sam Loyd and His Chess Problems.* Leeds, England: Whitehead & Miller, 1913.

353. Wiles, A. "Modular Elliptic-Curves and Fermat's Last Theorem." *Annals of Mathematics,* 141: 443–551, 1995.

354. Williams, W. T., and Savage, G. H. *The First Penguin Problems Book.* New York: Penguin Books, 1940.

355. Wyatt, Edwin Mather. *Puzzles in Wood.* Milwaukee, Wis.: Bruce Publishing Company, 1928.

356. Yan, Li, and Du, Shiran. *Chinese Mathematics: A Concise History.* Oxford: Oxford University Press, 1988.

357. Zaslavsky, Claudia. *Africa Counts.* Brooklyn, N.Y.: Lawrence Hill Books, 1973.

Solutions to Puzzles

abracadabra

To solve this problem, redraw the diamond and replace every letter on the outside top half of the diamond with a 1 and every other point in the diamond with a dot. This gives

```
            1
          1   1
        1   .   1
      1   .   .   1
    1   .   .   .   1
  1   .   .   .   .   1
    .   .   .   .   .
      .   .   .   .
        .   .   .
          .   .
            .
```

Next, replace each of the dots, starting at the top, with the sum of the two numbers in the northwest and northeast positions, so that the pyramid now starts

```
      1
    1   1
  1   2   1
1   3   3   1
```

After you have worked your way down to the very bottom, the bottom number in the diamond is the answer to Polya's problem: 252.

age puzzles and tricks

1. 18.

2. 16½ (Mary is 27½).

alphametic

1.
```
      67432 (EARTH)
        704 (AIR)
       8046 (FIRE)
    + 97364 (WATER)
    _____
     173546 (NATURE)
```

A = 7, E = 6, F = 8, H = 2, I = 0, R = 4, T = 3, W = 9

2.
```
     127503 (SATURN)
     502351 (URANUS)
    3947539 (NEPTUNE)
    + 46578 (PLUTO)
    _____
    4623971 (PLANETS)
```

A = 2, E = 9, L = 6, N = 3, O = 8, P = 4, R = 0, S = 1, T = 7, U = 5

3.
```
      862903 (MARTIN)
    + 1627342 (GARDNER)
    _____
     2490245 (RETIRES)
```

A = 6, D = 7, E = 4, G = 1, I = 0, M = 8, N = 3, R = 2, S = 5, T = 9

anagram

1. Wolfgang Amadeus Mozart.

2. Thomas Alva Edison.

3. William Shakespeare.

Carroll, Lewis

1. There is as much water in the milk/water mixture as milk in the water/milk mixture.

2. There are 30 ways of painting the cube. If the restriction that each face be painted a different color is dropped, there are 2,226 ways of painting the cube.

3. FOUR → FOUL → FOOL → FOOT → FORT → FORE → FIRE → FIVE

Alternatively:
FOUR → POUR → POUT → ROUT → ROUE → ROVE → DOVE → DIVE → FIVE
(Thanks to my editor, Stephen Power, for this one.)

4. Carroll didn't have an answer in mind when he wrote the riddle, though he later came up with: "Because it can produce a few notes, though they are *very* flat; and it is nevar put wrong end in front!" (Note the variant spelling of "never.") Other authors have come up with: "Because Poe wrote on both"

(Sam Loyd); "Because it slopes with a flap" (Cyril Pearson); and "Because both have quills dipped in ink" (David Jodrey).

clock puzzles

1. Approximately 27 minutes and 42 seconds (exactly 360/13 minutes) after 6.

2. The stopped clock, because it will give the right time twice a day, whereas the other is only correct approximately every two years.

3. The positions of the hands shown in the illustration could only indicate that the clock stopped at 44 min. 51 1,143/1,427 sec. after eleven o'clock. The second hand would next be "exactly midway between the other two hands" at 45 min. 52 496/1,427 sec. after eleven o'clock. If we had been dealing with the points on the circle to which the three hands are directed, the answer would be 45 min. 22 106/1,427 sec. after eleven; but the question applied to the hands, and the second hand would not be between the others at that time, but outside them.

cone

The simple rule is that the cone must be cut at one-third of its height.

de Morgan, Augustus

43 (the only number that, when squared, gives a number between the years of de Morgan's birth and death).

domino

No! Suppose it *were* possible to totally cover the modified chessboard with nonoverlapping dominos. In any complete tiling, every domino must cover exactly one white square and one black square. Thus the modified board must have exactly the same number of black and white squares. But the two removed squares, from diagonally opposite corners of a chessboard, must be same color. Since there can't be the same number of white squares and black squares on the modified board it must be impossible to tile the modified board with nonoverlapping dominos.

hundred fowls problem

Using the two equations to eliminate C, gives the equation $7R + 4H = 100$. R must be less than 15 (since $7 \times 15 = 105$). Trial-and-error shows that values of R that allow whole number values of H are 4, 8, and 12, from which it follows that the three possible solutions to the problem are 4 roosters, 18 hens, and 78 chicks; 8 roosters, 11 hens, and 81 chicks; and 12 roosters, 4 hens, and 84 chicks.

kinship puzzles

1. Your son.

2. Sons of two men who married each other's mothers.

3. The party consisted of two girls and a boy, their father and mother, and their father's father and mother.

measuring and weighing puzzles

The innkeeper first filled the 5-pint and 3-pint measures, then turned the tap on the barrel and allowed the rest of its contents to run to waste. He closed the tap and emptied the 3-pint into the barrel; filled the 3-pint from the 5-pint; emptied the 3-pint into the barrel; transferred the 2 pints from the 5-pint to the 3-pint; filled the 5-pint from the barrel, leaving 1 pint now in the barrel; filled the 3-pint from the 5-pint; allowed the company to drink the contents of the 3-pint; filled the 3-pint from the 5-pint, leaving 1 pint now in the 5-pint; drank the contents of the 3-pint; and finally drew off 1 pint from the barrel into the 3-pint. He now had 1 pint of ale in each measure!

missing dollar problem

There is no missing dollar (of course!). Adding $27 and $2 (to get $29) is a bogus operation. They paid $27, $2 went to the dishonest waiter, and $25 went to the restaurant. You have to subtract $27 minus $2 to get $25. There never was a $29; it has nothing to do with anything.

Mrs. Perkins's quilt

The following diagram shows how the quilt should be made.

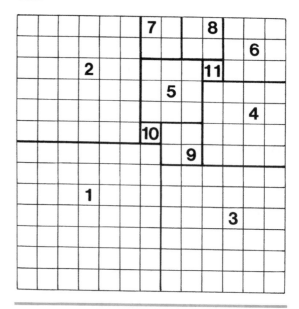

The solution of Mrs. Perkins's quilt.

In *Amusements in Mathematics,* Dudeney writes: "There is, I believe, only one solution to this puzzle. The fewest separate squares must be eleven. The portions must be of the sizes given, the three largest pieces must be arranged as shown, and the remaining group of eight squares may be 'reflected,' but cannot be differently arranged."

nine rooms paradox

One of the two customers who is initially placed in room A, whom we refer to as the "first" customer, is later transferred to room I and treated as if he were also the tenth customer!

postage stamp problems

Dudeney gives the solution to his "The Four Postage Stamps" problem as follows: "[T]he four stamps may be given in the shape 1, 2, 3, 4, in three ways; in the shape 1, 2, 5, 6, in six ways; in the shape 1, 2, 3, 5, or 1, 2, 3, 7, or 1, 5, 6, 7, or 3, 5, 6, 7, in twenty-eight ways; in shape 1, 2, 3, 6, or 2, 5, 6, 7, in fourteen ways; in shape 1, 2, 6, 7, or 2, 3, 5, 6, or 1, 5, 6, 10, or 2, 5, 6, 9, in fourteen ways. Thus there are sixty-five ways in all."

probability theory

1. There are four possibilities:

Oldest Child	Youngest Child
1. Girl	Girl
2. Girl	Boy
3. Boy	Girl
4. Boy	Boy

If your friend says "My oldest child is a girl," he has eliminated cases 3 and 4, and in the remaining cases both are girls ½ of the time. If your friend says "At least one of my children is a girl," he has eliminated case 4 only, and in the remaining cases both are girls ⅓ of the time.

2. There are six possible bullet configurations (B = bullet, E = empty):

B B B E E E → player 1 dies
E B B B E E → player 2 dies
E E B B B E → player 1 dies
E E E B B B → player 2 dies
B E E E B B → player 1 dies
B B E E E B → player 1 dies

One therefore has a ⅔ probability of winning (and a ⅓ probability of dying) by shooting second.

Pythagorean square puzzle

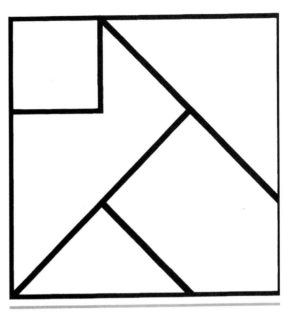

Pythagorean puzzle solution.

river-crossing problem

The missionaries (M) can avoid being eaten by the cannibals (C) if the crossings of the river and the returns are arranged as follows. Crossing #1: 1M + 1C; return #1: 1M; crossing #2: 2C; return #2: 1C; crossing #3: 2M; return #3: 1M + 1C; crossing #4: 2M; return #4: 1C; crossing #5: 2C; return #5: 1C; crossing #6: 2C.

word puzzles

(1) $T_1W_4E_1L_1V_4E_1$.

(2) Dreamt.

(3) Tremendous, horrendous, stupendous, and hazardous.

(4) I am.

(5) skiing

(6) "Are" by an adding an "a" becomes "area."

(7) underground.

Category Index

Algebra
abstract algebra
algebra
algebraic fallacies
algebraic geometry
array
associative
Bieberbach conjecture
binary operation
coefficient
commutative
cubic equation
determinant
diagonal matrix
discriminant
distributive
field
fundamental theorem of algebra
gradient
Hankel matrix
hundred fowls problem
idempotent
invariant
invariant theory
Jordan matrix
Lie algebra
linear algebra
linear system
matrix
monomial
non-Abelian
quadratic
quadric
quartic
quintic
resultant
ring
scalar
solution
tensor
Toeplitz matrix
trace
transpose
trinomial
unimodular matrix
vector
vector space
weak inequality

x
zero divisors

Analytical geometry
abscissa
algebraic curve
analytical geometry
Cartesian geometry. *See* analytical
  geometry
Cartesian coordinates
coordinate
coordinate geometry. *See* analytical
  geometry
octant
ordinate
origin
polar coordinates
quadrant
rectangular coordinates. *See* Cartesian
  coordinates
slope
x-axis
y-axis

> *See also:* **Geometry, general terms and
> theorems** and **Graphs and
> graph theory**

Approximations and averages
Banker's rounding
arithmetic mean
average
ceiling
difference equation
extrapolate. *See* interpolate
floor function
geometric mean
interpolate
iterate
iteration
mean
median
mode
Newton's method
rounding numerical analysis
round-off error
significant digits
universal approximation

Arithmetic
arithmetic
binary
casting out nines
cryptarithm
denominator
digimetic
direct proportion
division
fly between trains problem
four fours problem
fundamental theorem of arithmetic
greatest common divisor
greatest lower bound
harmonic mean
least common multiple
multiple
negative base
number system
pandigital product
period, of a decimal expansion
place-value system
product
proportional
quotient
ratio
reciprocal
Russian multiplication
St. Ives problem
skeletal division
subtraction
wheat and chessboard problem

> *See also:* **Number theory**

Biography
Abbott, Edwin Abbott (1838–1926)
Abel, Niels Henrik (1802–1829)
Abu'l Wafa (A.D. 940–998)
Agnesi, Maria Gaetana (1718–1799)
Ahrens, Wilhelm Ernst Martin Georg
  (1872–1927)
Alcuin (735–804)
al-Khowarizmi (c. 780–850)
Archimedes of Syracuse (c. 287–212 B.C.)
Apollonius of Perga (c. 255–170 B.C.)
Atiyah, Michael Francis (1929–)
Babbage, Charles (1791–1871)

373

3m

DATE DUE

MAR 0 7

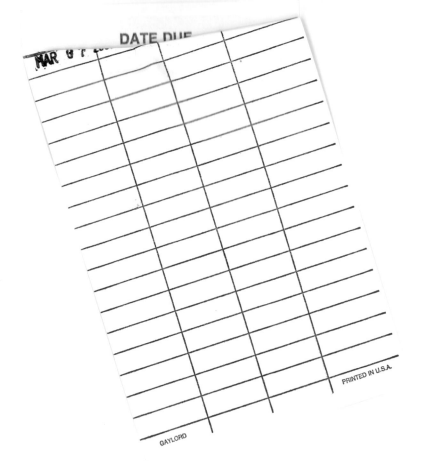

PRINTED IN U.S.A.

GAYLORD